# Hints & Kinks

18th EDITION

## For the Radio Amateur

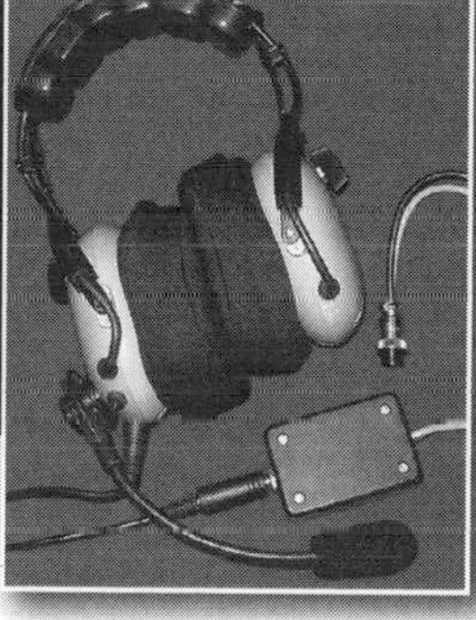

**Editor**
Steve Ford, WB8IMY

**Composition**
Shelly Bloom, WB1ENT
Jodi Morin, KA1JPA

**Cover Design**
Sue Fagan, KB1OKW

**ARRL** *The national association for* *AMATEUR RADIO®*
Newington, CT 06111-1494
www.arrl.org

Printed in the USA



ISBN: 978-0-87259-520-0

Eighteenth Edition
Fourth Printing

# Contents

# Foreword

Welcome to the 18th edition of *Hints and Kinks for the Radio Amateur*. This book is a compilation of material published in *QST* magazine's popular "Hints and Kinks" column between January 2005 and December 2011.

Between these covers you'll find handy advice on a broad range of topics from antennas to vintage radio restoration. The great thing about "Hints and Kinks" is that all these tips come from your fellow amateurs, people just like you who've discovered creative solutions to everyday problems. That's why "Hints and Kinks" has held an honored position for decades as one of the most-read sections of *QST*.

If you have hint or kink to contribute, by all means send it to our Hints and Kinks editor. You can reach the editor by e-mail at **h&k@arrl.org**.

73 . . .
David Sumner, K1ZZ
ARRL Chief Executive Officer
January 2012

**Please note**: The information in this book was current when it was originally published in *QST*. Call signs, postal addresses, e-mail addresses and Web sites cited in articles may now be out of date.

# US Customary to Metric Conversions

## International System of Units (SI)—Metric Prefixes

| *Prefix* | *Symbol* | | | *Multiplication Factor* |
|---|---|---|---|---|
| exa | E | $10^{18}$ | = | 1 000 000 000 000 000 000 |
| peta | P | $10^{15}$ | = | 1 000 000 000 000 000 |
| tera | T | $10^{12}$ | = | 1 000 000 000 000 |
| giga | G | $10^{9}$ | = | 1 000 000 000 |
| mega | M | $10^{6}$ | = | 1 000 000 |
| kilo | k | $10^{3}$ | = | 1 000 |
| hecto | h | $10^{2}$ | = | 100 |
| deca | da | $10^{1}$ | = | 10 |
| (unit) | | $10^{0}$ | = | 1 |
| deci | d | $10^{-1}$ | = | 0.1 |
| conti | c | $10^{-2}$ | = | 0.01 |
| milli | m | $10^{-3}$ | = | 0.001 |
| micro | μ | $10^{-6}$ | = | 0.000001 |
| nano | n | $10^{-9}$ | = | 0.000000001 |
| pico | p | $10^{-12}$ | = | 0.000000000001 |
| femto | f | $10^{-15}$ | = | 0.000000000000001 |
| atto | a | $10^{-18}$ | = | 0.000000000000000001 |

### Linear
1 metre (m) = 100 centimetres (cm) = 1000 millimetres (mm)

### Area
1 $m^2 = 1 \times 10^4$ $cm^2 = 1 \times 10^6$ $mm^2$

### Volume
1 $m^3 = 1 \times 10^6$ $cm^3 = 1 \times 10^9$ $mm^3$
1 litre (l) = 1000 $cm^3 = 1 \times 10^6$ $mm^3$

### Mass
1 kilogram (kg) = 1 000 grams (g)
(Approximately the mass of 1 litre of water)
1 metric ton (or tonne) = 1 000 kg

## US Customary Units

### Linear Units
12 inches (in) = 1 foot (ft)
36 inches = 3 feet = 1 yard (yd)
1 rod = 5½ yards = 16½ feet
1 statute mile = 1 760 yards = 5 280 feet
1 nautical mile = 6 076.11549 feet

### Area
1 $ft^2$ = 144 $in^2$
1 $yd^2$ = 9 $ft^2$ = 1 296 $in^2$
1 $rod^2$ = 30¼ $yd^2$
1 acre = 4840 $yd^2$ = 43 560 $ft^2$
1 acre = 160 $rod^2$
1 $mile^2$ = 640 acres

### Volume
1 $ft^3$ = 1 728 $in^3$
1 $yd^3$ = 27 $ft^3$

### Liquid Volume Measure
1 fluid ounce (fl oz) = 8 fluidrams = 1.804 $in^3$
1 pint (pt) = 16 fl oz
1 quart (qt) = 2 pt = 32 fl oz = 57¾ $in^3$
1 gallon (gal) = 4 qt = 231 $in^3$
1 barrel = 31½ gal

### Dry Volume Measure
1 quart (qt) = 2 pints (pt) = 67.2 $in^3$
1 peck = 8 qt
1 bushel = 4 pecks = 2 150.42 $in^3$

### Avoirdupois Weight
1 dram (dr) = 27.343 grains (gr) or (gr a)
1 ounce (oz) = 437.5 gr
1 pound (lb) = 16 oz = 7 000 gr
1 short ton = 2 000 lb, 1 long ton = 2 240 lb

### Troy Weight
1 grain troy (gr t) = 1 grain avoirdupois
1 pennyweight (dwt) or (pwt) = 24 gr t
1 ounce troy (oz t) = 480 grains
1 lb t = 12 oz t = 5 760 grains

### Apothecaries' Weight
1 grain apothecaries' (gr ap) = 1 gr t = 1 gr a
1 dram ap (dr ap) = 60 gr
1 oz ap = 1 oz t – 8 dr ap – 480 fr
1 lb ap = 1 lb t = 12 oz ap = 5 760 gr

**Multiply** →

Metric Unit = Conversion Factor × US Customary Unit

← **Divide**

Metric Unit ÷ Conversion Factor = US Customary Unit

| Metric Unit = | Conversion Factor × | US Unit |
|---|---|---|
| (Length) | | |
| mm | 25.4 | inch |
| cm | 2.54 | inch |
| cm | 30.48 | foot |
| m | 0.3048 | foot |
| m | 0.9144 | yard |
| km | 1.609 | mile |
| km | 1.852 | nautical mile |
| (Area) | | |
| $mm^2$ | 645.16 | $inch^2$ |
| $cm^2$ | 6.4516 | $in^2$ |
| $cm^2$ | 929.03 | $ft^2$ |
| $m^2$ | 0.0929 | $ft^2$ |
| $cm^2$ | 8361.3 | $yd^2$ |
| $m^2$ | 0.83613 | $yd^2$ |
| $m^2$ | 4047 | acre |
| $km^2$ | 2.59 | $mi^2$ |
| (Mass) | (Avoirdupois Weight) | |
| grams | 0.0648 | grains |
| g | 28.349 | oz |
| g | 453.59 | lb |
| kg | 0.45359 | lb |
| tonne | 0.907 | short ton |
| tonne | 1.016 | long ton |

| Metric Unit = | Conversion Factor × | US Unit |
|---|---|---|
| (Volume) | | |
| $mm^3$ | 16387.064 | $in^3$ |
| $cm^3$ | 16.387 | $in^3$ |
| $m^3$ | 0.028316 | $ft^3$ |
| $m^3$ | 0.764555 | $yd^3$ |
| ml | 16.387 | $in^3$ |
| ml | 29.57 | fl oz |
| ml | 473 | pint |
| ml | 946.333 | quart |
| l | 28.32 | $ft^3$ |
| l | 0.9463 | quart |
| l | 3.785 | gallon |
| l | 1.101 | dry quart |
| l | 8.809 | peck |
| l | 35.238 | bushel |
| (Mass) | (Troy Weight) | |
| g | 31.103 | oz t |
| g | 373.248 | lb t |
| (Mass) | (Apothecaries' Weight) | |
| g | 3.387 | dr ap |
| g | 31.103 | oz ap |
| g | 373.248 | lb ap |

# About the ARRL

The seed for Amateur Radio was planted in the 1890s, when Guglielmo Marconi began his experiments in wireless telegraphy. Soon he was joined by dozens, then hundreds, of others who were enthusiastic about sending and receiving messages through the air—some with a commercial interest, but others solely out of a love for this new communications medium. The United States government began licensing Amateur Radio operators in 1912.

By 1914, there were thousands of Amateur Radio operators—hams—in the United States. Hiram Percy Maxim, a leading Hartford, Connecticut inventor and industrialist, saw the need for an organization to band together this fledgling group of radio experimenters. In May 1914 he founded the American Radio Relay League (ARRL) to meet that need.

Today ARRL, with approximately 175,000 members, is the largest organization of radio amateurs in the United States. The ARRL is a not-for-profit organization that:

- ♦ promotes interest in Amateur Radio communications and experimentation
- ♦ represents US radio amateurs in legislative matters, and
- ♦ maintains fraternalism and a high standard of conduct among Amateur Radio operators.

At ARRL headquarters in the Hartford suburb of Newington, the staff helps serve the needs of members. ARRL is also International Secretariat for the International Amateur Radio Union, which is made up of similar societies in 150 countries around the world.

ARRL publishes the monthly journal *QST*, as well as newsletters and many publications covering all aspects of Amateur Radio. Its headquarters station, W1AW, transmits bulletins of interest to radio amateurs and Morse code practice sessions. The ARRL also coordinates an extensive field organization, which includes volunteers who provide technical information and other support services for radio amateurs as well as communications for public-service activities. In addition, ARRL represents US amateurs with the Federal Communications Commission and other government agencies in the US and abroad.

Membership in ARRL means much more than receiving *QST* each month. In addition to the services already described, ARRL offers membership services on a personal level, such as the ARRL Volunteer Examiner Coordinator Program and a QSL bureau.

Full ARRL membership (available only to licensed radio amateurs) gives you a voice in how the affairs of the organization are governed. ARRL policy is set by a Board of Directors (one from each of 15 Divisions). Each year, one-third of the ARRL Board of Directors stands for election by the full members they represent. The day-to-day operation of ARRL HQ is managed by an Executive Vice President and his staff.

No matter what aspect of Amateur Radio attracts you, ARRL membership is relevant and important. There would be no Amateur Radio as we know it today were it not for the ARRL. We would be happy to welcome you as a member! (An Amateur Radio license is not required for Associate Membership.) For more information about ARRL and answers to any questions you may have about Amateur Radio, write or call:

ARRL—The national association for Amateur Radio
225 Main Street
Newington CT 06111-1494
Voice: 860-594-0200
Fax: 860-594-0259
E-mail: **hq@arrl.org**
Internet: **www.arrl.org/**

Prospective new amateurs call (toll-free):
**800-32-NEW HAM** (800-326-3942)

You can also contact us via e-mail at
**newham@arrl.org**
or check out *ARRLWeb* at **www.arrl.org/**

# Common Schematic Symbols Used in Circuit Diagrams

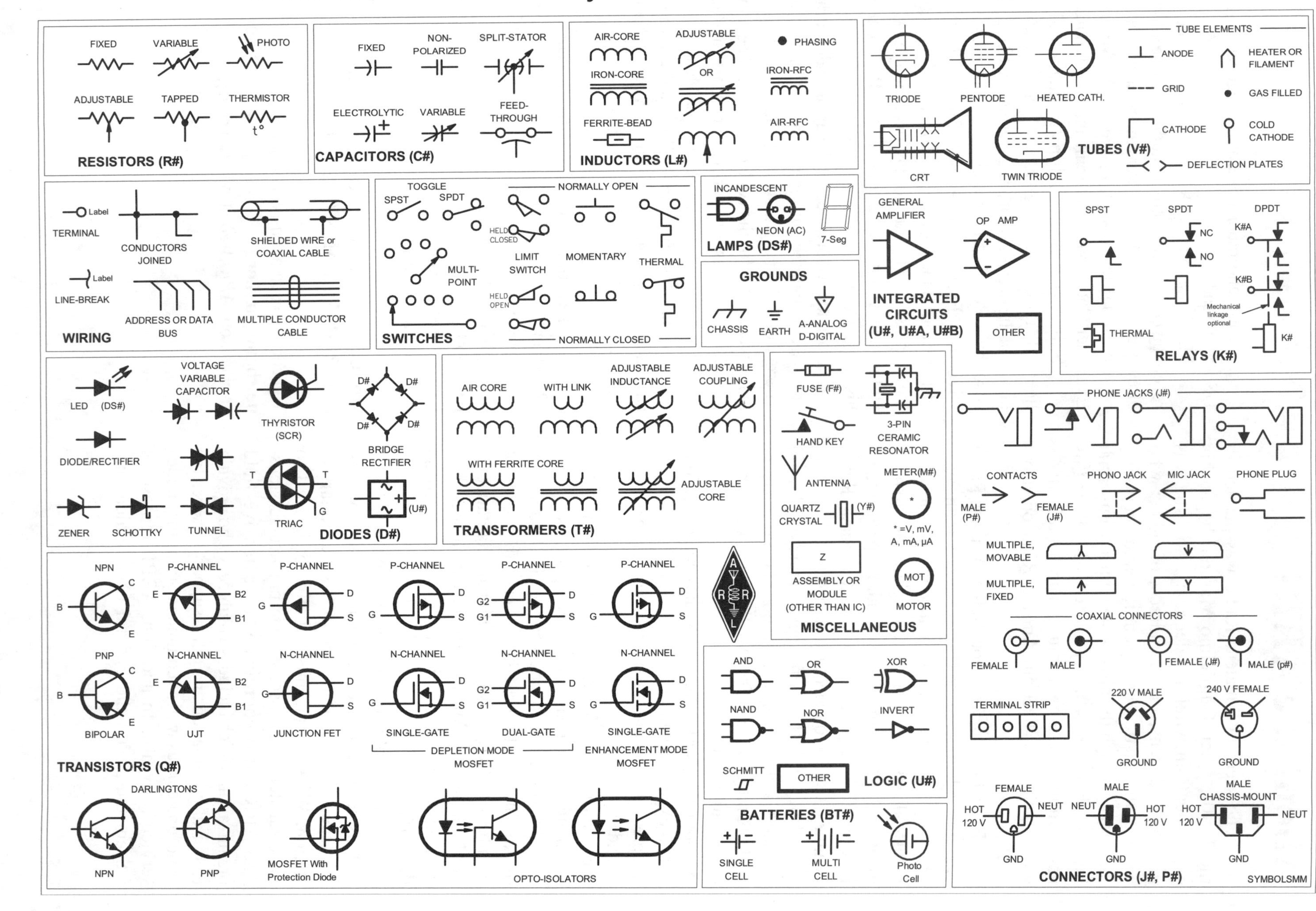

# Chapter 1 Equipment Tips and Mods

## KEEP YOUR ICOM IC-706/MKIIG COOL!

◊ The ICOM IC-706 series of radios is very popular for mobile and portable operation. A lot of heat is generated in this small package, however, and with moderate-to-heavy use, the radio and heat sink can get very hot. The remedy for this is additional airflow around the heat sink. This project had three goals: First, I needed to solve the heat problem. Second, the solution should not be obtrusive. Third, it should require no mechanical changes to my IC-706MKIIG. There are about a "zillion" surplus computer fans available for very little money, so I figured that there had to be one that was perfect for adding to my IC-706MKIIG. I decided to use a 12-V 50-mm square fan available from All Electronics (**www.allelectronics.com**, part number CF-203, $2.50 each). The trick was mounting it to the IC-706G heat sink. The answer turned out to be simple. (Refer to Figures 1, 2 and 3.)

Figure 1—The parts used to add a fan to the heat sink of an ICOM IC-706MKII2G.

I used plastic screw anchors to mount the fan to the heat sink. The anchors I used are 1¼ inches long and are normally used with #10 or #12 screws. In this application, the anchors fit tightly between the heat-sink fins and work perfectly with #8 machine screws. I needed to cut about ¼-inch of the length from the wall anchors. I used one-inch-long #8 machine screws to mount the fan. Figure 3 shows how to mount the fan. Orient the fan so that it blows away from the IC-706 heat sink. Rotate the fan assembly as necessary for the anchors to align with spaces between heat-sink fins. Then just tighten down the #8 screws. This fan draws 70 mA, so I took the necessary +12 V from the accessory connector on the IC-706G. Pin 8 (gray wire on the supplied connector) is +12 V and pin 2 (red wire on the supplied connector) is ground. The maximum current available at this connector is 1 A. To protect the radio in case the fan shorts, I connected a 33 Ω, ½ W resistor and a 0.25 A fuse in series with a fan lead. The resistor drops the fan voltage a bit, which makes it run very quiet. The fuse is a ¼ A pico fuse (**www.mouser.com**, part number 504 MCR ¼, $1.03 each). Both the fuse and resistor are covered by heat-shrink tubing. How does this modification work?—absolutely outstanding. My IC-706MKIIG runs cool even during heavy use. Fans of other sizes can obviously be used, though I think that a 50-mm square fan is about the largest that would fit. A 40-mm fan may actually be a better fit, but you get the idea.—*Phil Salas, AD5X, 1517 Creekside Dr, Richardson, TX 75081;* **ad5x@arrl.net**

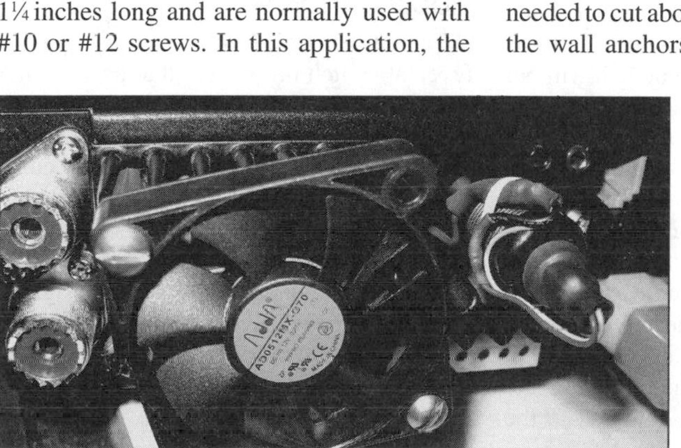

Figure 2—A rear view of the ICOM IC-706MKII2G with the fan mounted to the heat sink.

Figure 3—A top view of the ICOM IC-706MKII2G with the fan mounted. Notice the screw anchor between the heat-sink fins.

## ANOTHER TAKE ON TS-850S BATTERY REPLACEMENT

◊ After reading N1FB's fine article on TS-850S battery replacement in the October column, I decided I had better replace my battery before I ran into the leakage problem. I am a little faint-hearted when it comes to removing cables (because of some unpleasant experiences in the past), so I tried to devise a less-invasive way to replace the battery. I decided to position the new battery in a holder under the access hatch for the crystal filter switch.

In the end, I opened the case as directed by KO4NR, but I put the case on its right side instead of its left, thus placing the battery higher. I then moved the front panel away from the board that holds the battery. (Please note I have not removed any cables.)

I carefully bent the battery up away from the board and cut the battery tabs with diagonal pliers. Next, I bent the remaining portions of the tabs perpendicular to the board and carefully tinned them with solder.

I cut and tinned two wires to reach from the two battery pins to where the DRU-2 would sit. There, I "superglued" a piece of unclad RadioShack circuit board to the metal chassis, purely for insulation. Next, I mounted a piece of hook-and-loop material to secure the battery holder with a battery in it.

Now, battery replacement is as simple as opening the cover over the crystal filter switch.

I used a CR2477N battery, which is the highest capacity 3 V battery I could find. I got my parts from Mouser (**www.mouser.com**). The battery holder is 614-HU2477N-1 and the battery is 614-CR2477N. I suggest a very low wattage soldering iron with a pointed or conical tip. Please feel free to e-mail me if you have any questions.—*Harry M. Hammerling, W6HMH, 32115 Road 416, Coarsegold, CA 93614;* **harrymh@ hammertech.com**

## CONVERTING KENWOOD HS-5 HEADPHONES FOR STEREO

◊ My favorite headphones for Amateur Radio have always been the Kenwood HS-5. I've

Figure 5—The Kenwood HS-5 headphone cover.

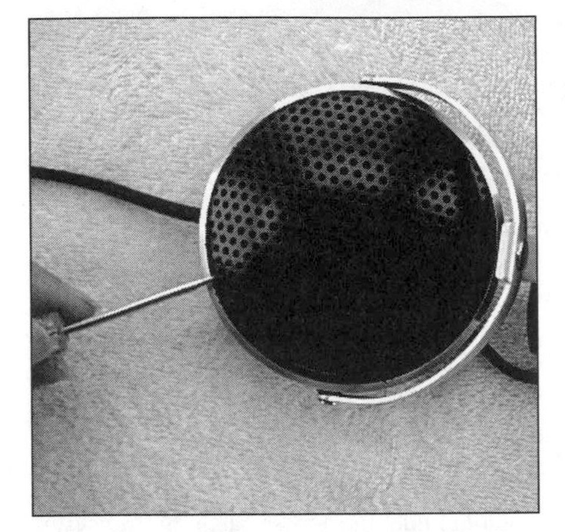

Figure 6—Gently remove the cover from the headphone.

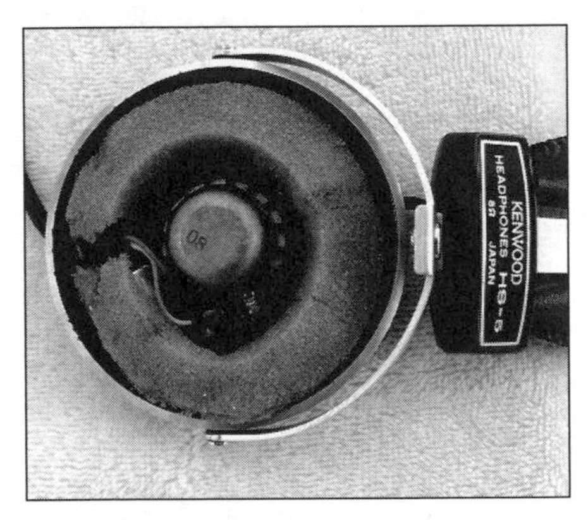


Figure 7—Foam covers the headphone wiring.

Figure 8—Removing the foam reveals the wiring.

had mine for almost 20 years and Kenwood still sells them. They are, however, monaural headphones. My latest transceiver, a Ten-Tec Orion, requires stereo headphones.

Troy, W6HV, mentioned on the Ten-Tec e-mail reflector that he had replaced the cord on his HS-5 with a stereo cord and that it worked with the Orion. I decided that I was going to replace my cord as well. Here's how you can do the same.

I unscrewed the black cover on the mono plug and discovered that instead of two wires there were actually three: black, red and white. The black wire was connected to the sleeve and was the ground. The red and white wires were both connected to the plug tip and carried the audio.

To gain access to the wiring inside the headphone, one must remove the perforated back cover of the earpiece (Figure 5). It turns out that this perforated piece is just a decorative cover over a plastic housing that in turn snaps into the earpiece with two tabs 180° apart around the circumference. These tabs snap into metal hooks on the body. They can be released without damage by gently prying the cover over them (Figure 6). Be careful how you apply the pressure, as the decorative chrome ring scratches easily. There is no wiring in the cover; once it is pried loose, it comes off easily. With the cover removed, you can see a piece of foam covering the wiring (Figure 7). This foam was not glued down but had become stuck to the wiring over the past 20 years and was a bit brittle. I was able to remove it without damaging it.

With the foam removed, the wiring is easily accessible (Figure 8). In addition to the red, white and black wires coming from the cord, there was a two-conductor cable with red and black wires that goes to the other earpiece. The black wires from both the cord and the other earpiece are soldered to one terminal of the speaker. The red wire to the other earpiece and the red-and-white wire from the cord are all soldered to the other terminal.

The modification inside the earpiece for stereo operation simply involved removing the red wires from the speaker and connecting them to each other. That way, the white wire carries the audio for the earpiece with the cable connected to it, and the red wire carries the audio for the other earpiece.

That left the plug to be changed. I already had a ¼ inch stereo plug in my parts box. As it turned out, the threads that hold the black cover on it matched the threads that held the black cover on the original mono plug of the HS-5. In my case, that cover has a built-in rubber strain relief, so I reused it, just swapping the plug.

It was obvious that the black wire would go to the sleeve of the stereo plug, but in what order should the red and white wires be attached to the tip and ring terminals? The standard for a three-conductor stereo plug is that the left channel audio is on the tip and the right channel audio is on the ring (or middle). It's been my experience that on any stereo headphones with a cable coming from only one earpiece, that earpiece is on the left. On the HS-5, there is no difference between the earpieces, and you can wear them with the cable on either side, as best suits your setup.

I chose to follow the convention and have the cable on the "left" earpiece. That means that the white wire connects to the tip and the red wire (the one that goes through to the "right" earpiece) goes to the ring.

After verifying that this setup works, I replaced the foam and snapped the cover back in place. Since I reused the plug cover, the cable looks the same as it did before. The only noticeable difference is the stereo, rather than mono, plug (and the minor scratches that I put in the chrome ring in my early attempts to remove the back cover from the earpiece).

I can now use my (old) favorite 'phones with my (new) favorite transceiver. Most of my other transceivers also accept, but do not require, stereo 'phones, so I can still use the HS-5 with them.—*Mark Erbaugh, N8ME, 3105 Big Plain-Circleville Rd, London, OH 43140-9474;* **n8me@arrl.net**

## HEIL PROSET HEADPHONE MODIFICATION

◊ The newest Heil ProSet features a phase-reversal switch on one ear that changes the center-focused sound (in phase) to a spatial-diversity focus (out of phase). According to the Heil Web page, **www.heilsound.com/proset.htm**, the out-of-phase focus helps in pileup conditions. I've modified my ProSet and find that the out-of-phase focus is in truth very beneficial in pileups as multiple signals are more easily differentiated.

The modification is simple: Mount a RadioShack Subminiature DPDT slide switch (#275-0407) as low as possible in the right ear speaker housing (see Figure 9). I bent

Figure 9—A photo of Heil ProSet headphones with VE6LB's speaker-phase switch in place.

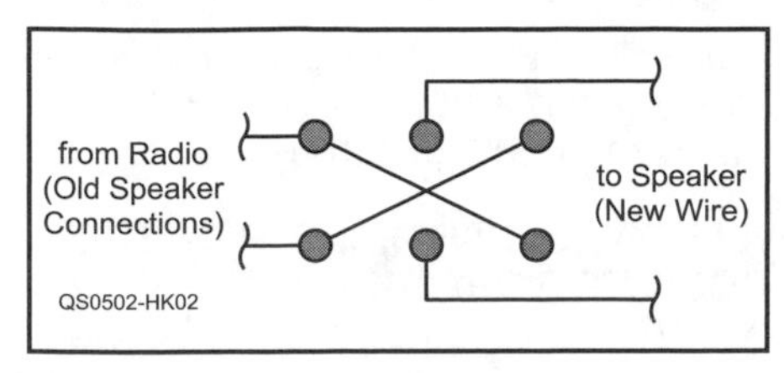


**Figure 10—VE6LB's wiring diagram for the speaker-phase switch.**

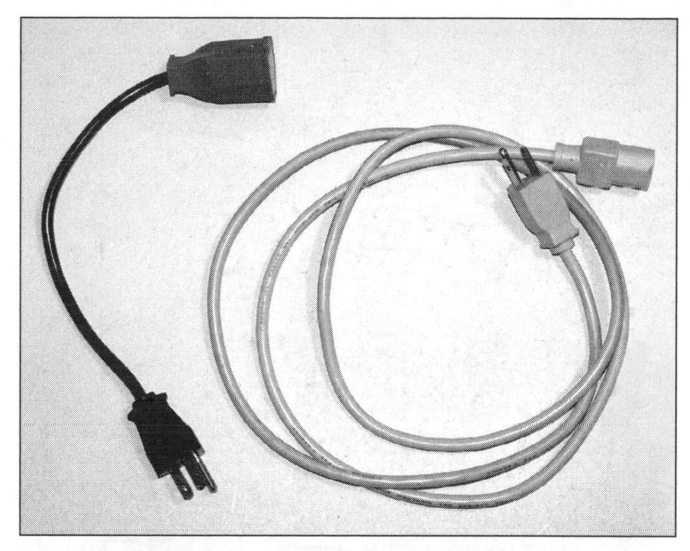

**Figure 12—A pigtail and the power cordset it's made from.**

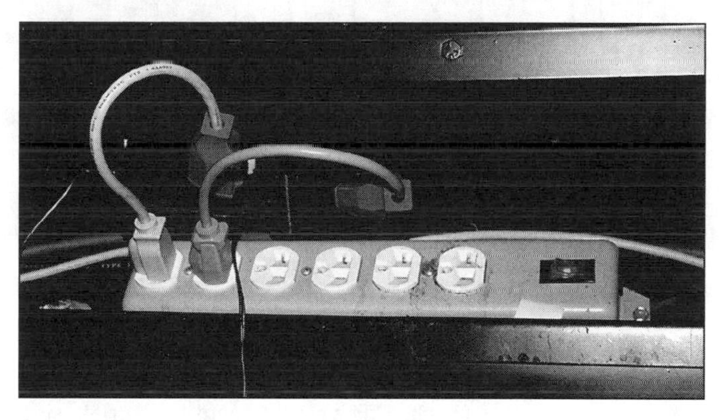

**Figure 13—The newly made pigtails hard at work.**

over the lugs on the switch to allow clearance to the speaker and wired it as shown in Figure 10.—*Gerry Hohn, VE6LB, 72 Woodacres Cres SW, Calgary, AB T2W 4V6, Canada;* **ve6lb@rac.ca**

## ANDERSON POWERPOLE GROUNDS

◊ With enjoyment of going portable QRP. Removing and attaching the grounds from the backs of my rigs was time consuming. So I use an Anderson Powerpole connector for the grounding braid. I mark these connectors with green heat shrink. Anderson does make green Powerpoles.—*Dwight Agnew, AI4II, 9335 King George Dr, Manassas, VA 20109;* **Ai4ii@arrl.net**

## (WALL) "WARTS" ON PIGTAILS

◊ How many of us have filled a power strip made for five to six plugs with only two or three "wall "warts" as shown in Fig 11? The solution is short extension cords, but at least one manufacturer wants $25 for five extensions in a national catalogue. A visit to the parts bin yields the perfect item—left-over ac power cords from discarded printers, computers and other equipment. Simply cut off the proprietary female plug, leaving the factory-made male three-prong plug and about a foot of cable (Figure 12). Install a female plug of your choice (wall "warts seldom have three-prong plugs) from the local home center, and space-hogging transformers are tamed (Figure 13). Instead of $25, I recycled left-over cables with $5 worth of parts.—*73 de Bill Smith, K1ARK, 3032 N Strawberry Dr, Fayetteville, AR 72703;* **k1ark@arrl.net**

*Bill has a fine idea here, and his idea led me to another way to skin this cat. Because wall "warts" seldom have three prongs, simple lamp extension cords do the job too. The typical three-outlet tap on such cords can hold two "warts," three if the spacing of the outlets permits it. Such cords often sell for less than $1 on sale. Finally, some outlet strips are constructed with the prong sockets perpendicular to the strip (Figure 8). If they're closely spaced, you need several outlets for two "warts," but properly spaced they can accept a row of "warts."*—Ed.

**Figure 11—Some outlet strips are more friendly to wall warts.**

## A TS-140S/680S RECEIVING PROBLEM

◊For a few years now, when I've used my Kenwood TS-680 transceiver, I would often reach a point when the received signals and background noise level would slowly drop until the S-meter barely registered. This suggested to me that perhaps a relay or two wasn't making proper contact. Usually, I could clear this condition by merely changing bands (up or down) and then back, letting the relays wipe themselves clean. This Christmas, I finally had some free time to wade in and sort out the problem. I'm happy to say that I was successful (and learned something new). The '680 is exactly the same radio as a '140 but with 6-m capability added, so what follows applies to both radios.

Refer to the accompanying block diagram and to the schematic in the owner's manual, if you have them available. The received signals go from the ANT(enna) connector to the FILTER UNIT circuit board. At that point, there is an RF choke, L32, in shunt to ground (to drain off antenna static charges) and a direct path through the SWR sensing circuit to a set of low-pass filters.

The appropriate filter is selected using small OMRON G6E-series relays at the input and output of each filter. The signals then go through the TR relay to the input of the SIGNAL UNIT circuit board.

From here, they go through a relay-switched 20 dB attenuator, an IF trap, a parallel RC network (designated R3C2) and then through a set of diode-switched high-pass filters to the first mixer (Q18 and Q19).

You'll see that the circuit has been designed so that a dc path exists from ground through L32 to the antenna connection and

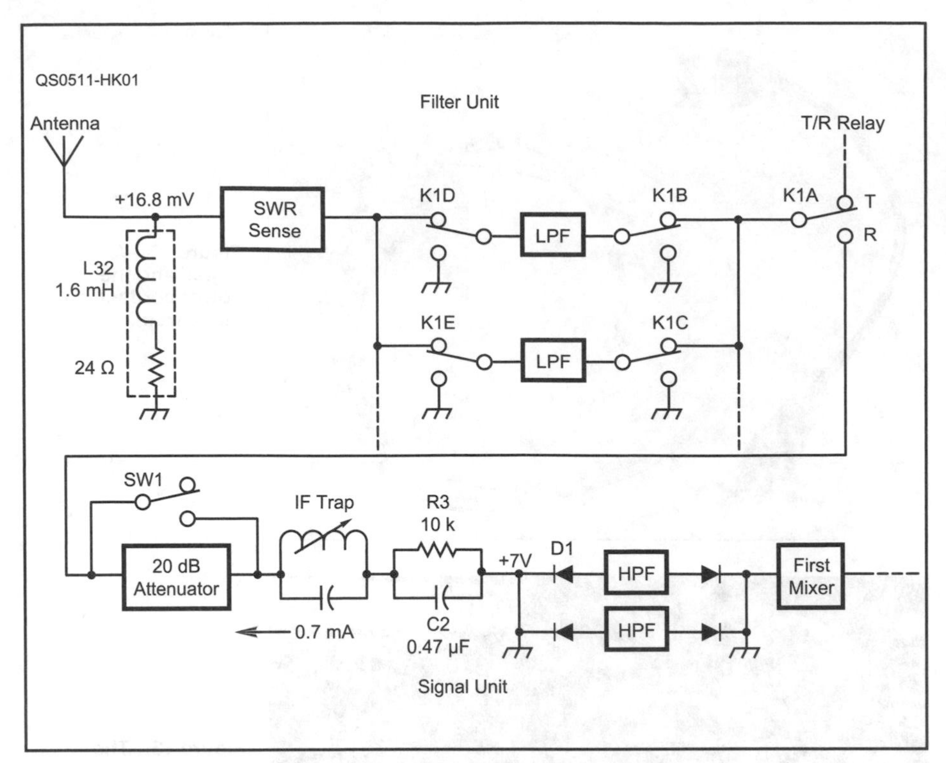


**Figure 14—A simplified block diagram of the TS-140S/640S receiver.**

right through to the junction of R3C2 and D1.

In receive mode, there is about +7 V dc at this junction. R3 is a 10 kΩ resistor. Assuming that all other components have little resistance, about 0.7 mA of current should flow back through the dc path to ground at at the "cold" end of L32. This current is what is called a dc *wetting* current.

Relay contacts carrying very little current (as in a receiver) eventually become oxidized. Allowing a small wetting current to flow through the contacts prevents this oxidation and keeps the contacts clean and presenting little resistance. Apparently, telephone and audio engineers knew this many years ago, but it doesn't seem to have been passed down to us Amateurs (or we never learned it!).

I measured the winding resistance of L32 as 24 Ω. Therefore, 0.7 mA of current flowing through it should produce a reading of +16.8 mV across it (measured with a DVM at the antenna connector). Mine read nothing (0 V). No voltage means no wetting current, leading to oxidized relay contacts and unreliable receiver performance. Where did the wetting current go? This radio isn't too difficult to disassemble. Using proper anti-static-charge precautions (that is, a grounded wrist strap and conductive mat), once the case halves are are removed, only four screws join the chassis halves. Then the chassis halves "hinge" apart, just like opening a book.

With the front of the radio facing you, removing the cover plate on the extreme left exposes the FILTER UNIT so that it can be checked. The circuit board on the right side is the SIGNAL UNIT. Unfortunately, most of the components that I was interested in checking were on the underside of that circuit board, so it had to be removed.

Once I could see the underside of the SIGNAL UNIT and got myself oriented, I found where R3C2 were supposed to be. Well, C2 was there, and the solder pads for R3 were there, but R3 *wasn't there!* It had never been installed. Mystery solved. Not having any surface mount resistors, I found and installed a 10 kΩ, ¼ W resistor from the junk box. It fit with lots of room to spare.

After a careful re-assembly (all screws accounted for) came the moment of truth. The radio powered-up fine. I checked for voltage at the antenna connector and found +16 mV! We were back in business. Subsequent testing showed all functions working normally without interruption and the receiver even *seemed* a bit hotter (a satisfied mind adds 6 dB to reality). I went through each of the bands and gave the wetting current a few seconds to clean up the relays. That's all it took.

The problem that I had with this radio certainly illustrates the need for (A) "wetting" the contacts of relays that are used to carry low-level signals and (B) buying the service manual with the rig. I hope that not too many TS140s or 680s left the factory without resistor R3. If you're a current owner of one of these radios (I bought mine in 1989) and are experiencing the problem which I did, perhaps I've given you a good lead.—*Ken Grant, VE3FIT, 5 Windrush Tr, West Hill, ON, Canada M1C 3Y5;* **ve3fit@rac.ca**

## A FREQUENCY INDICATOR ON THE ROCK-MITE QRP RIG

◊ For QRP enthusiasts who like just the basics you can't beat the Rock-Mite designed by Dave Benson, K1SWL ("The Rock-Mite — A Simple Transceiver for 40 or 20 Meters," Apr 2003 *QST*, pp 35-38).

The Rock Mite looked like an easy to build project, and when I saw the $27 price, I said: "I need one of those for 20 meters." So it went on my 2003 Christmas list, and Santa — in the form of one of my offspring — put it under my tree. It went together easily, as anticipated, and by mid-January I was on the air making contacts. Yes it takes a little doing to be a "sitting-duck" with no knobs to turn or tweak, other than a push button, but I soon learned that this little guy worked quite well. I get frequent comments on how well it keys and how strong the signal is, and once you develop the patience to wait on your frequency for the contacts, you get along just fine.

On my 20 meter version I added two push buttons on the top of my package of choice — the Altoids tin that serves as a key — one for dits and one for dots. This made a complete station you can fit in a shirt pocket. I just used two of the same type push buttons as on the front panel (for keying speed and frequency shift control), and added knobs. It works fairly well using the first two fingers straight out with an up and down motion. I also added a jack for a key in case you get tired of having your fingers do the talking! Having had so much fun with the 20 meter version, Santa again favored me with a 40 meter version in 2004. Figure 1 shows my Rock-Mite arsenal.

After much operating, I found one thing that was a little annoying — trying to figure out on which frequency I was transmitting. As users know, the Rock-Mite normally operates on the QRP calling frequencies of 7040 kHz on 40 meters and 14060 kHz for 20 meters. It can also be shifted down 700 Hz for a second frequency on each band. You toggle between the two frequencies with a single push of the control button. I found it handy to know when you were actually on the calling frequency and contrary wise when you were hearing a station that was probably on the calling frequency. (I say probably because in a direct conversion receiver you get both the sum and difference heterodyne product of incoming signals.)

My answer to this was to mount a miniature LED on the case to indicate transmitting frequency. At the junction of R9, R10 and D5 in the Rock-Mite, the voltage comes and goes depending on whether Q2 is conducting or not. My approach to adding the light is shown in Figure 16. I selected LEDs about ⅛ inch in diameter. Using a multimeter, I

Figure 15 — The author's Rock-Mite collection.

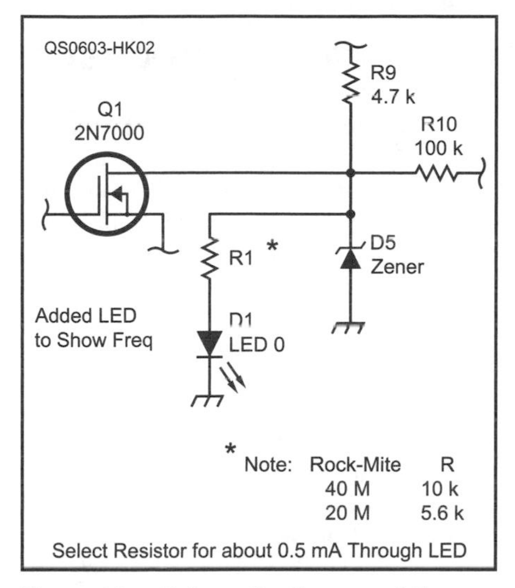


Figure 16 — Schematic diagram of the Rock-Mite frequency indicator circuit. Operation — Lamp steady ON in receive and blinks OFF at key down, transmitter frequency7039.2 kHz; lamp steady OFF in receive and blinks ON at key down, transmitter frequency

Figure 17 — Details of LED lamp (lower left) mounting.

Figure 18 — D cells ready for portable operation.

experimented to find the lowest current that would give a usable indication. I found this to be about 0.5 mA. I then selected the resistors to give the necessary voltage drop. This resistor in series with the LED and connected to the parts junction mentioned earlier in the Rock-Mite did the trick.

Figure 17 shows how I mounted the lamp in place with a glue gun. One end of the LED goes to ground (watch polarity) and the series voltage dropping resistor was covered with a sleeve and connected to the top of R9 in the Rock-Mite. Using this method you make no alterations to the Rock-Mite circuit board.

Now when I am transmitting on the calling frequency the light is "off" during receive, and flashes "on" with keying. When I transmit on the alternate frequency, 700 Hz lower, the light is continuously illuminated when receiving, and flashes "off" with transmit keying. Now you not only have an RF transmitter — you have ship-to-ship flashing light capability! It works quite nicely and is a pleasure to know the frequency you are operating on.

I run my Rock-Mites from eight AA or D cells, and it works great either way. The D cells are an exceptional choice since they provide months of operating time. I put two plastic four D cell holders back to back with a switch and terminals as shown in Figure 4.

The Rock Mite brings me back to 1947 when I started as a ham and my ear was the filter. Don't worry too much about the low power of about a 0.5 W or 500 mW. When you think about it, with all other things being equal, the signal you receive from a 0.5 W rig compared to that from a 100 W rig is only 23 dB or about 4 S-units lower. I can say from experience, I have had reports from S3 to S9 plus and have worked from the Chicago area to all coasts and eastern Canada. It is a fun rig, and I think if you use a Rock-Mite you will find the frequency indicator lamp a worthwhile addition. — *A. Robert Patzlaff, W9JQT, 422 W Maple, Hinsdale, IL 60521;* **bobpatz@comcast.net**

## BATTERY POWER FOR THE FT-897D

◇ I recently purchased a Yaesu FT-897D MF/HF/VHF/UHF transceiver with the intention of using it as my portable and backup radio. I wanted a radio that would give me 100 W with a regular power supply as well as operate on battery power with an output of 20 W. What I was not pleased about was the cost of the batteries and charger, either from Yaesu or aftermarket suppliers. I also did not want NiCd or NiMH batteries because they should be fully discharged before recharging. I preferred using a gel cell type battery. This would also leave the section where the batteries are mounted available for the internal power supply, reducing the number of pieces of equipment I would need to transport for a portable outing.

After studying the schematic of the radio I knew it could be modified to my specifications. I removed the bottom cover and the power amplifier (PA) shield to allow access to the dc power input section so I could take some ohmmeter readings to verify what I saw on the schematic. Sure enough, it could be done, but exactly how was the next problem. When I modify a radio I do not like to alter the capabilities built into the radio. I much prefer to add to them.

After removing the top cover and a bit of snooping, I found the circuit board I was looking for under the BATTERY SELECT switch (see Figure 19). I carefully disconnected the internal speaker and the battery select switch connector wires and set the top aside. Again using the ohmmeter I confirmed I was looking at the right PC board. I had also noted that the radio used the standard 6-pin Molex dc power connector. It only made use of four of the six pins, leaving two available. Using the unused pins, I brought in an extra wire to carry the battery power and attached them to the B connector pins on the board as shown in Figure 20.

I replaced the battery switch and internal speaker connectors and reinstalled the top cover. Again I used the ohmmeter to double check that I had a good ground through all the new wiring and that the red wire did indeed go to the dc power input on the PA board. There are two dc input points on the PA board. In Figure 20, the connections on the left are the input for an external 13.8 V dc supply for 100 W operation and come directly from the Molex connector at the back of the radio. The wires on the right come from the white connector seen in the Figure 1 and feed dc power from the internal battery select switch to the 20 W section of the PA board. The ohmmeter indicated the new red wire had a good connection to this point.

You can also see the two new wires installed in the rear Molex connector in Figure 19. I had a spare six position male Molex connector and took my time installing the red and black wires to make sure I had the correct polarity. I also have only those two wires in this male connector. Even if I use it with an external dc supply I'll still only get 20 W output. Finally after making sure all connections were good I reinstalled the PA cover in the battery compartment and then the bottom cover. This addition does not alter the radio capabilities. You can still install two batteries, charge them while they are installed and operate from either internal battery as before. The only unusual thing you'll notice is that with an SLA or gel cell battery, even fully charged, the battery LED on the front of the radio will flash as you transmit. This lets you know when your internal battery is going dead. The internal packs provide a full charge voltage of 13.8 V since they use an 11 cell pack. Go below 11.7 V and the radio begins to get grumpy and shuts off shortly below this voltage. So using a 12.6 V battery to begin with and transmitting can make the voltage float below 12 V so the radio thinks the battery is going dead.

Testing the installation went well and I made contacts with KØAAA while he was ice fishing, K7RX while he was giving his new TS-480 a workout, and finally CO8LY. All with 20 W while using my portable Cushcraft MA-5V antenna. — *Johnny Knight, WB4U, 2104 Irby Rd, Monroe, NC 28112;* **wb4u@earthlink.net**

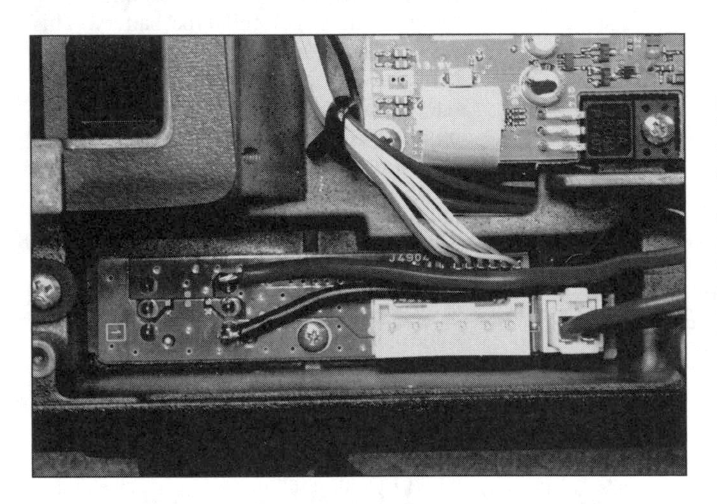

Figure 19 — Target area below the BATTERY SELECT switch under the top cover. Two unused pins in the power connector and some extra wire and we have external battery power on the B battery switch position.

Figure 20 — DC power input on the PA board. Wires on the left are connected directly to the dc power input Molex connector seen in Figure 19. The red and black wires on the right go to the battery switch PC board.

## A PORTABLE GENERATOR EVEN WHEN ALMOST HOME

◊ While working on a tower project that was located some distance from the house, or any other source of convenient ac power, we decided to use a small portable generator instead of a long cascade of extension cords. In addition to having more heat from the soldering iron due to reduced cable loss, we found that the exhaust from the generator provided a better source of heat to use with our shrink tubing than the usual soldering iron. — *Glenn Kurzenknabe, K3SWZ, 23 Carriage Rd, New Cumberland, PA 17070;* **bgkurz@epic.net,** *and Dave Smith, W3SOX, 1367 Kiner Blvd, Carlisle, PA 17013*

## A SIMPLIFIED LINEAR AMPLIFIER SOFT START CIRCUIT

◊ If you're like me, you searched until you found a filament transformer for your linear amplifier project with specifications that exactly matched the tube manufacturer's filament parameters. Then, worrying about cold start inrush current and loss of tube life, you concocted a circuit of ICs, relays and resistors to limit the current momentarily.

Well, it turns out that there is a much simpler and better solution. Just pick a transformer with a voltage somewhat higher than the rated filament voltage and put a resistor in series between the transformer and the tube and forget it. But just how to do that correctly and make the numbers fit together? Read on.

The equations governing start and run for the filament circuit are as follows:

$$V_{oc} = I_{start}(R_w + R_x + R_c) \quad \text{(Eq 1)}$$

Conditions for start

$$V_{oc} = (F_{fil} + I_{fil}(R_w + R_x))/(R_w + R_x + R_c) \quad \text{(Eq 2)}$$

Conditions for run

where:

$V_{oc}$ = Open circuit voltage of the filament transformer.

$I_{start}$ = Cold start "in rush" current.

$R_w$ = Winding resistance of the transformer (including correction for the influence of the primary, typically about 1.17 times the dc resistance of the secondary winding).

$R_x$ = Series limiting resistor placed in series with the transformer and tube filament.

$R_c$ = Cold resistance of the tube filament.

$F_{fil}$ = Manufacturer's rating for operating filament voltage.

$I_{fil}$ = Manufacturer's rating for operating filament current.

The two independent variables here are the choice of starting current, $I_{start}$, and the open circuit voltage of the filament transformer, $V_{oc}$.

For instance, the manufacturer's specs for the 8877 tube are5.0 V at 10.5 A. There are lots of 6.3 V transformers around. Let's see how one would fit with the equations. The selected transformer open circuit voltage turned out to be 7.20 V and the effective secondary winding resistance was determined to be 0.040 Ω. Read on to find out how we determine such small resistance values.

One more measurement is the cold resistance of the 8877 filament. That turned out to be a nominal value of 0.06 Ω.

Using Equation 2, the series limiting resistor, $R_x$, is computed to be 0.17 Ω. This should put exactly 5.00 V on the hot filament. Using Equation 1 we see that the calculated starting or inrush current will be 26⅔ A. How does this compare with the inrush current from a 5 V, 10 A transformer? Well, for a 5 V transformer that will handle 10.5 A, you are dealing with an open circuit voltage of about 5.70 V and an effective resistance of about 0.045 Ω. This predicts an inrush current of 54 A. So, selection of a 6.3 V transformer that will handle the required current has cut the inrush current in half. Not bad and well within the manufacturer's guidelines. You want less inrush current? Pick a yet higher voltage filament transformer and measure and work the equations.

What's the penalty involved here? The resistor will dissipate 18 W so the transformer needs to handle 75 VA rather than 53 VA. Not a bad trade-off.

The benefit to this circuit is that as the tube ages, its resistance goes up and the applied voltage will go up accordingly. This will improve the emission of the aging cathode. This is especially true of thoriated tungsten filaments. The benefit is less with indirectly heated cathodes such as the 8877.

How did this work out on the bench? I jury rigged some 3 W wirewound resistors from the junk box in parallel. The first reading at turn-on was around 1.0 V and final steady state filament voltage came out to be 4.81 V, nicely within the manufacturer's recommendation of 4.75 to 5.25 V. In practice this would be six, 1 Ω, 3 W resistors connected in parallel; the better the tolerance of the resistors, the more sure the result.

It does not pay to go outside of these limits for filament voltage. According to Eimac's book, *Care and Feeding of Power Grid Tubes,* running low voltage on the filament causes a chemical reaction called "poisoning" the cathode. And running excessively high filament voltages prematurely exhausts the emission of the cathode.

One does not measure the low values of resistance mentioned here directly with the usual digital multimeter. The procedure is to put the resistance to be measured, $R_x$, in series with a 3 or 4 Ω power resistor. By making the measurement quickly you will not need a power resistor with more than a 15 W rating. If you are slow, you better have a 50 W resistor and connect this to a power supply. The power supply is not critical. Everybody probably has a 12 V supply or 12 V battery these days. Measure the voltage across each part of the series circuit and calculate the result as follows:

$$R_x = (V_x \times R)/V_R \qquad \text{(Eq 3)}$$

When you have installed this simplest of circuits, you can sit back after you have turned on your homebrew linear amplifier and know that you are enjoying the best of all worlds in "care and feeding" of your power grid tube.

As a postscript to this article, let it at least be mentioned that there are transformers available, useful for tube filament service, that have multiple windings and multiple Voltages. Some of these are probably candidates for use but the math gets a bit more esoteric to make things come out right. — *Ralph Crumrine, NØKC, 1621 Sunvale Dr, Olathe, KS 66062;* **n0kc@arrl.net**

## IF ALL ELSE FAILS — RESET

◊I have RadioShack HTX-202 and -404 transceivers. Every once in a while I get an ERR 1 error message on the screen. I was tipped off on how to fix the problem. With the radio off press both the F and D buttons, and turn on radio. That will get you back on the air, although you will have to reprogram all your memories. — *Stuart Ballinger, WA2BSS, 11 Lown Ct, Poughkeepsie, NY 12603-3321,* **wa2bss@hvc.rr.com**

## KENWOOD TS-940 BATTERY REPLACEMENT

◊The Kenwood TS-940 was a very popular radio and there are still thousands in daily use. All Kenwood rigs in the TS-940 and similar series contain two lithium coin batteries, one for memory and one for the timer. The rig will retain most of its functions with a bad timer battery but when the memory battery goes soft the rig will not function. If the main display in your 940 "goes nuts" and the radio will no longer transmit or receive, you probably need to replace the memory battery.

The replacement batteries from Kenwood parts suppliers are a bit pricy. I recently bought a pair and with shipping the best price I could find was $25.00. One supplier wanted over $35.00. On examination of these batteries I found they were standard CR2032 (memory) and CR2450 (timer) batteries. The difference was that the expensive batteries had soldering tabs attached. Both of these batteries solder to posts on internal circuit boards. As an experiment I tried to use solder to attach tabs on an old battery but quickly found the battery case metal would not flow solder.

Figure 21 shows a standard CR2032 adapted for installation in my second 940. I slid the battery into a piece of ½ inch diameter heat shrink tubing, a snug fit, then tucked ground lugs into the top and bottom. Then I heated the tubing until it held the lugs tightly against the top and bottom of the battery. I prefer lugs with serrated teeth as illustrated; they seem to grip the battery surface better. Use good quality heat shrink tubing. *Be careful not to overheat the battery!* Now the battery, which is available locally for about $3 to $4, is ready for installation in the radio. The memory battery is located on Digital Unit A in the 940. You will have to remove the speaker bracket and the top shield under it to gain access. Once you have removed this top shield, the battery will be easily visible.

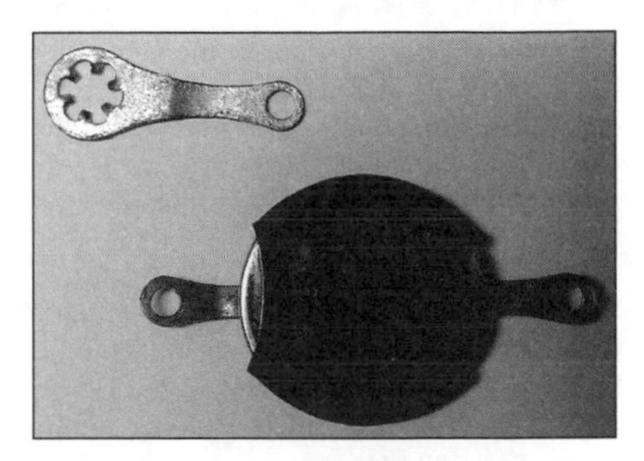

**Figure 21 — A pair of toothed-washer solder lugs are held tightly in place against a button-cell battery to make a replacement battery for the Kenwood TS-940 (and probably many other radios).**

The "timer" battery in the 940, located behind the LCD, is physically a bit larger but the same method can be used to secure the solder lugs using a larger diameter heat shrink tubing. I replaced both batteries with a savings of nearly $15.

Once soldered into place the battery is mechanically secure, although you might consider placing a small piece of double-stick foam tape under the battery if your radio might require frequent movement or transportation. With these batteries, my 940 has given hours of reliable service.

Kenwood recommends these batteries be changed every 3 years for the timer and 5 years for the memory backup. Replacement of these batteries requires some elementary disassembly of the radio. This and certain keystrokes to reboot the memory after battery replacement may vary from one model to another. For the 940, after memory battery replacement, hold in the A=B button while you turn the power on. — *James Viele, W8JV, 161 Fox St, Hubbard, OH 44425-2122;* **w8jv@yahoo.com**

## CYLINDER CRYSTALS FOR 10 METERS

◊ If you are looking for inexpensive 10-meter quartz crystals, try out the new cylinder crystals on your next 28-MHz project. These very tiny (3 mm diameter by 10 mm long) crystals are high quality components in spite of their bargain 83-cent price.

Mouser Electronics (**www.mouser.com**) offers three 10 meter frequencies: 28,000, 28,636, and 29,491 kHz (part numbers: 695-CSA309-28, 695-CSA309-286, and 695-CSA309-294). Frequencies are specified for an 18 pF load capacitance. The frequency of these cylinder crystals can be shifted upward by as much as 25 kHz by adding a small series capacitor. A 5 pF series capacitor increases the frequency by about 12 kHz. A 1 pF capacitor results in an increase of approximately 22 kHz.

Recently I rebuilt a QRPP 10 meter transmitter[1] using a 28,000 kHz cylinder crystal in series with a homebrew compression capacitor.[2, 3] This transmitter tunes from 28,002 to 28,023 kHz, covering most of the 10 meter CW DX window. Although I am waiting for some sunspot activity to give this rig a fair on-the-air trial, I am very pleased with the bench tests.

To get the most out of cylinder crystals, I suggest:

1) Watch out for crystal heating, evidenced by excessive turn-on frequency drift. If this occurs, reduce the value of the feedback capacitors in the crystal oscillator.

2) To get the maximum frequency shift when using a series variable capacitor, do everything possible to minimize stray capacitance at the junction of the crystal and the variable capacitor. Solder the crystal directly to the variable capacitor — don't make that connection via a printed circuit board. I find that a homebrew compression capacitor is far superior to any commercial capacitor for this application. Try a half scale version of the referenced compression capacitor.

3) Short lead lengths greatly reduce the likelihood of unreliable oscillator behavior. Tiny components — such as these cylinder crystals — make it fairly easy to keep leads short. — *Lew Smith, N7KSB, 4176 N Soldier Trail, Tucson, AZ 85749*; **n7ksb@arrl.net**

## DIRTY TRIMMERS DISABLE YAESU FRG-8800

◊ In 1986 I bought a new Yaesu FRG-8800 receiver for its general coverage. I've owned it continuously since then and have treated it well. Nevertheless, twice it has failed totally, with no audio and a flashing or blank display. With some effort, I traced each failure to a dirty trimmer capacitor across a crystal. In each case, merely adding a drop of contact cleaner and wiggling the trimmer corrected the problem. I suspect that others may have encountered the same problems and may benefit from details.

The first failure involved a flashing frequency display in addition to no receiver audio. This was caused by a dirty trimmer, TC01, in a shielded box on the PLL circuit board. It is accessible by removing the receiver top cover. The service manual specification for TC01, after warm-up, is 18 MHz (within 20 Hz) at nearby test point TP14. Lacking a frequency counter, I found it adequate simply to return the cleaned trimmer approximately to its original position and to verify the result by tuning WWV in a sideband mode.

The second failure was more complete: no audio and no information on the lighted display. This was caused by a dirty trimmer, TC02, in another shielded box on the PLL circuit board. Similar application of contact cleaner and trimmer wiggling brought the receiver instantly back to life. The service manual specification for TC02 adjustment is 5.996544 MHz at nearby test point TP09. Alternatively, I checked the receiver's clock display over time to verify reasonable adjustment of TC02 after cleaning.

It is worthy of note that those two trimmers are the only ones on the PLL circuit board, according to the service manual's parts list. It is disappointing that both trimmers ultimately failed, each time with baffling symptoms. The only other trimmers in the FRG-8800 adjust the BFO frequency on SSB and are located on the "main" board, accessible by removing the receiver bottom cover. If there are SSB problems, those components will be prime suspects. — *Paul Kirley, W8TM, 7805 Plainfield Rd, Cincinnati, OH 45236;* **w8tm@arrl.net**

## MFJ ANTENNA ANALYZER ON/OFF SWITCH

◊ If you have been the owner of an MFJ HF/VHF antenna analyzer for any length of time, you have probably experienced the agonizing feeling when you went to use it and found that the batteries were dead because the ON/OFF switch was accidentally pushed ON at some time. The switch is exposed on the front panel, which makes this scenario quite likely. It only happened to me once after buying a fresh set of AA cells before I found a simple solution.

I glued a rubber grommet on the front panel around the power switch and the problem went away. See Figure 22. You could use any ring-shaped device as long as it is higher than the switch protrudes from the front surface. The batteries last a lot longer now. — *73, Ed Denton, W1VAK, 14 Holland St, Falmouth, MA 02540*; **w1vak@verizon.net**

**Figure 22 — W1VAK glued a rubber grommet to the front of his MFJ Antenna Analyzer to prevent the unit from being turned on during transit or storage.**

## CIGARETTE LIGHTER ADAPTERS

◊ I recently purchased a solar-powered car battery charger and wanted to use it to charge my 2 meter radio. For this I needed a way to connect two cigarette lighter plugs. Visiting my local hardware store, I discovered that a standard copper T fitting for ¾ inch tubing works perfectly. (The fitting itself has ⅞ inch openings.) See Figure 23. The extra opening on the T can be used to monitor voltage, or, with a scrap of two-sided circuit board, current.

A standard ¾ inch straight coupling is too short to hold two cigarette lighter plugs, but it can be used to build a female connec-

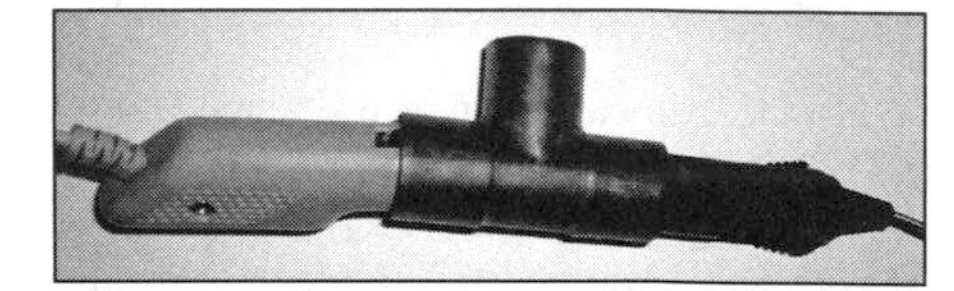

**Figure 23 — K2PNK used a ¾ inch copper pipe T to join two cigarette lighter plugs. The center pins of the plugs make contact in the center and the T provides the ground connection. If you are not using the side tap on the T, you might consider closing it with a pipe cap to prevent fingers or loose objects from falling into the T and shorting the center pins to ground.**

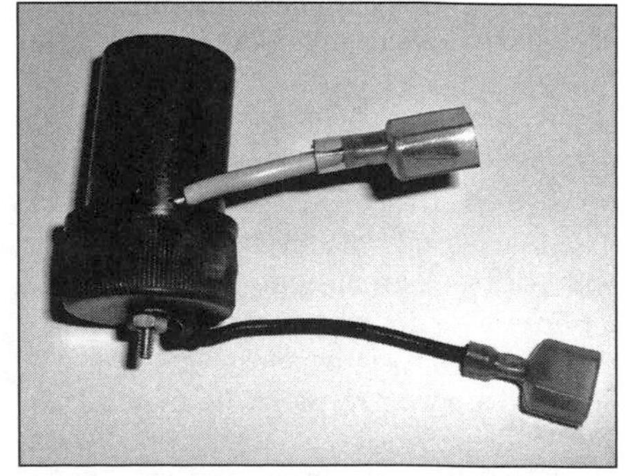

**Figure 24 — K2PNK made a cigarette lighter receptacle using a ¾ inch copper pipe coupling. This photo shows how he used a small machine screw in a soda bottle cap for the center pin connection. The ground lead is soldered to the pipe coupling and the cap is glued in place using epoxy.**

[1]L. Smith, N7KSB, "An Experimental ½ Watt CW Transmitter," *QST*, Nov 1994, p 84.

[2]L. Smith, N7KSB, "A Compression Capacitor For QRP Transmitters," *QST*, Feb 2002, p 67.

[3]L. Smith, N7KSB, "A Simple 10-Meter QRP Transmitter," *QST*, Mar 2000, p 43.

tor for such plugs by fitting it with a plastic end cap with a screw terminal in the center. A screw-on soda bottle cap and some epoxy seems to work. Figure 24 shows a completed connector. — *Arnold Reinhold, K2PNK, 14 Fresh Pond Pl, Cambridge, MA 02138*; **k2pnk@arrl.net**

## TEN-TEC POWER SUPPLY CONNECTOR HINT

◊ Having recently purchased a nice Ten-Tec Triton 544 from eBay, I needed to make an additional dc power cable. A look through the Mouser catalog showed that they had exactly what I needed. The plug for the radio and the power supply are the same; just the pins inside are different. The plugs are AMP Universal MATE-N-LOK, four position connectors, Mouser part no. 571-14807020. The pins for the power supply end are male, part no. 571-3505471. For the radio end, the pins are female, part no. 571-3505501.

The pins will take an AWG 12 to 14 wire. I crimped and soldered the pins on, and it works great. Total price for two plugs and eight pins is approximately $1.50, not including shipping. The best part about ordering from Mouser may be that they do not have a minimum order. The Web address for Mouser is **www.mouser.com**. — *73, John Best, Sr, WA1YIH, 162 Meadow Rd, Topsham, ME 04806*; **wa1yih@yahoo.com**

## REPLACING DEFECTIVE METER MOVEMENTS

◊I suspect that all of us have used some type of equipment (such as a piece of test equipment) that incorporated an analog meter movement. If the meter movement fails and the equipment is sufficiently old, such that an identical replacement is unavailable, we are usually inclined to believe that we must "junk" the entire piece of equipment. There is a tendency to believe that each meter movement is a unique device. In one sense, it *is* unique. In another, it is a common device.

All analog meter movements use the same principle to project information. Two magnetic fields interact to cause a repulsive force. For most meter movements, these opposing fields are caused by a permanent magnet, and current through a coil. The resultant repulsive force pushes against a spring, causing the needle to rotate. The spring tends to keep the meter needle pegged against a stop when there is no current. See Figure 25.

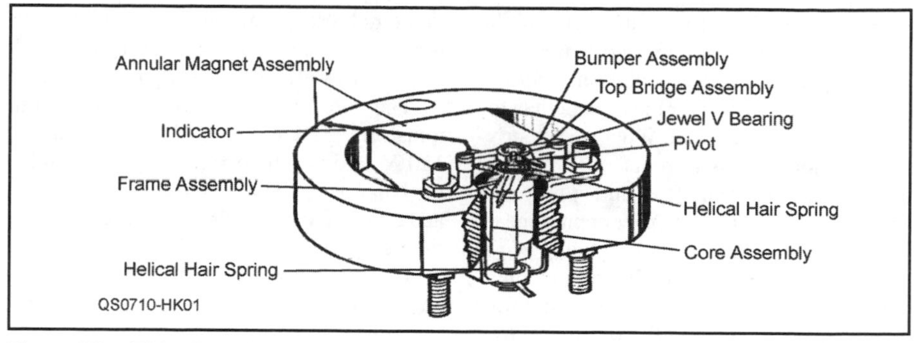


**Figure 25 — This diagram shows a typical D'Arsonval meter movement.**

So how does a meter movement in a voltmeter differ from one that is used in a tube tester? There are two ways they may differ: the electrical characteristics and the markings on the meter face. Are there special electrical characteristics that define that a particular meter movement must be used for a particular application? No! Since the meter — I will abbreviate the words "meter movement" to just the word "meter" must derive power from some source to enable the needle to move up scale, it is desirable to use as low a power meter as possible. For most applications, three related electrical characteristics define the meter. They are the meter current sensitivity (current for full-scale deflection), $I_{CS}$, the meter resistance, $R_M$, and the voltage across the meter to achieve a full scale indication, $V_{FS}$.

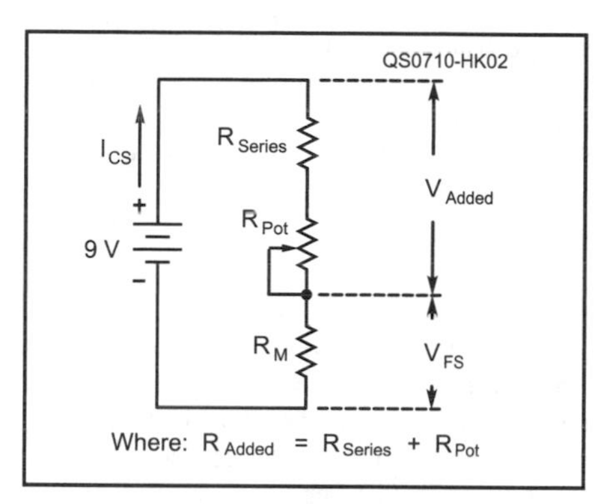


**Figure 26 — This circuit is used to determine meter electrical parameters.**

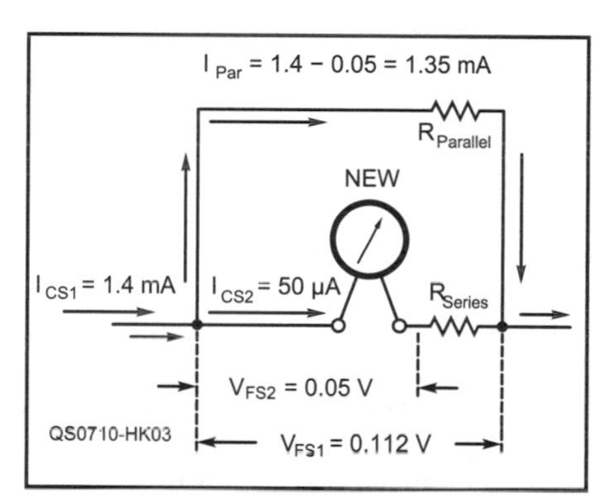


**Figure 27 — This diagram illustrates the process of calculating the new meter parameters.**

To determine these parameters for an existing movement, we can either refer to documentation for the device or measure the parameters. The documentation may be information printed in a data sheet, printed on the meter movement or printed in a user's manual. If it becomes necessary to measure the parameters, the meter must be somewhat functional. Primarily, a full-scale needle deflection must be achievable.

To determine the electrical parameters of the meter, we will use the circuit of Figure 26. A somewhat high-voltage source is used. In this instance, a 9 V transistor radio battery is a high voltage compared with the $V_{FS}$ for a meter movement (perhaps 30 mV to 200 mV). By adjusting the potentiometer such that a full scale needle deflection is noted, one can then measure (with a high impedance VOM), the voltage across the meter movement, the voltage across the resistor string, and after disconnecting the battery, the resistance of the resistor string, $R_{Added}$. With these data, we know $V_{FS}$ and can compute $I_{CS}$ plus $R_M$.

$$I_{CS} = \frac{V_{Added}}{R_{Added}} = \frac{8.878\text{ V}}{6.34\text{ k}\Omega} = 1.4\text{ mA}$$

$$R_M = \frac{V_{FS}}{I_{CS}} = \frac{0.112\text{ V}}{1.4\text{ mA}} = 0.08\text{ k}\Omega$$

One may be inclined to suggest that a VOM could be used to determine the meter resistance. In some instances, VOM use can damage a meter movement, because the ohmmeter position may put too large a current through the meter.

Two associates from the Anne Arundel Radio Club (Anne Arundel County, Maryland), Oscar Ramsey and Tony Young, work as volunteers at the Bowie Radio And Television Museum, Bowie, Maryland. They obtain and test electron tubes and sell them to generate income for the Museum.

Oscar mentioned that the meter movement failed in one of the tube testers. He found data that stated that the meter current sensitivity was 1.4 mA and the meter resistance was 80 Ω. From these data, $V_{FS}$ may be determined.

$$V_{FS} = I_{CS} \times R_M = 1.4\text{ mA} \times 0.08\text{ k}\Omega$$
$$V_{FS} = 112\text{ mV}$$

We can also calculate the power dissipated in the meter coil:

$$P_{FS} = 156\ \mu\text{W}$$

It is appropriate to state that without

knowing the Thevenin equivalent of the source feeding the meter movement, the following circuit and calculations will enable a lower power meter movement than the original to be substituted. What we are striving to accomplish is to replace the original meter with a different meter and appropriate resistors such that the replacement appears electrically the same as the original.

Some limitations on the replacement unit are that its full scale voltage and power be less than or equal to the original. For a replacement, let's choose a contemporary meter. The electrical parameters typically are/were $I_{CS}$ = 50 µA, $R_M$ = 1000 Ω, $V_{FS}$ = 50 mV and $P_{FS}$ = 2.5 µW. If we employ a circuit as shown in Figure 27, and select appropriate resistor values, we should be able to "fool" the original meter drive circuit into thinking that the original meter is still connected.

$$R_{Series} = \frac{V_{FS1} - V_{FS2}}{I_{CS2}}$$

$$R_{Series} = \frac{0.112 - 0.05}{50 \times 10^{-6}} = 1{,}240\ \Omega$$

(A standard carbon film resistor value is 1,200 Ω.)

$$R_{Parallel} = \frac{V_{FS1}}{I_{Par}} = \frac{V_{FS1}}{I_{CS1} - I_{CS2}}$$

$$R_{Parallel} = \frac{0.112}{1.35 \times 10^{-3}} = 82.96\ \Omega$$

(A standard carbon film resistor value is 82 Ω.)

From these calculations, we note that the original full scale current, $I_{CS1}$ (1.4 mA) and full scale voltage, $V_{FS1}$ (112 mV) are still delivered to the "load" of the new meter and two resistors.

The "deception" is complete!

A simple computer program to perform these calculations is available. Send a stamped, self-addressed envelope to W. Rynone, PO Box 4445, Annapolis, MD 21403. A floppy disk with a compiled computer program will be returned.

*Acknowledgments:* Mr Oscar Ramsey tested the meter movements to determine their electrical parameters. He also tested the replacement meter movement and corresponding circuitry in an electron tube tester that had a defective meter. My thanks also to Professor David Harding for reviewing the text and suggesting changes. Mr Ramy Benha of Simpson Electric kindly supplied line drawings of basic meter movements. — *William Rynone, PhD, PE*

## ISOLATION TRANSFORMERS

◊Getting rid of a ground loop was very difficult with my Yaesu FT-920 transceiver and the attached amplifiers. The exciter and peripheral equipment were operating on 120 V and the amplifiers on 240 V. The problem turned out to be uneven current on the 240 V line coming to the radio shack, which was about 75 feet from the main breaker box. Even a ground rod in a 35 foot well next to the rig did not help.

The cure turned out to be an isolation transformer. By dropping the 240 V line to 120 V to drive the exciter and peripherals, the balance on the 240 V line became even, and eliminated phase shift and the current on the neutral line. Result — no ground loop. — *73, Bill Trippett, W7VP, 15525 NE 195th St, Woodinville, WA 98072*; **w7vp@arrl.net**

## PACKET RADIO AND THE YAESU FT-221

◊The Yaesu FT-221 dates to the late 1970s and is a solid-state, all mode 2 m rig whose still fresh design complements the FT-301 HF series. Commanding a premium price on eBay, but owned by many Amateur Radio operators that prize its SSB and CW abilities, it makes a very adequate rig for packet radio at 1200 baud with its 15 W (input) FM power level.

Most who employ the FT-221 for packet radio will inject ASFK through the front panel mounted, 4-pin mic jack and take audio from the nearby ⅛ inch phone plug. A more effective connection may be had using the rear panel's 5-pin DIN accessory socket. Connecting a TNC here allows the user to preserve the microphone and headphone connections for normal use. This socket is identified in the Yaesu manual as the interface for a touch-tone keypad, but since receive signals are available, it is ideal for packet.

### *Interface Details*

The owner's manual is not altogether clear on the signals found at each of the socket's pins. I cross-referenced the schematic and confirmed the functionality as summarized in Figure 28.

Audio-in and audio-out work well with TNC signal levels without modification. In my testing, both a Timewave PK-232 and an MFJ-1270 produced good results transmitting and receiving local APRS packets.

Audio-in and audio-out levels are not affected by the Mic Level or AF Gain controls, so a user may mute the packet burst by turning down the volume. Another way is to insert an unused ⅛ inch plug at the headphone jack.

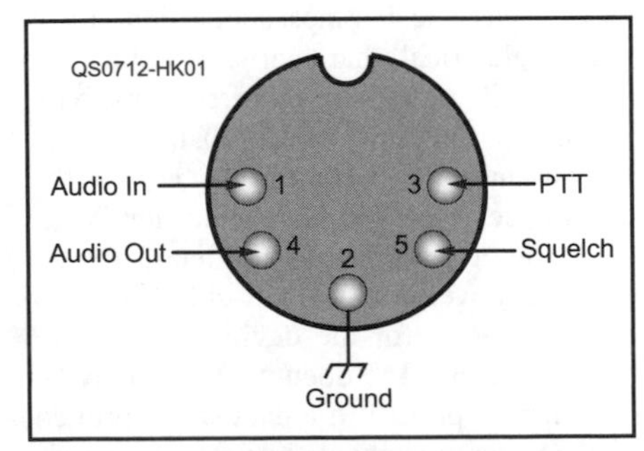


**Figure 28 — Connections at 5-pin DIN plug to fit FT-221 accessory socket. (Note that this is a mirror image of the socket's**

### *General Observations on the FT-221*

The FT-221 VFO is subject to some drift during warm-up, but is sufficiently stable for the generous bandwidth of an FM voice channel.

The Yaesu design provides for a dozen crystal-controlled channels. Those with particular sensitivity to drift might consider going "rock bound," which would also permit rapid switching from, for example, APRS monitoring to TELPAC.

Various CCTSC solutions are available to permit repeater access. Communications Specialists carry two versions of their externally programmable encoder and several companies offer kits for internal installation, using smaller but cumbersome DIP switches.

Mutek once sold a very well regarded front-end for low signal work and it appears they may once again be made available on a custom-order basis.

The typical failure mode for 1970s vintage Yaesu equipment is Relay 1 (RL1), which weakens over time and fails to properly connect various circuits. These relays are still manufactured by Matsushita (Japan) but are not presently imported to North America. They are available directly from Yaesu at a premium price.

### *Conclusion*

Since late winter 2007, I have been using two FT-221 radios for remote weather station telemetry to the county EOC, and simple APRS beaconing and monitoring. As these rigs are thirty years old, typical maintenance was required, including replacement of RL1. They have functioned flawlessly, however, in 24/7 operation with observed two-way results comparable to other stations in the neighborhood, using either a Cushcraft Ringo with Belden 9913 or a RadioShack discone antenna with plain old RG-8 coax. — *73, Joseph Ames Jr, W3JY, 10 Andrews Rd, Malvern, PA 19355*; w3jy@arrl.net

## SCRATCH-PROOF THE IC-706/703 FLIP STAND

◊ I frequently use the flip-up stands on both my ICOM IC-706MKIIG and IC-703 transceivers when operating from home and when operating portable. During portable operation (usually when I visit my Dad) I put the radios on top of a nice piece of furniture. At home, my IC-706MKIIG normally sits on top of my FT-1000MP MKV (I use it for 6 meters from home). I usually put a cloth under the flip-up stands just to make sure I don't scratch anything. (The stand is smooth, but it is hard metal.)

I wanted something a little better. While shopping at my local hardware store, I found a package of small rubber grommets designed for ⅛ inch diameter cable. These grommets are very small, and are only about ⅜ inch in diameter. These looked perfect

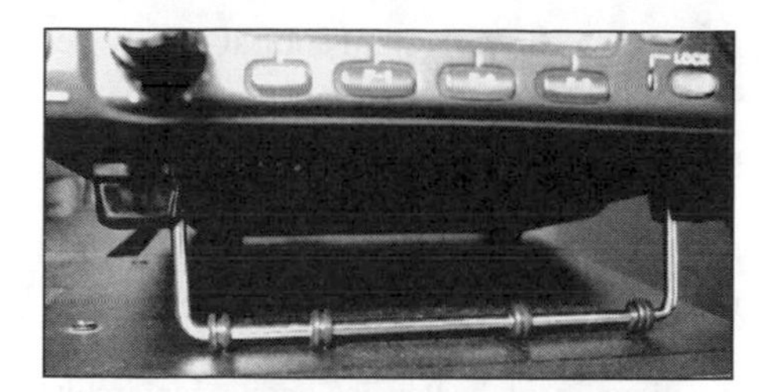

Figure 29 — Using protective rubber grommets on IC-706/703 flip stand bail.

for my application. I removed one of the mounting brackets holding the flip-up stand (two screws), and pushed four of these small grommets onto the stand. I arranged the grommets so they wouldn't interfere with any part of the bottom of the IC-706/703 when the stand was folded down. See Figure 29. Now I no longer worry about scratching anything with the flip-up stand. — *73, Phil Salas, AD5X, 1517 Creekside Dr, Richardson, TX 75081,* **ad5x@arrl.net**

## YAESU VX-5R TIPS AND TRICKS

◊ The Yaesu VX-5R is the most configurable radio I've owned. The user can assign icons and names to each memory channel, change the display size, turn the key beep on or off, scan modes and much, much more.

The radio gives you control over many features that gobble up precious battery power, such as the backlight duration, Busy/TX LED enable/disable, timeout timer and battery save modes.

It can be configured using the keypad (although time-consuming), or with a PC using a simple interface. I prefer the latter, for several reasons. For one thing, nothing beats a full-size keyboard when typing all of your favorite frequencies, CTCSS tones / DCS codes, memory names, call signs, etc. You can then save this configuration to disk or CD for later use. You'll be glad you did, should the radio ever need a CPU reset or other reprogramming.

The file can be shared with a friend or club members, or add it to the ever growing list of files available on the VX5 Web groups. These files allow a user to instantly load all area frequencies (including CTCSS/DCS) into their rig with just a few mouse clicks. At least two free software packages are widely available on the Internet (*EVE* and *VX5 Commander*).

For example, if you're planning a trip to Chicago, simply program the radio with the Chicago file and you instantly have all of the area repeaters loaded into your rig. Once you get back home, reprogram the rig and you're right back to normal operation.

There are two schematics on my Web site if you'd like to build your own interface. This is an easy and inexpensive project. You might even find (as I did) that this type of interface works to control other radios in your shack. Visit yokshs.fortunecity.com and click on the SCHEMATICS tab.

Recently, I discovered that even the S meter display characters are definable. The default character is a double greater-than symbol (>>), which can be tough to read from an arm's length and I find double characters to be less than intuitive.

You may choose from several choices in the S meter set menu, or you may even create your own using the Custom setting. Note that each segment can be a different character when using the Custom setting.

I set up the display using the numbers one through eight for the S meter and now the display is much easier to read, especially at an arm's length. — *73, Kyle Yoksh, KØKN, 125 N Chambery Dr, Olathe, KS 66061, AMSAT #35249, VUCC Satellite #150,* **k0kn@amsat.org**

## AUXILIARY 9.6 V BATTERY FOR THE YAESU FT-50

◊ If you're like me, you want to extend the service life of your equipment to its maximum. My old Yaesu FT-50 (**www.yaesu.com**) handheld transceiver still works fine but the design is obviously becoming rather dated when compared to Yaesu's latest products.

My friend, Tom, K4TCH, was anxious to show me his new FT-60s features and I must say it made me a bit envious. The overall shape and size of the FT-60 is much more convenient. The dual-tone multifrequency (DTMF) buttons are backlighted and the battery is a NiMH 1400 mAh unit that is more than double the power available in my radio.

Still, I can't justify discarding my FT-50 since it still works as well today as it did the day I bought it, so I began trying to think of ways to increase its utility. The number one discrepancy that I sought to address was the limitation imposed by my old NiCd 600 mAh battery packs. Besides being prone to "memory," they require frequent charging. The short battery capacity has caused me to "ration" my transmitting time, which has a major effect on my operating style. Often, I'm tempted to elaborate on some topic, but with the limited battery capacity always in the back of my mind, I usually try to not say much more than is necessary.

A few weeks ago, I discovered a dozen 9.6 V, NiMH, 1600 mAh battery packs in an eBay auction for $1.99 each. They were designed for remote control toy cars and were sold under the brand name, West Coast Choppers. I bought the lot of them and decided to hot-glue two together to form a single 3200 mAh pack with a 3 foot lead to my handheld transceiver.

The first task in making this happen was to remove the existing NiCd cells from one of my expired FT-41 Yaesu battery packs. To open the pack (which is glued securely together at the factory) I placed it into the freezer overnight to assure that it was quite cold and brittle. Even at these low temperatures, the plastic is much more durable than the glue that holds the pack together. I simply removed it from the freezer and, while it was still cold, smacked it smartly with a mallet on all four sides. The pack popped apart at the seams without damage.

Next, I removed the cells and the contact plates from the pack. (I would have reused the contact plates but they are made of stainless steel and the contacts are spot welded, rendering any solder connections to them difficult, if not impossible.) I then cut two new contact plates for the positive and negative poles of the battery pack, using a material that would readily accept solder. (In this case, a square cut from the side of a tuna can.)

I soldered one each to the stripped conductors on one end of a 3 foot piece of red and black zip cord. The contact plates (with wires attached) were then glued into their

Figure 30 — The complete system: charger, battery packs and FT-50.

respective positions inside the empty FT-41 battery pack. To verify the proper polarity, I tested the voltage present on the contacts of a good FT-41 battery pack for comparison. The zip cord was routed out of a hole drilled at the side of the pack, which still allows the radio to stand in an upright position.

Finally, the empty FT-41 pack was glued back together with super glue and the trailing zip cord was connected to the new West Coast Choppers pack with a pair of polarized Anderson Powerpoles. See Figure 30 for a view of the completed battery system.

Now when I leave the house, I have the choice of two options. I can use one of my standard FT-41 packs with its 600 mAh NiCd cells or if I plan to be out for a while, I can just snap on my modified, empty pack and drop my new, oversized battery into a pocket.

While the connecting wire can be a nuisance when performing some physical activities, it is an acceptable compromise when extended operating time is required. If it becomes too much of an inconvenience I'll probably buy a new Yaesu FT-60 like Tom's! For now at least, I can enjoy a normal QSO without having to ration my transmit time and I'm no longer particularly concerned when I need to switch to the full 5 W output to reach a distant receiver. Now if I could just figure a simple way to make those DTMF buttons light up! — *73, Johnny Angel, W4XKE, 120 Rhododendron Circle, Crossville, TN, 38555,* **w4xke@arrl.net**

## FT-817 POWER CONDITIONER KIT CONSTRUCTION MODIFICATIONS

◊ Phil Salas, AD5X, wrote an article ["Input Voltage Conditioner — and More — for the FT-817," *QST,* Jun 2005, pp 53-55] that described an external modification to add fusing, reverse-voltage protection and over-voltage protection in one small box for the Yaesu FT-817 low power (QRP) transceiver. Brian Riley, N1BQ, improved this by offering an inexpensive kit (**www.wulfden.org/ft817.shtml**) for these functions. Here are some simple mechanical changes to supplement Brian's well-written instructions. Fumble-fingered tinkerers, take note.

There are three primary changes to address, described in the approximate sequence encountered while I built this kit.

1. Input voltage selection switch
2. Internal power connection point
3. Input power connection

### Input Voltage Selection Switch

A simple double-pole double-throw (DPDT) slide switch is mounted with two 4-40 by ½ inch bolts to hold it in the case. Builders may find the inside dimensions very crowded, despite the very low parts count. Substitution of shorter bolts, such as ¼ inch, frees up valuable volume in the box. It may be necessary to slightly move or reposition the switch. Enlarge the mounting holes in the switch's ears and the plastic box to give some wiggle room.

### Input Power Connection Point

This is the point inside the assembly's box where the power leads attach to the voltage regulation circuitry. Once again, the fit is tight. There is less room for those of us who squeeze or cram components in.

These leads connect to the Powerpoles' internal connection tabs — the wire insertion end, not the springy tongue that holds the mating connectors in place. These hold the wire leads in the plastic shells. Once inserted in the shell, they cannot be removed easily and this is the advantage. Carefully cut off about a ¼ inch of the plastic on the inside end of each of the red and black shells. Cut all four sides of each of the two shells. A small, fine-toothed hacksaw blade works for me. As with any tool, use proper eye and hand protection. Use care when using a soldering iron, which may melt the shell into an unusable blob.

If you anticipate a space problem, do the cutting before inserting the tabs in their shells. It is okay to insert the tabs in the shells, but do not solder the Powerpole leads to the other components of the circuit yet. After cutting off this ¼ inch by ½ inch volume and when installed in the case's cutout slot, there will be extra volume gained to compensate for construction technique.

### Input Power Connection

Anderson Powerpoles provide polarity protection. Once made properly [that is, consistently with the ARES® norm; see "More Power to You," *QST*, Mar 2006, pp 31-33], cables and equipment become almost totally interchangeable. The kit instructions provide a simple diagram for measuring and cutting a slot for the two plastic shell assemblies, with suggested orientation. Once again, good construction technique ensures a good fit. However, other techniques (such as mine) mandate additional cutting or nibbling with a nibbler tool that can munch its way through the thin, soft plastic case very easily and quickly.

Measure carefully! Fortunately or not, extra cutting allows slew or slop room for repositioning the mated Powerpole pair, which allows forgiveness for cramped components inside the case. At this point, attach the top cover as instructed to hold the pair in place in its "adjusted" position and use hot-melt glue to fill any gaps. Use sufficient glue to fill the gaps on all sides except the top cover space, if any. Squeeze extra glue into the case's cavity for extra holding and support. The space between the Powerpole pair and top cover can be filled with thin sponge rubber or other filler material. Hold in position carefully until cool; it'll only be a minute or two.

### Finished Result

When the job is complete, the electrical integrity is maintained, the mechanical fit fits and, cosmetically, the case looks just as it should. This set of changes does not add to the functionality of the Power Conditioner circuit or kit, but it compensates for construction challenges present in the design. — *73, Eric Falkof, K1NUN, 2 Hickory Hill Rd, Wayland, MA 01778,* **k1nun@arrl.net**

## KEEPING POWER SUPPLIES COOL

◊ While operating Radioteletype (RTTY) recently, I noticed the heat fins of my Astron RS-35A power supply were quite hot to the touch. Although this excellent supply was still operating fine, I know that heat is an enemy of electronics — and RTTY operation generates a lot of heat.

I decided to mount a couple of small muffin fans to the heat sink. These are typically available at hamfests for $2-3 each, so I knew this would be a cheap fix. The fans draw just a watt or two. While there's probably nothing new in this story so far, I did two things that might be of interest:

First, I used two fans and connected them in series. (Be sure to use fans with the same current drain. Otherwise the one with the larger drain won't get enough current.) Connecting them in series kept the noise down in the shack and still allowed for plenty of cooling. I chose to mount them blowing in over the heat sink, but I suppose you would get equal results if you blew the air away from the power supply. Be sure to purchase and install the small metal protective grills for safety.

Second, I gave some thought to how I would mount them on the heat sink. I didn't want to drill any holes — what about hot-melt glue? So I tried that, and it worked beautifully. (Careful, the hot-melt glue sets up immediately — you're mounting it to a heat sink after all!) So everything seemed just fine until I pulled the supply out to take a picture for this article and bumped one of the fans causing it to pop off. So v2.0 of this concept used a combination of household GOOP adhesive (**www.amazinggoop.com**) for strength and hot-melt glue for instant set-up. This approach seems to be working just fine.

The final result is that the fans are working perfectly and the power supply stays just a cool as a cucumber even in the "heat" of RTTY contests. — *73, Mark Klocksin, WA9IVH, 1725 Wilmette Ave, Wilmette, IL 60091,* **wa9ivh@arrl.net**

## RADIO DISPLAY PROTECTOR

◊ The plastic displays on radios can be

scratched easily. PDA screen protectors will prevent accidental scratches to these plastic faces. Once they are scratched, there's little chance of removing the scratch, and forget about a replacement screen.

Use a scissors to cut and fit a protective sheet to the radio's display face. Also, Duke found that his Kenwood D-700 display is slightly curved requiring two sheets and a bit more care in cutting and trimming. Patience with the trimming and subsequent installation makes a very acceptable job and the vertical line on the D-700 face, where the two sheets join, is barely perceptible. Fingerprints are pretty hard to avoid on the adhesive surface but don't seem to show up once the protector is pressured in place. Above all, use clean hands (I used an alcohol pad on mine) and don't use rubber gloves. Most rubber gloves come lightly powdered and that powder will definitely show up on the adhesive surface! If the protector is "rolled" on, air bubbles are minimized and they can be worked to the side with the applicator card. For the bubble that just won't be moved, use a fine needle and puncture it (very gently) and it will disappear with a little pressure. — *73, Duke Knief, W4DK, P O Box 1000, Etowah, NC 28729,* **w4dk@arrl.net** and *Milton Garb, W6QE, 1426 Delamere Dr, Rowland Heights, CA 91748-2429,* **w6qe@scsxc.org**

## ASTRON RS-12A FIX

◊ Not long ago I discovered that I was getting a spark whenever my desk top ground braid touched the case of my well-used Astron RS-12A power supply. Thinking that there might have been a difference in the ground potential between the ac safety ground and the ground braid that was tied to a nearby cold water pipe, I called Astron's technical support for advice.

The technician there took me on a different tack. He suggested the ac ground was not tied directly to the chassis due to the paint. I opened up the supply and neither the solder lug on the ac safety ground, nor the earth ground post, appeared to be directly touching bare metal.

Taking my trusty Dremel Tool I removed the paint inside the chassis (see Figure 31) and tightened everything up. I also substituted the single nut on the earth ground with a wing nut and two washers for ease of use.

I retested the setup and found no more sparks. Evidently the chassis had been floating above ground potential due to the faulty safety ground. I reassembled my station and included the power supply on the ground braid. All seems well now.

I would suggest a check of all ground points on all your equipment to make sure you have a good mechanical and electrical bond from the chassis to a known good earth ground. It could improve your signal, prevent sparks and perhaps save your life. — *73, John Powell, KF6EOJ, 8325 Otto St, Downey, CA 90240-3924,* **jpowell@csulb.edu**

JOHN POWELL, KF6EOJ

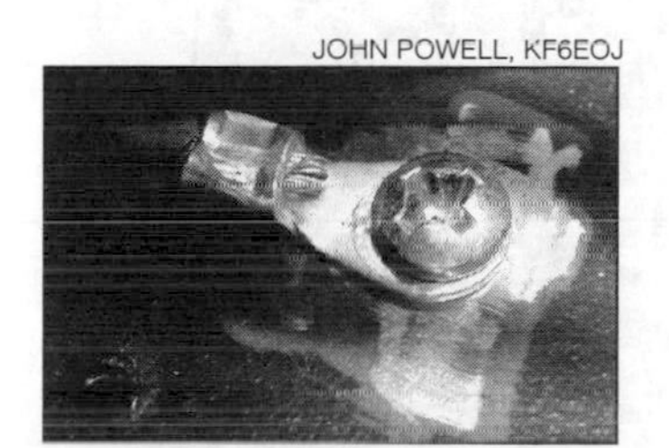

**Figure 31 — A close-up view of the repair to the RS-12A ground connection.**

## IC-706 CONNECTOR PROBLEMS

◊ I have had an ICOM IC-706MKII transceiver for a few years. It developed a problem — when I would transmit, the radio would shut down. This first showed up on 2 meters. I tried a few things without luck until I finally determined the problem to be the Molex power connector. The '706 generates 20 W on 2 meters so it probably draws 30-40 W from the battery. I took the connector off the rig, cleaned the Molex and used a small pointed scribe to bend the connections. The problem has disappeared. Recently, I was listening to Rich, K1OF, who had a strange hum on his signal. I broke in and suggested he disconnect and reconnect the Molex. There again the problem went away. He uses the ICOM IC-706MKIIG model, which outputs 50 W and draws 70-90 W from the battery. I also determined my ICOM '706 is very sensitive to input voltage. I used the power cord supplied, which in my opinion is too small a wire size. I suggest a minimum wire size of #10 but #8 is a better choice. — *73, George Peters, K1EHW, 41 Barbara Dr, Norwalk, CT 06851,* **k1ehw@arrl.net**

## HOT SUN — COLD SOLDER

◊ About 3 years ago I bought an ADI AR-447 70 cm transceiver, which worked great until just after the warranty expired. Does this sound familiar? I had the radio mounted in my car, which was always parked in the sun during the day. In El Paso, Texas that means that the interior of the car can reach temperatures above 250° F. With some of the silver-based and lead free solders used in manufacturing today this can be dangerous to your electronic equipment. When I got off work one day my ADI AR-447's display presented me with garbled characters. The radio worked fine. I just couldn't tell what frequency I was on.

I sent it off to be repaired and it came back with a letter stating that it was unfixable. I put it in my junk box thinking that I could use parts from it someday. It stayed there for 2 years.

While searching for parts for a project recently, I pulled the AR-447 out of the junk box and opened it up to see if there were any parts that I could use. I was looking at the parts when I noticed that much of the solder looked like crushed tin foil. Upon further study I discovered that most of the board had cold solder joints.

I got out the soldering iron and solder and resoldered every joint and connection I could find. I put the radio back together and it worked — and it is still working.

Over the years I've heard other hams complain of losing the use of equipment from the same cause, an unreadable dial. Many hams have had their radio's display fail and their radios have become unusable. Further research also indicated that in many cases the radio had been mounted in a vehicle that spends a fair amount of time sitting in the sun. If you have a radio that has display problems, check it out. It may just be cold solder joints. If your radio comes back to life after some serious resoldering and you plan to put it back in a vehicle, park in the shade and find a way to vent the vehicle to keep the temperatures down. — *73, John P. Conlon, WB7NPF, 10033 Mercedes St, El Paso, TX 79924-3816,* **aconlon1@elp.rr.com**

## MFJ CUB DIAL LABEL

◊ I recently built a 40 meter MFJ Cub in the ARRL Laboratory. It's a great little radio to operate, but lacked dial calibration. An idea came to me after labeling a CD-ROM: Use the little donut left over after peeling off the CD label.

First, remove the tuning knob. Cut the donut to fit and place it around the tuning shaft. Rotate the tuning shaft fully clockwise. Position the tuning knob a little past the 4 o'clock position and tighten the knob to the shaft. Calibration marks can be carefully made by using a receiver with a digital readout and a fine point pen (you could also use an accurate signal generator). Removing the knob made writing the numbers on the dial a lot easier (see Figure 32). ARRL staff members who have built the Cub now use

**Figure 32 — Bob's Cub transceiver sporting its new, calibrated label.**

this method for dial calibration. Someone who is skilled with label making software could produce a label with a neater appearance. — *73, Bob Allison, WB1GCM, ARRL Test Engineer,* **wb1gcm@arrl.org**

## HELP YOUR ICOM IC-27/37/47 REGAIN ITS VOICE

◊The most common problem leading to the "junking" of many of the above workhorse FM transceivers is their mechanically weak volume control potentiometers (pots). Turning the radio on and off via the volume pot's tandem rotary switch over the transceiver's useful life invariably results in an open or intermittent wiper. The net result is little or no receive audio, leading owners to believe their receivers are completely dead.

When I owned my first IC-37A in the mid-1990s, I experienced this very problem and purchased an exact replacement potentiometer/switch from ICOM. The replacement pot was mechanically identical to the original (a bad sign). Performing the repair was also a tedious job, requiring a complete removal of the front panel, unbundling a wiring harness and removing a daughterboard containing both original volume and squelch controls. Three hours and a splitting headache later, I vowed to never do this type of repair again. Until I sold that IC-37A at a hamfest in 2003, I always turned it off and on with my dc supply and never did anything with the volume control except turn it up or down a bit.

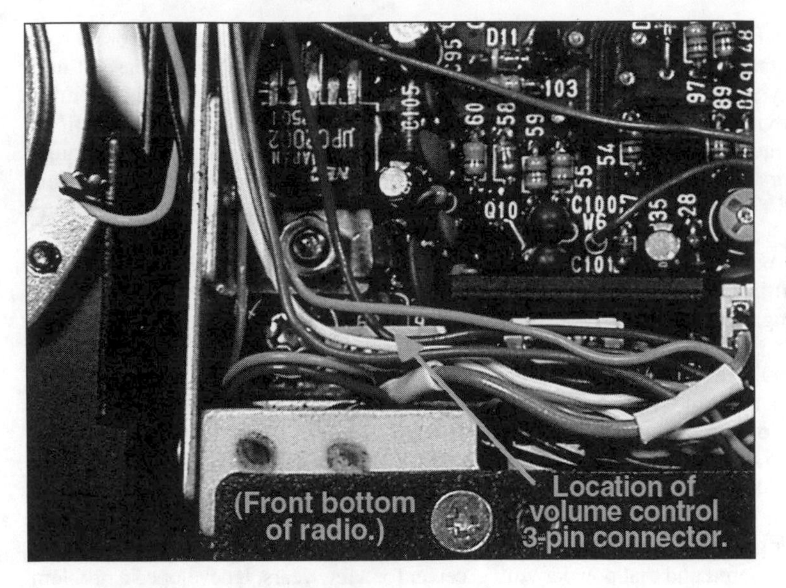


Figure 34 — Volume pot connector and wire locations.

In 2008 I got a good deal on several 222 MHz radios on eBay. Two of them were IC-37As and both had the volume pot issue! I learned that ICOM no longer sells the replacement volume potentiometer/switch for these radios. It turns out that there *is* a simpler way to get your radio "talking again." You have two choices.

As shown in Figure 33, a miniature, 10 kΩ audio taper, shafted potentiometer can be mounted vertically on insulating foam, double-stick tape in the top-left-rear of the transceiver (where the optional voice synthesizer would normally mount). I drilled a hole to pass its ¼ inch diameter shaft, centered 1 inch to the right and 1⅛ inches forward from the back left corner of the transceiver's top cover. To guarantee that the pot would "stay-put" during knob twists, I also placed dabs of

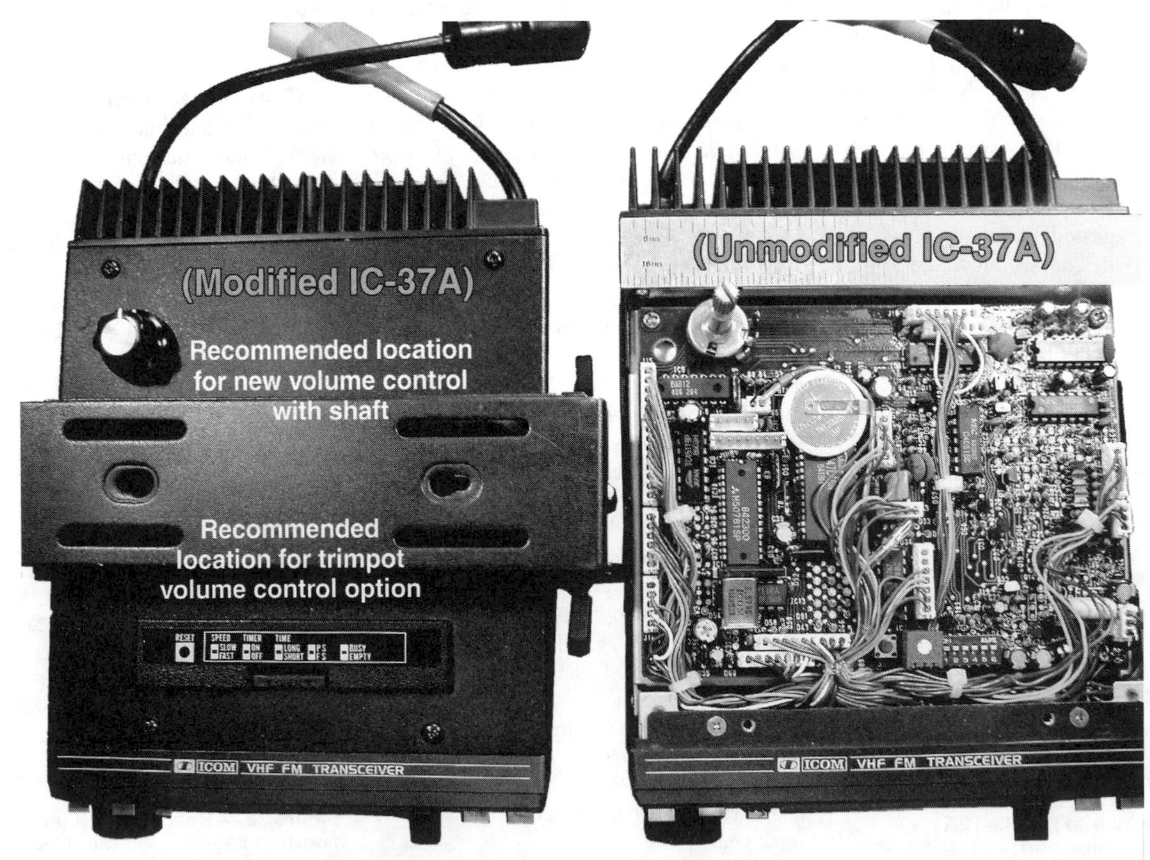


Figure 33 — Placement options for the new volume potentiometer.

RTV adhesive between the pot's circular body (left and right sides) and the power amplifier housing behind it. Alternately, a small trimmer pot can be used in the other location shown if you don't want to drill holes. Option two still maintains volume adjustment via a "diddle-stick." The trimmer pot's is located under the plastic access cover between the processor RESET button and scan options dip switch. (A dab of RTV adhesive is used to secure the trimmer pot in place.)

Figure 34 shows where to splice in your new volume control. Unplug the connector at the point shown on the bottom side of the transceiver. On it will be a purple wire, a white wire and a black wire. The black wire (ground) goes to one end of your new pot. The white wire goes to the middle (wiper) tab and the purple wire goes to the remaining tab. Carefully clip these three wires, leaving their excess lengths within the wire bundle from which they emanate. If you desire 1 inch more lead-length to splice to, the front panel will need to be removed allowing clipping of these three wires directly at the original volume control's daughterboard.

I used two cuttings of #28 twisted pair Kynar wire-wrap wire to connect to my new volume control pot, doubling up on the ground (black) wires. Otherwise, three twisted wires or two skinny runs of shielded cable/coax are necessary to keep crosstalk down. This way, the length of the run to the new pot is not critical. Plug the three pin connector back into place after insulating your splices. The turning on and off of the transceiver is maintained by the original volume pot's tandem rotary switch. (That function never breaks!)

Now you'll get many more years of enjoyment out of a great family of VHF/UHF FM transceivers with *no* further unexpected silencing of your receive audio. — *73, Tony Bogusz, W9MT, 5129 N Oketo Ave, Harwood Heights, IL 60706-3632,* **w9mt@arrl.net**

## REVIVING A YAESU VX5 TURN-ON BUTTON

◊My Yaesu VX5 handheld transceiver developed the peculiar problem it would not turn on reliably. I mentioned it to Paul Danzer, N1II, who had a similar problem. He bought a VX6 to replace his. He gave me his defective VX5 saying I could take it apart to see if I could cure the ills. I did that with no luck. I had a second VX5 at the office that had a similar on/off problem. I went home and did a factory reset on both VX5s. The problem was cured on both. I reprogrammed the Norwalk repeater into memory on one of them to test it. Doug Troughton, N2RDF, was on. I explained what I had done, and to my surprise he said his VX5 has the same problem. I bet this could happen to the VX6 or VX7 also. To reset the VX5: with the power off, hold down the 4, VFO and MR buttons at the same time then push power-on. This is tricky because the power-on is the problem to start with. Eventually it will come on. You then press FW. You do lose all programming using this method. — *73, George Peters, K1EHW, 41 Barbara Dr, Norwalk, CT 06851-5306,* **k1ehw@arrl.net**

## AUDIO FILTER IMPROVED CW PERFORMANCE

◊ I use an SCAF-1 audio filter from Idiom Press (**www.idiompress.com**) with my ICOM IC-706 MKII transceiver for CW. This filter is an enhanced version of the one designed by Denton Bramwell, K7OWJ, which appeared in the *1999 ARRL Handbook.*[2] It consists of an adjustable switched capacitor low-pass filter followed by a two stage active high-pass filter. Overlap in the two filter responses creates a band-pass filter. As supplied, the low-pass filter cutoff frequency is adjustable from about 740 Hz to 3600 Hz and the high-pass cutoff frequency is fixed at about 300 Hz. This is great for phone operation, but is far from optimum for a CW signal at 750 Hz as there is no way to filter out unwanted signals and noise between 300 Hz and 750 Hz.

A simple solution is to replace six 0.1 μF capacitors (C9, C10, C11, C14, C16, C25) in the high-pass filter with 0.047 μF capacitors. This raises the low frequency cutoff of the unit to about 600 Hz and provides a much better band-pass characteristic for CW (and digital) signals. The filter is still usable with voice, but some of the low frequency voice components are attenuated. — *73, Frank Getz Jr, N3FG, 685 Farnum Rd, Media, PA 19063-1611,* **n3fg@arrl.net**

## GARBAGE BAG TIE RESCUES PTT BUTTON WOES

◊For the January sweepstakes, I had borrowed an ADI AR-247 222 MHz radio to coordinate contacts (QSOs). In the middle of the contest, the PTT button on the microphone stopped working. Fighting a momentary panic, I carefully unscrewed the back of the microphone and surmised that the foam pad behind the button was compressing. I quickly rummaged through a junk box and found a polyethylene garbage bag tie that seemed to be the right thickness. I cut a small shim and placed it between the PTT micro switch and the foam pad. After reassembling the microphone, I was back on the air, coordinating QSOs.

Most interestingly, another ham who had lent me the manual from his AR-247 reported to me that he had the same problem and had resorted to using a microphone from another rig. I gave him the remainder of my "shim stock" to fix his microphone. — *73, Michael Davis, KB1JEY, 533 Tennis Ave, Ambler, PA 19002-6016,* **michael.davis@alumni.duke.edu**

## A QUICK RELEASE MOUNTING BRACKET FOR THE YAESU FT-857D

◊Recently I decided to add HF capabilities to my 1999 Ford F-150 pickup. Being a frugal ham, as most of us are, I wanted to use the radio both in the truck and in the shack. I eventually selected the Yaesu FT-857D. The only disappointing thing about the FT-857D is it did not have a quick release mount.

In the truck, I was using a Yaesu FT-7100 radio, which uses the Yaesu MMB-60 quick release mount. My first thought was to see if the same mount could be used for the new radio. Unfortunately, the mounting arms are too short for the FT-857D. Searching the close-out clearance items on several of the online Amateur Radio stores, I found a Yaesu MMB-67 quick release mount for the discontinued FT-100, so I ordered it to try. When it arrived I found that the slide mount itself was the same as for the FT-7100 and the mounting arms were just the right length to hang the new radio at the proper distance below the mount. As Murphy would have it, the mounting holes did not match plus the FT-857D is approximately ¼ inch narrower. It was time for some good old ham ingenuity.

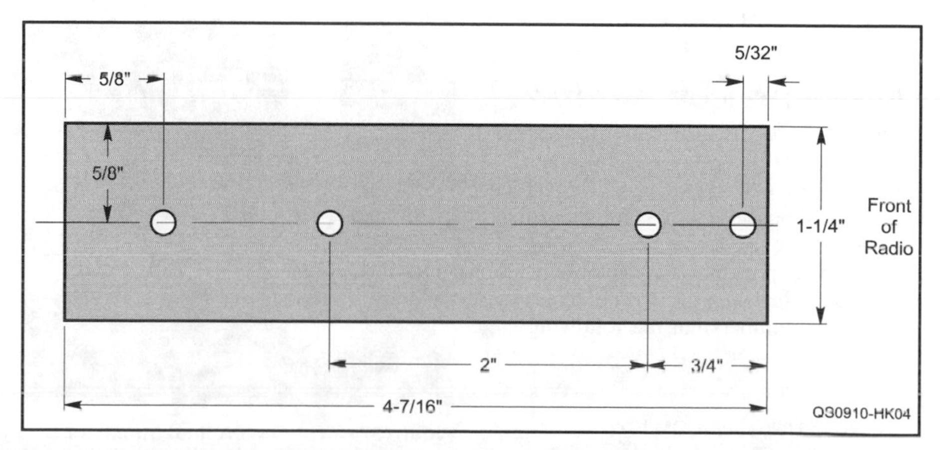


Figure 35 — Mechanical layout of the adapter bracket.

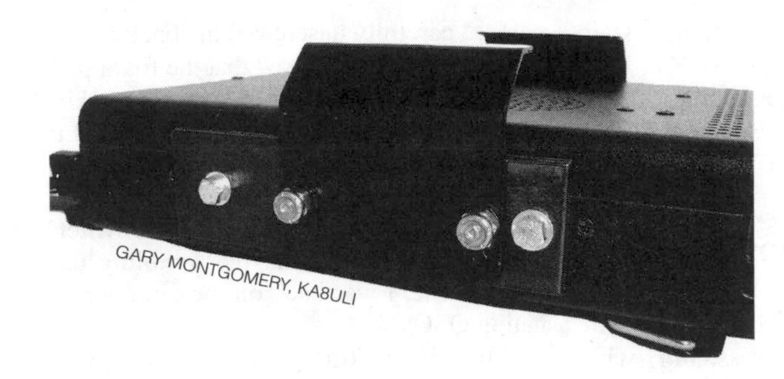


**Figure 36 — The new mounting arms and brackets attached to the radio and ready to slide into the truck's mount.**

After careful measuring, I determined that making a pair of adapter brackets would allow me to use the MMB-67 mounting arms on the new radio. To make the brackets I used two pieces of flat steel 1¼ inch wide by ⅛ inch thick by 4⁷⁄₁₆ inches long; four hex head metric bolts, M4 × 12 mm; four split-ring lock washers, 4 mm; four flat washers, 4 mm; four countersink head bolts, 8 × 32 × ½ inch; four 8 × 32 nylon insert lock nuts.

Figure 35 shows the dimensions of the adapter bracket and location and size of the four holes needed. Note that I used a larger drill bit to countersink the holes for the bolts that hold the adapter bracket to the mounting arms on the back side so it mounts flat against the radio cabinet when installed. I then primed and painted the adapters to match the radio and assembled the adapters and arms together using the countersink head bolts with the nylon insert lock nuts. I used the upper holes in the mounting arms to position the radio at the height I desired.

Figure 36 shows the mounting arms with the new adapters mounted to the radio using the hex head bolts with split lock washers and flat washers. (*Caution:* Be sure you do not use bolts longer than 12 mm as they will touch the circuit board inside and possibly damage the radio.) I can now use either radio in the truck. Additionally, I use the MMB-67 at my operating desk, so I can use either radio inside as well. For only a few dollars and a couple of hours work, I now have a great rig that I can use in the shack or on the road. — *73, Gary Montgomery, KA8ULI, 206 Hernandos Loop, Leander, TX 78641*, **ka8uli@arrl.net**

## MORE ON THE ICOM IC-27/37/47

◊ Lynn Bisha, W2BSN, wrote to me about the recent hint concerning audio problems in the IC-x7 series of mobile rigs.[1] He learned that it is still possible to obtain replacement volume and squelch controls from ICOM. Lynn had to bypass the parts department and contact technical support in order to get the correct part numbers but the following are direct replacements:

- Volume pot/switch, p/n 7210000250
- Squelch control, p/n 7210000230

Thanks Lynn for doing the legwork on this.

## HOMEBREW TRANSISTOR SOCKET

I recently built a DC40A QRP rig (**www.qrpkits.com**) and installed it in an Altoids tin. I love it. Making contacts with about 1 W while being crystal bound is one of the more exciting ham radio experiences. I also like to play with old gear and recently restored and put a Heathkit DX-60 on the air. A DX-60 was my first "good" transmitter 40+ years ago.

Unfortunately, the way I connected antennas, etc I ended up blowing the final 2N7000 FET in my DC40A — twice — and it is a real pain to remove the circuit board from the Altoids tin, remove the blown FET, reinstall one, etc. How I wished for a socket. Then it hit me. I have some old IC sockets around. Carefully using my wire cutters I destroyed one to get the pins. Then I simply soldered three of the pins into the DC40A circuit board (see Figure 37).

Now I can blow FETs to my heart's content knowing the next one will be easy to replace. — *73, Martin Huyett, KØBXB, 7735 Big Pine Ln, Burlington, WI 53105*, **huyettmeh@tds.net**

MARTIN HUYETT, KØBXB

**Figure 37 — The DX40A final transistor mounted in the recycled IC socket pins.**

[While I think Martin is being facetious, this is a point of shack safety. Never operate any transmitter or amplifier without an appropriate load. It only takes a 1 microsecond error to produce many dollars of repairs. — *Ed.*]

## WEATHERPROOFING YOUR AUTOMATIC ANTENNA TUNER

◊ As an avid Amateur Radio operator, I sometimes use an automatic antenna tuner to operate more than one band with the same antenna. My mobile unit uses an LDG RT-11 autotuner to feed a pair of "Hamstick" style antennas on several different bands. One antenna covers the lower bands and another antenna covers the higher ones. An excursion into operating "fixed portable" with a telescopic vertical had me looking for another solution.

In order to minimize the coax losses when feeding a vertical monopole away from its resonant band it is necessary to reduce the distance between the tuner and the antenna to a minimum. The use of ladder line is not feasible when the antenna is fed almost at ground level. In order to do this, it is necessary to place the tuner almost directly at the base of the antenna and thus reduce the coax run from the tuner to the antenna to nearly nothing. The coax from the tuner to the transmitter can then be any appropriate length since the mismatch has already been corrected.

Owning an LDG Z-100 automatic tuner already, I looked for a way to mount it at the base of the antenna and yet protect it from the elements. A phone call to LDG gave me the necessary specifications for a 50 foot extension of the control cable so the only thing left was to find a workable enclosure for the tuner itself. I discussed the requirements for such an enclosure with my spouse, Audrey. Without a word, she rummaged through a kitchen cabinet and produced a semi flexible plastic container that had a snap-on lid and fit my Z-100 and its cables to a T.

I drilled four small holes in one end of the container into which I fitted two short coax jumpers, one for the antenna and one for the radio. I also made up and installed a short 4-conductor cable to connect the stock control harness to the 50 foot extension. The fourth opening was used for a similarly short insulated wire to connect the grounding stud on the Z-100 to the radial system of the antenna. Once these four cables were in place, I sealed the drilled openings with hot glue. Silicone caulk could be used just as easily, provided it will stick to the container. I did not try that because the hot glue was at hand so you are on your own there (see Figure 38).

With the jumpers connected to the tuner, the extension cable and coax were run to the transceiver and the antenna erected. Now, testing was in order. The pressing of the TUNE button on my IC-706MKIIG did exactly what it was supposed to do. The Z-100

[1]T. Bogusz, "Help Your ICOM IC-27/37/47 Regain Its Voice," *QST*, Jul 2009, p 56.

GEOFF HAINES, N1GY

**Figure 38 — A detailed view of the autotuner and its wiring mounted inside its "raincoat."**

GEOFF HAINES, N1GY

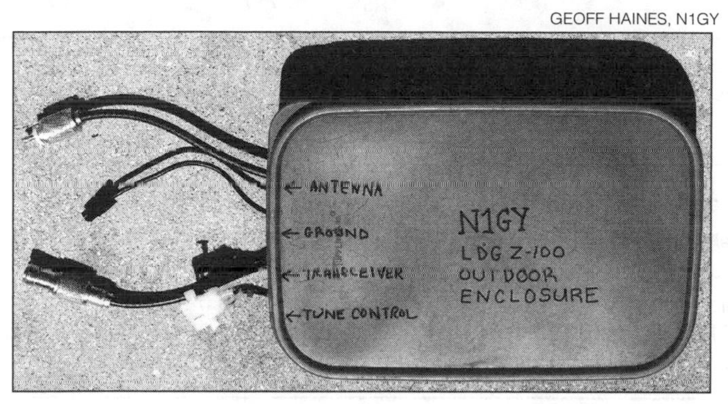


**Figure 39 — The autotuner all "buttoned up" and ready for heavy weather.**

ran through its paces and signaled a good match. Now if the afternoon showers come while I am operating "fixed portable," the only thing I have to worry about is keeping me and the radio dry. The tuner is cozy in its own little raincoat (see Figure 39).

Total cost, even if you had to buy the container new at the discount store, would probably not exceed $10. That does not include the extension control cable, of course. The container was already here and I had enough UHF connectors, coax, wire and weatherproofing on hand for the project. The only thing I had to buy was the 4-conductor cable and Molex connectors to build the 50 foot extension cable.

This project has enabled me to comfortably operate "fixed portable" from the beach, at Field Day and many other events where a vertical was the only feasible antenna. As long as the container can handle the physical size of the tuner with room for the connecting cables, any automatic tuner could be protected in this way. — *73, Geoff Haines, N1GY, 904 52nd Ave Blvd W, Bradenton, FL 34207*, **n1gy@arrl.net**

[*Note*: The plastic container, being airtight, will tend to accumulate condensation inside. If the weatherproof tuner is used for an extended period, you should either punch a weep hole in the bottom to allow water to drain or periodically open the top to vent the interior. — *Ed.*]

## IC-706MKIIG SO-239 PROBLEM

◊The IC-706MKIIG radios are made to be disconnected and reconnected to an antenna for mobile and portable use. Repeated reconnections may cause a problem — at least it did for me.

I have owned my ICOM IC-706MKIIG for about 2 years now. When I attached my antenna's PL-259 to the radio's HF SO-239 connector tightly, I would lose my received signal and the radio would not transmit. If I loosen the PL-259 slightly signals would be received and I could transmit as well.

I tightened the screws that hold the SO-239 to the radio thinking that I had a bad ground at this point — no luck. I made up a new cable with a new PL-259 and still no resolution. In the time that I have owned this radio I have used it less than a dozen times mobile. The HF SO-239 connector gets connected and disconnected more than other radios resulting in more "wear and tear" on them.

I decided to go into the radio. I removed the top cover so I could see what the SO-239 was connected to and then I saw what was happening. The center insulator along with its center pin was loose and moving in the SO-239 metal housing. When a PL-259 was inserted, the insulator would move out of the metal housing and touch a metal part of the chassis. Also, the wire going to the center of the SO-239 was hanging by only about one strand because of this movement.

In order to make a repair, I had to remove the top cover and five screws that hold a circuit board to access the SO-239 wire connections. I could only lift this circuit board slightly and when I moved the SO-239, the center wire broke right off. At this point, I decided to pull the SO-239 from the chassis. Be sure to unsolder the ground wire before removing it. Now I could see that the insulator in the SO-239 was actually loose and would slide back and forth with the insertion and removal of a PL-259.

As this is a special type of SO-239, I decided to "lock" the insulator in the housing by using a couple of drops of "thin" CA (cyanoacrylate), a type of glue which I use in my model airplane construction. CA adhesives are available in thin, medium and heavy; the thin is the consistency of water and flows ready into the joint by capillary action then cures almost instantly (*protect your eyes and fingers*). [Be aware of the drying time for any glue you use. Make sure it is completely dry before inserting the PL-259. — *Ed.*] I then pulled the two wires from inside the chassis out through the SO-239's hole and did my soldering safely on the outside of the chassis.

I am pleased with the repair and now the HF SO-239 connection works perfectly. While I had my CA out, I added a couple of drops around the insulator in the VHF/UHF SO-239 connector just to be on the safe side (without removing it from the chassis).

Also, when I solder all of my PL-259s, I always make sure that no excess solder gets on the outside of the center conductor; if it does, I take a small file and remove it. *Any excess solder on the center pin would cause excessive insertion force into an SO-239 possibly causing this failure. — 73, Karl Schwab, KO8S, 30752 Ridgefield Ave, Warren, MI 48088-3174,* **ko8s@arrl.net**

[*Note:* You should check with ICOM to determine if these changes will affect your radio's warranty. — *Ed.*]

## RIG STATIC PROTECTION

◊Paul, AA8OZ, was telling me about the problem he has had with his Elecraft K1 transceiver. Paul uses a wire antenna held aloft by a kite and he recently had a problem caused by static build-up on the antenna that zapped one of the components in his rig. After repairs were made, Paul installed a 1 kΩ resistor across the antenna terminals in the radio to drain any static.

When Paul told me about this I thought my rigs should also be protected as I also operate portable using wire antennas.

A little research told me that resistors between 1 kΩ and 10 kΩ or even higher would make a suitable drain resistor without affecting receive or transmit modes. I decided to install the resistor outside of the radio and not have to find room in the case.

I looked through my junk box and found a 2.2 kΩ resistor and a T-adaptor. I turned on the solder station and soldered the resistor from the center jack to the shell effectively bridging the antenna connections with the 2.2 kΩ resistor. For the K1 and KX-1 I needed an additional adaptor to connect the SO-239 to the rig's BNC connectors. [Note: This method should only be used with low power transmitters. — *Ed.*]

Now when connected to a random wire, I can be sure my rigs are protected from a static build-up from either wind or precipitation. [Note that this method is not lightning protection. If there is thunderstorm activity in your area — shut down *immediately*. — *Ed.*] — *73, Richard Arnold, AF8X, 22901 Schafer St, Clinton Twp, MI 48035,* **af8x@comcast.net**

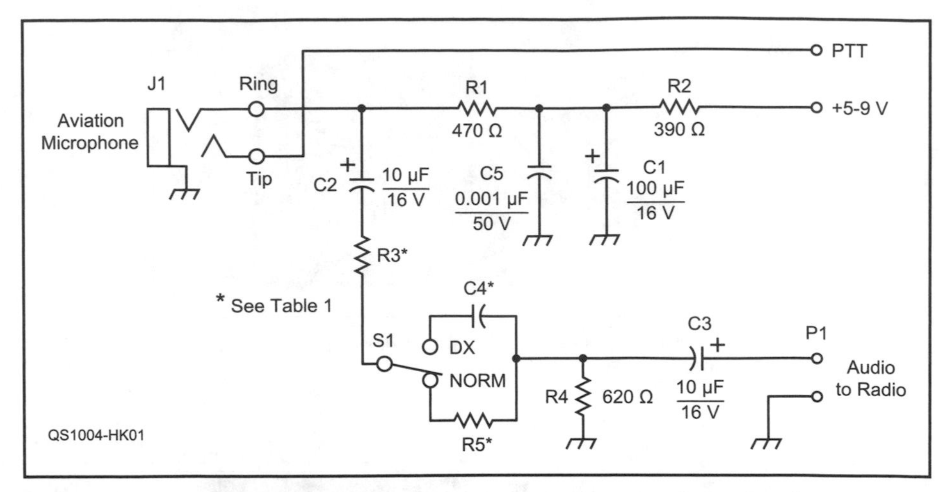


**Figure 40 — Schematic of the aviation headphones adapter needed to interface the headphones to a ham transceiver.**

## ADAPTING AVIATION HEADSETS TO HAM RADIOS

◊ Aviation headsets are widely used for commercial and private aircraft. These are rugged, expensive sets with a frequency response well-suited for Amateur Radio. If you have one you probably have wondered if it could be used with your amateur equipment. The short answer is yes, but…

Here's a way to do it. The headset earphones may be either 300 Ω stereo or 150 Ω monaural (some older sets are series connected 600 Ω). The plug fits a standard ¼ inch phone jack (two or three conductor to match). Stereo versions usually have a mono/stereo switch. If not, a RadioShack 274-360 stereo-to-mono adapter will match a mono radio jack.

Most ham radios these days are designed for 8 Ω headphones. If your radio works with a Heil Pro Set type headset (200 Ω), you are good to go. If not, a RadioShack audio output transformer (276-1380) will do the job. Hook it up backwards (that is, the 8 Ω side to the radio's phone jack) and use half of the 1000 Ω primary (= 250 Ω) for the headphones. The match will be close enough.

### Adapting the Microphone

Modern aviation headsets use an electret circuit that mimics the old-style carbon microphone. A dc voltage, in the range of 5-16 V at a few milliamperes, is required. Nominal output is 400 mV into a 150-500 Ω load. A PL068 microphone plug is standard and matches a 0.210 inch three conductor jack. Tip is PTT (push-to-talk) and the ring is audio.

The microphone cannot be used with ham radios without an adapter circuit. Most circuits I have seen use a small transformer to better match the microphone impedance to the radio. A simple resistive circuit will also work well and provide the reduction in microphone audio output to match ham radios. Modern ham radios can provide the needed bias voltage at the microphone jack (ICOM and Kenwood provide 8 V, TenTec 9 V, Yaesu 5 V, etc). Radios without such a voltage can use a 9 V battery. For ICOM radios use pin 2 for the bias.

The circuit shown in Figure 40 (the parts list is in Table 2) does the job. R1 provides the load for the aviation microphone. R1 plus R2 supply the dc voltage from the radio (or battery). For Yaesu radios with 5 V, eliminate R2 and C1 and connect R1 directly to the +5 V pin. C2 isolates the circuit from the microphone voltage and C3 isolates bias, if any, from the radio microphone pin (ie, ICOM radios). C5 reduces the chance of RFI via the bias supply.

Most radios are designed for microphone impedance of 600 Ω or so. This is provided by R4, which also worked with my vintage Collins KWM2A. The divider consisting of R3, R5, C4 and R4 reduces microphone audio output to an appropriate level. C4 creates a selectable "DX" rising frequency response. R5 reduces "normal" response by 2 dB to keep average output similar to that of the "DX" response.

By selecting different values for R3 (and related R5 and C4) we can adjust the microphone output to match just about any microphone/radio combination. Table 1 has values to match a popular Heil cartridge, other dynamic and electret microphones (as used by ICOM, Yaesu, Kenwood, etc). Values are also shown for older ICOM radios such as the IC-745/751/781 and early 706s, which require even higher output. Feel free to make further adjustments to these values to match your radio.

The jack is a Switchcraft S12B. The PTT connection can be ignored if not needed. You will also need a plug to match your radio's microphone jack. A little RadioShack 1 × 2 × 3 inch plastic box (270-1801) will accommodate the components. If you also need a 9 V battery, a larger 270-1802 box will work.

Aviation microphones are designed to be RFI resistant and I have not had problems with this circuit. A 0.001 µF capacitor across the circuit output to ground could be added if necessary. The finished product is shown in Figure 41.

A spectrum analysis of my headset (a SoftComm C-20) and the circuit shows

**Table 1**
**Aviation Headphone Adapter Microphone Equivalents**

| *Microphone Equivalent Output* | *R3* | *R5* | *C4* |
|---|---|---|---|
| Heil HC-5 (–68 dB) | 18 k | 3.5 k | 0.01 µF |
| Dynamic microphone (–62 dB) | 9.1 k | 1.8 k | 0.02 µF |
| Electret (–54 dB) | 3.5 k | 680 | 0.05 µF |
| Older ICOM radios (–46 dB) | 1.5 k | 300 | 0.12 µF |

**Table 2**
**Aviation Headphone Adapter Parts List**

| | |
|---|---|
| C1 | 100 µF, 16 V (not needed for Yaesu) |
| C2, C3 | 10 µF, 16 V |
| C4 | See Table 1 |
| C5 | 0.001 µF, 50 V |
| J1 | 0.210 3-conductor jack (Mouser 501-S-12B) |
| R1 | 470 Ω, ¼ W |
| R2 | 390 Ω, ¼ W (not needed for Yaesu) |
| R3 | ¼ W (see Table 1) |
| R4 | 620, ¼ W |
| R5 | ¼ W (see Table 1) |
| P1 | Plug to match radio microphone jack |
| S1 | SPDT miniature toggle switch (Mouser 108-1MS1T2B3M1QE-EVX) |
| T1 | 1000 Ω center-tapped to 8 Ω audio transformer (RadioShack 276-1380) if needed, see text |

**Figure 41 — A completed headset/adapter. The black box contains the adapter circuit for the aviation headset.**

the frequency response is flat up through 3200 Hz with a sizeable drop off thereafter. The "DX" frequency response is similar to Heil HC-4 cartridges.

The noise-canceling aviation microphone and 23 dB headphone isolation are very helpful in noisy environments and contests. The around-the-ear muffs are comfortable, especially with glasses. During on-the-air tests I have received numerous compliments about the audio quality. Since aviation headsets are designed to a specific standard, other brands and models should have comparable performance. — *73, John Raydo, KØIZ, 4901 NW 79th St, Kansas City, MO 64151-1099,* **k0iz@arrl.net**

## SPECTRUM SCOPE FOR YOUR K3

◊The K3 is my second radio. I obtained it for the joy of building and using it. It is my traveling radio and my standby radio. As a matter of fact, it would be my only radio if it had a spectrum scope. I'm so into watching those signals pop up on the band. I can see at a glance if 15 meters is open. I got used to having one of these gadgets long ago and I'm addicted.

What is a spectrum scope anyway? It's a wide band receiver that monitors the IF of another receiver and presents the signal to a display that sweeps across the IF bandwidth and displays any signals as amplitude spikes that are encountered along the way.

Digital control of the oscilloscope and sweep generator allow you to display the relative frequencies, change the width and sweep rate. The recommended spectrum scopes for the K3 are actually software defined receivers operating at the K3's 8.215 MHz IF frequency. Using a software defined receiver allows you to integrate the display into a digital oscilloscope that can, using data from the K3, combine all kinds of extra stuff into the display like filter bandwidth, second receiver data, "click and tune," etc.

Hmmm. Don't I have an 8.215 MHz general coverage receiver around here somewhere? Is there one with a built-in spectrum scope? That FTDX-9000D monster sits on my desk and watches tolerantly when I use my K3. I can just hear it sigh as that lightweight Elecraft borrows its antennas. Would my 85 pound '9000D like a job? It receives 8.215 MHz and has a spectrum scope with a 2.5 MHz bandwidth. (A bit wider than the digital, software defined 192 kHz.)

The other two requirements:

1) The K3 has the KXV3A transverter and IF output module

2) The transceiver operating as a spectrum scope has a separate, switchable receive-only antenna input. (Don't, I say, don't use the transceiver antenna jack. Okay? Accidentally transmitting into the IF channel is not healthy for your radio.)

So here's how to put it together: Use a BNC to UHF cable to attach the K3 IF output to the transceiver's receive-only antenna input. Okay, so I lied. It's not exactly free. You need a BNC to UHF cable. Tune the transceiver's general coverage receiver to 8.215 MHz and enjoy your new spectrum scope (see Figure 42). You did turn the K3 on didn't you?

There was enough drive from the K3 IF for a satisfactory display on 40 meters and below, but not quite enough above 20 meters. That's when I went out to the garage and hooked up my old Heathkit active antenna — just to "nudge" the IF output. The active antenna amplifier is peaked at the 8 MHz and the gain control is a handy display amplitude control. Any wideband or tunable preamp could be used.

The Elecraft IF output seems to "roll off" at 1 MHz bandwidth. The IF center frequency is offset by the mode at which the transceiver is operating. Where this would normally be handled digitally with computer control, you could tweak the IF receiver frequency to center up the signals on the display.

I really liked the spectrum scope on the Kenwood radios, especially the CRT system used with the TS-950SDX, not that blocky digital look. Obviously, you're not going to take this setup out to your cabin in the woods, but if the K3 is just exchanging enigmatic looks with the heavyweight on your desk, this little experiment will cost you (practically) nothing. — *73, Harvey S Laidman, W8DX, 22918 Crespi St, Woodland Hls, CA 91364-2807,* **w8dx@arrl.net**

## YAESU FT-530 ALKALINE BATTERY PACK REPAIR

◊The FT-530 is a wonderful dual-band handheld transceiver that I have had for years. The NiCd pack finally died and I tried to keep it alive by breaking open the NiCd pack and recharging each cell individually. After doing this for over 3 years, it is now ready for replacement.

Thankfully, I had the foresight to buy the alkaline battery pack when I purchased the FT-530. Unfortunately, the original design of the pack had three batteries on each half of the plastic shell and, to make a connection between the two halves, there was a spring-tab that was pressed against the docking tab on the other half-shell.

This design never worked well. It never made a good solid electrical connection and, while six fully charged NiMH batteries should have shown close to 8.4 V (1.4 V per

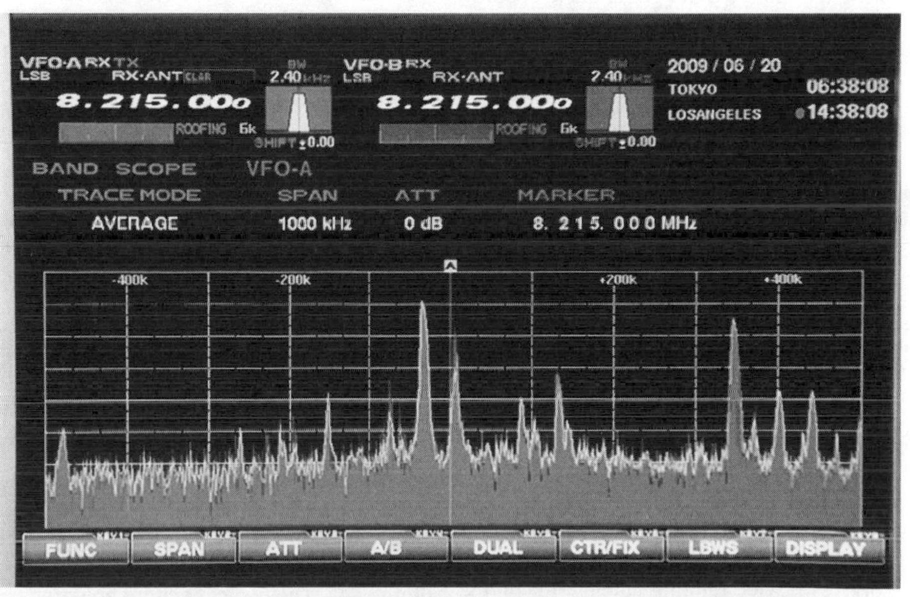


**Figure 42 — A view of the FTDX-9000D's bandscope display of the K3's IF channel.**

PHIL KARRAS, KE3FL

**Figure 43 — The alkaline battery pack opened and with two batteries removed to show the desoldering braid jumper.**

cell), it never did. I was always attaching the pack to the FT-530 only to find absolutely no voltage at all. I'd have to take it apart and try cleaning the tabs, then bend the spring-tab so it would sit with a bit more force. After two or more tries it would eventually work — but not well.

I figured that someday I'd come up with a real connection between the two halves. The day came when I wanted to use the rig but found that, once again, the tab was not making a good enough connection even though the batteries were just fine.

I pulled out both battery connection sections and tried to solder a length of desoldering braid size 4 (60-4-5) onto both the spring-tab and docking-tab, only to find that while it soldered with no problem to the docking tab (the one with the battery spring) it refused to solder to the spring tab from the other battery pack half. While the metal used was not aluminum, it acted like it and refused to accept any solder. [The spring-tab is probably nickel plated. Some careful filing should remove the plating. With the plating removed you should be able to solder to the spring-tab. — *Ed.*]

Since I had already cut off part of the spring tab there was no going back. A solution had to be found. The original design of the pack was a simple press fit. The problem with the original design was that it was never able to apply enough force or surface area to insure a good electrical connection. In looking at the situation, I noticed that I could probably fit some of the desoldering braid behind the positive battery plate connected to the spring-tab (see Figure 43). With the battery tip pushing on the plate, there should be more than enough force to insure a good electrical connection and, in fact, there was. — *73, Phil Karras, KE3FL, e-mail via* **cs.yrex.com/ke3fl**

## TUNE-UP FOR THE TEN-TEC 238B TUNER

◊The very popular Ten-Tec 238 series of antenna tuners utilize a very efficient L network, with wide ranging antenna-matching capabilities. This tuner was reviewed in the February 2003 issue of *QST*.[1] Now into their third generation, the latest "C" version offers improvements in metering.

I personally own the model "B" and have achieved excellent results with it. Here are two simple modifications that can improve the 238's operation. The changes involve the addition of a switch to control the meter lamp and installation of two electrolytic capacitors to damp the response of the SWR meter.

The switch, a miniature SPST toggle variety, is mounted in a hole drilled in the rear of the chassis, directly above the 13.5 V dc meter light connector. The wire from the power connector is disconnected and soldered to one terminal of the switch. An added wire is soldered from the remaining switch terminal to the vacant terminal of the power connector. (I also replaced the RCA-style phono connector with a coaxial power connector, which seems to be more conventional these days.) [It is a good idea to replace an RCA power connector on any unit so equipped with a coaxial one. This avoids the risk of accidently shorting the 12 V on the center pin to the chassis while plugging it on. If the 12 V source is a wall wart this is probably not that important. If the source is the accessory 12 V output of some radios, it can destroy a PC board trace. — *Ed.*]

The meter response damping is achieved by the installation of two parallel capacitors (220 µF and 470 µF) directly across the meter terminals (see Figure 44). Be sure to observe the correct polarity. Their combined value is 690 µF, with a voltage rating of 16 V.

[1]J. Parise, W1UK, "*QST* Reviews Five High-Power Antenna Tuners," *QST*, Feb 2003, pp 69-75.

**Figure 44— The two capacitors soldered in parallel to the meter board.**

This value was arrived at by trial and error and "tames" the jerky response of the meter indicator very nicely. Since the detector portion of the metering circuit is not involved, there is no difference in meter reading, other than the time required for the meter's needle to settle. Also, there is no more pointer slap when using CW.

Now, tuning is much easier, with a smooth meter response and I no longer have to unplug the wall wart power supply when I've finished operating. — *73, Steve VanSickle, WB2HPR, 3010 Tibbits Ave, Troy, NY 12180,* **wb2hpr@arrl.net**

## FAN SPEED CONTROL FOR THE W6JL 50 W AMPLIFIER

◊As a young ham I really disliked the background noise of amplifier fans. Aging made me more tolerant, but hearing aids suddenly brought back my irritation. After splurging $28.36 on the parts for W6JL's prize-winning 50 W 40 meter amplifier, I noticed from the schematic that the fan was always on. I decided to add a fan speed control.[1] In the spirit of the Homebrew Challenge II, I kept my budget to $2.83 (10 percent of the original cost). Figure 45 shows my design.

J3 on the original amplifier schematic is the PTT/QSK line. It connects to R5 of the fan controller. The only other change to the original circuit is connecting the minus lead of the fan to D1 of the controller. The controller has three speeds: low, medium and high. When the amplifier is idling the fan runs at low speed because both Zener diode D1 and resistor R7 reduce the fan voltage. When the PTT line goes low for more than about ½ second, the fan switches to high speed. The negative side of the fan is connected directly to ground by Q3. When the PTT line is released, Q3 turns off 15 seconds later. Q4 keeps the fan running at medium speed for an extra 15 seconds. Q4 shorts R7 to ground, so only D1 is limiting the fan speed.

Q3 and Q4 use nearly identical circuits for timing. When the PTT line goes low Q1 and Q2 turn on. R1 charges C1, so R1 determines how quickly the fan starts up. When the PTT line is released, C1 discharges through R2. That time is about 15 seconds because I chose a value of 10 µF for C1. Q4 uses C2, R3 and R4 for timing. Because C2 is about twice as large as C1, the "hang time" is twice as long. I set R3 to half the value of R1, because C2 is twice the size of C1; that way they have about the same startup speed.

It is very easy to modify the components to match your operating needs. D1 and R7 can be chosen for different fan speeds. C1 and

[1]D. Huff, W6JL, "Homebrew Challenge II Winner #1 — The Lowest Cost Entry," *QST*, Jun 2010, pp 30-33.

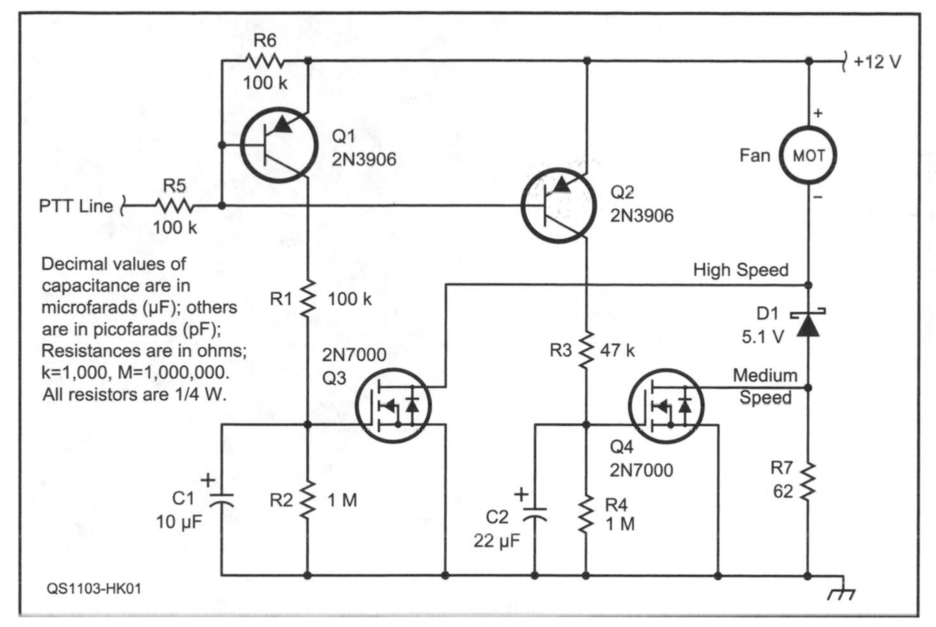


**Figure 45 — The schematic diagram of the three-speed fan controller. All resistors are ¼ W.**

C2 can be chosen for different "hang times" and R1 and R3 can be selected for very slow or very fast startup times. The values in the schematic work great for SSB with the amplifier as described in the article. — *73, Andrew Mitz, WA3LTJ, 4207 Ambler Dr, Kensington, MD 20895*, **arm@gnode.org**

## IC-229H REPAIR

◊As I was repairing one of my hamfest finds, an ICOM IC-229H, the major problem appeared to be in the microphone. This model came with the HM-56A DTMF encoder microphone. Intermittently, when the radio was powered up, it would go into transmit mode.

The problem turned out to be leaky aluminum electrolytic capacitors in the microphone. I have worked with this type of capacitor in the past. When they leak, they tend to cause all manner of problems.

Soon after replacing the capacitors, I began to intermittently lose audio. On further inspection, I found that the electrolyte from one of the capacitors had damaged the PC board. ICOM used a double-sided board in the HM-56A and plated-through holes are used between the top and bottom side of the board. [Plated-through holes are holes drilled through the board that are copper plated and tinned to provide a "jumper" between the surfaces. — *Ed.*] One of these plated-through holes was intermittently open.

After cleaning and tinning the plated-through holes (top and bottom), a short piece of small gauge wire was inserted and soldered in place. I keep some short pieces of #18 AWG stranded wire on my workbench. The small strands are perfect for this type of repair. After the repair, my ICOM works like a champ. [When experiencing problems with dual or multilayer PC boards it is always wise to check the plated-through holes for continuity. — *Ed.*] — *73, John Myers, KD8MQ, 510 W Harrison St, Alliance, OH 44601-1617*, **kd8mq@arrl.net**

## FT-857 AIR VENT MOUNT

◊I wanted to mount my Yaesu FT-857 transceiver's control head someplace on my car dashboard and not obstruct my radio, GPS or any other dashboard function. I bought the Yaesu YSK-857 separation kit but the kit does not come with a means to attach the control head to the dashboard. I looked into other kits that are specially designed to attach to car cup holder or the air-conditioning vents, but this was added cost.

The following design was a solution I found that would not add any cost. It requires two binder clips and four #10 nuts, screws and washers. You'll also need a screwdriver, needle nose pliers and ⅛ inch drill.

First, remove the chrome clip handles from the binder clips to permit access for marking the hole locations. The second step is to align the binder clips on the control head mounting frame and mark the location on the clips where to drill the holes. Next with the ⅛ inch drill bit, drill the holes.

**Figure 46 — Binder clips attached to the mounting frame.**

**Figure 47 — Controller head ready to be mounted to the air vent louvers.**

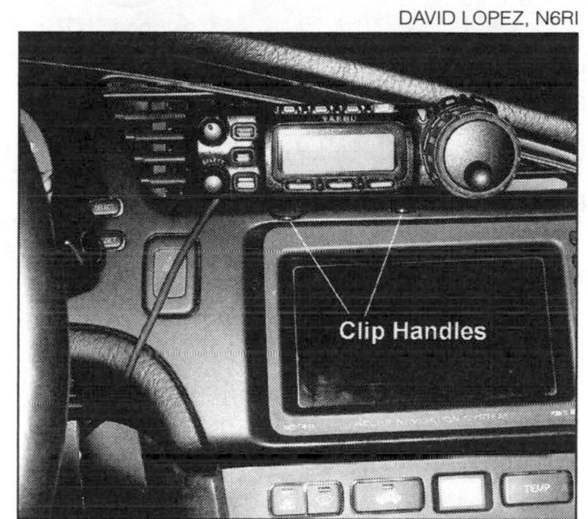


**Figure 48 — Mounted radio controller head with clip handles attached.**

Use the hardware to attach the binder clips to the back of the mounting frame (see Figure 46) Note: Do not over tighten the screws flattening the curvature of the binder clip. This curvature is what provides the force to hold the radio control head on the air vent louvers.

Once the binder clips are securely fastened to the mounting frame (see Figure 47) we are ready to reattach the lower clip handles to the binder clips. These clip handles are used to attach the completed assembly (control head and mounting frame) to the vent louvers. With the clip handles attached to the binder clips it's easier to clip or remove the assembly to the louvers. These clip handles can be removed once the assembly is attached to the vent for a cleaner look. I keep my clip handles attached so I don't lose them.

Now we are ready to attach the complete assembly to the air vent louvers. Simply squeeze the binder clip handle to open the clips slightly and slide the assembly over one vent louver. That completes the job as you can see in Figure 48 showing mounted head on car air vent. — *73, David Lopez, N6RI, 424 Las Riendas Dr, Fullerton, CA 92835*, **lopezd2@sbcglobal.net**

## TS-2000 TO SG-230 INTERFACE

◊I have lived in antenna restricted communities all my adult life. My HF antennas were shortened dipoles crammed into the attic and had limited band coverage. When I moved into another antenna restricted community, I sought a way to cover more HF spectrum. After doing a fair amount of

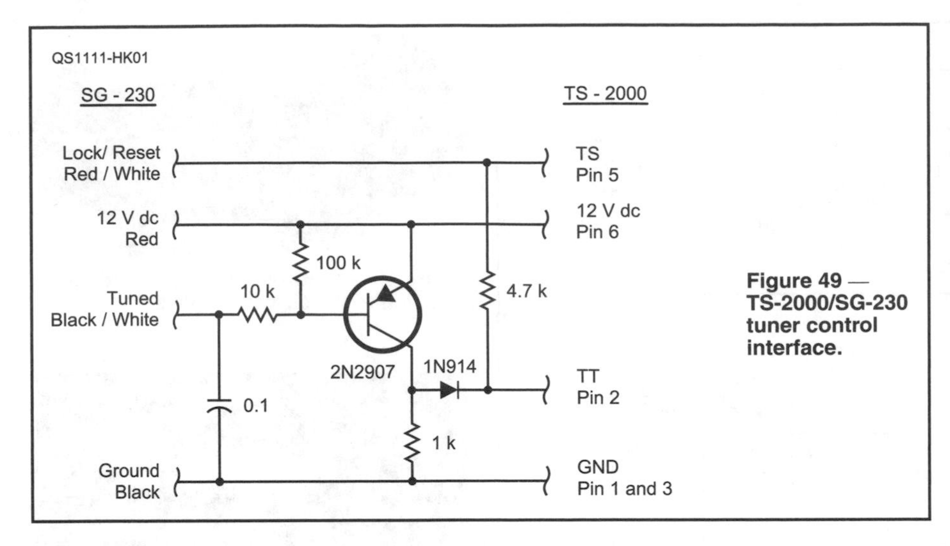


**Figure 49 — TS-2000/SG-230 tuner control interface.**

PHILLIP MIKULA, WU8P

**Figure 50 — The 10 cent connector cover adding a touch of USA to a Yaesu FT-2000.**

research, I decided to invest in an SGC SG-230 automatic antenna tuner that would cover all the HF bands and give me 200 W future capability. It feeds a delta loop located in an end gable of the attic. Fortunately the siding is not aluminum.

After installing the antenna and tuner in the attic, I wanted to interface the tuner and my Kenwood TS-2000 transceiver. The tuner requires 12-18 V dc at 0.9 A. This could have come from the station power supply, but the TS-2000 has an external antenna tuner connector that provides this power switched with the radio.

Aside from the power lines, the SG-230 has two control lines labeled RESET/LOCK (+12 V dc locks, momentary ground resets) and TUNED (goes low when coupler is tuned). Both these lines are optional, meaning the tuner will work without using these control lines. The TS-2000's external antenna tuner connector has two control lines marked TT and TS. Could these lines be utilized to control the SG-230? Some Internet searching turned up the service manual for the discontinued Kenwood AT-300 automatic antenna tuner. This answered how these control lines operate. These lines are bidirectional. TS stands for TUNING START and is activated when brought low. TT stands for TUNING TERMINATED and goes high when the tuner has completed tuning.

The TS-2000 disables its internal antenna tuner when it detects an external antenna tuner connected to the AT connector. But how does the radio do this? The AT-300 service manual did not reveal the answer, but again, some Internet searching did. (Thanks to Robert Lewis, AA4PB, for a nice technical description of the Kenwood tuner interface at **www.ham-kits.com/KWTuner/KWT%20technical%20description.pdf**). Upon power up, the TS-2000 sends signals over TS and TT and depending upon the tuner's response activates either the internal or external tuner mode. Connecting TS and TT through a 4.7 kΩ resistor fools the radio into selecting the external tuner.

Examining the signal polarity of each control line I determined that the TS signal matched the requirements of the SG-230's RESET/LOCK line and could be directly connected. The TUNED signal from the SG-230 was inverted from what the TS-2000 TT line requires. I designed a simple transistor switch to invert the signal. I combined all my requirements and created the circuit shown in the schematic (see Figure 49). I built it on a small perfboard using point-to-point wiring, installed it inline and insulated it with shrink tubing.

Now when I press the AT button on the TS-2000 the rig transmits a 10 W carrier, my SG-230 tunes, signals it is done and the TS-2000 switches back to receive — just the result I wanted. This circuit does not implement all the functionality of the original Kenwood AT-300 interface, but it does provide the basic tuning start/stop function between the TS-2000 and the SG-230 tuner. — *73, Roland Kraatz, W9HPX, 35185 Carnation Ln, Indian Land, SC 29707,* **w9hpx.4@gmail.com**

## TEN CENT CONNECTOR COVER

◊I have my microphone input and PTT line connected to inputs on the back panel of my Yaesu FT-2000 transceiver. I wanted to cover the front panel connector to keep out dust and protect the close spaced pins.

I found the screw on sleeve from a microphone connector in my junk box. I thought about what to use for the end of the cap. I looked around and the best answer was a good old USA dime. I soldered the dime on the inside and it worked out perfectly. It looks good and adds a little piece of USA to a very fine Japanese radio (see Figure 50). — *73, Phillip Mikula, WU8P, 6901 Hammond Ave SE, B, Caledonia, MI 49316,* **wu8p@comcast.net**

# Chapter 2 Batteries and Other Power Sources

## A CASE FOR EASY BATTERY TRANSPORT

◊ The Shasta-Tehama ARES group serves the California Department of Forestry (CDF) and the American Red Cross in our two counties. Covering more than 6800 square mile, this area includes some of the most rugged mountain terrain in all of Northern California.

Each year we travel miles along narrow and dusty logging roads to reach CDF command sites, CDF Equipment Staging Areas, and remote Red Cross shelter that house and feed fire victims. The only thing common to all these sites it the lack of electrical power.

The lifeblood of our remote stations is the current flowing out of donated PowerSonic PS-12260 sealed lead acid batteries. One 26 Ahr battery can easily power a remote station for more than a day or a packet station for an entire shift.

Without a handle, these 20 pound bricks are cumbersome to carry for any distance. Our solution is fabricated from salvaged .50 caliber ammo boxes. A 7/16 inch plywood insert with a cleat screwed and glued to the plywood retains the battery to one end of the case. The remaining space can be used to hold power cords, spare fuses, a distribution strip, and even a small power inverter.

It is easy to grab one of these battery cases in each hand and trudge up a slope to a strategic site that provides the critical communication between field operations and district offices.— *Gary Self, WA6MUU, 4350 Alta Campo Dr, Redding, CA 96002;* **wa6muu@arrl.net**

## ANOTHER SOLUTION TO EMERGENCY POWER

◊I was looking for an emergency power supply for my backup 2 meter radio, a Yaesu FT-2400 that I keep in my shack or take portable. I needed something that could go anywhere, that would stay charged and that would have enough power to keep me on the air for a good length of time if there were a power failure. What I tried and works great is a common portable automobile jumper battery that sells for $60 to $100. As an added feature they have a built in battery charger and will work while the charger is on making them a nice regular power supply.

All that I've seen have a cigarette lighter type outlet along with regular jumper cables. They also have indicator lights showing their level of charge. There are many different brands and some even come with built in 120 V ac outlets. They are available at almost any auto supply store or big box retailer. If that isn't enough, you can use them to jump your car if the battery goes dead. The battery, the Yaesu and an extra 2 meter magnetic mount antenna make up my go anywhere VHF radio when I have to stay connected in case of emergency. I have most of the local 2 meter repeaters programmed in to the radio and it's ready to go, as shown in Figure 1. — *Mayor Stewart Nelson, KD5LBE, 8 Deerwood, Morrilton, AR 72110,* **morriltonmayor@sbcglobal.net**

**Figure 1 — Emergency supply running a VHF transceiver on the bench.**

## MORE ON SMALL LITHIUM CELL REPLACEMENT

◊ At Dayton this year, I bought a NEC VERSA 4050C laptop computer. The CMOS battery was dead in it and so every time I turned it on I had to reset the memory for the clock and other parameters. I searched the Internet for some information on the battery type and came across **www.freelabs.com**.

The Web site is mainly about *Linux* usage, but has a lot of information on the 4050C. It seems the CMOS battery is a CR 2430 and has soldered leads coming off the battery, which are terminated in a *very* tiny plug.

I checked various battery suppliers on the Internet, and the cheapest price was in the $15 range with anywhere from $5 to $7 shipping and handling. Since this was about half the price I paid for the Versa, I was resigned to resetting the clock and everything else on start up.

While I was checking out the Hints and Kinks column in the July 2006 issue of *QST*, the picture in the upper right corner of page 57 caught my eye. The brief description of holding some tabs against a button cell made me stop and read the accompanying article. The idea of shrink-wrapping the connectors appealed to me and I checked my junk box for shrink tubing.

I picked up a CR 2430 from my local battery store for $3.90 plus tax and gave it a whirl. The battery location, according to the Freelab Web site was under the right hand portion of the laptop. What was missing in the information was the fact you have to disassemble almost the entire laptop to get to the battery. It's amazing how these things are shoehorned together!

I finally got to the battery and removed it. The leads coming from the tiny plug were indeed soldered to the battery with very thin flat metal strips on each lead. The battery also had a plastic film over it that looked suspiciously like clear shrink-wrap. Taking off the shrink-wrap, I was able to pry a lot of the metal strip up off the battery. Using a pair of small cutters, I clipped the leads off very close to the surface of the battery.

I slipped a piece of shrink-wrap over the new battery and, observing the polarity of the wires, pushed the leads under the shrink-wrap. Using a hair dryer, I shrank the wrap down to a nice tight fit. After numerous tries, I finally managed to plug the connector back into its socket in the computer.

The hardest part of the whole operation was taking the computer apart and putting it back together again. Thanks for a great money saving hint! — *Bob Cashdollar, NR8U, 1319 Granville Rd, Newark, Ohio 43055-2130*; **fdc1260@yahoo.com**.

◊ Several years ago I was also looking for an inexpensive way to replace a small lithium cell. I attempted to solder leads to the battery, with ugly results. I went back to the battery store (I can't recall if it was Batteries Plus or Interstate), and explained my situation.

To my pleasant surprise, they offered to "tack" leads onto a second new battery for me. I think they used some sort of small arc-welding setup, just for the purpose of attaching solderable leads to lithium batteries. There was no extra charge for this service. — *Regards, Joe Papworth, K8MP, 200 N Parkway, Delaware, OH 43015*; **k8mp@aol.com**.

## AUTOMATIC EMERGENCY POWER TRANSFER FOR YOUR HAM STATION

◊Living in the Midwest, adjacent to Tornado Alley, I have experienced many power outages. Most last only a few minutes to an hour or so. The ice storm of February 2002 lasted 10 days at my house, however. To keep my station on the air during these power outages, I use a 24 Ah, 12 V sealed lead acid battery. When the power goes out, however, I would have to disconnect the ac power supply from my radio equipment and connect the battery.

I worked on a design for an "uninterruptible power supply" using high current diodes to route the voltages. The problem is, if you place a diode between a battery and the transceiver, the voltage drop across the diode is often too much to keep the PLL circuits in the transceiver operating from a slightly discharged battery.

My solution is a simple high current relay to transfer the transceiver from the ac power supply to the battery when the ac power fails. When normal ac power returns, the relay is energized, connecting the transceiver to the ac supply. See Figure 2. The circuit also connects a trickle charge circuit to recharge the battery. With the 4.7 Ω current limiter, you cannot overcharge the battery. The diode prevents the battery from discharging through the ac power supply when the supply is switched off.

K1 is a DPDT relay with 15 A contacts. By paralleling the contacts, it has a 30 A capacity. I mounted the relay and its matching socket, along with the charging components in a project box.

I used a Tyco Electronics/Potter & Brumfield K10P-11D15-12 relay. The socket is a 27E895. A 3 A silicon diode (such as a 1N5400) and a 4.7 Ω, 1 W resistor make up the charging circuit. The project box is a RadioShack 6×3×2 inch box, part 274-1805. I bought the relay and socket from Electronics Supply in Kansas City, Missouri. My cost for the project was about $30.

I mounted RadioShack binding posts (part 274-661) on the plastic box cover as shown in Figure 3. I spaced them ¾ inch apart to allow the use of Pomona or General Radio double banana plugs for my connections to the power supply, battery and radio. [It would be simple to mount Anderson Power Pole connectors on the box, too. — *Ed.*] I used no. 16 AWG wire inside the box, but doubled all the runs for added ampacity. The wires to the battery should be at least no. 12, and should be as short as possible to reduce voltage drop.

When the ac supply is switched off (or the power goes out) my transceiver turns off but then returns to the same operating frequency when the relay transfers it to the battery. My transfer unit has been in service for some time and works great. I can forget it is even in my shack.

It would certainly be possible to build a more sophisticated power transfer circuit or buy a commercially available unit. The battery charger circuit could also be improved to include voltage and current sensing. This simple circuit meets my needs, however, and may help others. — *73, Roger Snowdall, WØKWJ, 8405 Everett St, Raytown, MO 64138*; **rogerw0kwj@aol.com**

Figure 3 — This photo shows how WØKWJ installed binding posts on a project box cover to connect his radio to an ac power supply or a battery.

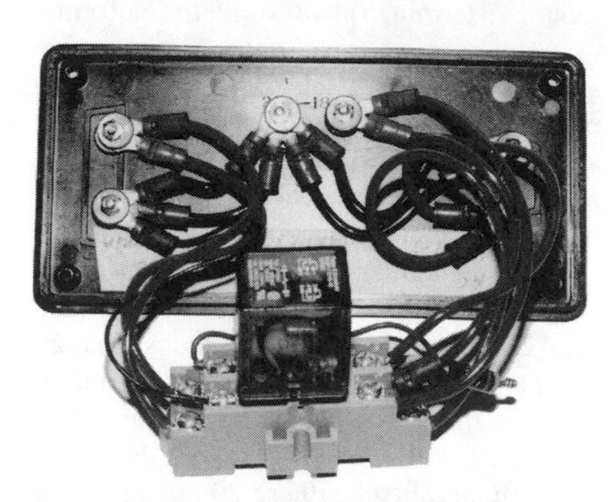

Figure 4 — This view shows how the relay is mounted on a socket, with wiring between the socket and the binding posts. Notice that the no. 16 AWG wiring is doubled for all wiring runs, to increase the ampacity of the wiring.

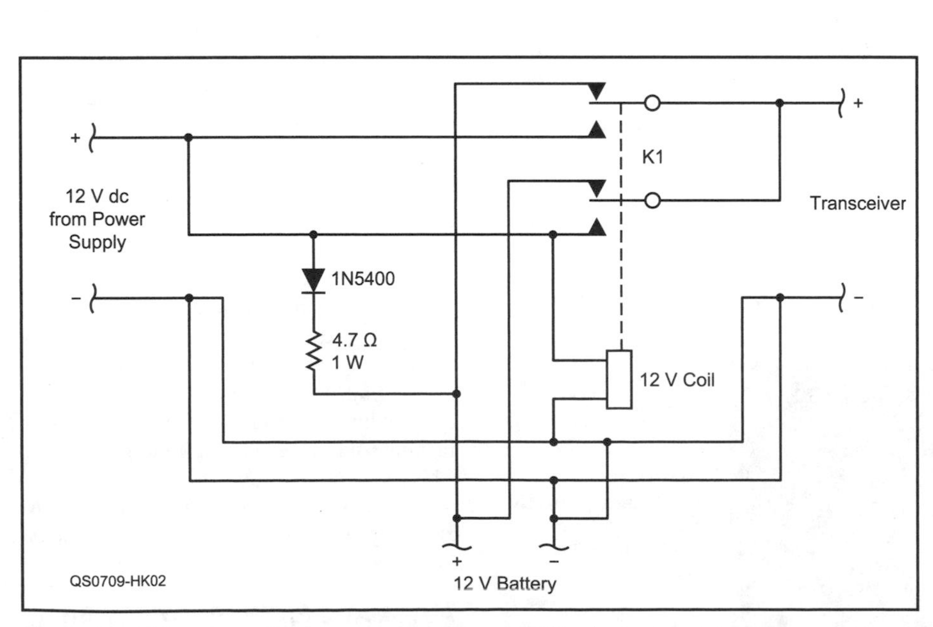


Figure 2 — This schematic diagram shows how a 12 V relay can be used to automatically switch between an ac power supply and a battery to power a radio. The diode and resistor form a trickle charger to keep the battery topped off and ready for the next power outage.

## ELECRAFT KX1 BATTERY PACK

◊After building an Elecraft KX1 I began to look for ways to improve this great radio. One trick is to take out the 1.5 V AA batteries and install six 3.7 V AA-size lithium-ion batteries. I just rewired the battery holders for three cells in parallel with the other three cells, and installed a 5 A protection board so the Li-ion batteries charge properly and are protected from an overload or short circuit. This gives me an 11.1 V, 1.5 Ah battery pack inside my KX1. See Figure 5.

If you decide to install a few 3.7 V Li-ion batteries in your KX1, please check out my "KX1 Pimp Pot" Web site: **www.wa3wsj.com/KX1.html** or go directly to the battery page at **www.wa3wsj.com/KX1-12vPower.html**. You must install a protection circuit board (PCB) so it is safe to charge these batteries, and also to protect against short circuits and other problems. Figure 6 shows the PCB with a wiring diagram for the batteries, radio and charger.

I use batteries and PCBs from **batteryspace.com**. See **www.batteryspace.com/index.asp?PageAction=VIEWPROD&ProdID=2779** for more information. I don't use the "Fuel Gauge" boards, but those add an interesting way to keep track of your battery pack charge condition.

I brought a set of leads from the battery out through a hole in the case and added a charge plug on the bottom of my KX1. See Figure 7. Now I can just plug in my charger and charge the batteries! The plug is the same type of plug used to power my KX1.

I use a Smart Charger for my six-cell

WA3WSJ

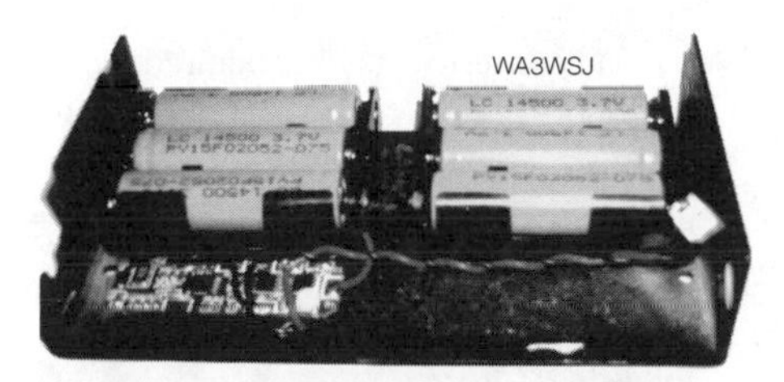

**Figure 5 — WA3WSJ modified the battery pack for his Elecraft KX1 by using 3.7 V Li-ion batteries and a protective circuit board (PCB) from Batteryspace.com. Ed connects two cells in parallel for increased ampacity with his 11.1 V battery pack.**

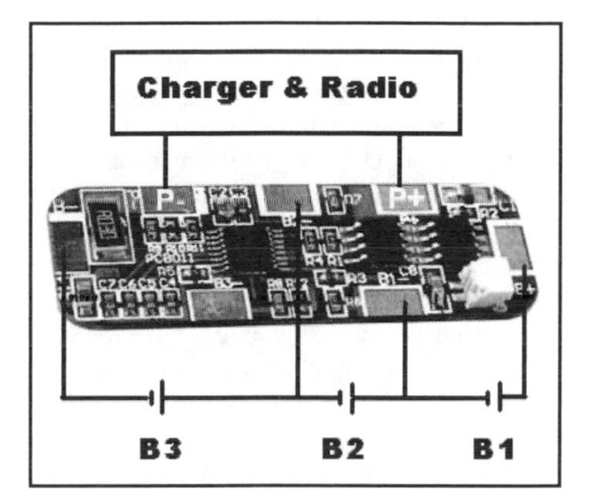


**Figure 6 — The protective circuit board used with the 3.7 V Li-ion batteries charges each parallel pair of cells individually, and also provides connections from the battery pack to the radio and charger.**

WA3WSJ

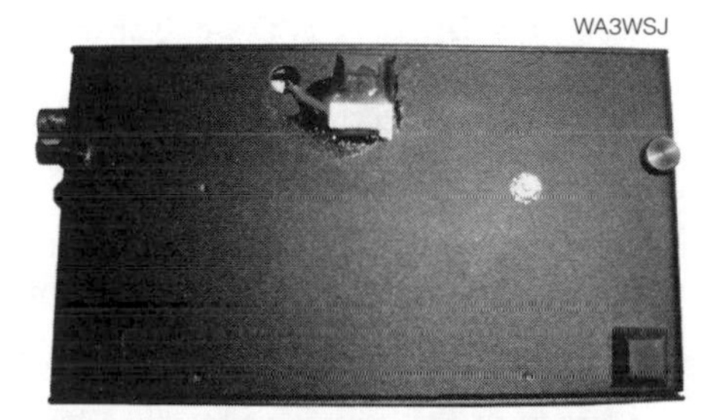

**Figure 7 — This photo shows how WA3WSJ brought a set of leads from the battery pack out on the bottom of the KX1 case to connect the battery charger.**

Li-ion batteries. It is set for any Li-ion battery pack at 11.1 V ( 3 cells) with 120 V or 240 V ac input for worldwide power support.

Here's another interesting idea to add to the KX1 battery conversion: A 2 A Poly Switch that resets when the high current load is removed. More details are at: **www.batteryspace.com/index.asp?PageAction=VIEWPROD&ProdID=2687**.

My friend Guy Hamblen, N7UN, uses a 14.6 V, 4 Ah Li-ion battery pack to power his K1. While on a hike, the power cable from the battery froze and cracked, shorting out the battery. One of these poly switches saved the battery.

I tried out my new "pimped-out" Elecraft KX1 on an Appalachian Trail Hike for the January 2007 Polar Bear operating activity. [For more information about the Polar Bear group and their Moonlight Madness activities, see **n3epa.org/Pages/PolarBear.htm**. — *Ed.*] My 2 ounce Bead-Wire Antenna quickly went up and I was on the air Saturday morning making Qs. While hiking on the AT in New Jersey, I had a dozen QSOs using my KX1 and the Bead-Wire Antenna. I checked my battery in the KX1 and it was still at 11.1 V. Plenty of power for more Qs!

My next KX1 Trick will be to put my "pimped-out" KX1 in a Pelican 1060 Micro Case for travel and hiking. — *72, Ed Breneiser, WA3WSJ, 775 Moonflower Ave, Reading, PA 19606;* wa3wsj@arrl.net

## EMERGENCY BATTERIES

◊ As EOC for Carroll County, Maryland, I know that hams can use just about any battery during an emergency. One of the things expected of an Amateur Radio operator is that he/she will find some way to keep a radio on the air. We often have to think outside the box to extend our ability to provide power for communications. What is one of the worst things an emergency radio operator might do before an emergency? — *Answer*: Not check the emergency power system and not charge all emergency batteries.

Recently when I went to check and top off my batteries, I found one 12 V, 7.5 Ah battery that measured about 11.8 V. I check and maintain my batteries every 6 to 12 months. A fully charged lead-acid, gel cell or deep cycle battery should lose less than 5% of its full charge in that amount of time. Further, the voltage should not drop below about 12.6 V. If the voltage falls below that level in 6 months it means the battery is bad, or is going bad, and it is time to recycle it.

During Hurricane Isabelle, in September 2003, almost every station I heard checking into the nets stated that they were on emergency power. So what is the second worst thing an operator can do during an emergency? *Answer*: It is to use emergency power when there's no need to do so! We are called on to provide communications during an emergency. It's a mistake to use batteries *before* they are needed, because we have decreased our ability to provide communication *after* the power fails!

I use my Astron (**www.astroncorp.com/**) ac power supply to power all my radios and my emergency lighting system. Both the Astron and my laptop computer are plugged into a 300 VA computer UPS (uninterruptible power supply). If ac power fails, I can change over the radios from the Astron to the batteries. Since I now use Anderson Powerpole connectors (**www.andersonpower.com/**), this is a very simple and quick thing to do.

I built an Anderson Powerpole Y-cable for my power system. See Figure 8. At the start of an emergency, I plug both the Astron and backup batteries into the power distribution system at the same time. This way I top off the battery, and I also won't lose any time in a switchover should commercial ac power fail. I simply unplug the Astron from my power distribution system if commercial power fails. I can then use the other connector for the next battery, once the first battery has run out of charge. Every emergency station should have at least two large deep-cycle batteries for emergencies.

Once commercial power fails, I'll print out anything needed from the computer and shut it down to save battery power. I then go back to using pen and paper. (I print only the things generated during the emergency that I need, not things that can be printed before losing power. Those should have already been printed out, of course.)

I have three methods to recharge my batteries once ac power has gone out. First, I have a solar panel, the one item I bought for the sole purpose of charging batteries. Second, I have a car, which, of course, has a battery recharged by its alternator. Last, I have a riding lawn mower. Like the car, the mower uses a battery for starting the engine and an alternator to recharge that battery.

In advance of an emergency you must determine the kind of emergency power you will need. My radios need a 12 to 15 V dc power system, and my battery system must

PHIL KARRAS, KE3FL

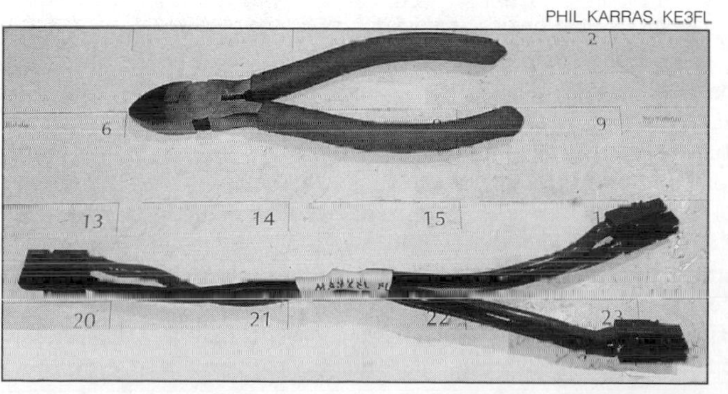

**Figure 8 — KE3FL's "Y-cable" system to supply emergency dc power. An Astron ac power supply tops off the charge on a rechargeable battery while commercial ac is available. Once power fails, the Astron is disconnected.**

supply power for perhaps seven days during an emergency. My computer and lighting system also run off 12 V dc.

### Stock Up or Charge Up

I also have a number of devices that use anything from 2.5 to 6 V dc. Most of these devices use AA batteries, and I own about 40 rechargeable NiMH batteries. You need to remember to either stock up on alkaline AA batteries or to charge up your rechargeable batteries prior to any emergency. Buying a bunch of NiMH batteries for emergency service means that you have to be sure to check and charge them every month or so, since they lose about 1% of their charge per day.

How can you be sure your batteries are charged prior to an emergency? As I said before, I charge up my lead-acid emergency batteries every six months to a year. At the very least I do this in the spring when ARES starts to support walkathons and things like that. Generally I take three batteries with me (not freshly recharged) for the event. I consider this a stress test for my batteries. Any that fail during the event are removed from service and replaced. All lead-acid batteries should still have over 90% of full charge after sitting for one year. Any battery that fails during a short duration, low-stress event like a walkathon is dying and needs to be replaced.

In my opinion, one of the worst things we can do is to keep our batteries always "topped off." Why? Because if we do this and an emergency hits, we assume that the battery is ready for use. The problem is that keeping it topped off could be hiding the fact that the battery is bad and it really won't be able to supply power for the duration of the emergency, or at least for its intended part during the emergency. Keeping your batteries topped off is not a panacea. You still need to check them and stress them every year to be sure they are still good.

I believe that my method of charging up these types of batteries every 6 to 12 months is almost as good as keeping them topped off. With my method, they will still have well over 90% of full charge, with the added advantage of reminding me that they need to be checked and stressed regularly. On the other hand, rechargeable NiMH (or NiCd) AA batteries need to be charged every one to two months.

Another emergency power recommendation is that whenever you buy a handheld transceiver, you should also buy an empty battery pack that accepts AA batteries. The newer handheld transceivers have sealed battery packs, which you cannot easily open, if at all, to check each cell and charge them individually, if needed.

If I had my way, I'd never buy another NiMH or NiCd pre-built battery pack. I'd only buy two empty battery packs and use my own rechargeable batteries. The main reason for this is that I can keep this kind of battery going for thousands of charges, while most of the pre-built packs I've used did well to last only three or four years. (That's about 36 to 48 charges, doing one a month.)

Your main batteries should be deep-cycle, gel cell, liquid acid or anything else that is made to be deep discharged. If you have, or can find, a supply of used UPS batteries, this is a good way to get usable deep-cycle batteries.

If you must, you can use the lead-acid batteries that start your car or mower. Just remember that once you deep discharge these batteries, you may have to replace them. They are not designed for deep discharge and, sometimes, even doing it once is enough to kill them. On the other hand, you could move the car or mower out of the garage, start it up and use jumper cables to charge up a discharged battery while you use your second battery to keep your station going.

When transporting batteries be sure all power connections are covered and protected. Also, be sure that all screws attached to the battery terminals have the long ends pointing to the inside of the battery so that the battery cannot be shorted if it just moves up against a metal object, like a tool kit.

Make sure you fill your gas tanks before an emergency, and also make sure you have some extra gas cans filled, if you can. Also try to keep enough gas in the car for after the emergency is over. You still have to get home, and you need to get to a working gas station to fill up again. — *73, Phil Karras, KE3FL, 3305 Hampton Ct, Mount Airy, MD 21771,* ke3fl@yahoo.com

## POWERING YOUR RADIO IN AN EMERGENCY

◊ Is your emergency receiver/transceiver emergency power ready? Is there compatibility between the power requirements of your emergency radios and the power outputs of your emergency power sources?

There are a number of ways to make your radios and emergency power compatible:

- Only buy radios that require 6 or 12 V batteries.
- Build power converters to match the 6 or 12 V battery voltage to the radio.
- Convert the radios to 6 or 12 V.

There are advantages and disadvantages to each:

*Advantages*

- The radios are easy to power during an emergency.
- One power converter can be used for more than one radio.
- Converted radios can be used with any standard 6 or 12 V source.

*Disadvantages*

- It's not always possible to find a radio with this power requirement.
- The power converter may not be compatible with other people's power sources. More than one converter may be required.
- The power converter can't be used with other radios, since it's built-in.

Which method you select depends on what you're after and how you are set up to supply emergency power.

In my shack I use 12 V batteries almost exclusively. (I have a few small lead-acid, gel cell 6 V batteries for my two smaller 6 V AM/FM/TV and AM/FM/SW receivers.) So I needed to find a way for all my radios to be powered by the 12 V backup batteries I use. Most of my equipment is 12 V ready since most ham gear is designed that way. However, a few years ago I used a Sangean ATS-803A and was very impressed with the radio; so much so that I decided to buy two of them.

I managed to get both a Sangean ATS-803A (**www.sangean.com**) and a Radio Shack model DX-440 (**www.radioshack.com**), which Sangean made for RadioShack. The only problem with these radios is that they use 9 V, from 6 D-cells. Since I do not normally buy D-cells, and since I use mostly 12 V emergency power cells, I decided to build a 12 to 9 V converter into both radios. To do this I bought a number of the 1.5 A LM7809, 9 V regulators.

The first step was to measure the power requirements of the radios. They were mostly identical; at 9 V the current requirements were from 100 mA (audio all the way down) to about 125 mA at full audio. This was excellent because an LM7809 in a TO-220 case can handle those current levels indefinitely without heat sinking.

I took the Sangean and DX-440 apart to see what kind of modifications would be needed. I found that I could put the regulator inside the radio without any problems. The LM7809 is a three terminal regulator that requires only an input and output capacitor to function. There was more than enough room for this small circuit with plenty of room for air flow as well.

I pre-built the circuit using dead-bug construction, meaning that I simply soldered the two capacitors directly onto the LM7809 chip with no circuit board being used (see Figure 9). [Refer to your chosen regulator's application notes for a schematic. Mouser Electronics (**www.mouser.com**) has many application notes available on-line. — *Ed.*]

As can be seen in Figure 10, there is plenty of room inside the radio to install the regulator and its required capacitors just off the board with plenty of airflow room as well. You can also see that there is only one other solder connection to the B+ line. I traced this and found it went to the 5 V regulator for the microcontroller. Thus there was no need to cut the trace and supply 9 V to this part of the circuit; the 5 V regulator just runs a bit hotter when it is supplied with 12-14 V as opposed to 9 V.

PHILIP KARRAS KE3FL

**Figure 9 — All that's needed to adapt a 9 V radio to a 12 V battery.**

**Figure 10 — The dead-bug installation of the power regulator directly onto the receiver's circuit board.**

It is important that you either trace the power circuits or have a schematic for the radio so you know if your modification(s) will cause any problems to the existing power regulators. In most of today's radios, there are frequently more than one power regulators.

There will normally be at least two, one for the microprocessor, usually either 5 or 3.3 V, and another for the radio/audio sections, usually 6, 9 or 12 V. All of this is subject to change as the manufacturers go to more and more digital circuits. Still, I imagine that there will always be the need for higher voltages in the output stage. In these stages (the audio amplifier in receivers and the final amplifier in transmitters) we use higher voltages since the higher the voltage, the lower the current requirements for the same amount of power. (P = EI) For example: If we have a 24 W load; to drive it with 6 V would require 4 A, but with 12 V only 2 A are needed. While this would be little problem for today's switching power supplies, it is more difficult to provide the higher currents with some of today's consumer batteries.

To do the same thing for other radios we need to check the power and voltage requirements for all operating modes of the radio. Commonly this includes listening at a low volume, listening at the highest volume and transmitting at the lowest and highest output power settings. Once you have that information, you can figure out if a simple regulator, like the LM7809, will work for your radio. My rule of thumb is that if the highest current needed is ⅕ or less of the maximum current output of the regulator, then the regulator can be used without a heat sink. If the current required is more than that, use a heat sink or a larger regulator.

So, take a close look at what type of power your radios need and what kind of power your emergency power system can provide. If there is a mismatch, take steps to adapt the radios to the available power sources before the emergency arrives. — *Philip Karras, KE3FL, 3305 Hampton Ct, Mount Airy, MD 21771,* **ke3fl@arrl.net**

## SOLAR QRP

◊ What could sound better to a ham than something for nothing? Well, solar power is free once you have the equipment to harness it. For those kindred spirits who enjoy operating low power portable, solar power may solve your battery power problems.

Solar panels meant to keep car batteries charged during periods of inactivity are available from auto parts stores and online for under $20. They are 12 V small capacity, 1.8 W or 125 mA units intended to trickle charge the vehicle battery by plugging it into the cigar lighter socket.

To modify one for my use, I first cut off the cigar lighter plug and replaced it with a dc barrel connector common to all my low power gear. I then made up a junction box with three plugs wired in parallel that I could plug the solar panel, radio and battery into, connecting all together.

Most large capacity solar systems use a charge controller to keep the batteries from being overcharged, but with a small capacity solar panel it is not necessary. I have been using this configuration for 2 years and have used gel cells, NiCd and NiMH batteries with good results. Be advised not to try to operate the radio using the solar panel alone; a battery must be connected. The battery acts as a buffer and voltage regulator and is a necessary part of the system. — *73, Richard Arnold, AF8X, 22901 Schafer, Clinton Twp, MI 48035,* **af8x@comcast.net**

## SAFEGUARD THOSE EXTRA HANDHELD BATTERY PACKS

◊Going off for a day with your handheld generally means bringing a spare battery pack. If you just toss one or two into your kit bag, you could be asking for trouble. Generally such battery packs contain a lot of stored energy in combination with exposed terminals. If the terminals short against a pair of pliers, or who knows what, you risk a ruined battery at a minimum — a fire or explosion could also result.

My batteries did not come with a protective cover, so I solved the problem on the cheap using a small, zippered, plastic "snack" bag as shown in Figure 11. While any size bag could be used, these seem a perfect fit for my handheld battery. Note that this is not a heavy duty solution and thus the bag should be inspected before each use to make sure it hasn't been torn. In any case, it's much safer than just tossing the pack in the kit and hoping for the best. — *73, Joel Hallas, W1ZR,* QST *Technical Editor*, **w1zr@arrl.org**

JOEL HALLAS, W1ZR

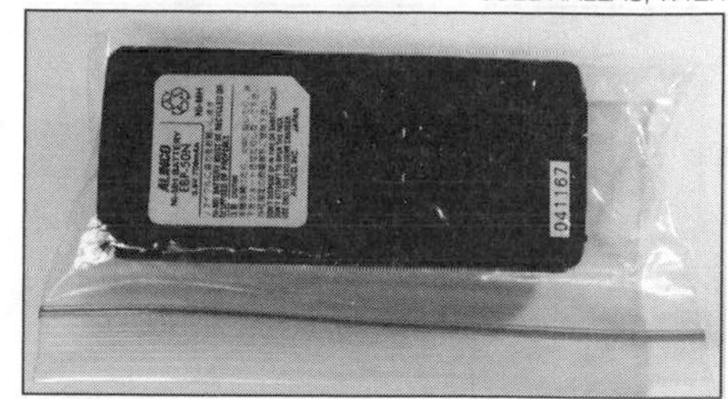

**Figure 11 — A spare handheld battery fits nicely into a snack sized zipper bag, insulating the exposed terminals.**

## GEL CELL MAINTENANCE

◊Being in the commercial fire alarm industry, I deal with a *lot* of gel-cell batteries, the most common size being the 7 Ah. Having been in the industry since 1982, I speak from experience when I tell you that the "expected" life of these batteries is around 3 to 4 years in average service. As a matter of fact, our fire codes *require* they be changed at 4 year intervals. Now, you may get more life from them, but you can only *depend upon them* for 3.5 to 4 years. Considering that the name of the game is emergency preparedness, I'd paraphrase my Uncle Bill's admonition about motor oil and motors, this way: "Batteries are cheap, lives aren't!"

I'd also comment on testing them — nothing works as well as a *real load*. Old automobile headlights draw about 3-4 A at 12 V dc making them a great test load (plus you can *see* the rate of change). I made my test load from seven 47 Ω, 10 W resistors in parallel (that's 6.7 Ω), which at 12 V dc yields a current draw of 1.79 A and at 24 V dc (typical fire alarm control panel voltage) gives you a drain of 3.6 A. All the new testers I've tried have been outshined by this tried-and-true method, which provides "real life" information. Lastly, *do not* keep those batteries on charge, all the time. You *must* cycle them (some chargers do a pretty good job of this).

As we learned in the Navy — never depend upon something that you haven't maintained *and* tested, properly! — *73, Tom Dailey, WØEAJ, 270 S Lafayette St, Denver, CO 80209-2524,* **daileyservices@qwest.net**

## BATTERY PROTECTION

◊I was just reading my May 2008 *QST* and saw the article on using a Y-cable for emer-

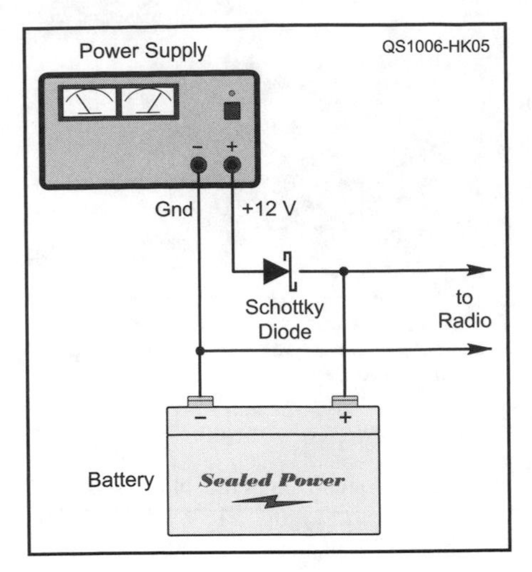


**Figure 12 — A schematic of the blocking diode arrangement used to prevent damage or discharge to a battery when connected to a load and power supply in a Y arrangement.**

gency battery power. While it is true that this will work on many linear power supplies such as the Astron, be warned that this is not safe for all power supplies.

I own an Alinco DM330MVT switching power supply (**www.alinco.com**), which has an over voltage protection circuit on the output. This circuit effectively shorts the outputs to ground when ac power is removed.

If you use a Y-cable to directly connect a battery to the Alinco DM330MVT, you will quickly drain the battery — or worse — when power is removed from the supply. (Don't ask me how I know this...)

Use of a blocking diode is required in this case (see Figure 12). Care will need to be taken that the diode is of a proper current rating for the application and one should be chosen with as low of a forward voltage drop as is possible, to reduce wasted power and generated heat. A diode with a forward current rating of 30 A or more is required for most 100 W transceivers. A suitable heat sink may be required for the diode. Even a 0.5 V forward drop at 30 A is 15 W of power. — *73, Russell Hoffman II, N3WDZ, 200 School St, Wabash, PA 15220-2718*, **n3wdz@arrl.net**

# Chapter 3 Mobile and Portable Stations

## PLAYING RADIO AUDIO THROUGH CAR STEREOS

◊ I operate mobile from a 1999 Honda CR-V with an ICOM IC-706MKII. Playing the IC-706 audio through the car's audio system has some distinct advantages: My car stereo system has much more power and fidelity than the speaker of the '706. Unfortunately, my factory stereo does not play cassette tapes, for which adapters that work very well are available cheaply, Neither does it—like many new cars and after-market stereos—have accessory inputs for an MP3 player or satellite-radio converter. I've tried wireless FM broadcast adapters, and they worked poorly. Thanks to demand from owners of MP3 players and satellite radios, devices now available connect auxiliary inputs to a stereo through its CD changer connector. The devices accept dual RCA jacks for stereo input and cost about $50. I ordered a PIE HOND98-AUX adapter for my CR-V from Logjam Electronics (**www.logjamelectronics.com**). After a fair amount of dashboard disassembly and re-assembly, I now can play the IC-706 through the stereo.

It works great! In the CR-V, the stereo works only with the key in the ignition, so when I'm stopped I disconnect the car stereo from the IC-706 front panel and use the built-in speaker.[1] I've documented the procedure at **www.geocities.com/ kf6iiu@sbcglobal.net/crvdash.htm** for anyone who is interested —*Wiley Sanders, KF6IIU, 984 Hawthorn Dr, Lafayette, CA 94549;* **kf6iiu@arrl.net**

## BENZOIN AIDS ADHESION FOR "A NOVEL MICROPHONE HOLDER"

◊ I read "A Novel Microphone Holder," in the August column. I appreciate the author's difficulty "convincing" commercial hook-and-loop material to adhere to the dashboard covering. I have used commercial hook-and-loop material for years, and have had success in keeping it attached to the surface to which I had stuck it.

I have been a registered nurse for numberless years, and have frequently encountered the problem of dressings that do not want to stick to the patient. In the hospital, we use a liquid known as "tincture of benzoin."

This is an alcohol solution of a plant extract (gum). When the alcohol evaporates, the resin is left behind, forming a tacky surface having the quality of enhanced adhesion. When I use it in the hospital, my intravenous site dressings *do not* fall off. I used it in my van several years ago to place hook-and-loop material on the dashboard to hold my radio's control head. It stuck then, and continues to do so today.

Tincture of benzoin is available over-the-counter, although it may not be obviously displayed on the shelf. You might need to ask the pharmacist where it is. It might cost 2 or 3 dollars for 4 ounces. [I found several sources of benzoin swabs on the Web.—*Ed.*] Use it sparingly, and it will last a long time and not make a mess. If you should make a mess, you can clean it up with a paper towel wet with rubbing alcohol. Simply be certain, beforehand, that the surface you are working on/cleaning up will tolerate rubbing alcohol without ill effect.

Thank you for both the interesting feature "Hints and Kinks" and the opportunity to contribute.—*Jon Seaver, RN BSN, N8SUA, PO Box 261, St Johns, MI 48879;* **n8sua@yahoo.com**

## A STRONG, FLEXIBLE MOBILE ANTENNA INSTALLATION

◊ In the August 2004 column discussion of homebrew wing nuts, I mentioned my mobile antenna setup that my wife and I (affectionately for me, probably not for her) call "the plumbing mobile" (Figure 1). I said it was a story that I would tell later. Now is the time.

Dwindling fuel supplies have led to lighter vehicles and made it difficult to mount an HF whip on a car. Metal body panels are thinner than the walls of a bicycle's frame tubes. They no longer provide a firm surface for an antenna mount. Automobile body quality is high; the gaps at doors and other body panels are now too small to accept thick metal antenna mounts. The metal parts of bumpers are hidden deeply under many layers of rubber and plastics so that we can no longer reach the metal to attach an antenna mount.

Manufacturers have supplied some worthy answers to these problems, such as the giant magnetic mounts and mounts that are secured behind the vehicle license plate.

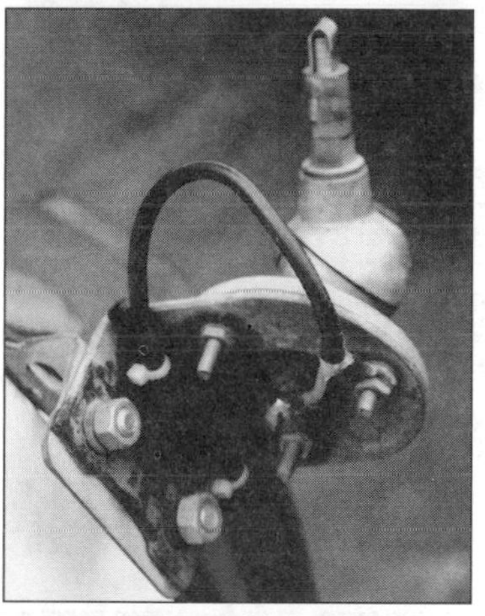

**Figure 2—A stock split-ball mount is mounted to the hatch lip via pre-drilled steel angle used to install garage-door openers. Two #12 × ½-inch self-tapping screws secure the angle to the hatch lip. The top photo is a view from the rear of the car with the hatch open. The bottom photo is a view from the front of the side of the car.**

**Figure 1—The "plumbing mobile" at rest.**

[1]About the SPEAKER-PHONES switch in the back of the IC-706 control unit: This switch configures the wiring of the front-panel jack for mono or stereo output. To patch the '706 into a car stereo, leave it in the "phones"

**Figure 3—Two braces (shown here as red and green) make the mast secure.**

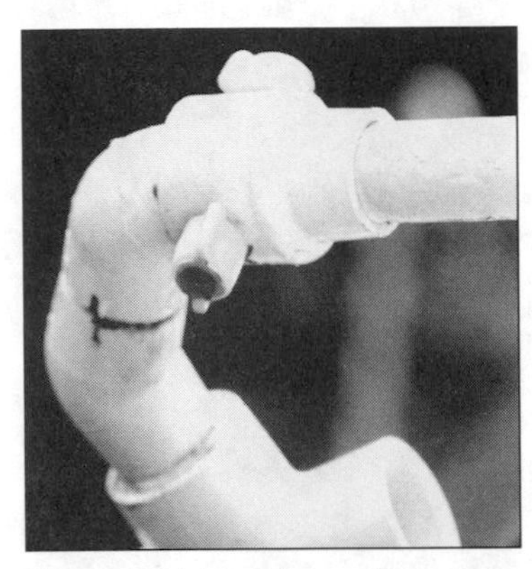

**Figure 4—A large hose clamp (see Figure 8) secures the ¾ PVC T to the roof rack at two places. This is the driver's side. A ¾ to ½ reducer, two short ½ nipples and two ½ 45° elbows reach up to the ½ PVC brace, which ends in a ½ to ¾ coupler to provide a loose fit over the elbow. #10 stainless hardware with homebrew wing nuts fasten the joint.**

My faithful Subaru wagon (1993) adds a few more problems to the mix. An antenna mounted near the end of the enclosed rear bumper would obscure the brake lights from the rear and could prevent the hatch from opening. The sides of the rear hatch are cut around the taillights, with no room for a thick bracket.

In addition, I wanted the option to mount either of two antennas on hand: A Hustler whip with a fold-over mast and a Spider antenna. Each of them has several resonators and a quick-release base. Because I live in a city, I must often park in enclosed garages with little overhead clearance (say, seven feet). Lastly, I wanted the installation to be removable in a few minutes.

The Spider requires that the resonators be mounted a certain distance above the vehicle roof. This determined the location of the antenna base and its mounting ball. It worked out well, with the ball just above the taillight assembly and the opening to fill the rear window-washer reservoir (see Figure 2, top). With the ball at this height, the folded Hustler antenna is low enough to fit inside parking garages and height barriers at fast-food restaurants. Thus, I can leave the Hustler in place and safely folded all the time I'm driving. (It would clear my garage door as well, but I park the car outside.)

The L-shaped mount is too thick to fit in the gap between the hatch and the car body, so I cut a short length of pre-drilled angle from the local home-improvement store that fits in the gap. If you don't find the material in the hardware or building materials departments, look near the garage-door-openers. This angle is often used to mount the openers to the ceiling. Thepre-drilled angle is easily secured to the vehicle body with two #12×½ inch stainless self-tapping screws and lockwashers (see Figure 2, top). Four screws would be stronger, but the hatch lip is not deep enough to permit four screws. The antenna-mount bracket is secured with hardware chosen to fit the mount. (I think mine are $\frac{5}{16}\times\frac{1}{2}$ inch stainless steel with lockwashers.)

This looked like a good mount until I tested it with the antenna in place. The length of the antenna gives it a lot of mechanical advantage over the short, narrow mounting area. When I simulated wind buffeting by moving the antenna somewhat, I saw that the angle and the vehicle body where it is mounted flex quite a bit from the stress. I think fatigue would soon cause a failure of the bracket or the body where it was mounted. A strong mount requires that all antenna torque from roll (side to side) and pitch (front to back) is removed from the mount assembly.

Because I formerly engineered roof-trusses and now enjoy how cheap and easy

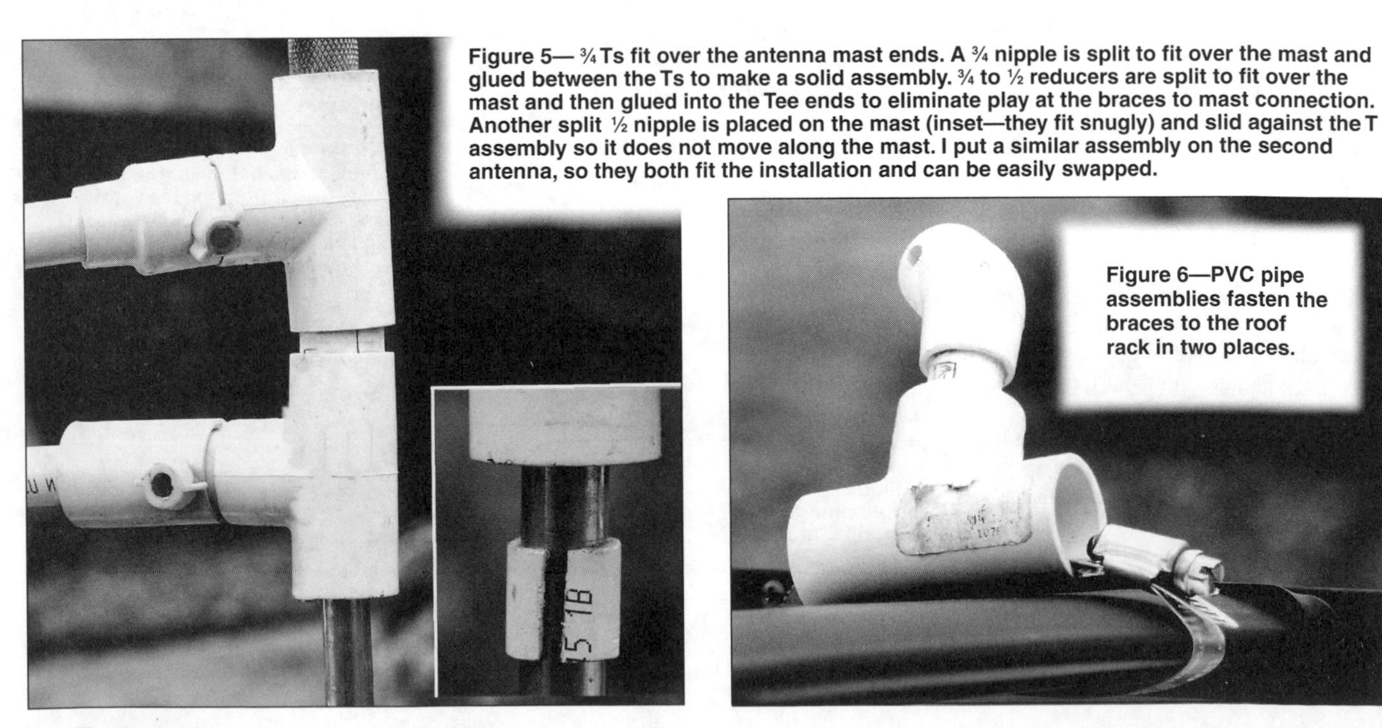

**Figure 5— ¾ Ts fit over the antenna mast ends. A ¾ nipple is split to fit over the mast and glued between the Ts to make a solid assembly. ¾ to ½ reducers are split to fit over the mast and then glued into the Tee ends to eliminate play at the braces to mast connection. Another split ½ nipple is placed on the mast (inset—they fit snugly) and slid against the T assembly so it does not move along the mast. I put a similar assembly on the second antenna, so they both fit the installation and can be easily swapped.**

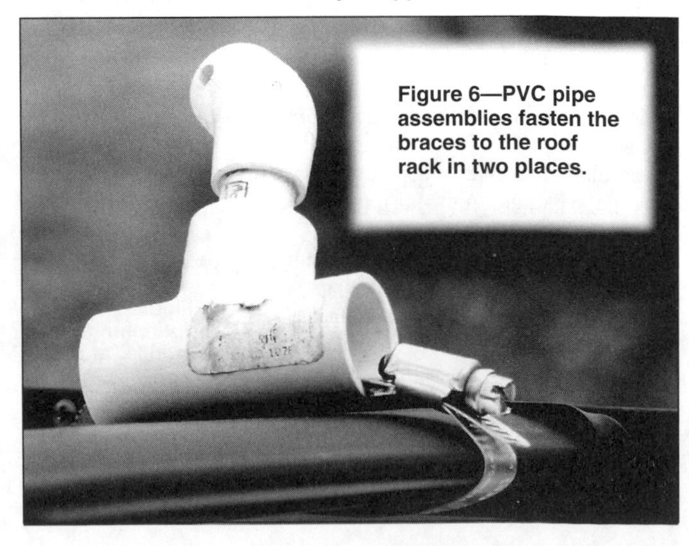

**Figure 6—PVC pipe assemblies fasten the braces to the roof rack in two places.**

Figure 7—An arm and slotted tee bracket keeps the folded antenna from "windmilling."

Figure 8—Only small white stubs remain when the antenna/brace assembly is removed.

it is to build things with PVC pipe parts, the answer sprang to mind immediately. One brace from the Subaru's roof rack braces against fore-and-aft motion. This is the red brace in Figure 3, while another from the rack rail on the other side of the car roof braces against side-to-side motion. This is the green brace in Figure 3. Together, they make the mount sufficiently strong. Yet PVC is normally fastened with glue; how could I make the mount removable?

The homebrew wing nuts from August 2004 Hints and Kinks are the answer. Four joints (Figures 4 and 5) fastened with #10× 1½ inch stainless-steel machine screws and matching nylon-insert homebrew nuts permit the antenna and braces to be removed or installed in a few minutes, and the parts all fit easily into the car.

### *Construction*

The braces are attached to the roof rack via 5 inch hose clamps and ¾ inch PVC T fittings. The hose clamp is backed open until the strap is free of the screw and its housing. The free end of the strap is then threaded through the top (straight piece) of the T, around the roof rack and back into the screw and its housing. The screw can then be tightened to secure the tee onto the roof rack (Figure 6). A ¾:½ reducer holds a ½ nipple in the upright of the T so that the 45° elbow fits on it and forms a stub pointing toward the mast.

In use, I found that the folded antenna had a tendency to "weathervane," so I added a vertical arm with a slotted T to hold the folded mast (Figure 7). I made the slot in the T about ⅛ inch narrower than the mast and then widened the slot with a file until I got the fit I wanted.

### *How Well Does it Work?*

My mobile station is not operated in motion, but mobile at rest, while my wife is in antique shops, shows or auctions. Therefore, the antenna is folded, except when I stand it up to operate. I've found, however, that the automatic antenna tuner on my rig is happy from 80-6 meters with the 40 m resonator and the mast folded. With a few minutes notice, most of the assembly can be removed (see Figure 8).

With 100 W out of the rig, North America is my playground. One rainy night near Boston, however, I was in a restaurant parking lot and had a wonderful ragchew with a ham in Argentina.

Other benefits: The car is easy to spot in a parking lot. Because my car is white, I made no effort to camouflage the white pipes, yet they can be painted to match the car if you desire. Paint makers have recently introduced spray paints that will stick to plastics.

This mount has been in place for almost two years, at high speeds (55-80 mph) on interstate highways and through one New England winter. Soon after I built it, my wife and I visited Acadia National Park in Maine. While parking one dim evening, I snagged a low tree with a 40 meter resonator. The resonator broke away at the mast, but the mast and braces were not damaged. That's strong enough for me.—*Bob Schetgen, KU7G, Hints and Kinks Editor;* **rschetgen@arrl.org**

## A SELF-CONTAINED IC-703

◊ The ICOM IC-703/703+ transceiver makes an excellent portable and backpack radio. One reason for this is that it is very stingy on current consumption when powered from a 9.6 V battery. When operating on an input voltage of less than 11.5 V, the IC-703 automatically drops into its "battery-saving mode," which switches off the display lighting and limits the transmit power to 5 W maximum. You can also set up the radio to be current stingy at higher voltages if you wish. I measured the receive current on my IC-703 at just over 300 mA when operating from a 9.6 V battery, and I measured a transmit current of just 1.7 A at 5 W output power. This is great, but I wanted a simple package that includes the IC-703 with an attached battery pack, for easy transport and portable operation.

### *The Battery Connection*

With the low current drain discussed above, I've found that I can get excellent operating time using 9.6-V 3-Ah radio-control battery packs. I buy mine from **www.batteryspace.com** (part number AS-HSC8I2TM). [This store has many interesting batteries, packs and chargers, but I expect their product and price lists change often. When I visited the site, the pack described was about $27. Resourceful readers could probably assemble a similar pack from the products offered, but notfor $20.—*Ed.*] These batteries have overall dimensions of 7×1.7×0.9 inches, weigh exactly 1 pound and cost only $20 each. Therefore, it appeared that these battery packs could possibly mount to the side of the IC-703, and maybe even look good!

My IC-703 includes the optional MB-72 carrying handle. This handle mounts on one side of the IC-703, and four rubber feet mount on the other side, which also has

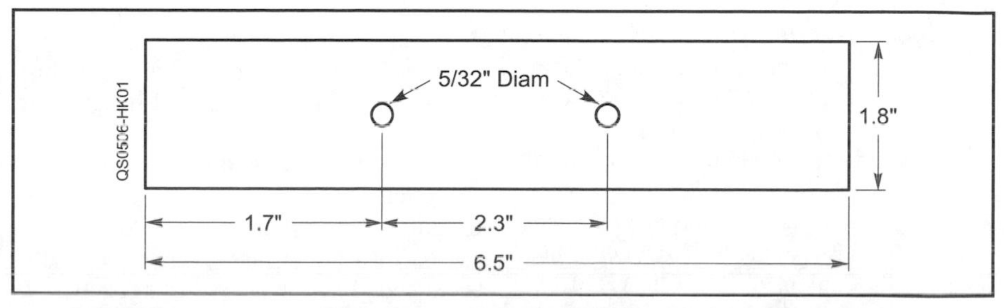


Figure 9—Battery Mounting Plate Dimensions. The height location of the holes is unimportant; but they must be 2.3 inches apart to match the mobile-mount holes.

**Figure 10—The new external battery, mounting plate, hardware and wiring. (The tie-wraps are not shown; they are visible in Figure 13.)**

**Figure 12—The clear-plastic battery mounting plate secured to the side of the radio.**

**Figure 11—A view of the fuses on the rear of the IC-703. The external battery and its power leads are at left. The pink square is the blade fuse.**

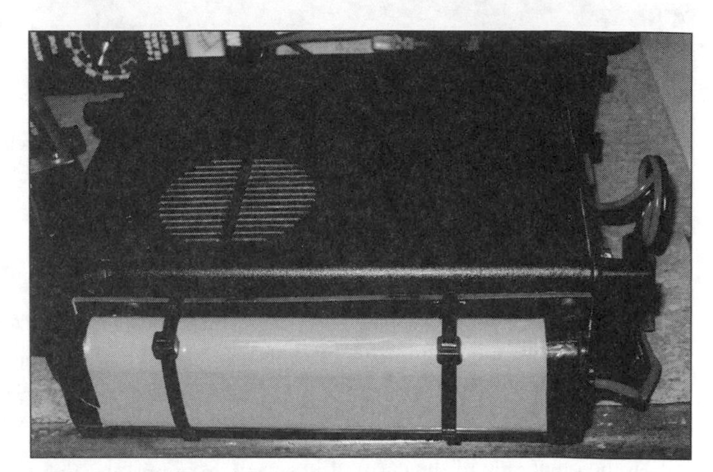

**Figure 13—A right-side-top view of the radio with the external battery secured in place. The black tie-wraps are visible on the green battery. The tie-wraps secure the battery to the mounting plate by passing around the battery pack then between the radio and the mounting plate.**

mounting holes for the optional mobile mount.

After thinking about this for a while, the solution became obvious. I simply made a battery mounting plate from a piece of sheet plastic purchased from my local hardware store, and attached this mounting plate to the side of the IC-703. PC-board material or aluminum will certainly work just as well. While you're at the hardware store, pick up two 4×16 mm screws, two ¼ inch rubber grommets and two 8 inch tie-wraps (unless you already have them).

**Figure 14—A front view of the radio with the external battery in place.**

Whatever material you choose for the plate cut it to the dimensions shown in Figure 9. If it's thick plastic, countersink the two mounting holes in the plate so that the heads of the mounting screws are flush with the surface of the plate. Figure 10 shows everything you need.

As you can see in Figure 11, I've changed the "RC" battery connector to an Anderson Power Pole connectors. This has become my dc connector of choice for all of my ham radio equipment. In addition, make sure to fuse the lines between the battery pack and the IC-703 because this battery can momentarily provide 30 A of short-circuit current! I removed the dual in-line fuses provided with the IC-703 and changed them to a PowerPole-to-fuses-to-PowerPole assembly for use when operating mobile (fuses recommended for both power leads). For this application, however, dual fuses are unnecessary. Therefore, I spliced in a mini-blade fuse holder and fuse as seen in Figures 10 and 11. I bought both the fuse and fuse holder from Mouser Electronics (**www.mouser.com**). The fuse holder is part number 5768-53002 ($1.25 each), and the fuse is part number 5768-97004 ($0.37 each). This in-line fuse takes up very little room and gives the necessary protection. If you do this, however, be very careful not to short the battery terminals while changing the connectors (remember the 30 A short-circuit capability of this battery!). Therefore, work with only one battery lead at a time, leaving the other still connected into the original connector. Clip the positive (red) wire from the battery connector first, crimp it to the in-line fuse holder and crimp the other side of the fuse holder to a PowerPole lug. Insert this lug into the red PowerPole connector housing. Next, clip the negative (black) wire from the battery connector and attach it to the other PowerPole lug. Insert this into the black PowerPole connector housing. You can do

Figure 15—The radio remains very portable with the added battery in

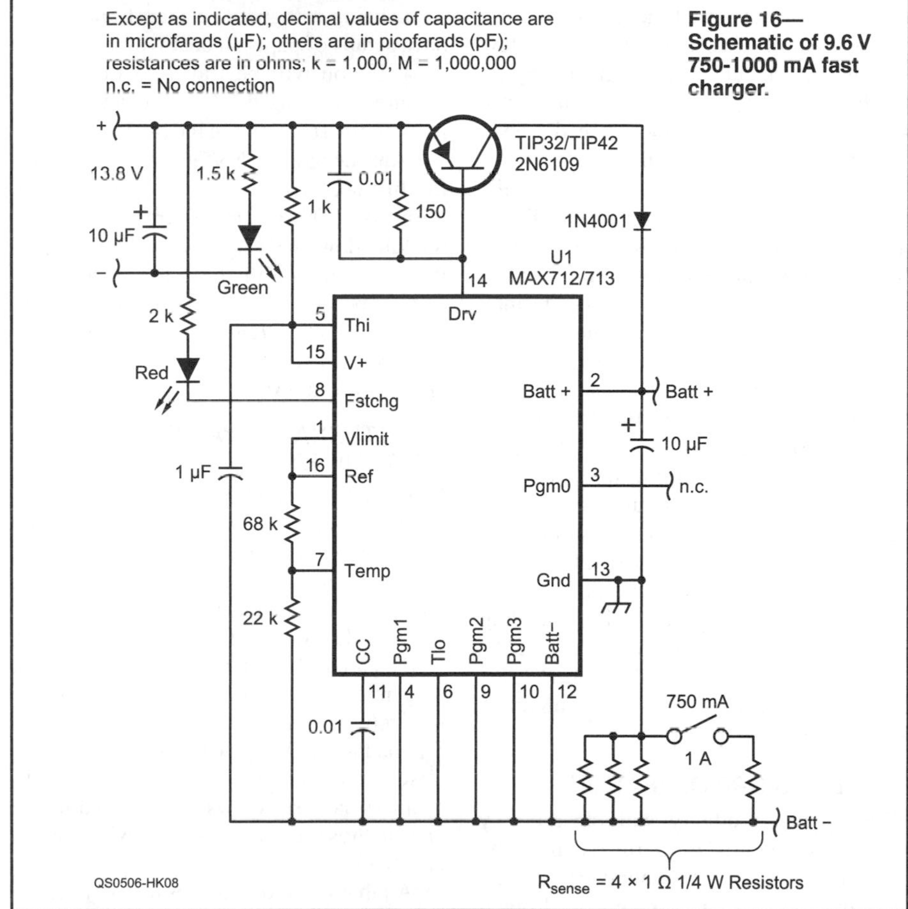


Figure 16—Schematic of 9.6 V 750-1000 mA fast charger.

this more easily and safely by simply making an in-line RC-to-fuse holder-to-PowerPole adapter cable. If you do this, you'll need to purchase a male battery connector from batteryspace.com (part number CN-TMML).

Now you can mount everything. First, pass the two metric screws through the mounting plate and through two ¼ inch rubber grommets and attach the plate to the side of the IC-703. The front and back edges of the mounting plate will rest on the MB-72 rubber feet. Figure 12 shows the plate mounted in place. Next, simply mount the battery to the mounting plate with two tie-wraps as shown in Figures 13 and 14 Now, the IC-703 and battery combination is easily carried using the MB-72 carrying handle. See Figure15.

### *Charging the battery*

There are several options for charging this 3 Ah battery. Batteryspace.com sells both normal (CH-1230-ULBX) and fast (CHUN-120) chargers for $6 and $28, respectively. The inexpensive charger takes 10-15 hours to charge this battery. The fast charger will charge the battery in 1.5-2 hours.

I use my homebrew smart charger that is described in Nov 2003 *QST*. I modified the circuit so that it can support either the higher current requirements of these NiMH RC batteries (1 A charge current) or those of smaller NiMH packs (750 mA charge current) by using a small SPST toggle switch. The revised circuit is shown in Figure 16. I also added a heat sink because the power dissipation at the 750 mA and 1 A charging current rates is more than the cover can easily dissipate alone. The modified unit with heat sink is shown in Figure 17.

Figure 17—The author's homebrew charger with the added heat sink.

An attached battery increases the flexibility of the IC-703 for portable applications. This article described a simple and inexpensive means of doing this, as well as making a recommendation on the type of battery to use. Build this modification and make your IC-703 portable operation even more enjoyable.—*Phil Salas, AD5X, 1517 Creekside Dr, Richardson, TX, 75081;* **ad5x@arrl.net**

## A LIGHTER PLUG MIC HOLDER?

◊ Do you need to install a mic holder in the car without drilling any holes? Here's your answer (see Figures 18 and 19). The large white piece is a ½-to-¼ nylon bushing (Watts-PL-1171Ace Hardware #48737) with a ¼×2¼ inch nylon hex-head plug (Watts PL-1065, Ace #48732) screwed into it. Drill a hole in the head of the plug to fit for a small brass screw (builder's choice) that will hold the mic clip to the plug. Drill a hole in the bottom-center of the mic clip to pass the screw. We'll use this instead of using the two existing holes; it keeps the clip centered. Finally, fasten the clip to the plug/bushing assembly.

In use, the ½ inch bushing fits per-fectly

Figure 18—W0CIA's lighter-plug mic holder at work.

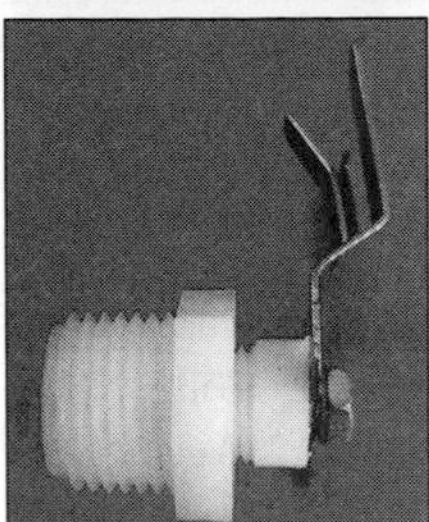

Figure 19—A side-view of W0CIA's lighter-plug mic holder reveals its construction.

in a car lighter socket (with some twisting to make it tight).—*David Quam, W0CIA, 754 Wagon Wheel Dr SE, Bemidji, MN 56601-8352;* **w0cia@arrl.net**

## EAR MUFFS (PROTECTORS) FOR NOISY ENVIRONMENTS

◊ With the arrival of warm weather, opportunities for public service trend upward. Many served agencies do not grasp the effect that operating environment has on clear communications (or unclear!). So, we might find ourselves in the shadow of a very loud fire department "pumper" truck. We might be in a chaotic shelter with many strident conversations in the immediate area or otherwise enjoying considerable ambient noise that competes with the traffic we are trying to pass or copy.

I have priced special-purpose headphones with noise attenuation properties, and my wallet was very unfavorably impressed. Seeking a more frugal (cheap) alternative, inspiration struck.

I am a graduate student in a Physician Assistant program, and therefore have considerable studying to do. One afternoon as I worked at the kitchen table, I found the dishwasher to be distractingly loud. I dug out a pair of shooting earmuffs (ear protectors) and found that they quieted the dishwasher effectively. As my mind wandered, I wondered if I might listen to some music while studying. I found that "ear-bud" earphones fit beneath the muffs and provided pleasing fidelity.

Is there a ham radio application? These same ear-buds give me "stereo" audio from my Alinco DJ-580, using one channel for VHF and the other for UHF.

In a noisy environment, slipping these earphones on, and fitting a pair of Silencio Magnum ear protectors over them gives me nearly perfect isolation from surrounding noise. Less noisy locations may be managed by a set of RadioShack "computer" headphones, at the price of more ambient noise intrusion. For a high-tech solution, I may use my Peltor "Tactical 6-S" electronic ear protectors. These have an amplifier in each ear cup, with circuitry that cuts out impulse noise (such as shooting-range blasts). While I certainly do not anticipate gun blasts while operating, the protectors reduce ambient noise by around 30 dB when off, and may be turned on, one ear at a time, by means of a volume control. Thus, I may turn on (at a low volume) the ear cup for the channel carrying the lower-priority traffic, and monitor my surroundings for conversations I need to hear, "Hey! Ham! Can you reach..."?

My local discount department store sells plain-vanilla ear protectors for around $7 and up. Silencio Magnum muffs may be around $50 (**www.gunaccessories.com/Silencio/default.asp**), although your mileage may vary. [Peltor Tactical 6-S electronic muffs are available from several Web vendors, and prices vary widely.—Ed.] The ear buds came from RadioShack for $5-$10, and the headset was $10-20, as I recall. It may take some research (or customizing) to find, or adapt, headphones or ear buds suitable for your handheld or mobile rig. Once that piece of the puzzle is in place, your next $20 puts you in business. While this solution may not be as elegant as David Clark noise canceling headphones, it won't set you back $200-$500, either!—*Jon Seaver, N8SUA, PO Box 261, St Johns, MI 48879-0261;* **n8sua@yahoo.com**

## TRAVELING QRP OPERATION

◊ Being retired military, with close family living permanently in Germany, plus the availability of $556.00 round trip tickets (Omaha to Frankfurt), recently allowed my wife (KB0JTJ), and me to return to my original family home — Deutschland. I made sure to pack my QRPp Rock Mite transceiver, a light 20 m dipole and a few tools. Soon after our arrival, the dipole was up on the roof, mounted on an FM-antenna mast, and we were on the air. Working CW from central Europe is usually lots of fun. Initially I had several good contacts to Italy, Scotland and Finland. Then disaster struck, with snow, ice and wind.Over the next week the dipole broke repeatedly, the roof was impassable, and less than optimal antenna positions made contacts hard to come by with the 500 mW signal from my RockMite. The 20 m band became very noisy, with few stations heard during the daytime. Late evening openings to the USA East Coast were brief, with lots of QRN and QSB, and no QSOs. It was a less than satisfactory experience, but a learning one.

For anyone doing similar winter travel with QRP operation in mind, I recommend the following:

1. Don't make antennas out of hook up wire (wind, ice/snow will ruin your day).
2. Converters to change the European electric supply from250 V ac to 120 V ac are noisy. The RadioShack one I brought to recharge my electric razor and operate the transceiver power supply was very noisy. Bring batteries: Two 9 V and two A batteries (50 mm tall by 17 mm diameter, 1.5 V), lasts a long time and is quiet.
3. Bring a logbook. I didn't and now I have scraps of paper to deal with.
4. If you will be traveling in Europe, be sure you bring the CEPT forms (download them from the ARRL Web site — **www.arrl.org/FandES/field/regulations/io/faq.html#CEPT**) and your *original* license with you. The Germans are great about tracking down new signals and knocking on your door. It didn't happen to me, but there is great peace of mind in being legal. Good luck. — *73, Dave Lohr, W0FBI, 78th Signal Battalion, Box 1096, APO, AP 96338;* **lohr@zma.attmil.ne.jp**

## AN ALTERNATIVE YAESU ATAS-120 MOUNT

◊ When I mounted my Yaesu ATAS-120 antenna on my van, I was concerned over the ease with which it could be removed. There was no provision to lock the antenna onto its mount and prevent accidental loss or theft.

I decided to remedy this myself. Having worked for a while in my younger years as an apprentice machinist, I'm reasonably handy with tools.

I had recently replaced the ordinary steel setscrews in my other mobile antennas with stainless-steel setscrews and decided to use one of those as my locking device on the ATAS-120.

All that was required was to drill the base of the antenna and tap the hole for the setscrew. Table 1 is a list of tools and materials; Figure 20 shows them.

Clamp the antenna in a vise. Use V-shaped

**Table 1**
**Antenna Mount Tools and Materials**

Center punch
Hammer
Electric drill
#21 drill bit
#10-32 tap and tap wrench
WD-40
#10-32×3/8 inch SS setscrew
Allen wrench for the setscrew

K9LDX

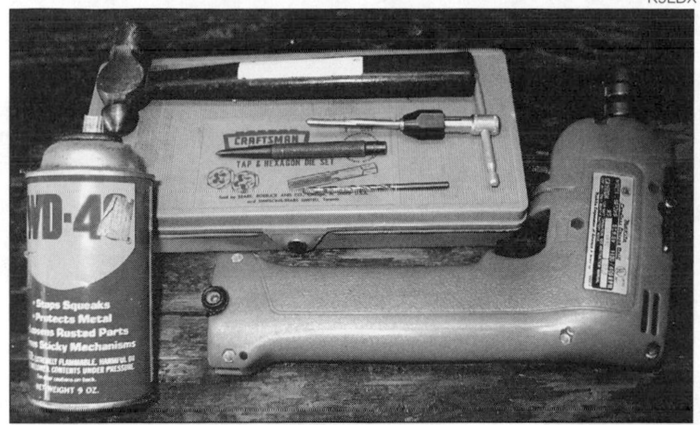


Figure 20—Tools and materials for the ATAS-120 modification. The setscrew is on the long end of the Allen wrench.

K9LDX

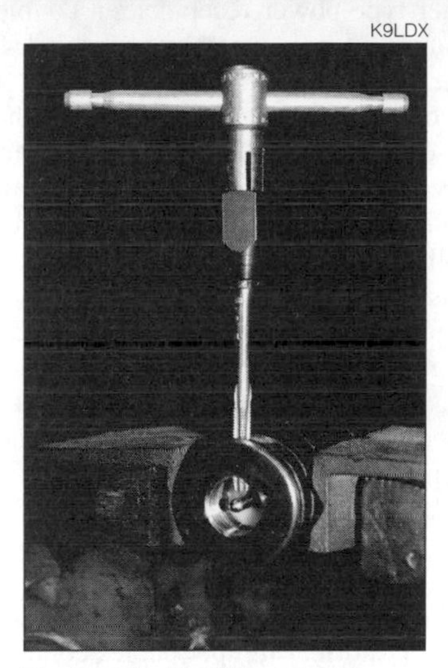

Figure 21—Tap the antenna base for the setscrew. Be careful not to damage the antenna center pin.

K9LDX

Figure 22—The modified antenna in place on the vehicle, with the setscrew visible.

vise inserts if you have them, and tighten the vise just enough to hold the antenna.

The ATAS has two flat spots (probably to accommodate a wrench) at the base. Center punch a mark in the middle of one of the flat spots about ⅛ inch from the lower edge. This will position the setscrew to contact the typical SO-239 antenna mount about midway.

Using the #21 drill bit, drill a hole at the center-punched location through one side of the antenna mount.

Tighten the #10-32 tap in the tap wrench and carefully start the tapping operation by pushing and turning the tap into the hole just drilled. Once you've made a couple of turns, give the tap a quick spray of WD-40 to lubricate the cutting of the threads. Continue turning the tap one-half-turn at a time, backing out the tap each time to break the chips raised by the cutting teeth. This prevents jamming of the tap as it gets deeper into the hole. When the tap breaks through into the center of the antenna, continue turning the tap far enough to clear the initial tapered portion of the tap through the hole. Watch out for the contact pin in the center.

Remove the tap and clean the tapped hole. Insert the #10-32 stainless-steel setscrew part way into the hole. Mount the antenna, and tighten the setscrew with the Allen wrench. Make sure you remember where you put that Allen wrench—you'll need it to remove the antenna. (I have several Allen wrenches of that size located in convenient spots.)

Optional: Insert a small plastic bead or piece of lead shot ahead of the setscrew to avoid damaging the threads of the antenna mount.— *Jim E. Augusteijn, K9LDX, 1542 Mellow Ln, Simi Valley, CA 93065-5745;* **k9ldx@arrl.net**

## SIMPLE HF MOBILE INSTALLATION

◊ The company that I work for furnishes a car for my travel, and I spend a lot of time in that car. I was looking for a quick and easy installation for my IC-706 that could also be removed easily. I "borrowed" one of my daughter's sport drink water bottles that would fit into the drink holders in the console of the car. I then used a drink koozie for a snug fit. I filled the drink bottle with plaster to keep the assembly from being top heavy.

The mount is an old cellular phone mount fastened to the top of the water bottle. I attached the 706 faceplate, speaker and microphone clip to the phone mount. See Figure 23. Now I have everything in one package, and it can be removed and stored in the trunk at a moment's notice. I hope this stimulates others to come up with some ideas, and hope to see you down the log on HF. — *73, Buddy Walker, AC5ZQ, 1058 Stallion Rd, Gilmer, TX 75645*; **ac5zq@arrl.net**

## A POWER SOURCE FOR PORTABLE OPERATIONS

◊ After I built my Radio Station in a Box, a portable station containing an ICOM 706MkIIG, an LDG autotuner and an MFJ switching power supply, together with antennas, laptop computer, coax and so on, I realized that without portable power it lacked that last piece to make it truly a go anywhere proposition.[1] I began looking for a suitable power source for truly portable power. Having seen several operators using automotive jump-starter battery packs during the 2004 hurricane season, I investigated these devices further.

[1]G. Haines, N1LGI, "Serious Portable Station," Up Front in *QST*, *QST*, Oct 2005, p 21.

Figure 23 — A plaster-filled water bottle and an old cellular phone mount form the basis of Buddy Walker's simple mobile installation. [Readers are cautioned against driving any vehicle with radios and other objects that are not securely fastened. The keyer-paddle shown here could become a dangerous missile in an accident. — *Ed.*]

Unfortunately, most of the ones I could lift were too small to be of much use and those that had enough juice were too heavy and too expensive. Some had too many power robbing accessories like air compressors or work lights. Something more suitable for my personal use would have to be homebrewed.

I already had a good candidate for the battery under my operating position desk. I use a type 27 deep cycle battery as backup power for my home station. It is quite heavy, but with a suitable luggage cart, I found it easy to move around. The kind of cart I used is the kind that some travelers fold up and strap to their suitcase for those long walks from terminal to terminal at the airport.

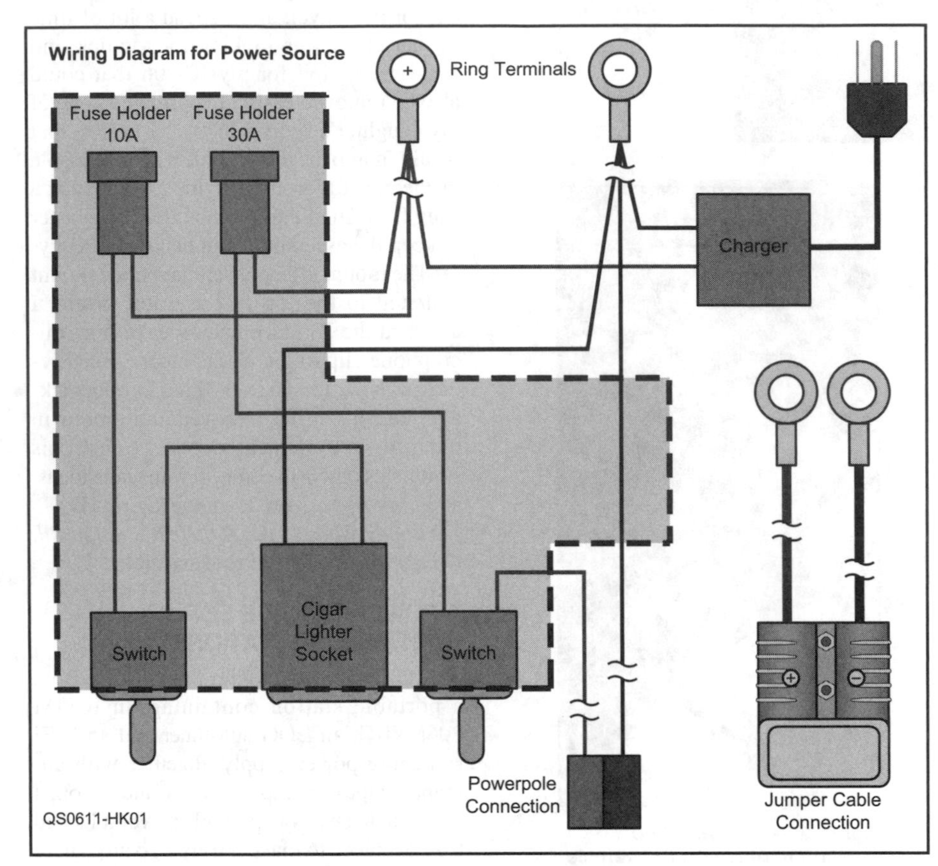


**Figure 24 — This is the wiring diagram for the portable power source.**

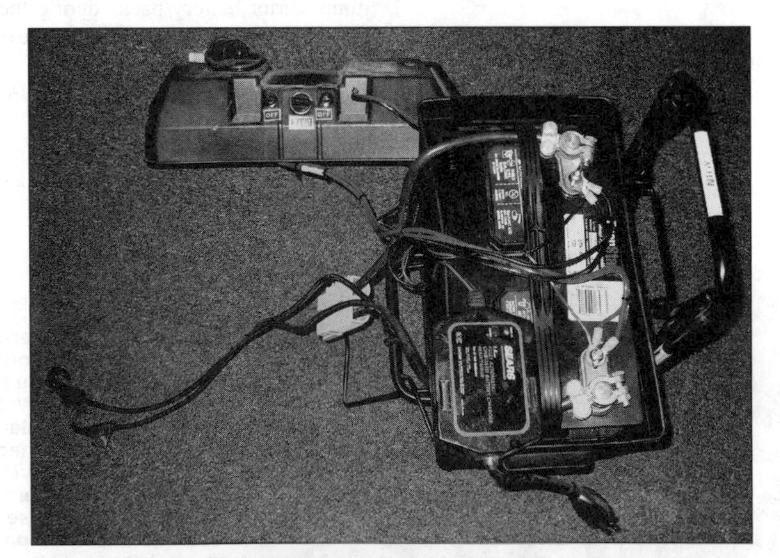

**Figure 25 — This photo shows the completed portable power source package. Blade fuses for the radio line and accessory jack are under the battery box cover. The battery box is attached to the luggage carrier. I also store spare fuses and extender wires to connect to the radio or other devices under the battery box lid.**

The battery is contained in a standard marine type battery box with a vented lid. This meant that I already had an enclosure for the various switches and connectors that I envisioned for my portable power source. I also had a dedicated charger/maintainer to keep it charged up and ready.

Putting these elements together was a one afternoon project. I purchased a cigar lighter type power receptacle, a couple of automotive toggle switches, some blade fuse holders and some 12 gauge wire in appropriate colors. Wiring it up as Figure 24 shows was easy. A pigtail lead with Anderson Powerpole connections provides the power feed for the radio. (Note that these are the small 45 A Powerpole bodies that many hams are using for their radio dc power connections.) The cigar lighter port could power my laptop or other accessories and both are controlled by the two switches and protected by the fuses in their holders. I mounted all of these components in the battery box lid.

While a deep cycle battery is not the best choice for jump starting a car, I have used it successfully for that purpose in the past. My normal jumper cables connect to my vehicle through a very large Powerpole Connector mounted in the grille of my Jeep Cherokee. (I use SB175 connectors, rated at 175 A for this purpose.) This works well and guarantees that the cables are connected correctly, at least at the Jeep end. Since I had an extra connector of this type on hand, and it even had lengths of heavy cable attached, I figured it was worth the effort.

My battery has dual terminals, wing nuts and the standard metal studs. I used a pair of replacement battery cable clamps to secure the jumper leads to the battery directly without disturbing the other wiring. These leads exit the battery box in the normal openings for heavy cables. The connector dangles out the front of the box where it is easy to connect my jumper cables just like I would on the Jeep.

I now have a portable power source with all of the desirable features of the commercial "jumper packs." Since I had many of the parts on hand already, it only cost about $20 to build. If you have a well stocked "junque box" you can probably do it for even less. Figure 25 shows the complete package.

Just a short thought about the weight of this power source. As I indicated, it is permanently mounted on a collapsible luggage cart. My portable station is also on wheels since it is part of the suitcase into which I built the station. Both units can be pulled along just like luggage at an airport or train station. Each package weighs less than 40 lbs. Lifting them into and out of my vehicle may require help, but I do not usually travel alone so that is not a problem. Once they are on the ground, they are easy to manage. — *Geoff Haines, N1LGI, 708 52nd Ave LN W, Bradenton, FL 34207*; **n1gy@arrl.net**

## BATTERY CLAMP FOR MOBILE RADIO POWER INSTALLATION

◊ I was getting ready to install a 2 m transceiver in my 1995 Ford Ranger pickup truck. In looking through my collection of *ARRL Handbooks*, I saw that it is recommended that the rig be powered directly from the battery, with fuses in both the positive and negative lines.

So, I wired the 2 m rig as shown in the *Handbook*, fusing both lines and ensuring that the length of unfused cable between the battery and my added fuse block was as short as possible. I wired the whole run with AWG no. 10 cable — probably overkill for a 50 W mobile rig, but why not? It wasn't much more expensive than using thinner wire. While I was at it, I gave all of the battery wiring in the truck a good check and am happy to report that all of the stock battery and ground cables are corrosion free and tight.

When I inspected the truck wiring, one thing gave me a fit — my truck had battery terminals made from relatively thin pieces of brass, instead of the standard big, lead blocks

Figure 26 — This photo shows the thin metal battery terminal used on many vehicles today.

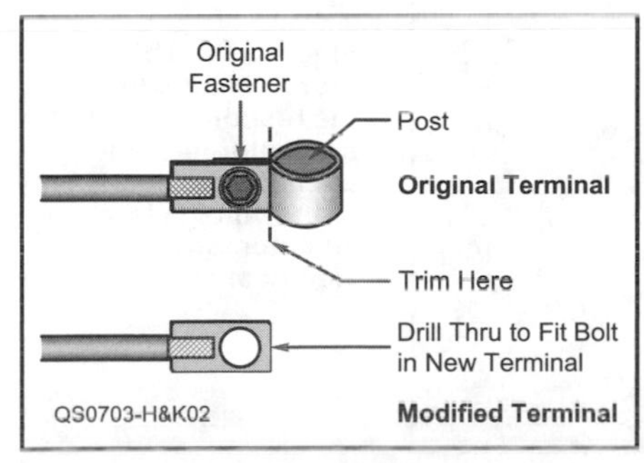


Figure 27 — This drawing shows how to modify the thin-metal battery terminal for use with the heavy-duty mil-spec terminal.

KD5BYB

Figure 28 — A new mil-spec battery terminal installed in Ben Hall's truck provides a very solid connection to the battery for both the truck's battery cables as well as the additional wires to power his

WA6MUU

Figure 29 — A discarded cell phone automotive power adapter is easily converted to an Anderson Powerpole-to-cigarette-lighter adapter.

vehicles used to have. I also discovered that my wife's Mitsubishi car had similar thin terminals. See Figure 26. I was scratching my head on how to make a good connection to the battery, until I came across a suggestion at an off-road truck Web page. The suggestion was to replace the stock terminals with military-specification terminals that have a long bolt, where additional power leads can be easily attached with eye terminals. McMaster-Carr sells these mil-spec battery terminals for about $4.50 each: **www.mcmaster.com/nav/enter.asp?pagenum=709**. They are shown on the Web page as Figure G, part number 7738K2 and 7738K1.

I simply cut off the circular portion of the original terminal, leaving the flat piece and drilled out the stock threaded section to fit the bolt on the new terminals. See Figure 27 for details. This made a very robust and clean-looking installation. See Figure 28. The black "goo" smeared over and around the terminal is Dura Lube Synthetic Grease that I picked up at a local Dollar Store. It is rated for chassis and wheel bearing use, but for a dollar a tube, I use it on my battery terminals to prevent corrosion.

The radio works great and I cannot hear any noise that I can attribute to the engine electronics in my truck. I'm very, very pleased. — *73, Ben Hall, KD5BYB, 102 Stoney Point Dr, Harvest, AL 35749*; **kd5byb@bellsouth.net**

## POWERPOLE AUTOMOTIVE POWER ADAPTER

◊ The number of discarded cell phone automotive power adapters has increased each year as cell phones are upgraded. I find several each week at garage sales for 25 cents or less. There are as many different models of power adapters as there are models of cell phones. I prefer working with the ones that have a threaded and fused tip and spring loaded negative terminals. They can be converted into useful Anderson Powerpole-to-cigarette-lighter adapters with a little work and minimum expense. See Figure 29.

The first step is to open the adapter body by sliding a screwdriver between the halves at the power cord and gently prying them apart. The circuit board and cord are not used and should be recycled through a proper facility.

Next, reassemble the shell and carefully mark the end where the Powerpole connectors will be mounted. The opening height depends on the number of Powerpole connector pairs that you plan to install. A width of 0.55 inches allows the shell to engage the Powerpole connectors at the roll-pin indentation. Separate the shell halves and clear the opening with a rotary hand tool, such as a Dremel tool, and pattern files.

Connect the Powerpole terminals to their respective adapter terminal. You can reuse the pilot LED by connecting a 3 kΩ resistor in series with the LED. See Figure 30.

A spot of hot-melt glue holds the LED and Powerpole connectors in place and the shell halves are brought together. — *73, Gary G. Self, WA6MUU, 4350 Alta Campo Dr, Redding, CA 96002*; **wa6muu@arrl.net**

WA6MUU

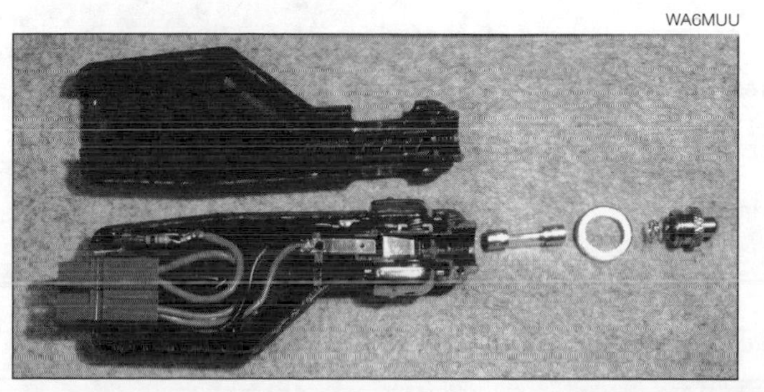

Figure 30 — This photo shows how WA6MUU turns a discarded cell phone automotive power adapter into an Anderson Powerpole-to-cigarette-lighter adapter.

## CURING ALTERNATOR WHINE

◊I recently installed an IC-208H transceiver in my truck. On the first day of using the new radio I got reports that I was transmitting very noticeable alternator whine. I could also hear it on receive and when the radio was quiet. I checked the diodes in the alternator, verified I had good grounds, and I even ran the truck with the alternator removed to be sure that the whine was indeed from the alternator.

The alternator produces ac, which is rectified into dc. The problem is that the rectified dc is not filtered adequately. The dc output will have a small ac signal riding on it. That ac signal will have nine cycles for each revolution of the alternator. Suppose your engine is idling at 600 RPM and the drive pulley ratio to your alternator is 1:3. At that engine speed your alternator is turning 1800 RPM, or 30 rotations per second. Each rotation gives you 9 cycles of ac. Do the math and you get a 270 Hz sinusoid (not a perfect sinusoid but close enough). Cruise down the road at 2000 RPM and you get a 900 Hz sinusoid riding on your dc power supply.

I tried an off-the-shelf filter from an auto parts store. It did very little to cure the problem, so I decided to build a filter. The first filter I built worked *very* well. The problem is that not everyone has the tools required to build that filter so I decided to figure out a filter design that could be built in less than an hour by anyone with basic tools, have a cost under $20, and handle a current of at least 12 A with acceptable voltage drop. Table 1 lists the materials I used.

### Directions

The ½ inch Quick Link will be used as the inductor core. (Quick Links are intended to quickly join sections of chain.) A fellow ham, Dave, KC1LT, suggested using a shackle. I went to get a shackle and came across this quick link. I went with the quick link to make more efficient use of project box space. See Figure 31.

Wrap the hook-up wire around the closed side of the quick link, starting from the left as shown. Leave about 9 inches of wire free on the left end. Try to keep the turns as close

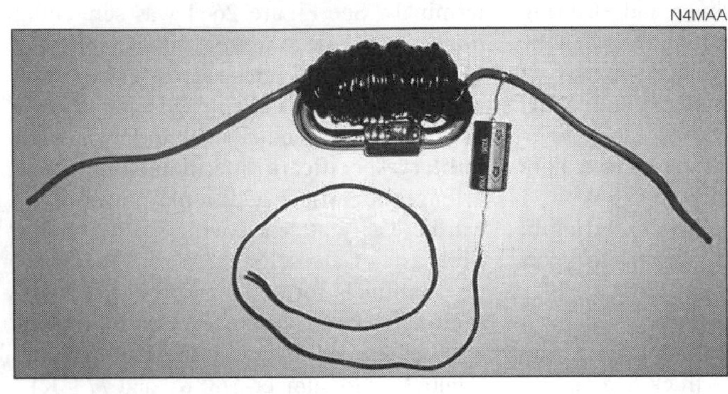


**Figure 33 — Solder the positive capacitor lead to one inductor lead as shown here.**

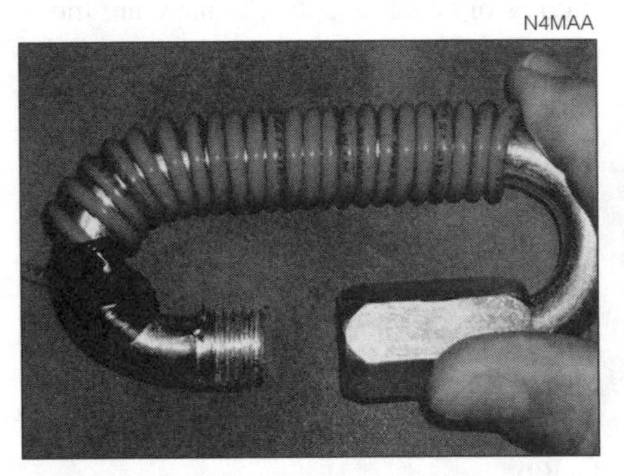


**Figure 31 — This photo shows the inductor being wound on a Quick Link form.**

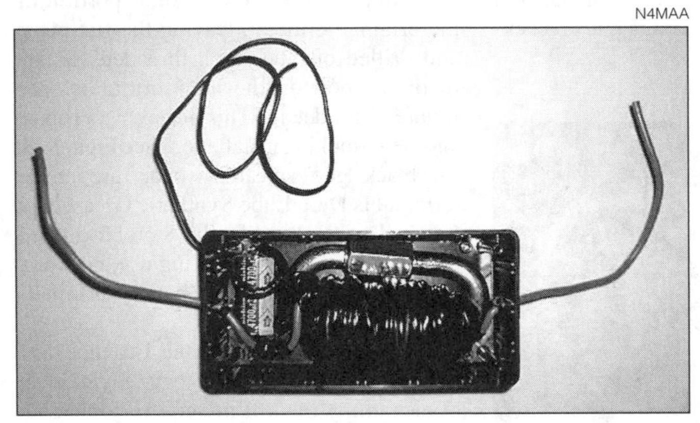


**Figure 34 — Secure the inductor and capacitor inside a**

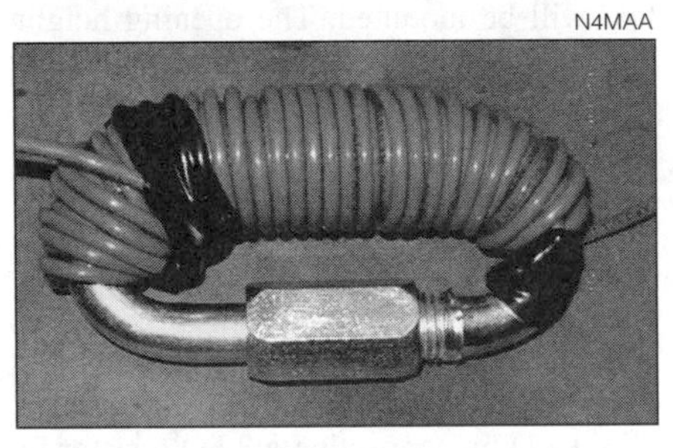


**Figure 32 — Here is the completed inductor, ready to be wrapped in electrical tape or some other method to hold the turns tightly in place on the form.**

**Table 1**
**Parts List**

|  |  |
|---|---|
| ½ inch Quick Link from Lowe's | $2.98 |
| 6 × 3 × 2 inch project box from RadioShack | $3.79 |
| 20 foot roll, 12 gauge red hook-up wire from RadioShack | $4.99 |
| 4700 μF 35 V capacitor from RadioShack | $5.29 |
| 18 inches of black 16 gauge wire | |
| Electrical tape | |
| GOOP or similar glue | |
| 3 zip ties | |
| Total: | $17.05 |

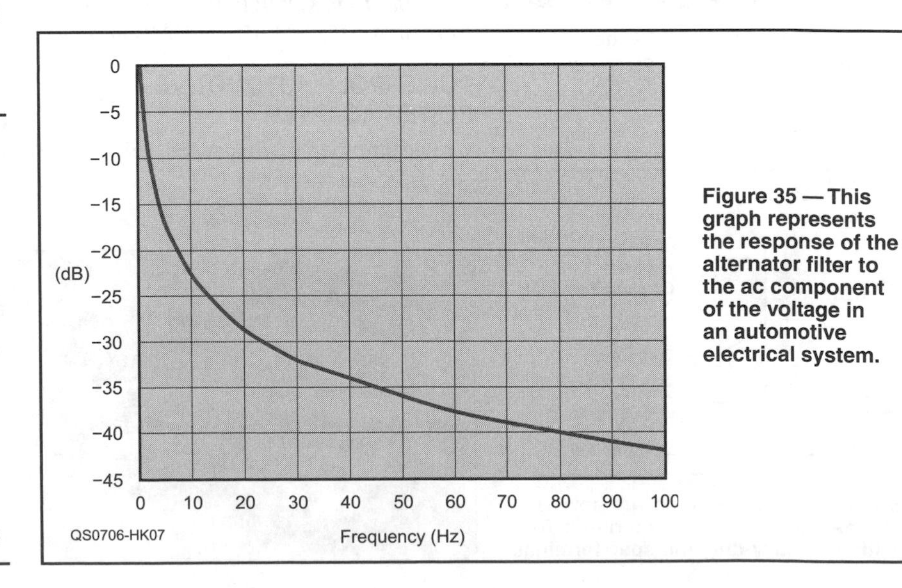


**Figure 35 — This graph represents the response of the alternator filter to the ac component of the voltage in an automotive electrical system.**

together and tight as possible. On the last layer, space the turns so that you have about 9 inches of wire left on the right end. Use all 20 feet of wire. See Figure 32.

Wrap the coil in electrical tape and close the quick link. About 1.5 inches from the right end of the inductor, strip ¼ inch of insulation off the red wire and solder in the positive capacitor lead. Make sure you observe the capacitor polarity. In Figure 33, you can see the negative arrow on the capacitor pointing down. Solder the 18 inch piece of black wire to the negative lead of the capacitor.

Cover the soldered connections and capacitor leads with electrical tape. Drill a 3⁄16 inch hole in both ends of the project box for the red wires. Drill a ⅛ inch hole in one end of the project box for the black wire. Run the wires through the holes. Put a zip tie on each of the three wires to limit how far the wires can be pulled out of the box. Make sure to leave a little slack in the wires inside the box. Using GOOP or some other thick, strong adhesive, glue the capacitor and inductor into the project box. Leave the cover off until the glue dries. Figure 34 shows the components inside of the plastic box.

### *Measured Filter Response*

I measured the filter response using a low frequency signal generator and an oscilloscope. At 25 Hz, the filter has better than 30 dB of attenuation. In other words for frequencies above 25 Hz, the noise power has been knocked down by more than a factor of 1000. See Figure 35.

### *Installation*

The filter was installed in a Honda CRV owned by Ron, KB1KRG. Ron is very active on the WB1GOF repeater, and one of the first people I talked with on 2 meters. Ron uses an IC-V8000 with a magnetic mount antenna and a cigarette lighter power plug as his mobile 2 meter rig. We installed Anderson Powerpole connectors on the filter and in the IC-V8000 power leads. On the filter the black wire goes to ground, the red wire on the capacitor side of the inductor goes to the radio and other red wire goes to the battery. This proved to be a very convenient testing setup. We could very easily remove and install the filter and listen to the difference in alternator whine on an HT. The filter worked very well. With the filter installed there is no audible alternator whine on Ron's signal. In a permanent installation, mount the filter using two-sided tape or a good adhesive, and ground the filter to the chassis. The radio should be grounded to the chassis as close to the radio as possible.

I am new to practical electronics. By answering basic questions and making suggestions, several folks contributed to this project. In particular, Dave, KC1LT, was very helpful. I hope this information is useful. If you use this design to build a filter or if the information presented was useful please send an e-mail to **KB1MVX@comcast.net** and let me know it was worth the effort to write this Hint. — *73, Jim Perkins, KB1MVX, 27 Nathan Dr, Clinton, MA 01510*; **kb1mvx@comcast.net**

## A QUICK GROUND CONNECTION

◊This idea for quick-connect antenna grounding was suggested by experience at the 2006 Vienna Wireless Society Field Day. Northern Virginia was doused by thunderstorms for a couple days preceding, and the same weather was promised through FD itself. At dinnertime Saturday the flashing and rumbling resumed and we decided to button up. That's when the problem reared its head. How do we ground the coaxial cables from all those high-flying Windoms, inverted V dipoles and long-wire antennas — and do so very quickly?

Our stations were on picnic tables with ground rods driven in near them. There was no good way to get both the shields and conductors of the antenna leads solidly connected to the rods, though.

Here is the solution whispered to me by the Radio Muse:

Get a 2-foot RG-58 patch cable with PL-259 connectors on the ends. Cut it in half. Strip off a couple of inches of the outer insulation, push back the braid, remove the foam, twist the shield down tight against the center conductor and hammer everything flat. Tin the whole tab thus created into a nice, big short. Add a barrel connector to attach the antenna feed line quickly and easily. Do the same with the other half of that piece of coax, and have a pair of grounding stubs.

When you prepare ground rods for Field Day or other portable operations, attach one of these stubs just below the top of each rod with two radiator hose clamps or other type of grounding clamps you can cinch down tight (for the best in strength and connectivity without the need for a kilowatt soldering iron). You may have to enlarge the short tab for the clamps to get a good grip. Just wrap stranded wire around the length of the tab and add more solder. — *73, Alan Bosch, KO4ALA, 5832 20th St N, Arlington, VA 22205*; **ko4ala@arrl.net**

## TRAVELING WITH HANDHELD RADIO "RUBBER DUCK" ANTENNAS

◊Do you travel with your handheld? Many of us do, and if you like to keep your equipment in the best of shape, this idea is for you. Before you go on your next trip, visit your local drugstore and purchase a toothbrush holder such as the one in Figure 36. Many of the new handheld antennas fit into a toothbrush holder, and can pack into your bags quite nicely this way. This protects the antenna and its connector from unnecessary bends and/or breaks. — *73, Nathan Cufo, KA3MTT, 6323 Cinnamon Ridge Drive, Burlington, KY 41005*; **ka3mtt@arrl.net**

## PORTABLE BATTERY POWER

◊Great ham minds think alike! The portable battery operation setup described in the November 2006 *QST* Hints & Kinks column looks a lot like mine. You got it down to the same type of automatic battery charger! However, I might offer one more tip.

Picking up a big car/marine battery can be a challenge for some of us. The weight is a concern even for the normal maintenance and that inevitable battery replacement, where you have to return the old battery to the store for a core credit.

I use two of the smaller Lawn and Garden Tractor batteries in one marine-battery storage box. It is a pretty snug fit but they will fit in there. I can use one battery at a time or can switch them to parallel use for higher-demand loads.

This dual battery system also helps by having one battery available to run the rigs while another one can be charging. This configuration allowed me to keep my shack operational for the days following hurricane Katrina until utility power was restored. My vehicle was turned into a recharge station for the batteries during my commutes to work and back home. This setup will also run one of those RV fluorescent light fixtures for quite a long time. You would be amazed how much emergency lighting one of those lights will provide to a dark setting.

Keep up the good work! — *Joseph Maurus, KA5TWS, PO Box 45714, Baton Rouge, LA 70895*; **jmaurus@brgov.com**

[Yes, I realize that Lawn and Garden Tractor batteries are not designed for deep discharge applications, so these batteries should not really be used for this application. The concept of packaging several smaller, lighter batteries in a box, with a charger as described here and in the November 2006 Hints & Kinks column has some merit, however. While we are at it, it is also not a good idea to mount switches inside the battery box, where hydrogen gas could accumulate and a spark from throwing the switch could have explosive results. Mount the switches on an external panel or in a separate box, perhaps attached to the battery-box lid. — *Ed.*]

KA3MTT

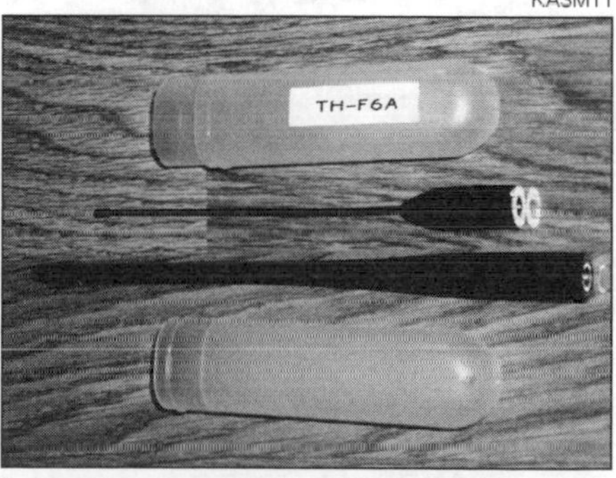


**Figure 36 — A toothbrush holder is a convenient storage container to safely pack a rubber duck antenna. If you carry more than one hand-held radio, use stick-on labels to identify which antenna is inside each holder.**

## PTT SWITCH ON SHIFT LEVER

◊Connecticut, along with many other states, has passed a new cell phone law and there appears to be some confusion regarding Amateur Radio operators using their radios while their vehicle is in motion. Police officers may see a ham operator using a radio while the vehicle is in motion and think the operator is on a cell phone.

I received a catalog from Ham Radio Outlet, and noticed an ad for a visor-mounted boom microphone with a PTT switch that is mounted on the shift lever. This allows me to operate while in motion without having to hold a microphone up to my face and give the appearance of being on a cell phone. It eliminates the chance of being stopped and accused of breaking the law. This might save me bothersome time and trouble. Even without a law against handheld cell phone operation, it makes mobile operation safer.

I mounted my microphone on the dash of my vehicle, where the mike hanger was, and found it to be excellent to use while driving. See Figure 37. The PTT switch must be held on to transmit so you can't leave it on by accident. You do not have to search for the mike hanger and take your eyes off the road. It's not perfect but a good alternative, and a lot safer.

The microphone system I purchased was made by Pryme Products some time ago, and I was told they still have some in stock. They told me that if they do receive requests for more they will be able to supply them in the future. — *73, Harry H. Abery Jr, AB1ER, 18 Dalewood Rd, Newington, CT 06111*; **ab1er@arrl.net**

AB1ER

**Figure 37 — The boom microphone is attached near the top of the dash, where the microphone clip had been attached previously. The PTT switch is attached to the shift lever.**

## EASY MOBILE TRANSCEIVER AUDIO PUNCH

◊ I recently bought a new AM-FM radio for my car. On the rear of the radio are jacks for the input of audio from external devices, such as CD players, TV, audio and anything else you might want. So I took the audio from my FT-100 and put it into one of the input jacks of the car radio. If your radio has jacks on the front, it's even easier.

Now I have four speakers working for me, plus the bass and treble are adjustable in the car radio. I can ride with my windows down and have lots of audio to overcome road noise.

You may have to check your owner's manual for instructions on how to remove your car radio. Some radios need a special tool to remove it. — *73, James Casper, K2KAA, 46 Lynus Dr, Fairmont, NC 28340;* **k2kaa@arrl.net**

## ANOTHER WAY TO BOOST TRANSCEIVER AUDIO IN YOUR CAR

◊ I suffer from significant hearing loss and find it difficult to hear my transceiver when it is mounted in the dash because the speaker is mounted on its top side, like most modern transceivers. My goal was to come up with a cheap solution for monitoring the transceiver through the factory installed speakers fed by my car's AM/FM radio.

Inspired by my wife's purchase of a similar device to listen to her Apple iPod in the car using an external device FM modulator called an iTrip, I thought the same technology could apply to mobile amateur radio operations. See Figure 38, which shows the adaptor I purchased, a Monster Cable RadioPlay Car Stereo Wireless FM Transmitter (MBL-FM XMTR).

I use the transceiver's external speaker jack to feed the adapter, which I set on an unused FM broadcast frequency. [Note: Editor Larry Wolfgang, WR1B, asked ARRL General Counsel Chris Imlay, W3KD, for a clarification about the legality of using a "Part 15" device to "broadcast" Amateur signals on FM Broadcast frequencies. Chris says that such use is permitted under the Rules.]

I paid about $25 for my audio adapter and replaced the input stereo jack with a mono jack. I have also successfully tested another kind of audio adapter designed to feed audio from an external device into a vehicle stereo via the cassette deck. — *73, Hank Kenealy, K1DOS, 9517 Kirkfield Rd, Burke, VA 22015;* **henry@kenealy.com**

K1DOS

**Figure 38 — MBL-FM XMTR audio-FM adapter.**

## DRESS UP YOUR HF ANTENNA MOUNT

◊ For mobile HF in my current vehicle, I use a "Hamstick" style HF antenna. Between the ball mount on the side of the SUV and the antenna I installed a pair of old style springs. Because of their age they were rusty and looked pretty nasty. They still worked fine, but their appearance left a lot to be desired. To neaten up the installation and to hide the ugly parts, I went to a local performance truck parts shop and purchased a shock absorber boot made of flexible vinyl. I tie-wrapped this to the top of the upper spring, since it fit neatly down on the top of the ball mount. To keep the bottom of the shock boot from wrinkling up, I tie-wrapped a ring made from a slice of 2 inch diameter PVC pipe just inside the bottom of the boot. See Figure 39.

N1GY

**Figure 39 — Rubber shock absorber boot covering spring mount for mobile antenna.**

These boots come in several colors to match or contrast with the paint color of the vehicle, so the color combination is up to you. If desired, you could construct a disk of plastic or other material to lock between the bottom of the spring and the antenna mount to support the shock boot, but I did not see the need in my installation. — *73, Geoff Haines, N1GY, 708 52nd Ave Ln W, Bradenton, FL 34207,* **n1gy@arrl.net**

## BATTERY PACK VOLTAGE CONTROL

◊ Portable operation has become quite popular and with the new, almost pocket size rigs, you can carry your entire station in a very small container. It would be ridiculous to carry a small bantamweight rig and use a home shack type key and a heavyweight automotive battery.

My preference for a paddle is the Palm Mini Paddle (**www.mtechnologies.com/palm**), but there are a number of other small lightweight paddles on the market. As for power, I prefer to use either NiCd or NiMH battery packs, both types rated at 1.2 V per cell and rechargeable.

I have made up battery packs using 10 cells of each type as well as a 10 cell alkaline pack. Both the NiCd and the NiMH packs measure a little over 15 V straight off the charger. A set of new alkalines measures about the same.

The low power rigs that I use specify operating voltages of 9-15 (Elecraft K1), 7-14 (Elecraft KX-1) and 12-14 (MFJ QRP-Cub). Since the voltage of fresh 10 cell packs exceeds the upper limit of all these rigs, I needed to reduce the voltage until the pack was partially discharged, then add more voltage to keep it at optimum.

I discussed this problem with Paul, WØRW, a well-known pedestrian mobile operator in Colorado. Paul's solution is to use two silicon power diodes in series to drop the voltage (0.7 V for each diode) and a switch to bypass the diodes as the voltage comes down. He uses the low voltage alarm in the KX-1 to let him know when to switch out one of the diodes.

With the switch in the center off position, voltage is reduced through both diodes (1.4 V). In position "A" the first diode is shorted out and in position "B" both diodes are bypassed allowing voltage direct from battery.

RICHARD ARNOLD, AF8X

**Figure 40 — A dummy AA and a full battery pack.**

My own solution was to make up a couple of *dummy batteries* out of ½ inch wooden dowels cut to the length of AA cells. These are then drilled through lengthwise and fitted with a long 6-32 bolt and nut to act as a conductor. See Figure 40. Placing two of these dummy cells in the pack reduces the voltage to 12.4 V. One cell at a time is added to restore the voltage as the pack is exhausted. See Figure 41. — *73, Richard Arnold, AF8X, 22901 Schafer, Clinton Twp, MI, 48035,* **af8x@comcast.net**

## CORRODED CONNECTIONS MAKE MOBILES MALFUNCTION

◊ Not long ago my HF mobile station developed some strange symptoms. After a couple of months of inactivity, when I turned on the radio all functions would operate normally until I keyed-up the microphone, at which time the display would go haywire, blink once or twice and the rig would turn itself off.

Like any good do-it-yourself ham, I thought "bad connection." I proceeded to check the remote cable and the power and ground connections for the radio. I then made a quick voltage check at the rig end of the power cable, which showed the expected 12.6 V.

To my disappointment, it wasn't anything simple like a bad ground connection, so I turned my attention to the battery terminals under the hood. They looked clean and tight, and a voltage check showed the same 12.6 V. To be sure, I decided to clean them too, disconnecting the negative terminal first and then the positive terminal; I cleaned both using a battery terminal brush available from most auto supply stores. When I reassembled the connections, I did it safely, putting the positive terminal on first and then the negative terminal.

To my disappointment the radio displayed the same symptoms as before — when I keyed-up the microphone, it still turned itself off! Then, I noticed the positive battery cable going into a power distribution block on the side of the engine compartment. When I removed the plastic cover it was clear that the point where the positive battery cable connected to the power distribution block (Figure 42) could loosen or corrode.

Cleaning both sides of the crimped-on lug, along with the copper stud, eliminated a high resistance connection that was introducing just enough of a voltage drop to cause my HF radio to shutdown. For extra insurance I applied a light coating of an antioxidant product that is available commercially from several different manufacturers to both battery terminals, the crimped-on lug and the copper stud, and then retightened the connection securely. My HF mobile now works perfectly; no blinking and no shutting down unexpectedly. Happy DXing! — *73, Andy Vavra, KD3RF/VE2DXY, 11 Collins Ln, Schwenksville, PA 19473,* **kd3rf@arrl.net**

## CONTROL HEAD MOUNT MADE FROM MINICEL FOAM

◊ I recently purchased a mobile transceiver and a separation kit. I then pondered what would be the best way to mount the control head in my Subaru wagon. Because I would like to use the radio in a different vehicle sometimes, I wanted the installation to be easy to set up and take down. I also hoped to avoid cutting and drilling in the car.

In the Subaru, there is an oval depression in the horizontal surface between the gear shift and the ashtray. It typically becomes a collecting place for little odds and ends. I wanted to support the control head in that unused space.

As a whitewater canoeist, I have used Minicel foam to outfit my boats with customized, contoured seats and knee pads. It occurred to me that I could use the same material to fill that space in the car console and mount the control head. I had several pieces of the gray, closed-cell foam in my bin of boating gear.

RICHARD ARNOLD, AF8X

**Figure 41 — A dummy cell voltage reducer at work.**

ANDY VAVRA, KD3RF/VE2DXY

**Figure 42 — The power distribution block with an arrow indicating the location of the corrosion.**

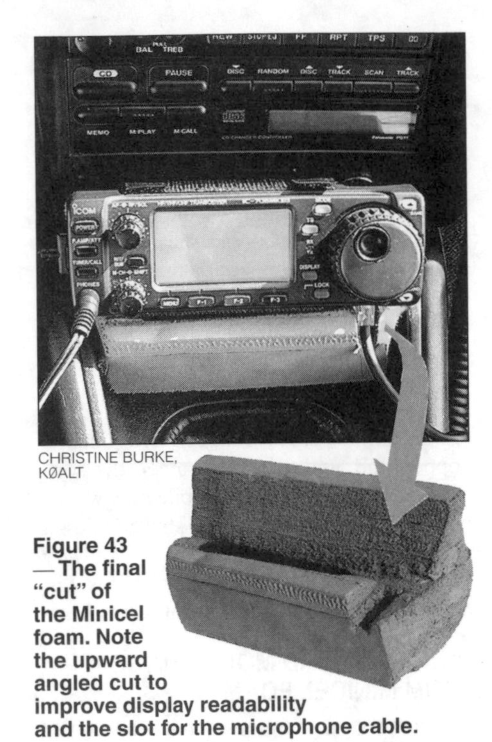

CHRISTINE BURKE, KØALT

**Figure 43 — The final "cut" of the Minicel foam. Note the upward angled cut to improve display readability and the slot for the microphone cable.**

The trademarked name for the material is Minicel. You can purchase it from stores selling kayaking supplies locally or on the Internet, in thicknesses of either ⅝ inch or 3 inches. The 3 inch pieces are usually the firmer M200 version of the product, which is the easier one to carve and sculpt. The ⅝ inch pieces are likely to be the slightly softer, more flexible L200 version.

Tools for working with Minicel can be found around the house. You can cut it into an approximate shape with a long, thin, serrated bread knife. You can also laminate several thicknesses together using contact cement (read the directions carefully). After you apply a thin coating to each surface, don't mate the two surfaces until they are both dry.

After I had constructed my rough shape, I sculpted it to fit snugly into place (see Figure 43). That involved more work with the bread knife and also some shaping with a rough sanding sponge. There were various curves and ridges to fit into. I also carved out a notch for the microphone connection and cable. The result is not beautiful, but it is very functional. A more meticulous person, using a variety of sanding and carving tools, could make a more presentable piece.

There was no need to attach the mount to the car, because it fit snugly into place. I did add a couple of pieces of hook and loop fastener to make sure that the control head stays put. — *73, Christine Burke, KØALT, PO Box 346, Silt, CO 81652,* **k0alt@arrl.net**

[Whenever using any type of "no-holes" mount, be sure that the mount and its device are secure in case of accident. Remember, it it's "real easy" to pull it out then it's real easy for it to go flying about. — *Ed.*]

## CUP HOLDER MOBILE MOUNT

◊ Have you ever wondered what motivated auto manufacturers to include cup holders? The answer, of course, is Amateur Radio.

The auto industry gave us a method for moving a mobile rig between cars without drilling holes in the passenger compartment or using separate mounts for each car. The cup holder, your front panel separation kit and a few items from your big box home improvement center will do the trick.

Purchase a small piece of 2 inch black PVC pipe, two end caps for the pipe and a flexible 2 inch to 3 inch pipe coupler. You can usually find a 2 foot section of the pipe so you do not need to buy the standard 12 foot piece. Cut a 3¼ inch piece of the pipe and glue an end cap in place as shown in Figure 44. The larger bottom part of the flexible coupler is cut off so that the remainder will give a snug fit into the upper part of your cup holder. The modified coupler is placed on the pipe assembly so that the end cap will reach the bottom of the cup holder when the flexible coupler is seated in the top part. For my installation, best fit was achieved when the coupler was placed all the way onto the pipe assembly.

Lead-shot fishing weights can be added inside the pipe for extra stability and the coupler can be secured with a cable tie if desired. Figure 45 shows the modified coupler on the pipe assembly.

**Figure 44 — A view of the pipe and end cap assembly.**

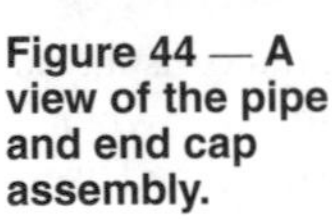

**Figure 45 — Next attach the modified pipe coupler to the pipe and end cap assembly.**

FIGURES 4, 5 AND 6 BY LYLE KRUSE, KØPHT

**Figure 46 — The finished mobile mount in use in the author's car.**

The front panel of the rig is secured to the remaining end cap and placed on the pipe assembly. Do not glue the top end cap in place in order to allow future removal of the front panel bracket. The completed unit is shown in Figure 46. — *73, Lyle Kruse, KØPHT, 2009 White Cloud NE, Albuquerque, NM 87112-3716,* **lwkruse@earthlink.net**

## EASY ANTENNA MOUNTS FOR A HONDA RIDGELINE

◊ As a recent returnee to Amateur Radio, I wanted to use my 2006 Honda Ridgeline as my mobile ham station. I wanted to mount some antennas in a more "permanent" fashion than using a magnetic mount antenna. Unfortunately, this model of the Ridgeline does not have a roof rack and is a leased vehicle, so I was limited in my mounting options (for example, I could not cause visible damage to the vehicle by drilling holes).

The Ridgeline has six tie-down points in the bed of the truck that look like cleats on a boat. The tie-down points measure 1.75 inches long by 3.5 inches with a 1 inch square opening in the center (see Figure 47). Using the two tie-down points located at the top of the front of the bed on each side of the truck as the base for an antenna mount makes it easy to route the coax in through the center sliding window at the back of the cab.

I decided to develop an antenna mount that would clamp onto a tie-down point. To make my antenna mounts, I used 0.25 inch thick, 2 inch wide, flat aluminum stock. The local hardware store conveniently sells 3 foot lengths of this stock supplying material for several mounts.

The simplest type of mount is made by cutting a 3 inch (or longer) piece from the stock (we'll call this the upper plate). Measure 0.5 inch in from each of the long sides of the upper plate and scribe a line. Now measure and mark lines across the width of the upper plate 1 inch in from one end of the upper plate and another line at 2 inches in from the same end of the upper plate. You should now have a 1 inch square

FIGURES 3-6 BY RANDY KULZER, N2CUG

**Figure 47 — Honda ridgeline tie-down point.**

box marked on the upper plate. This scribed box corresponds to the 1 inch opening in the tie-down. Drill two 0.25 inch holes at opposite corners of the 1 inch box marked on the upper plate.

Next, cut another 2 inch square piece from the stock (we'll call this the lower plate). Clamp the lower plate underneath the upper plate with one end of both the upper and lower plates and the sides of both the upper and lower plates aligned. Using the two holes you just drilled in the upper plate as a template, drill two more 0.25 inch holes through the lower plate. Figure 48 shows the upper and lower plates for the antenna mount (along with mounting hardware).

To install the antenna mount, hold the upper plate on top of the tie-down. Slide the lower plate underneath the tie-down. Insert one of the 0.25 inch bolts through the holes in both the upper and lower plates and screw a nut on the bolt that extends through the lower plate. Install the second bolt through the remaining hole in both plates and screw a nut on the bolt. As you can see from Figure 48, for extra security I also used a 0.25 inch flat washer between the bolt head and the upper plate and flat and lock washers between the lower plate and the nut. Depending upon how you drilled the holes in opposite corners of the upper plate, the lower plate may align only one way. Figure 49 shows a side view of the assembled antenna mount.

You should now have an inch or more of the upper plate extending past the end of the tie-down. If the antenna you want to use requires a counterpoise, then simply drill a hole of appropriate size for the antenna mount. For example, in the previous figures, you can see where I drilled a hole large enough for a double female SO-239 "bulkhead" connector. I verified the tie-downs are grounded to the Ridgeline chassis using a multimeter between the tie-down and the outside of the cigarette lighter/power receptacle in the back of the center console. Note that, depending on your specific antenna, the counterpoise might not give acceptable SWR readings.

Figure 50 shows the simple antenna mount installed on the tie-down. — *73, Randy Kulzer, N2CUG, 2235 Allenwood Rd, Wall, NJ 07719,* **rkulzer@usa.net**

**Figure 48 — Close-up of antenna mount hardware (disassembled).**

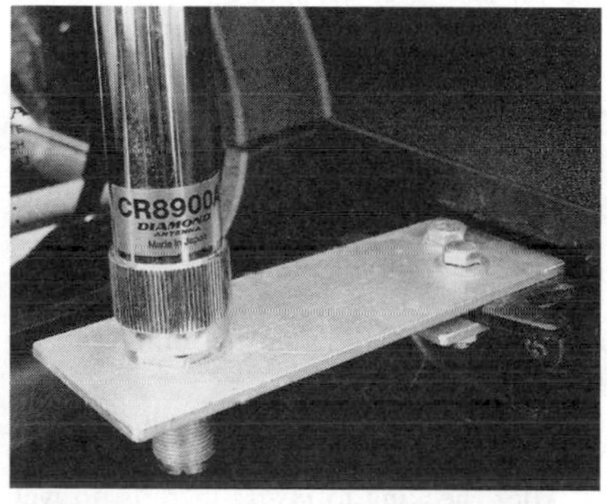


**Figure 50 — Antenna mount installed on tie down.**

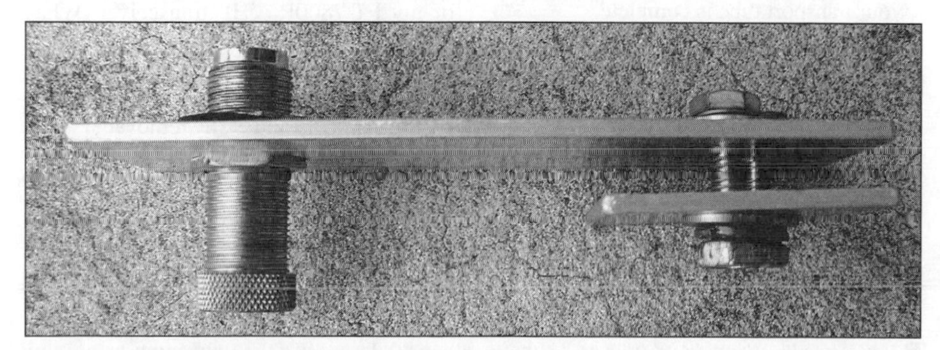

**Figure 49 — Side view of assembled antenna mount.**

## MOBILE CONVERSION OF THE MFJ-1762 6 METER YAGI

◊ The MFJ-1762 3 element, 6 meter Yagi (**www.mfjenterprises.com**) is a simple lightweight antenna easily converted to a travel Yagi. When the antenna was first received, I put it together exactly per the instructions and the SWR was 1.2:1 at 50.200 MHz. A couple of the holes needed to be deburred, but otherwise, all went well. The elements all lined up and it looked good. On-the-air tests for gain and front to back ratio were exactly as expected for a short 3 element Yagi.

I then proceeded to make it into a "travel" Yagi. I cut the boom at about 38 inches, just past the matching stub, halfway between the mast U-bolt mounting holes. I then fitted a 4 inch (2 inch either side) piece of tubing inside the boom splice, using the mast U-bolt to hold all tightly in place. Through the boom screws could be added for additional strength if you wish. I then cut all the elements at 36 inches and made 1 inch splices from ¼ inch tubing, for their reassembly on-site. I added a 5 foot RG-8X pigtail, with three snap-on ferrite split beads right below the matching stub.

Now it all fits in a 4 inch × 40 inch fishing rod carrying case. Reassembly on-site is fast. It is exactly what I need for my travels around the Caribbean with my ICOM IC-7000 transceiver and Alpha Delta HF dipoles. — *73, John Abbruscato, W5JON, 22107 Pine Tree Ln, Hockley, TX 77447,* **w5jon@sbcglobal.net**

## AUTO CLEARCOAT DISCOLORATION

◊ One of the great things about this hobby is how people pass along great ideas, and warnings of possible problems. A great idea is putting cotton balls loosely in your remote speaker. That cheap little 3 inch speaker in your car will sound incredible with the air space loosely filled. This hint is about an expensive lesson I learned, which can become a problem for a lot of hams with newer vehicles.

I own a 2007 Camry Hybrid. The special light green color is available only on the hybrid, but I've seen a close match on new VW Beetles. The color itself really has nothing to do with the problem, but made it very visible.

Not wanting to drill any holes, I selected a dual band mag-mount antenna from a well-known supplier. The base is about 3 inches in diameter and has a thin rubber boot that fits around the rim and covers the bottom. The combination of newer model paint and rubber on the bottom of the antenna created a major problem.

I happened to be washing the car, lifted the mag-mount base from the trunk and found a discolored patch where the antenna had been. Against the green paint, the patch was brown. This was not simple fading from being covered and not exposed to the sun. It would not clean off with Windex, or any other household cleaner I tried.

I took it to a body shop near where I work and showed it to my "guy" there. He tried lacquer thinner on it and the stain was unaffected. The next day I made arrangements to leave the car with him overnight. He ended up wet-sanding it and had to sand down well into the paint, almost to the metal, to eliminate the stain. It took $200 worth of his labor and paint to redo the entire top of the trunk to make it right.

He explained to me that many of the clearcoats used on newer cars are water based and that rubber and many other things will bleed into and through it, which is what happened to me. So rubber boots on mag-mount antennas and even the rubber top part of a trunk lip mount can and will stain the car right into the paint.

Both the auto manufacturer and the antenna manufacturer have never heard of this problem before. The auto manufacturer says, of course, that they don't make mag-mount antennas and so have no experience with them. The antenna manufacturer says that the usual complaint they get is that the rubber boot is missing. So neither of them has any information to help avoid this problem.

I'm now hoping the thin felt stick-on dots I put on the base of the antenna will prevent this from happening again. If you have a newer car, take a moment to look under any rubber making contact with the paint. It might also be worthwhile finding out if your clearcoat is water based, and what, if anything, they can recommend to keep your antenna boot from bleeding through. It'll be a nasty surprise if you're at your dealer trading in the car and remove the antenna to find a 3 inch round stain that won't go away. — *73, Bill Stewart, 1131 SE 18th St, Cape Coral, FL 33990*, **n4cro@hotmail.com**

## A MOBILE ANTENNA TRANSPORT TUBE

◊ Carrying any of the popular fiberglass shaft antennas (Hamstick, MFJ, Outbacker, etc) unprotected in a car or as a portable antenna for use in the field can easily result in damage. Here is a fast, easy and inexpensive way to protect them.

You can make antenna transport cases using PVC pipe, end caps and foam pipe insulation. The cost is minimal and the protection is great.

To make two PVC transport tubes obtain the following parts:

- One 10 foot section of PVC pipe 1.5 inches in diameter
- Two glue-on PVC end caps
- Two screw-on end caps and two matching glue-on threaded adaptors
- One 10 foot length of foam pipe insulation
- Four medium sized (1.5 inch) felt stick-on pads for chair/table legs
- One can of general purpose PVC cement

BOB PATTERSON, K5DZE

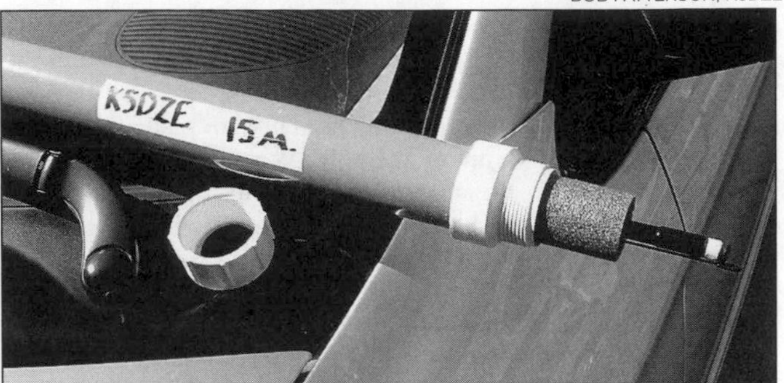


**Figure 51 — A completed travel tube for a 15 meter Hamstick.**

I recommend you read through this procedure carefully and look at Figure 51 before you start to cut and glue:

1. Cut a PVC pipe to fit the length of the antenna. If you use a "quick disconnect" for your mount and whip make sure you allow for this extra length when you cut the pipe. Measure twice — cut once.
2. Put a 1.5 inch self sticking felt pad on the inside of each glue-on cap and one on the inside of each screw-on cap. This will protect the antenna from hitting the two ends as it slides back and forth inside the tube.
3. Cement the glue-on end cap on one end of the tube. This is the bottom of your transport tube.
4. *Caution*: Do not screw the cap on before, during or after this step for about 10 minutes. This will allow the cement to dry completely, preventing the screw-on cap from being cemented in place. If you do get glue on the cap and it touches the threads, the glue will bond instantly and permanently and you will have to start over with new materials.

With the screw-on cap removed, cement the screw-on adaptor on the other end of the tube.

5. Insert the foam pipe insulation in the PVC tube and cut it to fit the length of the tube.
6. After checking to *insure* there is no wet PVC cement on the screw-on threads or the end cap, gently screw on the end cap to see how it all fits. That's it; the construction of your transport tube is complete.

### *Finishing Touches*

- Using a permanent marker, mark the tube side and screw-on end cap to identify the antennas the tube is for, such as "40 M, 20 M" etc.
- Mark your antenna tips for the band and overall length for storage. White tape wrapped around the antenna tip will indicate how far to slide the tip into the antenna shaft for each antenna.
- You can also paint your transport tubes, but if you do, be sure and use PVC approved paint or the paint will likely flake off. Camouflage paint or tape looks sharp, but it can also help you lose a tube or end cap in the field. Instead, try a bright color paint (orange?) or leave them white and add decals. Also, it's a good idea to write your call sign and name on the tube.
- A discarded shoulder strap, sling or web strap from a sports bag, backpack, etc, makes a good carrying sling or carrying handle.

**Note**: Do not glue the transport tubes directly together as a cluster for carrying as it makes the screw-on end caps hard to manipulate. If you have two or three transport tubes and intend to always carry them together, you might consider placing a couple of 3 or 4 inch pieces of PVC tubing between the transport tubes as spacers and use PVC cement to glue these in place to make a 'cluster' for ease of transport. After you cement these in place, you can run a strap thru these short pieces for your carrying handle. As always in ham radio, add, delete, change or modify these ideas to suit your needs. Enjoy. — *73, Bob Patterson, K5DZE, 110 Charles Givens Dr, Dry Ridge, KY 41035*, **k5dze@arrl.net**

## NO HOLES MOUNT

◊I'm fortunate to be provided a company vehicle every year, which is especially nice these days since a credit card for gas is included. But what's not so nice is that I'm not allowed to drill holes or otherwise mutilate the vehicle in order to mount radios, antennas, audio gear, etc. This prompted me to explore ways of mounting the control head for my FT-7800R VHF transceiver. What I've settled on is a simple and inexpensive method. My personal requirements were:

- Be able to easily see the display
- Have a quickly and easily removable setup
- Not cost an arm and a leg
- Use readily available materials
- Be within very easy reach for safety while driving

My new mount consists of a modified goose neck clamp-on light available at most discount stores. The light I bought cost less than $9. I removed the light components and

JOHN D. MERRITT, K4KQZ

**Figure 52 — A cheap solution to a no holes mount installation. Note how the unit is clamped to the center console and also the position of the microphone clip.**

JOHN D. MERRITT, K4KQZ

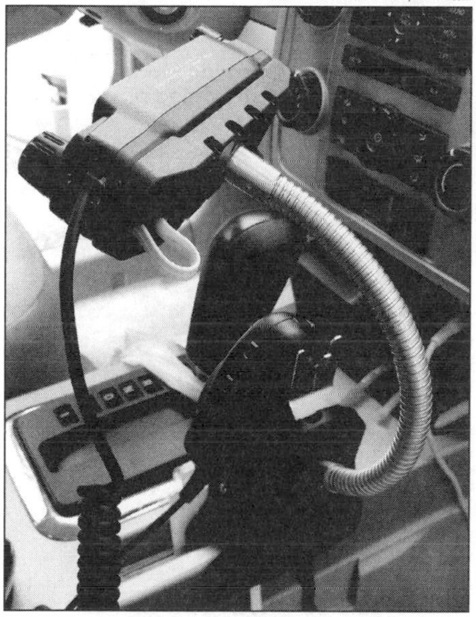

**Figure 53 — A rear view of the no holes mount attached to the center console and showing the microphone clip installed on the lamp base.**

drilled the mounting bracket so the goose neck would fit. I could have used the original cable that came with the radio mounting kit but decided to leave it intact and instead used flat 6 conductor standard phone cable. One end had to be cut off in order to snake the cable through the goose neck and I installed a new RJ-11 connector on the cable end that plugs into the radio. I also installed the microphone clip onto the clamp.

The result of my effort is shown in Figure 52. It may not work as well in other vehicles but it works great in mine. I haven't tried it in other vehicle models but I suspect there are spots to which the clamp could easily be attached. — *73, John D. Merritt, K4KQZ, 2430 Hidden Lake Cir, Columbia, TN 38401-5832,* **k4kqz@arrl.net**

## AN IMPROVED MOBILE CONSOLE

◊ It was getting to be that time — the time to get a new radio. I really liked the idea of removable control modules and settled on the Yaesu FT-7800R mobile transceiver. The separate control panel provides many possibilities regarding how to install the radio. The automobile was a Buick LeSabre, which has the car battery under the rear seat (some GM engineer must be a ham) and no front center console. What more could any ham ask for?

Well, I didn't want to drill any holes; I didn't want to fish wires up under the instrument panel and I wanted easy viewing of the radio control head and to be able to hear the radio easily (read external speaker).

While perusing the automotive department at the local discount store, I noticed the console (see Figure 54). This after-market center console immediately evolved in my ham mind's eye into a mobile mount that could easily be modified to meet all of my requirements.

A piece of masonite was cut and sanded to fit the tape/CD storage space. It is supported by the ridges used to separate the media. The center of the masonite was overlaid with a piece of perf board and holes located for mounting a 4 inch speaker. The perf board drill guide was taped in place and using a 1/16 inch drill bit, all 1000 holes were drilled. One might just cut a round hole and cover it with speaker cloth or a preformed speaker grill but the masonite provides a firm base to support items which might be stowed in the shallower compartment.

A ½ inch hole was drilled in the bottom of the speaker compartment as well as just behind the cup holders (more on this later). I had some scrap lamp cord and used this as speaker wire, knotting the end for strain relief. A mating plug was installed on the other end for use with the '7800R. Purists might want to use shielded audio cable. I also mounted a short screw in the center of the speaker grill to act as a handle to lift the speaker assembly out when needed.

Now about that extra hole. I fashioned a small aluminum clip (see Figure 55) to be mounted in the center of the console above the cup holders with two self tapping screws. A piece of aluminum about 2½ × 1 inch was slightly bent to form a hook at one end to hold the control head.

I just clamped the metal behind and slightly above the edge of an old hinge in a vise. I then hammered the aluminum over the edge to form the hook. This was then filed and wire brushed to remove any burrs and sharp edges. A small strip of black electrical tape was folded over the upper front of the plastic control head to further protect it.

Adjust the tension of the clip to firmly hold the control head.

JERRY SOBEL, KØMBB

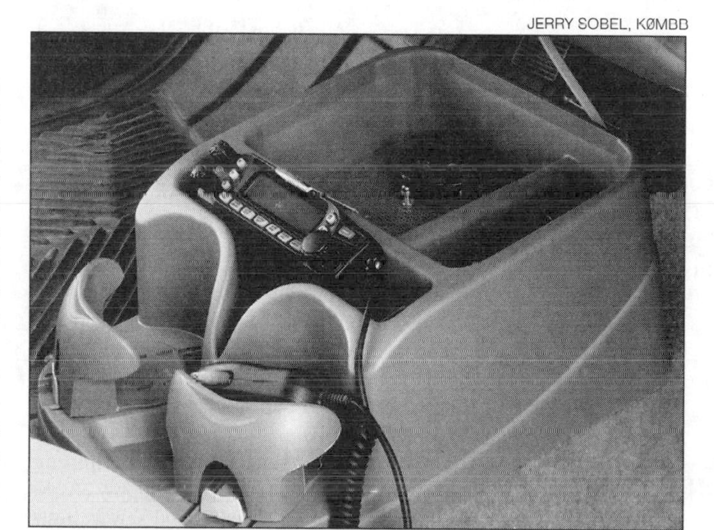

**Figure 54 — The completed center console mounted in place. Note the bracket securing the FT-7800 control head.**

JERRY SOBEL, KØMBB

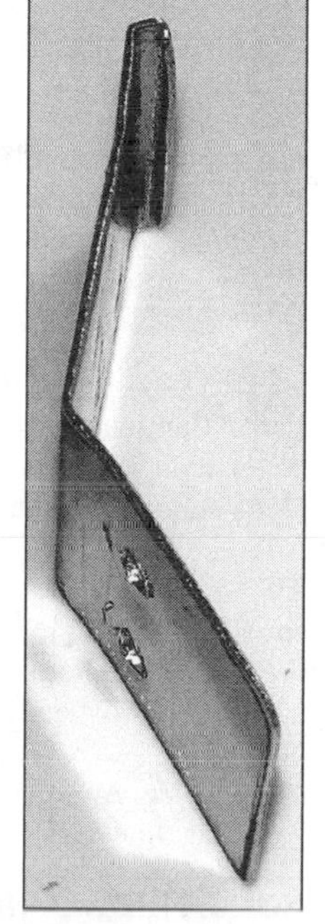

**Figure 55 — A close-up view of the aluminum bracket used to hold the control**

This "spring clamp" then secures the control head at an angle perfect for viewing and adjusting. Oh yes, the extra hole is for the control head cable to exit below the console and under the seat to the radio body on the floor behind the driver's seat.

I can see the control head easily and can use the somewhat shallower compartment on top of the speaker (nothing liquid please) and the cup holders as well. Fidelity from the 4 inch speaker is great and everything is easily transported to my other Buick LeSabre when switching cars. Best of all — the entire project was very inexpensive (read cheap). Changes and modifications to this design are numerous and I'd be interested in knowing what others come up with. — *73, Jerry Sobel, KØMBB, 10409 Broom Hill Dr, Las Vegas, NV 89134-7337*, **arsk0mbb@aol.com**

## STEERING WHEEL TABLE

◊ When parked at a campground I operate HF from the driver's seat of my motor home. Taking notes, using the control head and logging was not convenient. Since the steering wheel tilted I thought a flat surface, on the wheel, would be ideal.

ALLEN WOLFF, KC7O

Figure 56 — The underside of the table showing the added PVC caps and tubing "legs."

The local home improvement store had a small laminated table top that was slightly larger than my steering wheel. Using PVC plugs screwed to the bottom, spaced to accommodate the openings in the wheel, and PVC pipes on the plugs, the table slides snugly on the steering wheel (see Figure 56). Using hook and loop fasteners I attach the control head and a clipboard for easy operating. — *73, Allen Wolff, KC7O, 57 W Grand View Ave, Sierra Madre, CA 91024*, kc7o@arrl.net

## HONDA RIDGELINE CUP HOLDER MOUNT

◊As a recent returnee to Amateur Radio, I wanted to use my 2006 Honda Ridgeline as my mobile ham station. This was complicated by the fact that the Ridgeline is a leased vehicle so I was limited in my mounting options (that is, I could not cause visible damage to the vehicle by drilling holes). The Ridgeline does have two cup holders at the front of the center console that were potential mounting options.

Figure 57 — The Lido Products LM-800 EXP Cup Holder Mount with felt pads added for vibration control.

Figure 59 — The Lido Mount installed in the modified pipe connector.

Figure 58 — The PVC pipe connector modified to better fit the base of the Lido mount.

Figure 60 — A bottom view of the Lido Mount showing how it fits in modified pipe connector.

These are rather large cup holders (for when you just have to have that "big gulp") measuring 3¾ inches in diameter and 3 inches deep. Fortunately, these cup holders do not have a taper or ridge under the top edge but they do have rubber "lips" around the periphery of the cup holder to hold smaller diameter drinks.

It just so happened that I was flipping through an Amateur Electronic Supply (AES, **www.aesham.com**) catalog and saw various "no holes" mounts manufactured by Lido Products. In particular, the LM-800EXP Cup Holder Mount (**www.lidomounts.com/catalog/item/1478816/4979439.htm**) seemed perfect for my needs. As you can see from Figure 57, I modified the mount with some self-stick heavy duty felt strips as padding against shock and vibration.

The information in the AES catalog stated that this mount would fit cup holders from 2½ to 3¼ inches (in diameter). So I had a gap of ½ inch to close.

I first tried increasing the diameter of the two rubber coated clamps at the bottom of the mount using some 1¼ inch wide by 7/16 inch thick closed cell weather stripping. It turned out that the weather stripping wasn't stiff enough (it compressed too much) allowing too much movement of the mount.

If attempting to increase the diameter of the mount didn't work, then the next thing to try was decreasing the size of the cup

Figure 61 — The entire device as it appears in the Ridgeline cup holder.

ROBERT PATTERSON, K5DZE

Figure 62 — A camping privacy tent that could make a quick and easy portable shack for Field Day or any other portable operation.

RON TOYNE, WAØAJF

Figure 63 — The movable mount strapped in place in the author's F-150 truck. The width of the base board and length of the legs will vary according to your vehicle's interior layout.

holder. After doing some searching through the plumbing department at the local home center, I found a 3 inch white PVC S&D female adapter.

This PVC connector measures 3¾ inches across the top and has an inside diameter of 3¼ inches inside the lower part of the connector. Also, there is a ridge just below the threads against which the PVC pipe would butt.

I modified the pipe connector by grinding down the threads in the top portion of the connector. This allowed the larger portion at the bottom of the Lido mount to sit flush against the ridge inside of the pipe connector. I also ground two "channels" in the ridge inside the pipe connector corresponding to the two rubber coated clamps in the Lido mount. This would allow the clamps to sit flush against the side walls in the lower part of the pipe connector (see Figure 58).

The completed Lido mount and modified pipe connector is shown in Figures 59 and 60. The assembly as it sits in the Ridgeline is shown in Figure 61. Note that I purchased a second Lido LM-800EXP and modified another PVC pipe connector to hold my Nokia cell phone. The Lido mount conforms to the four hole AMPS standard for mounting plates. — *73, Randy Kulzer, N2CUG, 2235 Allenwood Road, Wall, NJ 07719,* **n2cug@arrl.net**

## A PORTABLE HAM SHACK FOR FIELD OPERATIONS

◊If you have ever spent any time operating in the field for Field Day or on a low power outing you have likely run into a cold wind, a light rain or night dew settling on everything to become a nuisance. To beat this, consider making a sheltered one or two person operating position for your field operations using an inexpensive pop-up "Privacy Tent" of the type normally used as a shower or a portable toilet in the field. The Outback Porta-Privy is a typical tent of this type that will work well for amateur field use (see Figure 62). It sells for $59 to $69 at Bass-Pro Shops, Cabela's and Emergency Essentials among others. Here are some features for this model that makes it a good portable ham shack:

- Pops up in under a minute — no assembly. (Just toss it out.)
- Generous 54 × 54 × 80 inch interior holds a 4 × 2 foot folding table for rig(s), laptop, logbook and up to two people.
- Weatherproof nylon lining keeps dew off equipment.
- Large ventilation screens.
- Removable waterproof floor.

Add a 4 × 2 foot lightweight folding table and a folding chair or camp stool and you have a nice operating position for Field Day, a low power outing, or an EmComm deployment. More information can be found at **www.k5dze.net**. — *73, Robert Patterson, K5DZE, 110 Charles Givens Dr, Dry Ridge, KY 41035,* **k5dze@arrl.net**

## PLATFORM MOBILE MOUNT

This describes an easy to build and install mobile mount. Just a board cut with a bevel and then carpeted, two legs with rubber ends and a nylon strap. Very simple and handy to use. Figure 63 shows the mount in my Ford F-150 pickup truck.

I used a standard 2 × 10 board 13½ inches long purchased at any lumber store (the length of this board should match the center console width of your vehicle). This board has a 30° bevel cut along one edge. The Kenwood TS-50's quick release mobile mounting bracket is screwed onto this cut edge.

I attached carpet to the board first and then mounted the bracket. Two ¾ × 1½ inch boards were cut and screwed to this edge to form the legs. The length of these two legs was determined after strapping the carpeted board to the center console with a ratcheted nylon strap commonly used to tie down a boat onto a trailer.

I mounted rubber furniture pads onto one end of each of the legs, held them in place to the board already tied down to the center console and marked the length they needed to be. After cutting them I screwed them onto the 30° beveled cut edge. It is important to get these "legs" the right length to give stability to the mount. Otherwise it would bounce up and down when on the road. — *73, Ron Toyne, WAØAJF, 1220 Hertz Dr SE, Cedar Rapids, IA 52403-3450,* **wa0ajf@aol.com**

## MOLESKIN PAINT FINISH PROTECTOR

◊In response to Bill's, N4CRO, hint about auto finish discoloration, I found myself cringing at the thought of my car's finish being ruined by a mag-mount or the rubber pad that is supposed to protect it.[2] The mag-mount padding I use has not marred the surface of the car in any way shape or form, despite fall and winter weather, spring rain and summer sun.

What do I use? I use plain old fashioned moleskin. It can be found in the foot care aisle of your local discount store or pharmacy and it's cheap, to boot. To make it stick better, when you remove the paper backing, heat up the sticky side with a hairdryer before placing it on your mag-mount. Carefully trim with a sharp knife or file and enjoy. — *73, Georges P. Godfrin, KJ4BNE, 112 Grant Dr, Laurens, SC 29360-3712,* **georgesrn@yahoo.com**

[2]B. Stewart, N4CRO, "Auto Clearcoat Discoloration," *QST*, Aug 2009, p 62.

## QUICK BICYCLE MOBILE SETUP

◊I enjoyed WA3LKN's article about bicycle mobile ham radio.[1] David has certainly created a first-class setup, but there's a simpler way.

I commute via bicycle to my job and like David want to enjoy ham radio while I ride. Having radio equipment attached to my bike isn't an option since it is unattended while I am at work. While riding I wear a daypack, which contains work-related items. It occurred to me that the pack's shoulder straps were sturdy enough to support a handheld transceiver. I wrapped a length of #14 insulated wire around the left strap and soldered the two ends together, forming a flat loop with the strap passing through it. I then duct-taped the loop to the rear of the strap, to hold it in place. A handheld transceiver is attached with its belt clip to the loop on the front side of the strap, so the radio faces forward when the pack is worn (see Figure 64). The handheld transceiver is easily attached or detached from the pack in a few seconds.

The loop is positioned on the strap so the top of the handheld transceiver is level with the top of my left shoulder when I wear the pack. This makes the handheld transceiver's controls easily reachable with either hand and puts the rubber duck antenna "in the clear" above my left shoulder.

I use a few variations of this basic arrangement. If I'm not feeling chatty and just want to read the mail on the local repeaters, the handheld transceiver alone fills the bill. Positioned about 6 inches below my left ear, I can easily hear it even in a noisy environment. No need for earphones or a headset, which could block other sounds and create a safety hazard.

If I want to talk, I add a speaker-microphone. I have to remove one hand from the handlebars to hold the speaker-microphone in front of my mouth and key the transmitter, but I do this only when I judge it is safe. When not transmitting, the speaker-microphone clips to the opposite (right) pack strap in a position where it is easy to hear with my right ear.

If I know I'll be doing a lot of transmitting, I use a small 12 V gel cell battery, which fits in the pack. I run a power cord from the battery out the top of the pack and over my left shoulder, where it plugs into the handheld transceiver's external dc jack. This provides much longer battery life and higher transmit power. By the way, the same setup works great for hiking or walking too. — *73, Joe Dickinson, WTØC, 8152 S Saint Paul Way, Centennial, CO 80122,* **wt0c@arrl.net**

[1]D. Pennes, WA3LKN, "Bicycle Mobile Ham Radio," *QST*, May 2009, pp 69-70.

JOE DICKINSON, WTØC

**Figure 64 — William, KCØYQN, the author's son, demonstrates the use of the Quick Bicycle Mobile setup.**

## FLEXIBLE MOBILE MOUNT

◊A great deal of my Amateur Radio work takes place in my vehicle. My 2001 Chevy Tracker has a gearshift console between the two front seats that meets the dash right in the center where radio gear would normally be mounted. I needed another solution for mounting a radio.

Being a non-smoker, I removed the ash tray and found it opened to the area behind the dashboard. The cavity was too small for my ICOM IC-706 transceiver, but it has a removable control head. I could mount the radio body behind the dash and feed the control cable through to the control head. I wanted something adjustable and the idea of an arm on which the control head could be mounted came to mind (see Figure 65). The arm shown in the photo and the accompanying diagram works very well.

KAYCEE JOHNSON

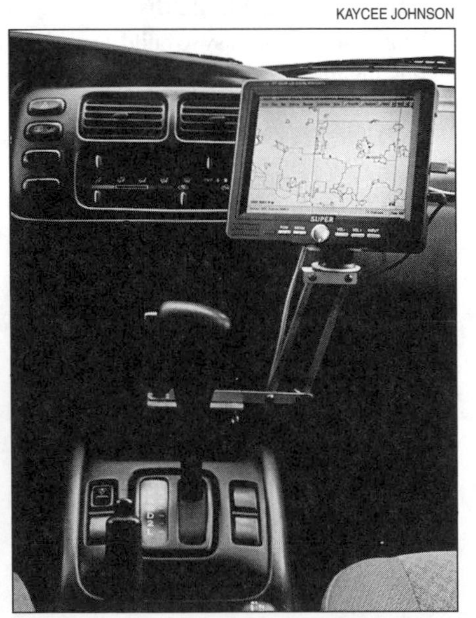

**Figure 65 — A view of the completed mount in the raised operating position.**

While examining the ash tray cavity, I noticed two ⅛ inch grooves for the ash tray runners. In these grooves I positioned a ⅛ inch aluminum plate just long enough to extend from the dash without interfering with the gearshift lever. This provided the mounting point for my device.

The construction is very simple (see Figure 66). A bench top drill press is the only power tool necessary. Otherwise, a hacksaw, file, screwdriver and combination wrench are all that are required. The materials are all stock parts found at any hardware store except the ⅛ × 4 × 8 inch plate, which was purchased at a local metal supplier for $3. The total cost was under $20 and took about 2 hours to complete.

### *Things to Think About*

Before building the mount, here are two things to consider. First, different vehicles have ash trays that operate differently. A modification of the base plate might be necessary. Once mounted, the rest of the design should be the same. Second, this mount allows a great deal of motion both horizontally and vertically. The display also pivots toward the driver, the passenger or any position in between. Care must be taken during construction to insure that the range of motion in any direction doesn't interfere with the operation of the vehicle. The bottom or top arms can be shortened or lengthened to fit your vehicle's controls and available space. Test the fit as you build.

### *Getting Started*

Cut the mounting plate to a length that, when inserted into the ash tray cavity, will not interfere with the gear shift. Using the full depth of the cavity provides the best support for the device. Next, cut a piece of aluminum L channel an inch or two longer than the plate's width. This will be the bottom arm. In my truck, a piece about 6 inches long gave the best range of motion without interfering with the gear shift or my passenger.

Place the bottom arm on the mounting plate with the widest side down and the other side facing the rear. Cut away about an inch of the vertical side nearest the driver for the horizontal thumbscrew. Place the arm on the plate with the tab nearest to the driver and pivot the arm on the plate to determine the range of motion and length of arm necessary. Outline the arm on the base plate. Drill a hole at the center of the tab and through the base plate. Using a thumbscrew, two wash-

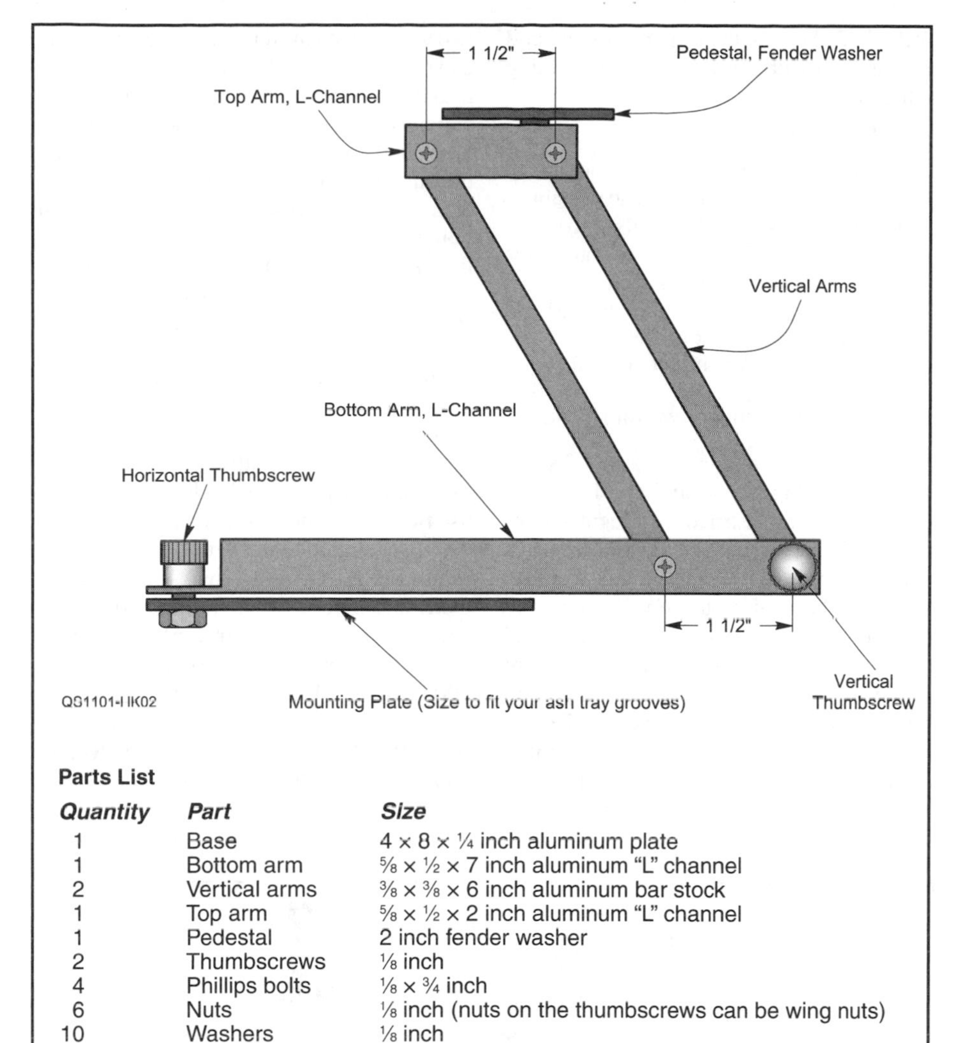


**Parts List**

| *Quantity* | *Part* | *Size* |
|---|---|---|
| 1 | Base | 4 × 8 × ¼ inch aluminum plate |
| 1 | Bottom arm | ⅝ × ½ × 7 inch aluminum "L" channel |
| 2 | Vertical arms | ⅜ × ⅜ × 6 inch aluminum bar stock |
| 1 | Top arm | ⅝ × ½ × 2 inch aluminum "L" channel |
| 1 | Pedestal | 2 inch fender washer |
| 2 | Thumbscrews | ⅛ inch |
| 4 | Phillips bolts | ⅛ × ¾ inch |
| 6 | Nuts | ⅛ inch (nuts on the thumbscrews can be wing nuts) |
| 10 | Washers | ⅛ inch |

**Figure 66 — Construction details for building the flexible mobile mount.**

ers and a nut, install the arm and check for range of motion.

The vertical arms are two identical pieces of ⅜ inch aluminum key stock. Round the corners of each end of the key stock so the corners don't bind when moved. Drill a hole in each end of both pieces of key stock. These holes must be positioned identically so the arms travel parallel to each other keeping the top of the control head mount level throughout the arm's range of movement. The vertical arms must be identical; otherwise, the mounted object will tilt.

Cut a 2 inch piece of L channel; this will be the top arm. In the narrow face, drill two holes, 1½ inches apart. Drill two corresponding holes in the vertical side of the passenger end of the bottom arm. Attach the two vertical arms to the top and bottom arms. Use three bolts, nuts and washers, and one thumbscrew, wing nut and washer. The thumbscrew will allow the upper arm to be moved and locked into position and the wing nut will rest against the channel, making a self-locking nut.

Finally, the pedestal. To allow it to rotate I bolted a 2 inch fender washer to the top arm using a bolt and two nuts. The two nuts are tightened against each other, fixing the bolt to the top arm. The nuts can also be adjusted so the fender washer can rotate with slight finger pressure, but can't turn on its own. This step, like the base plate, might require a little ham ingenuity depending on the mounting needs of your device. These directions should provide an adequate starting point to build your own mobile mount.

*73, Michael K. Johnson, NØVX, 19215 NE 129th St, Kearney, MO 64060-7945,* **n0vx@arrl.net**

## MOBILE LOGBOOK

◊At Dayton 2008 I kept seeing call sign tags and thinking, "I think I worked that guy at some point in time." Wouldn't it be nice to have a logbook on one of the new handheld phone devices where I could search the call quickly? If a call was found it would allow me to strike up a conversation showing him the contact in the log and possibly filling out a QSL for him if he wanted one.

After coming home from Dayton I approached all my techie friends and posed the problem to them with the following requirements:

- Have a mobile handheld logbook that would hold at least 50,000 contacts with time, date, band, mode and call.
- Have a search time of less than 10 sec-onds so that if I saw someone and entered the call, I could find them in the log before they disappeared.

We started out with a friend's Palm Pilot, but could not find any available programs that would meet the requirements. Most of the year passed with no progress. Then a coworker's cousin who was upgrading his phone let me have his old T-Mobile Wing. The nice thing about it is that when connected to a computer via a USB cable it acts like a thumb drive and you can transfer files back and forth between the Wing and the main computer. The Wing also has *Windows Mobile* and *Excel Mobile* installed.

I use *DX4WIN* as a normal record keeping program on my main computer. Mack, N4SS, suggested that I export the *DX4WIN* file as a .CSV file, which *DX4WIN* will generate. Then, still on the main computer, open the .CSV file with *Excel,* which accepts .CSV files. After cleaning up the *Excel* version of the *DX4WIN* file, export it to the Wing's *Excel Mobile*. It worked easily but we did have a few minor things to contend with.

We discovered that my version of *Excel Mobile* would only accept 16,384 contacts of the 41,000 we exported to it. The next suggestion was to divide the 41,000 into individual sheets, which would be set up in alphanumeric order so I could select the sheet I wanted to search from. The first try was with three sheets, 1-J, K-R and S-Z. This made each sheet cover about 13,000 contacts. The worst case search time on each sheet was around 10 seconds so, to reduce the search time, my plans are to divide it up even more.

During these tests the thought was to expand the Wing's memory with an 8 GB SD memory card. Unfortunately, we found that the Wing only stores data in the external memory card and does not actually run programs in that area. The Wing does have a fixed amount of onboard memory so the data being used by *Excel Mobile* had to be kept within certain limits. All manipulation of the *Excel* file is done on the main computer before exporting to the Wing.

Even a computing novice like me found *Mobile Excel* fairly easy to use. You can even do a zoom on the log page to improve readability. I am sure other devices out there will do the same thing but the Wing makes it easy. With technology moving at a high rate I am sure this solution will be outdated soon but the price of used Wings on eBay will come down, making them more available.

Maybe I will see you at Dayton this year and find you in my logbook, which has

21 years of contacts so far. My thanks go to Mack, N4SS, and Larry Griffie for all their help, and to Roark Jones for supplying the Wing. — *73, Mike Greenway, K4PI, 4055 Kings Hwy, Douglasville, GA 30135-3763,* **k4pi@arrl.net**

## MICROPHONE CLAMP

◊I use a wire clamp known as an Adel loop clamp for my microphone holder (see Figure 67). It is a metal strap with a protective rubber shell that will not scratch what it is hanging on. It bends to any shape hook you might need. It's cheap, it works, it's very available and can be used on any brand microphone.

LEE GEORGE, KE4VYN

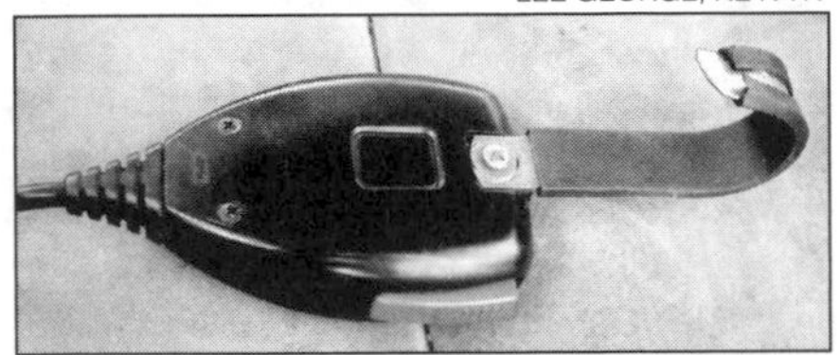

**Figure 67 — An Adel loop clamp attached to the back of your mobile microphone allows you to "hang" the microphone from your hand, leaving your fingers free to grip the wheel.**

If you're operating mobile, you can hang the clamp on your hand and still have most of it available to hold the wheel with your small finger keying the microphone. I have two fingers missing from my hand and only 50 percent use of what's left. Using the Adel clamp I don't have to hold onto the microphone. The hanger also easily hangs on anything so you're not distracted when trying to hang up the microphone.

With all the collateral damage hams are receiving from people using cell phones, this hint actually makes for safer travel. Your hand does not hold the microphone, so you can still grip the wheel. You do not have to hang up the microphone when finished and so your eyes never leave the road. Your primary fingers do not press the PTT, only your pinky.

If you do have to hang the microphone, the wide hook allows for hanging on most any surface. No looking for the hole. Finally, its home engineered look fits with most hams. — *73, Lee George, KE4VYN, 3420 Deerwood Cir, Vestavia Hills, AL 35216-4816,* **ke4vyn@msn.com**

## NO HOLES SUV MOUNT

◊After purchasing a GMC Yukon and trying to decide where and how to install the antennas, I found I couldn't use the same method as my previous SUV since the Yukon's trailer hitch is an integral part of the frame and rear bumper.

While searching for a mounting method, I shared my problem with Grady Ball, WB3JUV (he has the same vehicle) and we came up with the following no-welding solution.

The parts required are:

2 — Diamond K-400 heavy duty mounts
1 — ¾ inch × 8½ to 9 inch threaded stainless rod (**www.speedymetals.com**)
1 — ⅝ inch OD × 7½ inch round stainless tube (**www.speedymetals.com**)
2 — ¾ inch stainless nuts and lock washers
4 — ¾ inch stainless flat washers
1 — 3 inch stainless bolt, nut and flat washer
1 — truck mirror mount (two for heavier antennas)

Refer to Figure 68 and the Diamond K-400 mounting instructions. Begin assembly by discarding the flat grounding strips that come with the K-400 mounts. Loosely assemble the dual mount on the bench by inserting the threaded rod into the tube, add a flat washer and insert this into the bottom of the flange of a K-400 mount. Then add a flat washer, lock washer and nut. Do the same for the opposite end, mounting it into an inverted K-400, and loosely tighten the two nuts. Attach the dual mount to the vehicle and loosely tighten all eight mount setscrews. Make any final adjustments to both K-400s and finish tightening the top and bottom nuts.

Remove the factory socket bolt holding the antenna mount to the trunk base bracket from the upper K-400 and insert the 3 inch bolt with the inner bushings, washer and nut; this will be your ground post for the antenna.

Ground the mount by installing a stainless self-tapping screw into the chassis with as short a ground strap as possible and attach the strap to the 3 inch bolt. The shield from your coax will be attached here also (depending upon how the coax is attached).

The final step is to attach the mirror mount to the antenna platform. For heavier antennas use two mirror mounts, one upright and one inverted, for added strength. Presently I am running a HI Q 3/80 (**www.hiqantennas.com**) and before that a Tarheel 100 and a Tarheel 40A (**www.tarheelantennas.com**). I have driven well over 20,000 miles on my 2008 Yukon XL SLT since the installation without a problem. — *73, Anthony McAlister, KE9PH, 9141 S Paxton Ave, Chicago, IL 60617-3858,* **ke9ph@arrl.net**

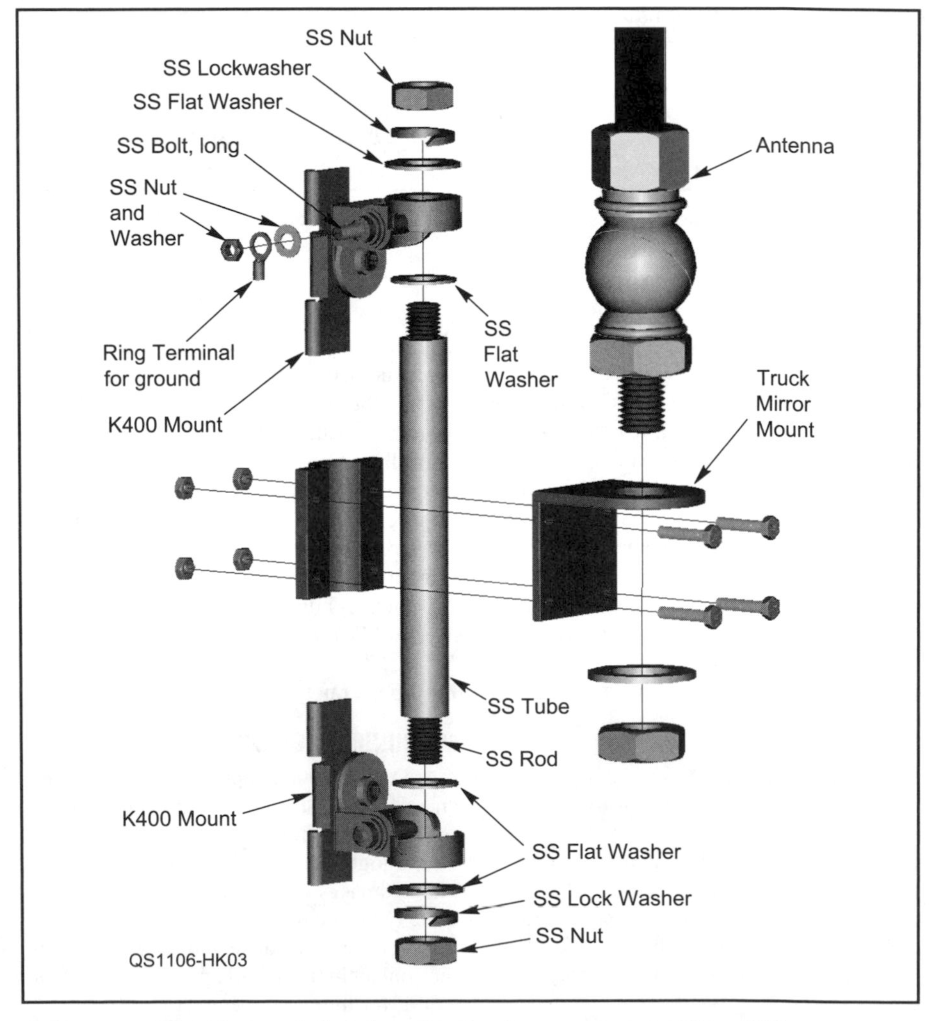


**Figure 68 — A detailed exploded view showing the components of the SUV mount.**

## PORTABLE WITHOUT THE PIZZA

◊Many hams have taken magnetic-mount antennas beyond the top of a vehicle, placing them in attics or using them for emergency or public service events to replace a flexible antenna. I hear operators say "I have it on a pizza pan." But is that pan 1 meter in diameter to provide the proper ground plane?

After building a 2 meter ground-plane antenna for a club presentation, I realized that four radials are sufficient and you don't need the rest of the pizza pan. I found a steel can lid an inch larger than my 4 inch diameter mag-mount. The lid center should be flat, without ridges, because the magnet has to capacitively couple to the ground plane. I soldered four ½-meter #20 AWG solid-copper radials to the edge every 90° cutting opposing radials (equally) to be 1 meter across (see Figure 69). If used on a supporting surface, smaller or stranded wire would do; otherwise, use #20 AWG solid or larger for freestanding applications.

KENNE YEE, K3YEE

**Figure 69 — With the radial adapter attached the mag-mount is freed from you car's roof and can be placed on any convenient surface or hung from a tree.**

My radials are flexible enough to bend for "pack and go" and then reshape for use. Try bending the radials down 45° to tune as with a ground-plane antenna. — *73, Kenneth Yee, K3YEE, 1713 Evelyn Dr, Rockville, MD 20852*, k3yee@arrl.net

## CONVERT TODDLER TRANSPORT TO TOWER TOTE

◊I have been operating from some nice places for the 10 GHz and Up contest for the past few years, but until recently I have had to use loaner stations. I finally got my own transverters to use so I don't have to worry about damaging someone else's equipment.

The problem has been toting all that equipment. Some locations require me to carry the equipment to a distant site. One such site is affectionately called "Sackrider Hill." This site is a good quarter mile hike traveling uphill over an unimproved path that is actually a rainwater runoff. Carrying about 50 pounds of equipment (depending on the battery used) up this hill isn't easy. Usually it requires making numerous trips back and forth between the site and your car.

I started thinking about an easier way to move my equipment to these distant sites. I think I came up with an interesting solution.

I had a toddler bike-carrier frame (see Figure 70) that attaches to a bicycle and totes your toddler in comfort as you ride along. I paid $30 for it at a bike shop 2 years ago. I thought of using it for carrying equipment around during special events or when I wanted to operate in some of the more scenic spots of Michigan. If you look around, you may find someone who is discarding theirs. Look for one that lets you remove the wheels. This way it sits on the ground adding more support.

JAMES FRENCH, W8ISS

**Figure 70 — The toddler transporter ready to be repurposed for hauling ham equipment.**

JAMES FRENCH, W8ISS

**Figure 71 — The completed portable antenna system ready to head for the hills.**

To convert the frame I took an 8-foot 2 × 4 and cut it so that I had two pieces that would fit across the front and the back of this carrier. Then I used U-bolts to anchor it to the frame. Next I drilled three holes for ¼ inch hex bolts, washers and nuts to attach my 3 foot tripod.

The frame has a cross piece that is bent up in the center with a hole for supporting the toddler enclosure. There I placed a 2 inch PVC cap with a ¼ inch hex bolt, washer and nut as a support for the base of the mast. Those 4 foot antenna supports you find on sale at Dayton and other swaps fit in this nicely and it keeps the mast from slipping out from under you.

Figure 71 shows the finished setup with a surplus DIRECTV dish and my 10 GHz Down East Microwave (**www.downeastmicrowave.com**) transverter. When this picture was taken the wind gusts were around 15-20 mi/hr and it wasn't moving. The pipe used as the mast is a surplus piece of galvanized gas pipe that I had been given.

So far it is a sturdy setup with plenty of flexibility for other things like the state QSO Party, a trip to a scenic operating position or on top of a mountain that you can only access by foot or bicycle. My plans are to add a battery box for an undetermined size battery and "maybe" an operating table or storage box. I may add outriggers to improve stability. The main idea was to keep it as mobile as possible but still be able to carry it on a car bike rack or stow it in the car if need be. The total cost for this including the frame and tripod has been about $65. — *73, James French, W8ISS, 1811 Horger St, Lincoln Park, MI 48146*, **w8iss@wideopenwest.com**

## RIG WARMER

◊ Some time back I made the mistake of putting my HT-2400 handheld transceiver in the glove box for safe keeping. I forgot about it and left it there over an atypically cold Ohio winter weekend. As you might guess the liquid crystal display "rainbowed" and turned black from the below freezing weather, which was an expensive lesson learned. Now the question was how to keep the weather from getting into the radio?

I found a foam lined aluminum case that fits both my handheld transceiver and mobile rig quite nicely. For those times I can't take either with me, I break open one of those hand warmers that hunters use and put it in the case with the radios. You can pick them up at any sporting goods store. They are non-toxic and produce enough heat to keep the radio(s) happy and protected from the cold weather.

The case fits easily under the front seat or anywhere out of site from prying eyes until I get back in the car. Once the hand warmer runs out of heat merely dispose of it. The trick is to not forget you left them under the seat in the first place. — *73, Gregg Gary, WB8YYS, 3775 Kauffman Rd, Stow, OH 44224*, **wb8yys@arrl.net**

# Chapter 4 Software and Computers

## A VOX CIRCUIT FOR PSK31

### *Design Goal: PSK31 Operation with a Laptop Computer*

◊ Many laptops today no longer feature legacy serial (EIA-232) ports. Their built-in sound cards, however, make for a very small terminal, a quality I specially appreciate in mobile operation. The VOX circuit described here generates the PTT signal to key the transmitter. In addition, I found that the sound card signal is rather large and has, even at small levels, a tendency to overdrive the transmitter. The transmit signal from the sound card is therefore attenuated by a factor of about 10 for reduced transmit drive.

### *Circuit Description*

The circuit uses three op amps (see Figure 1). First, the transmit signal is ac coupled. Then it is rectified with a gain of 10, and the resulting positive voltage is applied to the second op amp. The signal is integrated with a time constant of 0.27 seconds and at the same time amplified by a factor of about 5. The output of the integrator is approximately one half of the supply voltage on receive and close to ground on transmit. The third op amp is configured as a comparator, with its output close to ground on receive and near the positive supply voltage on transmit. A capacitor across the feedback resistor eliminates switching transients.

The output is then applied to an NPN switching transistor that keys the transmitter. The bias setup for the op amps allows for large part tolerances. It is for amplifier one and two one-half of the supply voltage and for the comparator one fourth of the supply voltage. None of the amplifiers are subjected to a common mode latch up. A diode at the collector of the NPN transistor catches relay-switching transients, while a diode in the positive power-supply lead prevents operation if the power supply polarity is reversed. The ac coupling allows for very low transmit frequencies, down to 200 Hz.

### *Implementation*

I used a quad low power operational amplifier IC, and I've tested several of the popular brands: A National LM-324 (bipolar), as well as a TI-274 (MOS) were handy. Either one works satisfactorily. In order to keep the package small, I used surface-mount parts soldered to a 9162 "surfboard." Discrete parts can be used just as well, but they will result in a slightly larger unit.

### *Performance*

The circuit was tested with three radios: Kenwood TS-570D and TS-50 and my mobile rig (a Yaesu FT-817. The VOX circuit was tested independently over a voltage range from 6 to 16 V, where it performed well. It simplifies the hookup to the transceiver (once the unit is built, of course). The Kenwood TS-570D has an auxiliary dc output to feed the VOX circuit, while I intended to use a 9-V battery when operating with the Yaesu FT-817. A PS2 mouse cable works great to provide the interface between the Yaesu FT-817 data output and the VOX box. The VOX box power consumption is minimal: 1 mA on receive and slightly more on transmit. For keying, modern transmitters typically sink very little current, so to some degree, the NPN transistor is overkill; the fourth IC could handle it very nicely. Should something go wrong, however, it is much easier to replace an NPN transistor than a whole surface-mount IC.—*Walter Kaelin, KB6BT, 1500 NE 15th Ave Apt 349, Portland, OR 97232-4417;* **kb6bt@arrl.net**

## MORE ON DATA SWITCHES

◊ I have connected a data transfer switch to my ICOM IC-7800. This enables me to use my computer monitor when I am working RTTY. This means that I do not need an extra monitor on my operating desk as mentioned in the Product Review of the IC-7800 (Mar 2007, p 62). A further advantage of this switch is that I can quickly view the spectrum over 3 kHz when the monitor is switched to the computer.

VGA monitor extension cables will be required to connect the 15-pin T switch and gender changers may be needed. The pole of the switch connects to the monitor and the computer. The video output from the IC-7800 connects to the switch outputs.

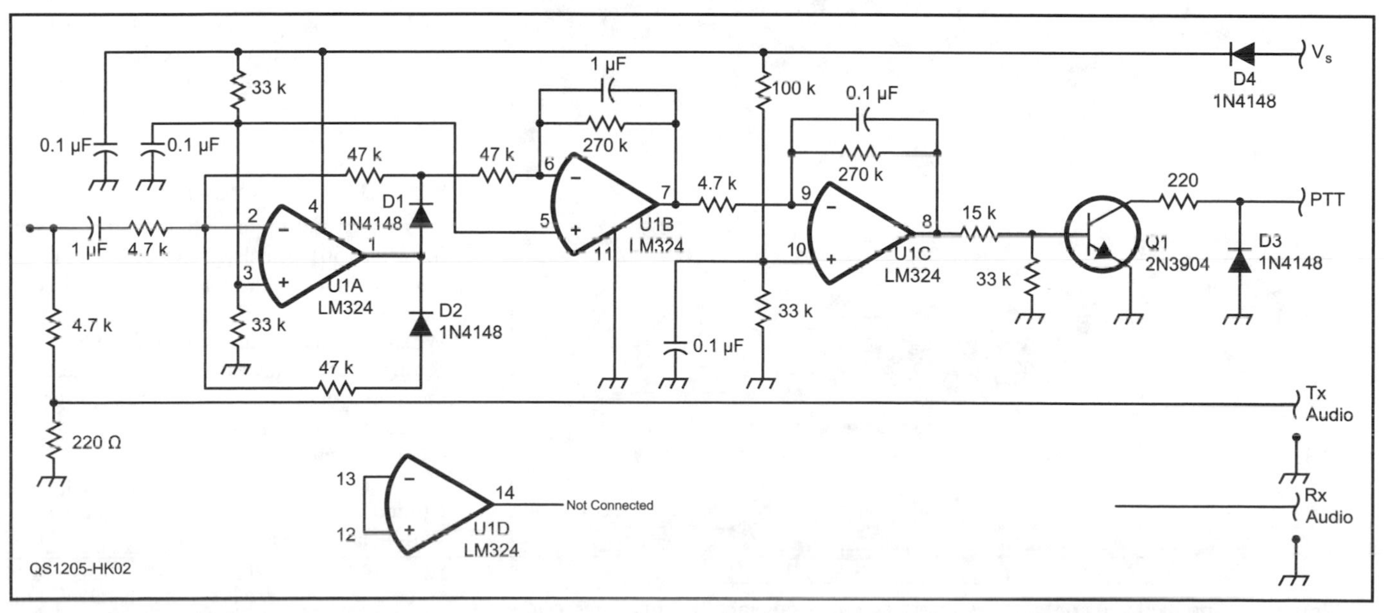


**Figure 1—KB6BT's PSK31 VOX circuit.**

FRANK E. WYER, G8RY

**Figure 2 — Using a data switch to interface a flat-screen monitor to the ICOM IC-7800.**

Figure 2 shows the switch to the right of the RIGblaster Pro in the Display position. *— 73, Frank E. Wyer, G8RY, 23 Sheriffs Ct, Burrough Green Newmarket, Suffolk CB8 9NJ, United Kingdom,* **frank@wyerg8ry.freeserve.co.uk**

## USB TO ICOM TRANSCEIVER INTERFACE

◊ Adapting equipment for Amateur Radio use is a long-standing ham radio tradition. So I was intrigued when I learned about a RadioShack Scanner Programming Cable that could be used to connect my ICOM IC-7000 to the shack computer and provide the hardware interface required by various remote control software programs. I now have computer-controlled radio at KQ4BY!

### The $29.95 Interface Solution

I discovered this programming cable while monitoring a long thread on transceiver control from the Yahoo IC-7000 Group (**groups.yahoo.com/group/ic7000/**). Since my shack computer has USB ports, when I read a message describing success with the Full-Speed USB to Scanner Programming Cable, part number 20-047, I was motivated to plunk down my $29.95 plus tax and bring one home to try.

The 20-047 cable is designed to interface RadioShack's popular scanners to a personal computer Universal Serial Bus (USB) port and allow the scanners to be programmed using software available from third party vendors. Fortunately for us ICOM owners, those scanners require the same voltage-level conversions as our transceivers! As you can see in Figure 3, the package contains the interface cable assembly, a stereo to mono converter (required for my IC-7000), a driver CD and instructions to install the drivers.

### Hardware Installation

I installed the drivers, per the instructions, on my Compaq laptop running *Windows XP*. The only trick is to make sure that Windows does not install the Microsoft-provided drivers for the USB serial converter and the USB serial port. So, after my computer reported that the installation was complete, I opened the *Device Manager*, found the "USB Serial Converter" device and the "USB Serial Port (COM5)" device and verified that the "Driver Provider" entries read "FTDI." With the installation complete and verified, I connected the radio end of the cable to the CI-V Remote Control Jack on the rear panel of the IC-7000.

### Adding Software

I had been wanting to try the *Ham Radio Deluxe* (HRD) CAT control software, by Simon Brown, HB9DRV, since reading about it in "Short Takes" [*QST*, Apr 2007, p 56]. So, I dug out the April issue, found the Web site (**www.ham-radio-deluxe.com**) and downloaded and installed the latest version. When I clicked on the program icon, the "Connect" wizard was displayed. From the "Company" tab, I chose "ICOM" and then selected "IC7000" from the "Radio" tab. I noted the default settings (COM Port = Auto-Detect, Speed = 19200, CI-V Add = 70, DTR = not checked, RTS = not checked) and clicked on "Connect." After a short pause, I noticed that the red and green lights on the interface were blinking and the Status window displayed the connect messages. A moment later, the HRD window changed to display the data from my radio! I disconnected and then reconnected to verify that the COM Port entry had changed to "COM5."

I also downloaded and tested the *TRX Manager* control and logging program (demo version), by Laurent Labourie, F6DEX, from **www.trx-manager.com**. I had to manually configure the interface settings and I used the information that I had recorded when I set up the HRD software. After that, the program connected to the IC-7000 and I was able to explore the functions of that software package also. In short order, I downloaded, installed and tested the *CI-V CAT* program/memory manager software, by Jr Spolestra, PE1BYW, from **www.tellina.nl/software** and the *IC-7000BKT ICOM IC-7000 CAT Control* software (free version), by Mauro Capelli, IZ2BKT, from **iz2bkt.altervista.org/ic7000bkt_en.htm**, as well. Like *TRX Manager*, I had to input the settings before the programs would connect and allow the software to control the radio.

### Hardware Test Complete!

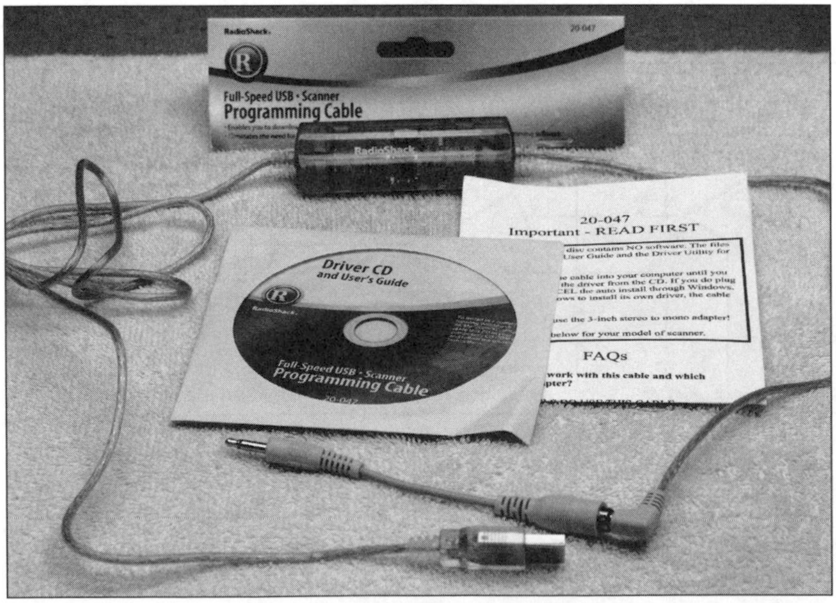

So, I consider my RadioShack interface cable to be a keeper! Now, all I have to do is figure out which of those excellent software packages to use. *— 73, Larry Keith, KQ4BY, 231 Shenandoah Trl, Warner Robins, GA 31088-6289,* **kq4by@rocketmail.com**

## SOUND CARD CAVEAT

◊ When I went to buy a new laptop, a Dell Inspiron 1501, I insisted that it had to have a stereo input sound card and port. I was assured that the 1501 did have such a card. When the computer arrived and I had finished setting up all the necessities, I started testing the sound card with a free audio program called *Audacity*. I soon found no way to record in stereo.

After a week on the phone to Dell support I found out that the card as supplied was a stereo card with a stereo input port but the software *drivers* supplied were only mono. Once the stereo *drivers* were installed and *Audacity* set to record in stereo it was able

LARRY KEITH, KQ4BY

**Figure 3 — The package contains the interface cable assembly, a 3 inch stereo to mono adapter, a driver CD and instructions to install the drivers.**

to record either microphone or line input in stereo.

So, for all those authors such as Rich, W3OSS ("Hendricks QRP Kits FireFly Transceiver," Sep 2007, p 65) who have laptops and can't get it to use stereo input, all they may have to do is find the correct *driver*(s) for the sound card in the laptop. I'd contact the manufacturer to find out for sure if the sound card is stereo, but if you can at least get the make and model of sound card used (Most manufacturers do not make their own sound cards. They use some brand and simply "plug" it into their motherboards.) then you stand a chance of finding out if the card is stereo or not. Then the manufacturer's Web site should have the stereo *drivers*. Good luck — *73, Phil Karras, KE3FL, 3305 Hampton Ct, Mount Airy, MD 21771-7201,* **ke3fl@comcast.net**

## WINDOWS SOUNDS ON DIGITAL FREQUENCIES

◊ As you listen on the various digital frequencies you will hear the familiar *Windows* startup and shutdown chimes. I just happened to find the following tip in AA5AU's "Getting Started on RTTY" Web page (**aa5au.com/gettingstarted/rtty_start5.htm**) that may be a useful reminder to *Windows* users on how to ditch the *Windows* sounds. One way to keep *Windows* sounds from keying your radio is to turn *Windows* sounds off. You do this by going to the *Windows* Control Panel and then to Sounds. Under Schemes, choose "No Sounds." This does not totally eliminate sounds generated by your computer. For instance, the beep sound used in many programs will still be generated. — *73, Dick Kriss, AA5VU, 904 Dartmoor Dr, Austin, TX 78746-5163,* **aa5vu@arrl.net**

## ECOHLINK IN A FLASH

◊ I like to be able to connect to our local EchoLink repeater when out of town, but I don't have a laptop to take with me and I don't want to install the software on other people's computers. To overcome these problems, I downloaded the EchoLink software and then I copied it from the program files folder on my computer's hard drive to my flash drive where it takes up 3.12 MB of space. This way, the software is on both the computer and the flash drive. When I use someone else's computer, I merely run EchoLink off of my flash drive. When running Echo-link on a borrowed computer there may be some interface problems when initially starting up the program but these should be simple to correct. — *73, Alex Bates, KI4UYF, 108 Mill Stone Dr, Dothan, AL 36305,* **ki4uyf@arrl.net**

## PSK31 FOR AUDIO MONITORING

◊ Many of the new rigs have the capability to listen to transmitted audio for setting up levels, compression, etc. The problem is that listening to your audio while speaking makes it difficult to judge differences in settings. To overcome this I use my PSK31 interface to connect to the *sound recorder* of my computer. This eliminates the problem of speaking and listening at the same time. When doing this, I actually state into the recording the various levels I have the transmitter set to, then reset them to another value and make another recording. This permits me to listen to the audio files and know what is going out on the air, with different audio levels and settings.

The PSK31 interface is also useful when someone asks how their audio sounds. I simply record it with the *sound recorder* and play it back for them using the PSK31 interface. I often do this and then e-mail them the file. This permits them to hear exactly what I heard without the effect of my rig's audio on the recorded audio I play back. — *73, Steve Ray, K4JPN, 104 Wendell Ct, Warner Robins, GA 31093-1035,* **sbralr@cox.net**

## RS-232 TRANSMIT CONTROL

◊ Here is a method of controlling a transmitter when using digital soundcard modes (see Figure 4).

I am using a PCMCIA RS-232 card in my laptop, as the computer does not have a built-in serial port. The circuit may have to be "adjusted" if your serial port is very different from mine. RS-232 outputs vary.

The output of the Request To Send (RTS) pin changes from about +6 V dc on transmit to –6 V dc on receive. The LED "rectifies" this voltage so there is only forward current (about 12 mA with my adapter) on transmit, sufficient to operate the relay. It also provides a "free" on-the-air light. If the light is not desired you can use a 1N34 diode for D1. The entire circuit will fit into the adapter hood for the connector.

Pin numbers are given for DB9 and DB25 serial ports for RTS control of the push to talk (PTT) transmit control. Other serial functions can be used by changing to the appropriate pins. This system works for me using the *MULTIPSK, EASYPAL* and the Digital Master portion of *Ham Radio Deluxe*. The needed parts are:

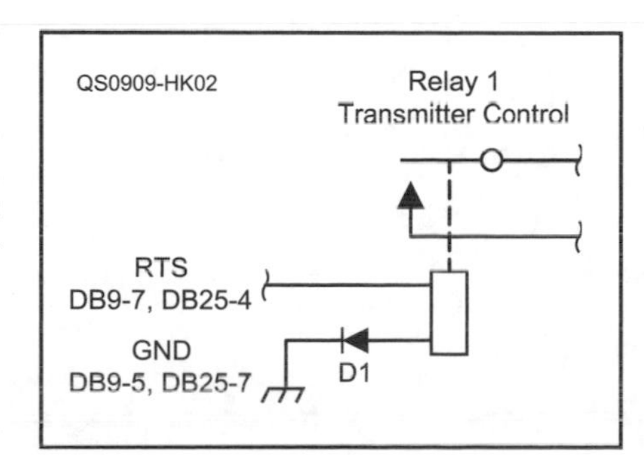


**Figure 4 — The diagram of the RS-232 transmit control circuit.**

D1 — 1.8 V dc red LED 20 mA (Radio Shack 276-0330) or 1N34 diode

RY1 — 5 V dc 250 Ω, 20 mA (Radio Shack 275-232) — *73, Scott McCann, W3MEO, 160 Shields Ln, Queenstown, MD 21658-1278,* **achess@juno.com**

## *WINDOWS* SOUNDS — ANOTHER APPROACH

◊ Rather than get rid of the *Windows* sounds that are produced by the operating system and other pieces of software that annoy the average digital operator, let them be heard — by yourself. Many people have several pieces of software on their computer that alert them to what is going on and don't want to mute the sounds in fear of missing an alert that may be of true importance.

The easiest way to fix this issue is to have two separate sound cards on your computer — one a default soundcard for all your *Windows* sounds and another that is dedicated to your HF rig. This instantly solves the headache of changing cables around and swapping settings everytime you want to run a digital mode. *Windows* and several digital packages allow you to select which sound card receives certain sounds. This is configured in the *Windows Control Panel* and in the settings of your digital software. Most computers that were built to run *Windows XP* and *Vista* have multiple USB ports. This allows the digital user to add an external sound card to a USB port. These can be found on the Internet for under $10. This is a simple fix that will keep the world from knowing when you have new e-mail. — *73, Dave Eagle, KB8NNU, 3780 Leeside Ln, Traverse City, MI 49686-8923,* **kb8nnu@arrl.net**

## MORE ON RS-232 TRANSMIT CONTROL

◊ In response to a hint on using RS-232 interfaces for transmitter control, Ralph Dieter, K1RD, wrote in to clarify some issues.

First, he points out that there is no guarantee that the signals will be + or – 6 V. In his experience he often sees +5 and –0.7 V interface signal levels with currents ranging from 40 mA down to 2 mA.

A second item that he points out is that the 1N34 diode is a germanium small signal diode, used mainly in receiver detectors. It typically has a peak inverse voltage (PIV) rating of 60 V. In some cases the inductive kickback generated by a 5 V relay may exceed this value. Your editor reviewed diode data and the 1N4445, a silicon signal diode with a PIV of 100 V, might be a good alternate.

Ralph's third point is that most RS-232 drivers are not designed to source large amounts of current. Your editor did a quick survey of drivers and found that typical source currents varied from 10 to 40 mA.

All in all, Ralph makes some noteworthy points. The bottom line of it all is that if you want to use Scott's circuit in your shack, you do need to do some homework on what the capabilities of your specific interface are and then modify Scott's design to accommodate them.

### ONLINE RINGTONE GENERATOR

◊On my commute in to work this morning I was reading through my May *QST* and read the submission from Greg about the CQ ringtone.[2] An alternative that might be easier for folks is to go to **www.planetofnoise.com/midi/morse2mid.php** and enter in whatever text you want converted to Morse code. You can adjust the speed, pitch and sound of the code and you can save and play it on your computer. It also provides a direct URL to a file, so if you have a browser enabled phone (such as the Blackberry referenced in the magazine) you can go to that URL, save the file and then make it your own ringtone. Depending on the features of your phone, you can have different ringtones assigned to different contacts in your address book. The Web site above is free and there's no cost to generating as many ringtones as you like. *— 73, David Levine, K2DSL, 11 Mackay Ave, Waldwick, NJ 07463-1909,* **k2dsl@arrl.net**

[2] G. Tyre, "CQ Ringtone," *QST*, May 2009, pp 65-66.

### MFJ ANTENNA ANALYZER HINTS

◊The January 2007 Hints & Kinks column suggested a way to protect MFJ Antenna Analyzers from accidental operation. I have another suggestion that doesn't deface the instrument. Simply put a coaxial type power connector in the external power supply socket. This disconnects the batteries. Any 2.1 or 2.5 mm ID connector with a 5.0 or 5.5 mm OD will do the job. I use a right angle plug, which provides a sort of handle for easy removal. — *73, John Roubie, K2JDD, 7205 Coventry Rd N, East Syracuse, NY 13057*

◊ When I bought an MFJ-269 Antenna Analyzer, I also purchased the matching MFJ-39C carrying case because I knew that this very useful instrument would get a lot of handling, both inside my shack and out. The carrying case is padded, and has a single large clear plastic window to allow full view of the analog and digital displays. Two punched holes are provided to fit over and under the TUNE and FREQUENCY knobs. If the case is fitted under these two knobs, removal of the analyzer from the case — to service the batteries, for example — is going to be difficult because the padded case is now under the knobs. In addition to this, the case completely covers the FREQUENCY range markings, something that I could just not live with! See Figure 1. Yes, I know I can read out the frequency range on the digital display, but that did not satisfy me. After considering several possible solutions to this dilemma, this is what I decided to do. [It looks to me like the MFJ-259 Analyzer and MFJ-29C case will have the same problem. — *Ed.*]

**Figure 1 — KO8S modified the carrying case by cutting an opening to show the FREQUENCY and TUNE labels on his MFJ-269 Antenna Analyzer.**

### IMPROVED COOLING FOR THE MFJ-267 DUMMY LOAD/WATTMETER

◊ The MFJ-267 Dummy Load/Wattmeter is a nice station accessory that provides in-line SWR, average and peak power readings up to 3000 W in two ranges — plus it includes a legal limit dummy load that can be switched in-line. While the dummy load is rated at 100 W for 30 minutes and 1500 W for 10 seconds, it takes a long time for the dummy load to cool off. This means you must wait quite awhile between tests — especially at higher power levels. For example, while recently running tests with my ALS-600 amplifier at 600 W into the dummy load, the MFJ-267 would get very hot after just a few repetitive 10 to 15 second tests. I would then have to wait a good 5 minutes for the unit to cool down enough for me to feel comfortable applying full power again.

I solved the heat problem by adding an inexpensive surplus fan to the dummy load. I chose the All Electronics (**www.allelectronics.com**) CF-252 40 mm fan. I removed one of the #6 sheet metal screws on the dummy load shield and tapped this hole for a #6 machine screw. I used a single 1.25-inch long #6 screw, two nuts and two lockwashers to mount the fan to the dummy load shield, as you can see in Figure 2. Orient the fan so that it blows air into the dummy load.

The MFJ-267 SWR/Wattmeter is active whenever 12 V dc is applied. The ON/OFF switch just turns the meter lamps on and off. So you can meter SWR, peak and average power regardless of the ON/OFF switch position. Therefore, I wired the fan's dc input to the on/off switch such that the fan comes on only when it is turned on. This eliminates fan noise during normal use of the metering functions, but lets you turn on the fan as desired when using the dummy load. The correct place to connect the fan's plus (+) wire is to the lower center terminal on the power switch.

The impact of the fan is dramatic! With it running I have no problem running my ALS-600 at full output for testing (usually 10-20 seconds at a time), and I no longer have to wait at all between transmissions for the dummy load to cool off. — *73, Phil Salas, AD5X, 1517 Creekside Dr, Richardson, TX 75081;* **ad5x@arrl.net**

**Figure 2 — Fan mounting details.**

### LURKING BEHIND THE GRAY DOOR

◊Soon after Hurricane Rita in 2005 my wife and I made the decision to buy a larger generator, since we had been without power for some time and our 5 kW generator wouldn't run the air conditioner in the sweltering heat. When the new 15 kW automatic generator arrived, I called Steve, an electrical contractor and a close friend, to wire it in for us.

Like everyone we have an electrical panel. Ours is in the back bedroom closet. It is seldom opened except to reset a breaker (or to turn one off) or to change a switch or outlet. The panel remains out of sight. Steve opened the door and removed the cover over the breakers. What was lurking behind that gray door sent chills down my spine!

Hiding in the darkness was a disaster just waiting to happen. One of the wires coming into the box from the pole was covered with green copper oxide. The added resistance had caused the connection to become very hot, melting and burning the insulation. After disconnecting the power, Steve removed the wire and the 200 A main breaker. The back of the breaker was burnt and crumbling. Underneath the breaker, the plastic panel had melted with a hole showing the metal back of the electrical box behind it. Pieces off burnt plastic lay in

TERRY D. COFFMAN, KC5NAC

**Figure 3—Plastic panel holding the breakers in the metal box. Notice that a hole is burned all the way through the plastic.**

DAVID KOH, KE1LY

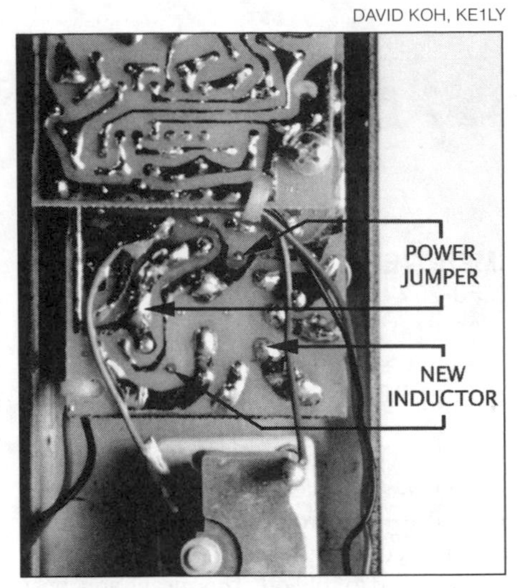


**Figure 4 — Bottom view of MFJ-207 PCB showing where connections for new inductor and new power switch are made.**

DAVID KOH, KE1LY

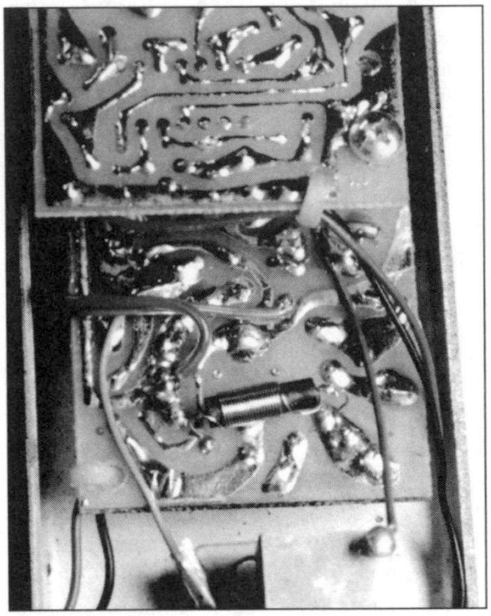

**Figure 5 — Finished view of MFJ-207 modification for 60 meter coverage.**

the bottom of the box. See Figure 3.

How long had it been like this? How long before it shorted out above the main breaker and our house caught fire? No one knows! We had not had any electrical problems. Nor had we smelled anything funny or had any clue that something was amiss. Had we not purchased a larger generator…who knows?

A loose connection was probably the culprit. This sort of thing could happen to anyone in a house or apartment. Everyone reading this article should check his or her electrical panel. Do not assume that everything is okay because the lights come on when you flip a switch. An electrician will charge you for a service call, which is a bargain compared to having your house burn to the ground. — *73, Terry D. Coffman, KC5NAC, 3040 FM 2109, Huntington, TX 75949,* **tdc9@valornet.com**

## ADDING 60 METERS TO THE MFJ-207 SWR ANALYZER

◊The 60 meter band requires a new level of care because of the unique operating requirements on that band, including channelization and a strict radiated power limit. Adjusting your antenna on-the-air is not viewed favorably, and this is where SWR Analyzers have become a real boon. An inexpensive and popular unit is the MFJ-207. Unfortunately, it does not cover the 60 meter band because of a gap in coverage from 5 to 6.5 MHz. The MFJ-209, MFJ-259 and MFJ-269 don't have this problem but these models are considerably more expensive.

Along with many MFJ products, a schematic was not supplied and is not readily available. Previously posted reviews and comments indicate that the SWR detector itself does not require tuning. Within wide limits it will respond to whatever frequency is generated by the exciter section. A bit of poking around revealed that the operating frequency is determined by a number of series-connected inductors. A bandswitch shorts out individual inductors. Since they are not inductively coupled there is no concern about creating shorted turns. An air variable capacitor does the tuning within each band. I wanted the modification to be simple and easily reversible. I also wanted to avoid sacrificing any other bands or changing the dial calibration for the existing bands.

The solution turned out to be straightforward. It involves creating a new band whose frequency is determined by adding a new inductor to the existing "C band," which presently covers 6.5 to 11.7 MHz, thereby pulling the frequency down to 3.7 to 6.4 MHz. The existing bandswitch is a 2-pole, 6-position affair. One half of the switch shorts out legs of the series-connected inductors and the other half serves to connect power to the circuit. The OFF position basically removes power and removes all shorts from the inductors. There is actually nothing connected to this position on either half of the switch. Adding a small inductor of 7.5 µH from the unused pin of the inductance side of the switch over to the C position pin will produce the new band. Obviously there is no dial calibration for this new band but the built-in frequency counter output allows for precise monitoring of the frequency (not a bad idea when working with this new channelized amateur band). When the new inductor is switched out of the circuit, it adds only a tiny amount of stray capacitance to the tuning circuit, not enough to change the existing dial calibration by even one percent.

Which position on the bandswitch accesses this new band? The OFF position! This leads to the final step of how to apply power to the circuit when using the new band. The simplest way is to solder a jumper from the unused pin on the power side of the bandswitch to the common trace connecting all of the other pins on that side of the switch. This means that the SWR analyzer is on all of the time — not a problem if you're using an ac adapter but not great for battery operation. Adding a conventional power switch means cutting a trace on the circuit board. Instead, I carefully soldered insulated wire leads (#22 to #26 works fine since supply current is only 45 mA) to the unused pin on the power half and to the common trace connecting all of the other pins on the power side of the bandswitch. See Figures 4 and 5. The other ends of the insulated wires are connected to a mini-toggle switch. When this switch is closed the unit is on and therefore energized in the new 60 meter position. When it is open the result is normal operation; that is, power is off in the OFF position and ON in any of the usual band positions. There is plenty of room to drill a small mounting hole for this switch near the accessory power jack. You could also leave the switch dangling from its leads routed through a case seam and mount it more permanently after the warranty period has expired.

I had a 10 µH choke from a RadioShack coil assortment. This had 25 turns on a 0.156 inch ferrite core. Removing 5 turns yielded an inductance of 7.7 µH. The closest common fixed value inductor is 6.8 µH, and this will also work. You can buy a small ferrite choke of this value from Mouser (part no. 542-77F6R8-RC) or from Digi-Key (part no. M8655-ND) for about a dollar (plus shipping charges). — *73, David Koh, KE1LY, 14 Shuman Rd, Marblehead, MA 01945-2744*; **ke1ly@nsradio.org**

## HOMEMADE OSCILLOSCOPE PROBES

◊ Following is a simple-to-make 100:1 resistive-input scope probe, which I prefer to the standard 10:1 capacitive probe for many (though not for all) purposes. Homemade resistive-divider oscilloscope probes can outperform standard low-capacitance probes

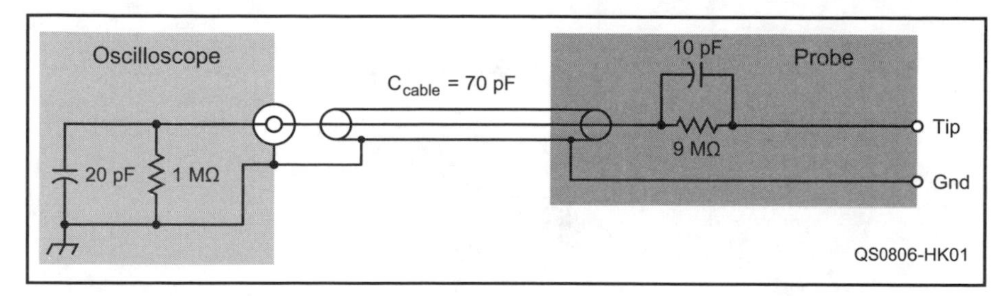


**Figure 6 — Circuit of oscilloscope input and traditional 10:1 low-capacitance probe.**

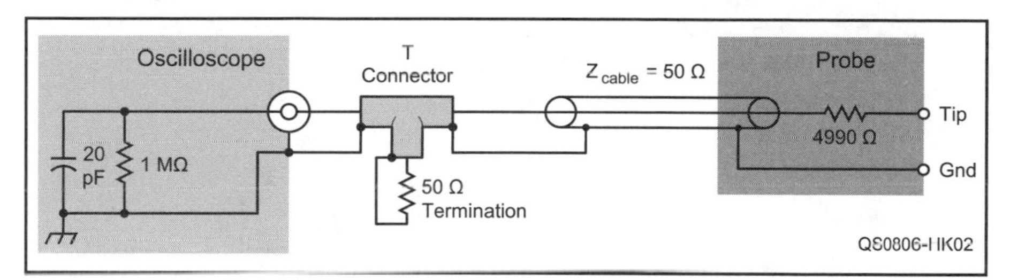


**Figure 7 — Circuit of oscilloscope input, 50 Ω termination, 100:1 resistive probe.**

in low-voltage applications.

I have been building my own scope probes for many years. These 100:1 probes have a 5000 Ω resistive input impedance, instead of the capacitive input impedance of conventional 10:1 low-capacitance scope probes. For modern low-voltage, low-impedance, high-speed circuitry, I find that the resistive probes work better than do the conventional 10:1 probes. Traditional 10:1 attenuation low-capacitance scope probes are resistive-capacitive voltage dividers as shown in Figure 6, giving an impedance at the probe tip of 10 MΩ resistance in parallel with a few picofarads of capacitance. The traditional probes were designed in the vacuum tube era, when resistances and voltages were much higher than they are in most modern equipment, and the high input resistance was essential.

Long ago, I found in a Tektronix catalog 100:1 probes intended for feeding signals into the 50 Ω input of a gigahertz-bandwidth oscilloscope, and adapted the design for lower frequencies. My probes are diagrammed in Figure 7. A 4990 Ω, 1⁄10 W resistor connects the probe input to the 50 Ω transmission line, and forms a 100:1 (actually 5040:50) resistive voltage divider between the input and the transmission line.

The other end of the transmission line is terminated in 50 Ω at the oscilloscope. I use a BNC T connector and a BNC 50 Ω load to accomplish this.

Theoretically, there are reflections from the oscilloscope input capacitance, but I have not noticed reflections in practice. Oscilloscopes fast enough to see these reflections normally have reflection-free 50 Ω inputs anyway. The resistor leads must be kept as short as possible, as their parasitic series inductance is in series with the resistor.

The probe is so simple, I sometimes hardwire one to a test point for best waveform fidelity. For general probing around, I build a probe into the plastic housing of a mechanical pencil. The coax comes out the eraser end, and one resistor lead sticks out the pencil-lead end. I omit the ground lead entirely, which makes for easier probing.

Using small-diameter coax cable, you can build probes into standard voltmeter probes, but the inductance of the probe tip degrades the high-frequency performance.

I don't bother with a ground lead when probing around, as the resistive input impedance suppresses ringing. I have found that these probes, without ground leads, show less ringing than the traditional low-capacitance probes show with ground leads. Built-in probes with ground leads don't ring at all.

You must keep the probe input away from high voltages, as just 20 V will take the resistor close to its dissipation limit. This is not a problem with modern circuitry running from lower voltages.

When you use the probe to check low-frequency signals, removing the 50 Ω load converts it into a 1:1 probe with a built-in low-pass filter. The filtering effect is the result of the cable and oscilliscope capacitance loading the 4990 Ω series resistance and exhibits a 1 MΩ resistive (in parallel with the cable and oscilloscope capacitances) input impedance at low frequencies. The input impedance is still 5000 Ω resistive at high frequencies, so the 1:1 probe is unlikely to make the tested circuitry oscillate. — 73, *Peter Anderson, KC1HR, 42 River St, Andover, MA 01810-5908,* **kc1hr@arrl.net**

## HANDY CONTINUITY TESTER

◊ Many of the problems the "handyman-ham" encounters around the home, shack, boat or automobile fall into the category of "Is there continuity?" Sure, one can pull out the multimeter and switch to the resistance mode to check the continuity of a circuit but a meter may not work in an awkward location. A garden variety code practice oscillator, such as described in many *QST* articles, offers a very useful alternative. Simply replace the connection to the key with a couple of tinned leads and use the tone from the oscillator as an indication of circuit continuity.

Quite often in transceiver hookups, we find that we have to either determine or verify the connections between a multicircuit plug or socket and the wires from the cable connected to it. Using the code practice oscillator to determine continuity is particularly valuable in this application because you do not have to be looking at a meter (or light) when attempting to also hold the wires on to the two ends of the circuit. You can keep your eyes on the connections and just listen for the tone from the oscillator as you search out the relationship between the socket pins and the wires. Of course, always check that the power is off when doing continuity checking. — *73, Ed Sack, W3NRG, 1780 Avenida Del Mundo Unit 40, Coronado, CA 92118-3002,* **esack@pacbell.net**

## MATCHING METERS TO THE MEASUREMENT

◊I recently needed an expanded scale voltmeter showing 12 to 15 V dc. Conservation of energy was not too important (it was part of a battery charging system) so I used the 0-1 mA meter I had on hand. The circuit consists of the meter (M1), a 12 V Zener diode (Z1) and a resistor (R1) in series. As voltage increased across the circuit the Zener diode (Z1) will conduct when the Zener voltage (12 V dc) is exceeded. This establishes the zero point. To set the full scale (15 V dc) point calculate as follows:

A) 15 V – 12 V = 3 V dc across R1 and M1 when there is 15 V across the whole circuit.

B) 3 V/0.001 A = 3000 Ω total resistance for M1 and R1.

C) 3000 Ω – 200 Ω (resistance of M1) = 2800 Ω for R1 (2700 Ω being the closest standard value.

The *ARRL Handbook* tells you how to determine your meter resistance or look it up in a catalogue; you can then do the calculations for your own meter and voltage range. If all else fails use a resistor substitution box or pot set at a very high value and adjust down until you get full deflection at the desired voltage. I calibrated the instrument by using an adjustable power supply and a digital voltmeter. You can make up a little chart giving new values indexed to the original meter scale numbers or use a scanner and computer to make a new meter face. — *73, Scott McCann, W3MEO, 160 Shields Ln, Queenstown, MD 21658,* **achess@juno.com**

## HANDY CONTINUITY TESTER

◊ Many of the problems the "handyman-ham" encounters around the home, shack, boat or automobile fall into the category of "Is there continuity?" Sure, one can pull out the multimeter and switch to the resistance mode to check the continuity of a circuit but a meter may not work in an awkward location. A garden variety code practice oscillator, such as described in many *QST* articles, offers a very useful alternative. Simply replace the connection to the key with a couple of tinned leads and use the tone from the oscillator as an indication of circuit continuity.

Quite often in transceiver hookups, we find that we have to either determine or verify the connections between a multicircuit plug or socket and the wires from the cable connected to it. Using the code practice oscillator to determine continuity is particularly valuable in this application because you do not have to be looking at a meter (or light) when attempting to also hold the wires on to the two ends of the circuit. You can keep your eyes on the connections and just listen for the tone from the oscillator as you search out the relationship between the socket pins and the wires. Of course, always check that the power is off when doing continuity checking. — *73, Ed Sack, W3NRG, 1780 Avenida Del Mundo Unit 40, Coronado, CA 92118-3002,* **esack@pacbell.net**

## ISOLATED IN-LINE RF TAP

◊ A couple years ago the Mid-MO ARC (**mmccs.com/mmarc**) was the recipient of a surplus Cushman CE-5 Service Monitor to help maintain our 2 meter repeater. The Cushman has been pulled into service in support of the Missouri Emergency Packet Network (MEPN) and its series of both 2 meter (local) and 6 meter (packet backbone) stations.

As received, the CE-5 included a high-band preselector that provides a narrow receive passband and added sensitivity for the receiver when operating in the VHF high-band. Unfortunately, no such preselector was available for the VHF low-band (6 meters). As a result, the CE-5 was deaf to the off-air signal from our 6 meter packet backbone transmitter, even with a short whip installed and the monitor directly (about 150 feet) beneath the 6 meter antenna.

I was able to force-feed RF to the CE-5 by direct-connecting the transmit coax (through a T-connector), but the level of RF input to the CE-5 would have been well over the rated RF input level to the front end. I had to find a better way.

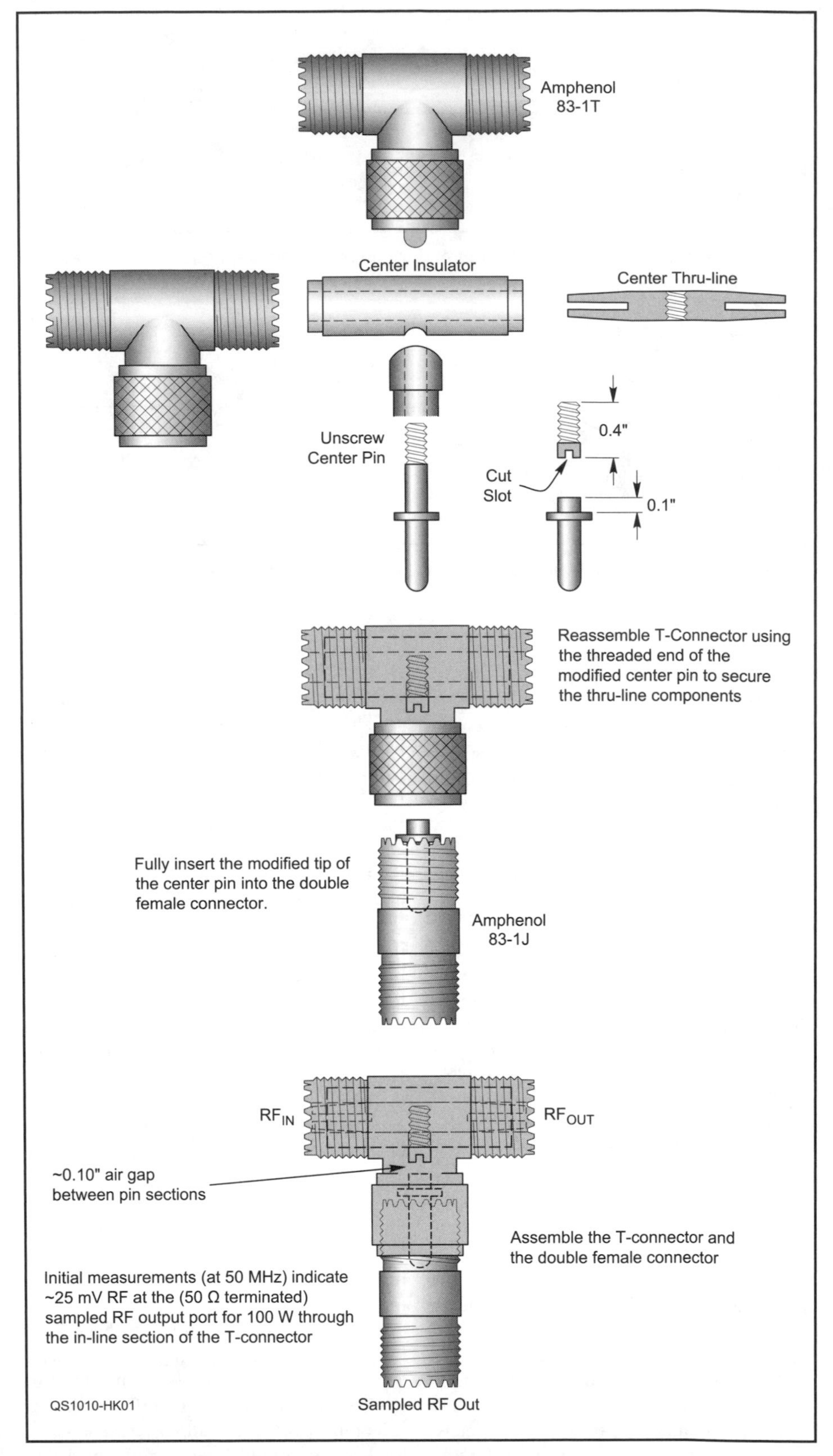


**Figure 8 — Construction drawing for the in-line tap modification.**

Then I remembered that about 30 years ago a good friend, Clarence "Coop" Cooper, KØLVR, used an "isolated RF tap" to feed RF from a signal generator *into* a device under test. If the RF tap that Coop used was suitable for getting RF *into* the signal line, it followed that it should be equally as good for getting RF *out* of the line as well.

To fabricate the isolated RF tap, I disassembled an Amphenol 83-1T T-connector into its five components (outer body, center thru-line, center T-pin and two clear center insulators).

Using my Dremel tool and a thin cutoff disk, I cut the center T-pin into two pieces as shown in Figure 8. Once I squared and deburred the cut ends of the T-pin, I cut a straight screwdriver slot in the threaded end of the T-pin to allow it to be firmly reattached when the T-connector was reassembled.

Using the cutoff disk in the Dremel tool allowed me to create a narrow (slightly less than 0.1 inch) air gap of removed metal between the two pieces of the T-pin when the T-connector was again reassembled and mated to the Amphenol 83-1J double female connector. This air gap provides more than ample RF coupling for picking RF off the thru-line.

Once the T-pin was modified as described above, I reassembled the T-connector, using all of the original parts *except* for the non-threaded tip of the T-pin. I used the screwdriver slot in the threaded end of the T-pin to ensure that everything was well secured.

I then inserted (fully) the non-threaded tip end of the T-pin into one end of the double-female connector and screwed that end of the double female connector into the body of the T-connector, ensuring that both connectors were firmly seated together. The resulting air gap allowed the CE-5 to be able to hear 6 meter RF from the transmitter, even at power output levels of 5 W or less. There was still not nearly enough RF to risk damaging the front end of the monitor at the 100 W level.

The Isolated RF Tap may be left in-line when not in use. I was not able to measure a significant degradation in SWR on the transmit line (through 150 MHz) either with or without the tap installed. — *73, Tom Hammond, NØSS, 5417 Scruggs Station Rd, Lohman, MO 65053*, **n0ss@arrl.net**

## RECALIBRATING WATTMETER ELEMENTS

◊Like many hams, I have a couple of wattmeters around the shack, but my standby is an old Bird 43. Recently I picked up a couple of elements from the local radio swap meet, one Bird and one Coaxial Dynamics. I found one registered about 12% high and one about 15% low. At first I put a sticker on the top stating the error but then decided that there must be a better way to make the two elements accurate.

I found an excellent article on the "repeater builder" Web site by Robert, WA1MIK, that gave full disassembly instructions for the element.[1] I realized that all I had to do was get the thin metal top label off the element and reset the calibration pot. As Mike noted in his article this is easier said than done. He recommended soaking the top of the element in a solvent. This did not work for me, so I borrowed my wife's hair dryer, which was able to soften the glue allowing me to pry off the cover. I then turned the pot to reset the calibration and put the cover back on. In less than 30 minutes I had the slug recalibrated. — *73, Harry M. Hammerling, W6HMH, 32115 Road 416, Coarsegold, CA 93614*, **harrymh@hammertech.com**

## IN-LINE FUSE FIX

◊While testing one of my mobile VHF/UHF transceivers on the bench, I found the power output was down 30% from normal and also varied very slightly. At first, I was concerned that either the power amplifier or driver was failing. As I continued testing, I noticed that one side of the in-line fuse in the power cord was noticeably warm, as was the connector at the far end.

I found the fuse holder used crimped connections. I soldered the wires taking care not to run any excess into the fuse connector. In addition, I took apart the power connector and also found crimped connections, which I also soldered. After reassembly, the power output returned to normal and was stable.

Most in-line fuse connectors are easy to disassemble. The typical T power connector takes a little more effort because the metal connectors have locking tabs. I found the tabs are easily compressed with a small, sharp tool and that the metal connector then slid out of its shell easily. The metal connector typically has ears and I recommend that you keep the solder from running past them.

I've had similar problems using cigarette lighter plugs with built-in fuses. Even when the current draw was less than half their rating, the plug body became quite warm. In those cases, I found that a compression spring at the plug tip was expected to apply enough pressure on the fuse to force a good mechanical connection at the top of the plug. It didn't. My solution was to solder the fuse into the circuit at the top of the plug. (I left as is the other end of the fuse where it slid into the tip with the spring.) After that, the cigarette lighter plugs were able to draw full rated current without noticeable warming or a material voltage drop. Cigarette lighter plugs are not the best solution to power equipment, but when you need to use them this fix should make them more reliable. — *73, Steve Glickstein, W4FMD, 3850 University Dr, Fairfax, VA 22030-2517*, **w4fmd@arrl.net**

## PEAK READING CIRCUIT FOR PASSIVE WATTMETERS

◊Many companies sell small peak-reading conversion boards for older passive wattmeters lacking this feature. These are easy-to-build circuits that usually require power, such as a battery or external wall wart supply. They can be wired to provide either the new peak-reading behavior or the standard unmodified wattmeter readings. The circuit is usually installed between the detector and the meter. The peak function is selected with a double-pole, double-throw (DPDT) switch. If you wish to also turn a battery on and off with the switch, a three-pole, double-throw switch is required — not an easy type of switch to find.

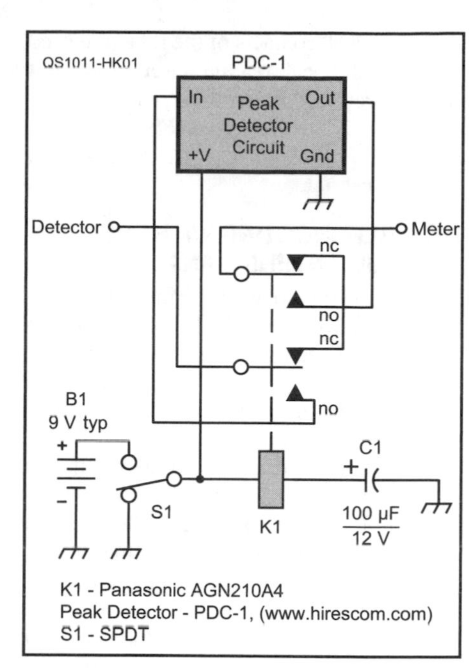


**Figure 9 — Schematic diagram of the peak detector circuit.**

Here is a simple circuit (see Figure 9) that reduces the switch requirement to a single-pole, double-throw (SPDT) switch. It allows the selection of peak or average readings and turns off the battery when the peak circuit is not in use. My circuit makes use of a latching relay. A latching relay is one that can be set to a certain state and then retains that state after the coil current is removed. They are polarized relays, meaning that to reverse the set state you simply reverse the direction of current in the coil. They typically require a short pulse of current to set or change state.

In this circuit, an SPDT switch is used to connect a latching relay coil to the battery or ground. There is a capacitor in series with the coil. The capacitor needs to be large enough so that the relay actuates. When the switch is turned on, voltage is applied to the coil, charging the capacitor and activating the relay. When the switch is set to the off position, the capacitor discharges through the coil. Because the current is flowing in the opposite direction from turn-on, the latching relay switches to the other state. The advantage of using a latching relay is that no power is used by the relay while the circuit is on. An inexpensive and tiny latching relay is specified in the schematic, but any latching relay of appropriate voltage can be used — you may have to experiment with the value of C1.

The DPDT contacts of the relay are used to switch the peak reading circuit in or out of the circuit between the sensor and the meter. — *73, Tony Brock-Fisher, K1KP, 15 Webster St, Andover, MA 01810-1109,* **k1kp@arrl.net**

## TROUBLESHOOTING WITH THE MFJ-259B ANTENNA ANALYZER

◊While trying to work some HF DX, I noticed my RF output had dropped to zero and the SWR had gone sky high. Moments before, everything was working fine. There were no indications of a problem, such as smoke, arcing or noise.

I did the routine checking of jumper cables, connectors and the coax switch and found nothing out of place. I then started isolating the devices in the RF line one by one. When I came to my desktop amplifier the problem disappeared.

I put the amplifier on the workbench and did a quick ohmmeter check. The reading was okay. I removed the cover and everything looked normal. I wondered how I could duplicate the fault without turning on the line voltage.

I decided to use my antenna analyzer as an RF source and a dry type dummy load. As soon as I turned on the antenna analyzer the SWR went to the top of the scale. The only thing between the input and output connectors is the TR relay. On my amplifier, this relay is an open frame type and easy to inspect. The contacts looked okay; there were no binding or burned contacts.

While watching the meter, I used a wood probe to move things around. Suddenly the SWR dropped down to normal. I found one of the SO-239 connectors was bad. It had worn out from normal use and would not grip the pin on the PL-259. It would pass a dc voltage from my ohmmeter but would block the RF signal. I replaced both input and output SO-239s and the problem was solved.

Using the antenna analyzer made a dangerous job safe. This same method could be used in the entire RF line in your station, from the transceiver to your antenna. — *73, Phillip Mikula, WU8P, 6901 Hammond Ave SE, Apt B, Caledonia, MI 49316-7651,* **wu8p@arrl.net**

## MEASURING "FOUR SQUARE" DUMMY LOAD POWER

◊It has been well established that the only meaningful way to monitor the performance of a "Four Square" array fed with a hybrid coupler is to measure the power dissipated in the dummy load.

To measure the "dumped" power, I have placed the directional coupler from a wattmeter in series with the dummy load located at the center of the array and used long leads (200 feet) to the meter placed in the shack. There seems to be no power difference between readings at the remotely located meter and measurements taken near the terminating resistor.

I used spare conductors in the 10 conductor unshielded control cable (with ferrite chokes at both ends) to connect to the meter. The wires are all #18 AWG. I wound as many turns as possible on two stacked MFJ ferrite chokes at the coupler end of the cable. At the meter end in the shack, I wound a few turns of the cable onto a snap-on ferrite bead. The original cable supplied with the wattmeter is unshielded.

This configuration is much easier, and less costly, than locating the dummy load in the shack connected to the hybrid coupler with a long run of coax. John Devoldere, ON4UN, describes a similar arrangement using an RF detector and a voltage comparator circuit. — *73, Ira Lipton, WA2OAX, 96 Hemlock Ln, Liberty, NY 12754,* **wa2oax@toast.net**

## PA MODULE REPAIR

◊I was working on a VHF transceiver whose symptoms were loss of output power during transmission or no output at all. With an oscilloscope, I was able to determine that the power amplifier (PA) module had RF input and proper dc voltage on the input pins but no RF output.

Not wanting to spend $80 to replace it, I decided that investigating couldn't make it worse. With the module mounted in place, I placed a small screwdriver between the heat sink and one end of the module cover and twisted the screwdriver until I heard a faint snap. I then did the same at the other end of the cover. The cover popped off revealing the components inside.

Since the problem was intermittent, I reasoned that the components were good and the problem was a faulty connection. I first confirmed that there was voltage on the two input pins (while transmitting) and the collector of the output transistor. Next, I checked the collector of the driver and found *no* voltage. The only thing between the collector and the input voltage pin is a stripline inductor. I moved my scope probe along the stripline until I located the break. I confirmed this by checking for power output with the probe bridging the gap.

The final fix was using a fine tip soldering iron and a dot of solder to bridge the gap (see Figure 10). To prove the fix, I placed the transceiver in a freezer for an hour and successfully retested. I snapped the module cover back in place, using a very small amount of silicon adhesive to hold it. — *73, Donald Larkin, W8RVT, 630 Garrison Rd, Apt B, Battle Creek, MI 49017-4545,* **w8rvt@arrl.net**

## CHECKING INFRARED LEDs

◊A trick to verify if an IR LED is working is to point your cell phone or digital camera at the LED output. You should be able to see it glowing in your camera's display if the LED is working and connected properly. Use a TV remote to verify that your camera will work this way. — *73, Stanley Slaughter, N3LU, 213 Berwick Rd, Columbia, SC 29212-1902,* **n3lu@arrl.net**

## INLINE FUSE POWER PROBLEM

◊In the February 2009 "Hints and Kinks" column, George Peters, K1EHW, discusses problems with the Molex connectors used on the IC-706MKII and IC-706MKIIG radios.[2] I have seen similar problems with Molex connectors. Also, I once owned an IC-706MKII. While I did not have trouble with the Molex connectors on that rig, I did have problems with the fuse holders in the supplied power cord. I had shortened the power cord to 3 feet and installed Anderson Powerpole connectors. I connected it either to my vehicle battery or my deep-cycle battery with #10 solid copper wire soldered to gold plated ring terminals on the battery end and Anderson

DONALD LARKIN, W8RVT

**Figure 10 — The tip of the dental pick indicates the repaired gap in the stripline inductor.**

Powerpole connectors on the other end.

While the radio worked fine from a 13.8 V dc supply or in my vehicle with the engine running, with the engine off or running from a 12 V deep-cycle battery, I was having problems even when the batteries were fully charged. Upon investigating, I found that there was a combined voltage drop of 1.75 V across the fuse holders at full power transmit (100 W FM into a dummy load). The voltage was dropping below the minimum operating voltage during transmit. I replaced the factory fuse holders with inline spade type fuse holders and 20 A spade fuses. The combined voltage drop across the new fuse holders is 0.5 V.

Better results could possibly be obtained by using the gold-plated fuses and holders used for high power car stereo installations. — *73, Martin Campbell, KBØHAE, 218 N Gertrude St, Burlington, IA 52601-2830,* **kb0hae@arrl.net**

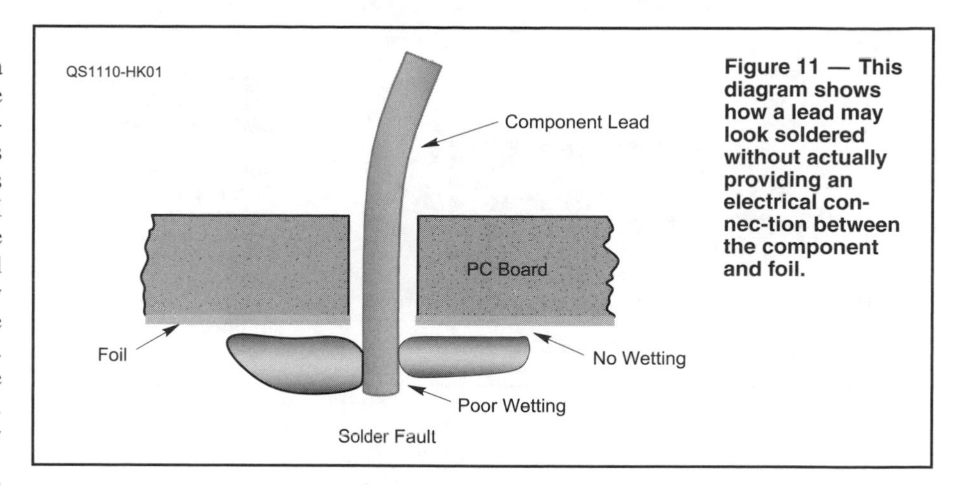

**Figure 11 — This diagram shows how a lead may look soldered without actually providing an electrical con-nec-tion between the component and foil.**

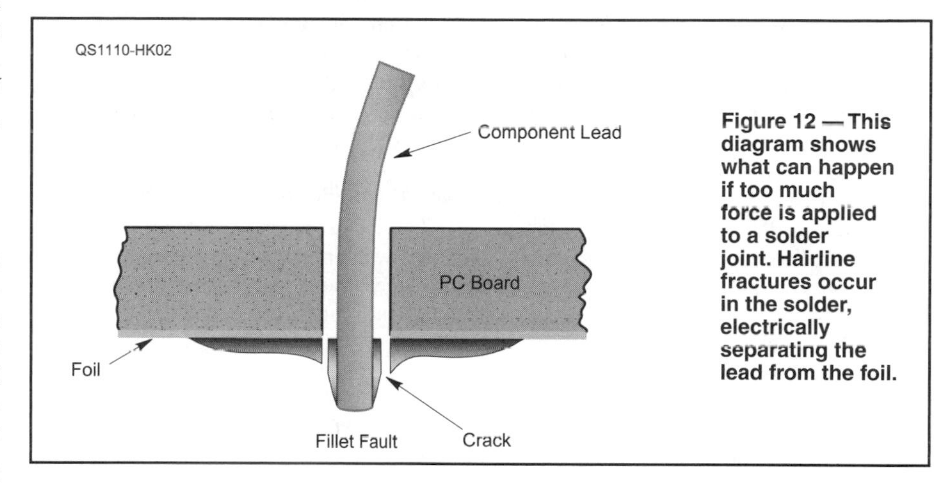

**Figure 12 — This diagram shows what can happen if too much force is applied to a solder joint. Hairline fractures occur in the solder, electrically separating the lead from the foil.**

## TROUBLESHOOTING WITH A PAINTBRUSH

◊I was pulling out my hair working on a pesky intermittent problem in a solid-state receiver. I had worked for over 3 hours trying to locate a loose connection on the circuit board. All the connections looked good when examined with a powerful magnifier but the signal was popping in and out at the slightest touch anywhere on the board. I walked away from it and after a while I had an idea.

I found a small paintbrush we bought for the grandkids and tried brushing each component. In about 5 minutes I had located the problem. It was a tiny capacitor that was not soldered to the board. Turning the board over I found that the solder had flowed right over the lead on the capacitor so it appeared to be soldered but was not. Resoldering the capacitor cured the intermittent problem thanks to a little kid's paintbrush. [For more troubleshooting information have a look at "Troubleshooting Radios" by Mal Eiselman, NC4L.][1] — *73, Bob Sumption, W9RAS, 61250 Cass Rd, Cassopolis, MI 49031*, **w9ras@arrl.net**

## SOLDER FAULTS

◊After building a new or kit circuit board, carefully inspecting it and bravely powering it up, if it fails to perform, troubleshooting is in the offing. Of course, if some parts burn and smoke, are in the wrong places or are left out, it may not be difficult to locate the defects. On the other hand, if everything looks good and it just doesn't work, some hard-to-see faults may be present. First use a magnifying glass to check for solder shorts. If there are none, you must delve a little deeper.

Probably the most common soldering failure is to properly "wet" the foil or component lead (see Figure 11). Note that the solder fillet around the lead is bulb-shaped, rather than being smoothly feathered at both the lead and foil. The solder may be adhering to the lead in a small area, but not the foil, or vice versa, so little or no electrical connection is made. There are several causes:

- Failure to apply enough heat to bring the foil and/or lead up to temperature.
- A film of dirt, oxide or coating may be present, making soldering difficult.
- The lead may be large and act as a heat sink. This can result in overheating the foil.
- Insufficient flux was present for good soldering.

### Broken Fillets

This fault may not appear until a finished board has been handled and installed, but it is very aggravating. It is difficult to see hairline cracks around component leads in otherwise good-looking solder fillets (see Figure 12). Note the cracks in what should be a well-soldered connection, where both the lead and foil have been wetted and the solder feathers out at the edges. Some causes for this fault are:

- Side pressure was applied to the lead, cracking the solder.
- The solder fillet is too small. The thickness of the solder should be at least equal to the lead diameter.
- Trimming the lead with a side cutter, when the cutter is resting on the board, can pull the lead hard enough to crack the solder.

### Surface Mount Device Faults

Looking at Figure 13, hairline cracks can occur at the terminals of surface-mount devices (SMD). They may be difficult to see, unless the device is broken at both ends and even then the break may not be obvious. This fault may be caused by:

- Bending of the soldered board, which stretches the surface-mounted devices, causing cracks in the solder or the device. If a 6 inch long board is arched a quarter-inch, it may damage SMDs.
- Failure to apply melted solder to both the SMD and the foil.
- Attempting to resolder SMDs, which often dissolves the metal end caps, especially when the solder does not contain silver. It is usually very difficult to reuse SMDs. Multilayered capacitors are especially susceptible, since each layer must be joined to the end caps.

[1]M. Eiselman, NC4L, "Troubleshooting Radios," *QST*, May 2009, pp 30-33.

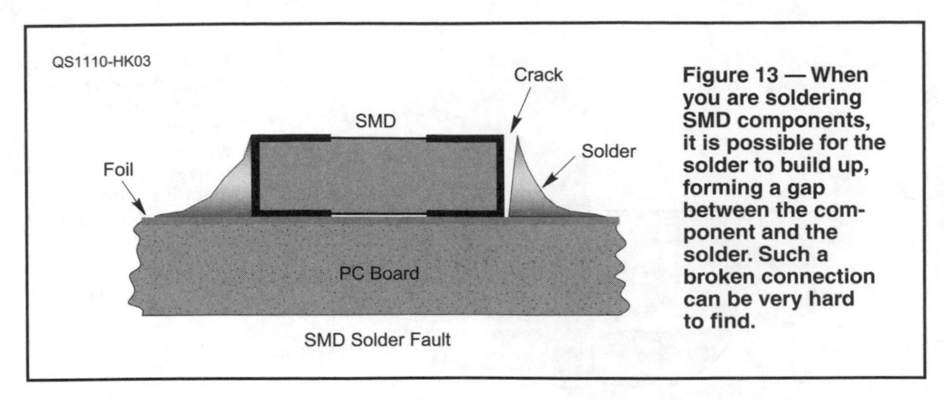


**Figure 13 — When you are soldering SMD components, it is possible for the solder to build up, forming a gap between the component and the solder. Such a broken connection can be very hard to find.**

### *Flat-pack Integrated Circuit Faults*

These ICs have their leads formed to lie flat on the foil pads. In order to avoid solder shorts between leads the foil is tinned first, flux is applied and the IC is held in place while the leads are heated, usually in groups. When a lead is not soldered firmly, it fails. Bad joints may be detected by carefully prying the leads sideways a bit to detect looseness. Carefully resolder loose connections.

### *Foil Cracks*

Hairline cracks in foil may be detected by shining a bright light on the foil side of the board and observing light leakage from the component side, provided that any solder resist lets light through. Foil breaks may be repaired by soldering a bit of wire across the break. Bridging the break with solder may fail if the board is flexed, stretching the solder patch.

### *Loose Pads*

Pads that have broken away from the board do not offer support to component leads and may be reattached with a bit of epoxy. If a pad is separated from its foil lead, a piece of small wire may be soldered around the lead and to the foil then glued with epoxy.

### *Missing Plated-through "Barrels"*

The inside surfaces of barrel holes are plated with copper on double-sided boards to connect the sides. If a barrel is missing, either the leads must be soldered on both sides or a piece of small wire may be installed on the component side foil and soldered to the lead before soldering the lead to the foil side. Care must be taken when removing leads and components to avoid pulling barrels. A solder-sucking tool is very handy for removing solder. If a multileaded component is to be removed, it takes a lot of care to avoid damage. If the leads are cut loose first, they can be removed one at a time with less board damage. — *73, Doyle Strandlund, W9NJD, 2849 N 035 W, Huntington, IN 46750-4012*

## AUDIO OSCILLATOR POWER TESTING

◊ I needed to do some testing on the power supply of a tube-type audio amplifier (fuse checked good but it showed dead when plugged in). Not wanting to apply 120 V ac or fiddle with my Variac, I simply set my audio oscillator for a low level 60 Hz sine wave output, connected it to the power plug of the amplifier and went on my merry way chasing down the problem without worry of a shock. Actually, there could be some non-trivial voltage developed across the secondary of the high voltage winding of the power transformer, so as harmless as this configuration might seem, some caution is still warranted.

The outcome probably deserves mention. When I removed the fuse to check continuity, I left it in the "cap" of the fuse holder as I always have in the past. The problem was that the small spring inside the cap that pushes the fuse into the mating contact was missing. Inserting the fuse into the amplifier never actually put it inline. Snipping a small piece of an ink pen spring and placing it inside the fuse cap fixed the problem. — *73, Jeff Bauman, W8KZW, 6647 Stonebridge E, West Bloomfield, MI 48322-3255*, **w8kzw@arrl.net**

## RESTORE DRAKE "B-LINE" -DIALS

◊ Recently, a friend brought a Drake "B-line" station to me for restoration. The receiver had moderate discoloration (Figure 1). The "phone" segment of the transmitter dial had the worst discoloration I'd seen in my 35 years of repairing and restoring radios. It was so bad that you could not see the black dial markings through the dark-brown pigment. I couldn't clean them because the discoloration had migrated into the plastic.

I had one good dial from a parts radio, but maybe there is a way to *replace* both dials at reasonable cost.

### *A Clear Idea*

I asked a local "plastic-house" to duplicate the dial, but they wanted a minimum order of 500 dials—*not* reasonable.

Then I remembered the clear discs on the top and bottom of the spindle in bulk CD packages. They're a little larger than the original disc, but when I put one into the radio, it fit!

Wow, what a concept! Maybe I could copy the dial onto clear sheet and somehow glue it to the clear CD. I went to the office-supply store, made about 10 copies of the clean dial and purchased a glue-stick. Unfortunately, the glue-stick adhesive turned white and made a mess.

Back at the store, I related my story to a woman and her young daughter, who said, "We use *spray* glue at school."

I bought a can and rushed home. It being winter, the basement was the only place to do the work (far from the furnace with plenty of ventilation). With flat cardboard to catch over-spray, I applied adhesive to the clear CD and clear sheet (printed side up, to receive the spray). I let them dry for few minutes and lay the sheet onto the CD. I used a soft rubber ball to roll any air bubbles from the center of the CD toward the circumference. When the glue was dry, I cut off the excess around the outside of the new CD-dial (see Figure 2). Now there is a new dial for one of the Drake radios.

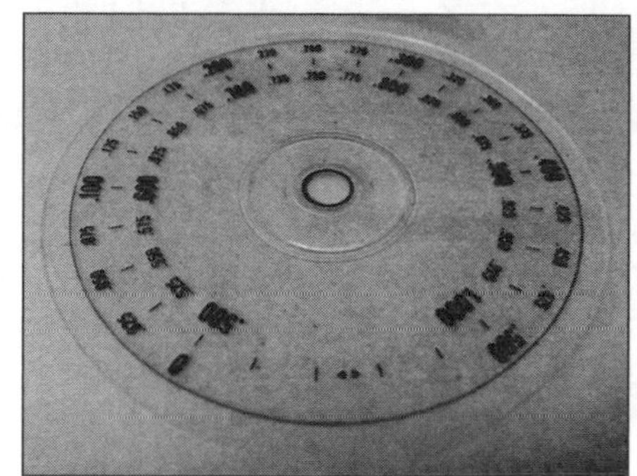

Figure 2—The clear sheet glued on the new dial.

### *A Computer Idea*

I had already scanned the "scrap radio" dial hoping to print its image on clear sheet. If you have a bad dial, use a computer graphics program to improve it or draw a new dial.

While I was looking for the spray glue, I walked by the printer supplies section and found another approach to all this. Clear CD labels come with "clear" sticky paper ready to apply!

### *Installing the New CD-Dial*

The CD-dial center hole is larger (⅝ inch) than the original Drake shaft (⅜ inch), so I placed the PTO in a vise on the bench with the dial shaft pointing upward. I laid the new CD-dial on a flat piece of spacer that lays loose against the back-plate of the PTO. The spacer is exactly the same size as the center of the CD-dial, so I used a very small amount of silicon glue to hold the CD-dial. I wanted to finish with the pre-bent washer installed, using the snap-ring to hold all in place. I made sure it would turn evenly, as I spun the dial shaft. The next step was to install the PTO assembly into the radio and make sure it would fit all around. It is larger, but I would have needed a lathe or a very sharp tool to cut it smoothly. As I mounted the PTO, there was a little rubbing of the new dial on the volume control. I made sure the three screws that held the PTO in place were loose, put leftward pressure on the unit, then tightened the screws. Finally, I installed some small flat washers under the PTO. They raise it enough to clear the volume control and let the PTO lamp be seen from the front. Wow, it fit! Next, the Drake dials always had a "white-lens" material behind the clear dial, along with a strip of Drake light-blue background, so the #47 lamp would light the whole dial with a very attractive light blue! Thank goodness for the white-lens material, which hides any imperfections in the glue or sticky material used to hold the calibration sheet on the new CD (see Figure 4).

Figure 4—The finished product at work.

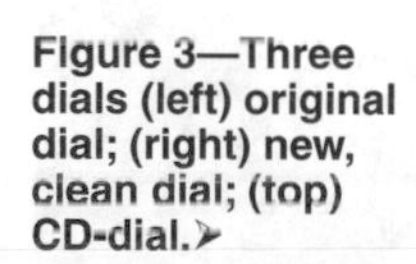

Figure 3—Three dials (left) original dial; (right) new, clean dial; (top) CD-dial.➤

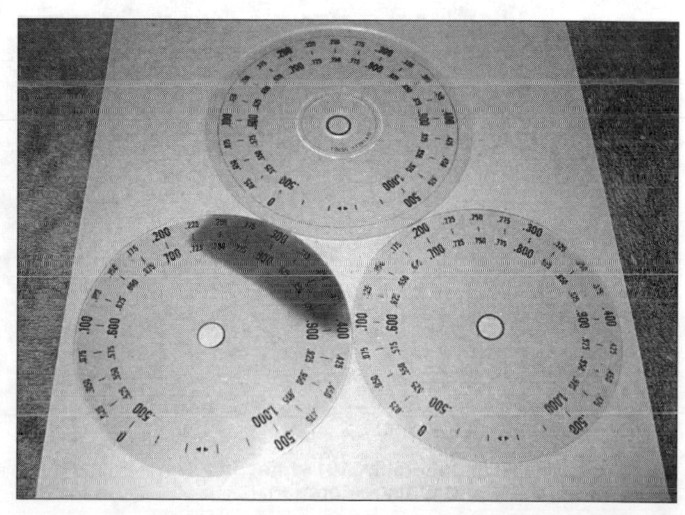

Figure 1—The receiver dial was only slightly discolored.

In closing, my friend came down to the basement expecting the worst. With the basement lights dimmed to show off the dials, his smile of pleasure was my reward.—*Jeff Covelli, WA8SAJ, 5368 Melody Ln, Willoughby, OH 44094;* **wa8saj@ncweb.com**

## RESTORATION HINTS

◊I have had quite a bit of trouble keeping the workbench clean during the initial clean up of older tube-type equipment. The clean up process can be very messy when removing old grease, grit and cigarette nicotine deposits. I urgently needed to contain the mess and not let it spill over into other projects. What I came up with is a sheet metal water heater drip pan with a pipe plug drain on one side (Figure 5). This allows containment of the cleaning process and a drain for residue. There's also enough room for an oven rack to hold the equipment off the pan bottom. Even though plastic drain pans are available, they may not be resistant to solvents, making them unusable as a cleaning tank.

The other issue I have to deal with is that I cannot pick up heavy items. I have a VA disability that prohibits heavy lifting and I am not about to give up my Hallicrafters HTs [the Hallicrafters HT series refers to vacuum-tube transmitters manufactured from the late 1940s to the early 1960s, not handheld transceivers — *Ed.*]. So I had to come up with a solution and ultimately found a hydraulic lift table on wheels that I can move around my work area. The lift table comes to nearly the same height as a work bench, which allows considerable flexibility in dealing with heavy items. Sliding heavy projects at waist height is very easy.

The lift table is a generic unit from Harbor Freight, item 93116-2VGA (**www.harborfreight.com**). The drip pan is from my local building supply dealer, City Lumber (**www.citylumber.com**). The price was slightly under $30 — *73, Jim Santee, KF7NE, 42162 Bagley Ln, Astoria, OR 97103-8416,* **kf7ne@arrl.net**

## PLEASE DON'T USE THIS TO CLEAN OLD RIGS!

◊ I just finished reading the ARRL's *Low Power Communication—The Art and Science of QRP,* and it's a great book, but in the section on vintage QRP rigs the author, Rich Arland, K7SZ, suggests using a liquid brass polish to clean and improve the gloss on some old radio cabinets. I believe he's referring to Brasso, which is a great cleaner but awfully harsh! I'd hesitate to use anything so strong on a vintage rig. Yet, sometimes there is some just plain old gunk, sticker glue, and such that must come off. A product I find invaluable is called Nevr-Dull (**www.nevrdull.com**). It is available in boating, auto and hardware stores. This is also a brass polish, but it's much milder than some others. It comes as a cotton "wadding" that's impregnated with the polishing agent. It's great for removing traces of old stickers, residue from a smoking household, polishing dull plastic, but it will take off decals or lettering, so be careful! It does not have the strong smell, and strong chemicals, as do the popular liquid products I've seen, and does a great job. 73—*Alexandra Carter, NS6Y, 450 N Mathilda Ave Apt T105, Sunnyvale, CA 94085-4235;* **ns6y@arrl.net**

*Nevr Dull is sold as a metal polish. Another good product for this use is Goo Gone (***www.magicamerican.com/googone.shtml***), which is sold in hardware stores. Goo Gone smells pleasant, but it leaves a shiny residue. You can easily remove the residue with soap and water.*—Ed.

## RESTORING ROTARY SWITCHES

◊ My Drake R-4B receiver, purchased new in 1968, recently became unusable because the bandswitch was worn out. Contact cleaner would no longer fix it, which is not surprising, considering that it has been turned many, many times.

Here's how I restored it. I cleaned the rotating contacts carefully with a cotton swab and contact cleaner. I then applied a very thin coating of Noalox (**www.idealindustries.com/**) with a toothpick. Noalox is a conductive anti-oxidant grease sold at home improvement stores in the electrical section.

It's important that you use only a small amount. I did one wafer at a time, so if I shorted one out I'd know which one it was. Sure enough I did short one. I blasted it with a can of pressurized contact cleaner, which removed the excess grease nicely. It's been a month now, and the switch works perfectly, just like a new one. — *73, Tom Webb, W4YOK, 3533 Teakwood Ln, Plano, TX 75075,* **sam9lives1@verizon.net**

## TEXTURED PAINT FINISH FOR RADIO CABINETS

◊ A way to generate a textured finish for a radio cabinet is to apply a coat of an automobile rocker panel coating (eg, 3M Rocker Panel Spray, 05910) after the cabinet is primed but before the color coat is sprayed. This technique avoids spouse ire that can occur when you heat the cabinet in the family oven before painting to make the paint wrinkle. Even better, make friends with the folks at a local auto body repair shop and let them paint the cabinet for you between other jobs. — *73, Michael Davis, KB1JEY, 533 Tennis Ave, Ambler, PA 19002-6016,* **michael.davis@alumni.duke.edu**

## DISPLAY SCRATCH REPAIR

◊It is possible to remove scratches from a plastic display. Now that I am retired and no longer in the electronics servicing business, I will reveal one of our shop secrets. A product by the name of BRASSO is commonly available in stores that sell household cleaning products. A few wipes with a soft cloth moistened with BRASSO will make scratches disappear as if by magic. Use it on electronics, watch crystals, your spouse's tableware, etc. — *73, Harold Wade, W4NVO, 102 Edisto Dr, Summerville, SC 29485,* **halbwade@bellsouth.net**

## FEET FOR HEATHKITS

◊When the plastic feet are missing from a Heathkit radio, it can be a little hard to find a suitable replacement. For a Heathkit GR-81 receiver, I substituted a stack of 37⁄64 inch flat faucet washers, secured by a long 6-32 flat slotted screw, flat washer, lock washer and nut. When the 6-32 screw is tightened, the head of the screw causes the bottom washer to become concave and will not scratch anything that the radio is placed upon.

This substitute foot will not peel off with age. Since the feet are largely hidden by the radio, no one will ever notice the substitution. Best of all, if you encounter a source for vintage feet, it is easy to unscrew the substitute feet and install the originals. — *73, Michael Davis, KB1JEY, 533 Tennis Ave, Ambler, PA 19002-6016,* **michael.davis@alumni.duke.edu**

**Figure 5 — The Hallicrafters SX-101 sitting in the water heater tray ready for the chemical cleaning part of its restoration.**

# Chapter 7 Construction/Maintenance

## A BACKYARD ASSAY FOR STAINLESS STEEL

◊ Hints and Kinks has previously suggested using a magnet to determine whether junk-box hardware is made of stainless steel: N1EDM, Jun 1999 *QST*, p 76. WB2SHR, Jun 1996 *QST*, p 70. N1EDM, Mar, 1994 *QST*, p 79. Some exceptions were mentioned, but there is another exception: nickel-plated brass nuts and screws. A magnet will not pick them up, but similarly, they will not rust. I have salvaged many steel shafts from old photocopiers for use in my machine shop. Some of the shafts are steel, some are stainless, and I used a magnet to sort them out. Yet, some of the magnetic shafts looked like they could be stainless, so I ground a little off the end to expose the base metal and laid them out in the backyard for a couple of days. The dew on the grass will tell you if the shafts are not stainless by causing them to rust. I became fond of stainless 30 years ago when I bought my first four-wheel-drive for plowing snow. Almost all the bolts and nuts are stainless now, and I still plow snow with the same 1970 CJ5.—*Pete Ostapchuk, N9SFX, 59425 Apple Rd, Osceola, IN 46561;* **n9sfx@aol.com**

## COPPER WIRE PROPERTIES IN YOUR HEAD

◊ Most hams have memorized a few simple conversions between decibels and power: 3 dB is a factor of two in power, 10 dB is a factor of 10. Adding decibels is equivalent to multiplying power factors, and so on.

To a high degree of accuracy, the same factors apply to the resistance of copper wire of various gauges. Just substitute AWG gauge for decibels and resistance for power.

The easy "zero point" to remember is that #10 copper wire has a resistance of 1 Ω per thousand feet. Then, for every three gauges up the resistance doubles, for every 10 gauges up the resistance is a factor of 10 higher and so on. The same rule applies for gauges below #10, but the resistance decreases.

For example, #20 wire is 10 gauges higher than #10, so the resistance is a factor of 10 higher, or 10 Ω per thousand feet. #4 wire is six gauges below #10, so its resistance is ½ × ½ = ¼ of #10 or 0.25 Ω per thousand feet.

These results are accurate to a few percent for either annealed or hard drawn copper, and either solid or stranded wire. They do not apply directly, of course, to alloys such as copperweld or other materials, such as aluminum.

If, however, you know the resistance of any gauge of wire made of a material other than copper, the "decibel rules" do apply to get you the resistance of any other gauge of that material. Simple, no?73—*Ned Conklin, KH7JJ, 2969 Kalakaua Ave #1004, Honolulu, HI 96815;* **ekc@forth.com**

## RE: RF GROUND—FOUR STORIES UP

◊ I would like to comment on the March Hints & Kinks item "RF Ground—Four Stories Up." I, too, live in an apartment but only three stories up. I have my shack in my bedroom, near the window.

Since I live in the very back of the complex and face a wooded area, I was able to put up an outside loop antenna. However, I knew I needed a ground and being three stories up, wondered how I could manage this.

My first thought was to run a heavy wire out the window to an outside ground. Not difficult, but rather unsightly and since my window is right above the windows of the apartments below, I figured those people might not appreciate the wire.

I then thought of the water heater/plumbing, but my complex, though not new, has had all the plumbing done in plastic pipe. Now what?

Well, it turns out that the air conditioner is in the same "closet" as the water heater and it does have copper pipe. It has two pipes in fact, that go straight down below the foundation and connect to the compressor outside—directly below my window. The compressor is connected to the building ground by a short, heavy length of wire. Problem solved. I ran a length of the heaviest "flat wire" speaker cable, tying both ends together, from a grounding strip to a good solid connection on the air conditioner and connected my equipment to the grounding strip. It works great and I have no RF in the shack.

So, for those living the "high life," check your plumbing; even if the building is old, it may have been retrofitted for plastic pipe, but do not discount the air conditioner.—*Chris Maukonen Sr, WA4CM, Systems Programmer, University of Central Florida, 4000 Central Florida Blvd, Orlando, FL 32816;* **wa4cm@arrl.net**

[Editor's Note: Chris found his solution in the air-conditioning ground wire, but similar arrangements may not work as well. To learn more, look at the discussions of RF Grounding in recent *ARRL Handbooks*, RFI books and "The Doctor Is IN" for November 2004.

## STILL MORE USB, EIA-232 AND AMATEUR RADIO

◊ The November 2004 column recommended several USB/EIA-232 converters. Another one may be more available locally. It's RadioShack's 6 foot USB-to-serial-port cable #26-183.

At the time of this writing, it sells for $39.99 in the 48 contiguous states. It is out of stock on the RadioShack Web site, but it remains in stock at some local stores.

The cable has gold-plated connectors (DB-9 at one end, USB at the other). ASIC converter circuitry in the cable (see Figure 1) draws its power from the USB connection. The supplied driver (Win98, 98SE, Windows ME, Windows 2000 or XP) links the USB (1.0) port to a new serial port, so operation is transparent to the user. Thanks to Stu Cohen, N1SC, for this information.—*Robert Schetgen, KU7G; Hints and Kinks Editor;* **rschetgen@arrl.org**

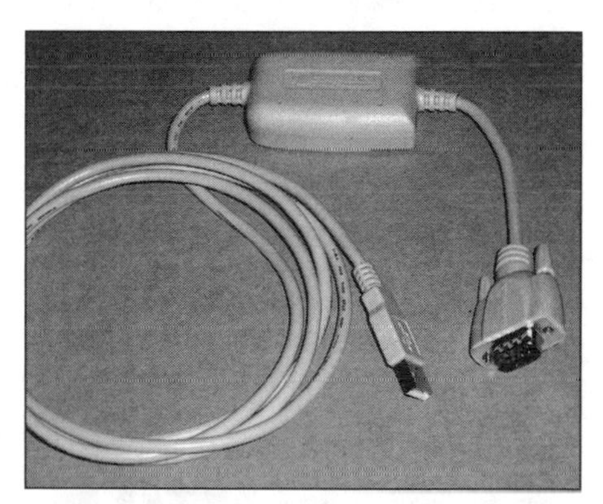

**Figure 1—The RadioShack USB to EIA-232 converter cable.**

## JUNK-BOX REFILLS

◊ If you are searching for a cheap source of components, don't overlook carport sales, yard sales and flea markets. With the ever-changing market in cellular telephones, there are few sales without an assortment of obsolete cell-phone chargers, batteries, hands-free headsets and antennas. Most of these sell for a dollar or less, and you can use them for a variety of projects and purposes.

For example, at one sale, I picked up a magnetic-base antenna for a Motorola cell phone. The cable was an RG-174 equivalent. The whip was probably λ/4 or ⅝ λ for some cell frequency, but it unscrews from the base. [Some are secured by a setscrew in the base—*Ed.*] With a new whip, it would be adequate for occasional use with a handheld transceiver—and it cost a whole quarter![1]

Similarly, wall chargers ("wall warts") from obsolete electronics are a source of parts (or power supplies) for small projects or they may be used to replace similar units that have failed.—*Bill Brown, KG5AR, 17420 Oak Forest Dr, Mabelvale, AR 72103-4500;* **kg5ar@sbcglobal.net**

## MORE ON WALL-WART REMOVERS

◊ If you don't want to make a pigtail or purchase an extension cord, there are commercial products available to move your wall warts away from the outlet strip. These are often available from professional audio supply houses. A Google search for "wall wart remover" or "outlet strip liberator" should yield several sources. Here are a couple of examples that I found: **www.hometech.com/power/wallwart.html** and **www.bitsltd.net/smartstrip/buycordscables.htm#liberator**. Many of these are small, flat receptacles that can be attached to a surface with double-sided tape. A short 2-conductor cord, plug and socket set then plugs into the outlet strip.—*Robert W. Lewis, AA4PB, PO Box 522, Garrisonville, VA 22463-0522;* **aa4pb@verizon.net**

### *A PowerSquid?*

◊ Power Sentry (A Fiskars Brands company; **www.powersentry.com**) has developed a new sort of outlet strip they call the PowerSquid (model #100596). As shown in Figure 2, the squid's "head" holds a switch and circuit breaker. A short cord from the head plugs into an ac outlet. Several "tails" of varying lengths exit the head, each ending in a three-conductor outlet. Each wall wart can have its own tail, where it blocks only one outlet. The tails and their outlets are made for heavy-duty use with power tools. I'm tempted to homebrew a similar device, but I can't do it for their price—$15). The PowerSquid is marketed through home-improvement and department stores.—*Ed.*

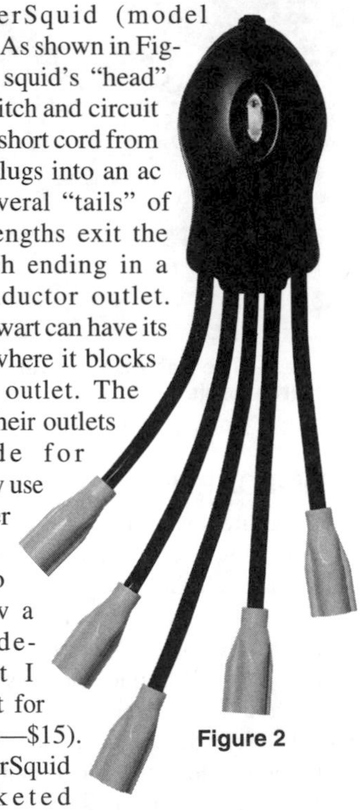

Figure 2

◊ In April's Hints & Kinks, Bill Smith, K1ARK, talks about making pigtails for wall-warts to make better use of power strips. I recently found these at a computer store chain (**www.microcenter.com**) for $1.39 each in the 10 inch length. They are made by QVS and are part number 320853.—*Scott Hernalsteen, W8CQD, 51325 Central Village Apt 203, New Baltimore, MI 48047;* **w8cqd@arrl.net**

## ALTERNATIVE FOR GROUND CABLE

◊ I was reading in the new *ARRL Handbook* about the best cable to use for grounding.[1] I would like to offer another suggestion. I live in a rural area and it isn't always easy or inexpensive to get good braided cable. As a substitute, I use automobile battery cables or jumper cables. They come with built-in clamps, are flexible and are usually long enough. Just be sure to get the good ones that are solid copper or copper clad aluminum. — *Stewart Nelson, KD5LBE, 8 Deerwood, Morrilton, AR 72110*

## A SIMPLE HANDHELD HOLDER

◊ A common problem for hams who own a handheld radio is "Where do I put it when it is not in use?" Here is a brief answer to that question. The design is ultra simple and requires only a minimum of hand tools; you may have the raw materials around your shack.

Measure the base of your handheld for width, length and depth. Depth is only important to keep the holder from pushing buttons you don't want pushed. Cut four pieces of wood trim stock of your choice to create a kind of miniature shadow box frame around the base of your handheld. Cut a base for this frame from any convenient material and attach as shown in Figure 3. Place stick-on pads on the bottom of this base to keep the screws from scratching the furniture and you are done.

Now you have a place to store your handheld after its time in the charger stand. The holder may even be pretty to look at; it all depends on how fancy you want to make it. —*Geoffrey E. Haines, N1GY, 708 52nd Avenue Ln W, Bradenton, FL 34207-2934;* **n1gy@arrl.net**

[1] *The ARRL Handbook for Radio Communications*, 83rd Ed, pp 3-7 to 3-8. Available from your ARRL dealer or the ARRL Bookstore, ARRL order no. 9493 in hard cover, or 9485 in soft cover. Telephone 860-594-0355, or toll-free in the US 888-277-5289; **www.arrl.org/shop/**; **pubsales@arrl.org**.

## CAST-OFF COAX CLEANS DRAINS

◊ An 18 to 24 inch length of old RG-59 cable (less connectors) makes an excellent emergency snake for clearing kitchen or bathroom sink drains. It's flexible enough to pass around bends in pipes but stiff enough to push through hair clogs and other blockages. It's cheap and effective!—*Arlan J. Brandt, K3NE, 4038 Bakerstown Culmerville Rd, Gibsonia, PA 15044-6311;* **k3ne@arrl.net**

## DEGAUSS YOUR TOOLS

◊ Sometimes tools become magnetic. This seems especially true of screwdrivers, but can happen to other tools as well. This may happen in a number of ways, but seems to be of particular concern when working around loudspeakers with their strong permanent magnets. If a tool fabricated from a ferrous metal, such as iron or steel, is exposed to a strong magnetic field, the molecules in the tool become aligned, forming a permanent magnet.

Sometimes this can be helpful, such as when trying to retrieve a dropped screw from the bowels of a piece of gear. Most of the time, however, it is just annoying. This is particularly true when you're doing something delicate in a small area and all of a sudden an unrelated screw ends up on the end of your screwdriver. Sometimes it gets in the way and other times it falls off, going beyond your reach.

An easy way to solve this problem is to degauss your tools. To degauss a tool, it is immersed within an ac magnetic field and withdrawn slowly. The ac magnetic field alternates the direction of magnetization as it is slowly withdrawn, resulting in a cancelling of the magnetic effects over the length of the tool. Specialty devices with a coil designed for this purpose have been around for a long time. The chances are you have one in your shop in the form of a soldering "gun."

Soldering guns of the type with two terminals protruding and connected to a copper loop on the business end can be pressed

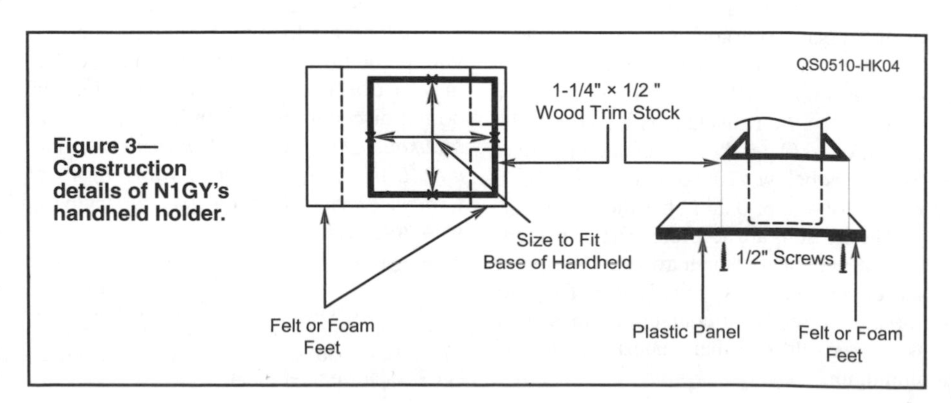

**Figure 3—Construction details of N1GY's handheld holder.**

into degaussing service. The copper loop is one portion of a transformer winding that encloses a magnetic field when the trigger is pulled.

To use a soldering iron to degauss a tool, slip the tool within the copper loop. Pull the trigger and slowly withdraw the tool from the loop. If the first application doesn't do it, a second pass usually will. If you have tools that won't fit in the loop, you can temporarily remove the loop and install a larger loop of 10 gauge wire. Just be careful you don't burn yourself — that tip gets hot even when you're not soldering! — *Hugh Inness-Brown, W2IB, 5351 State Hwy 37, Ogdensburg, NY 13669;* **w2ib@w2ib.com**

## INEXPENSIVE CABLE TIE MATERIAL

◊ A few years ago our local grocery stores began selling heads of leaf lettuce with a strip of hook and loop fastener wrapped around them to hold the lettuce together. The strips are about 12 inches long.

The material has the hooks on one side and the loops on the other, so the material sticks to itself. While the hooks and loops are very small, and the material doesn't have a lot of holding power, my wife Jean, WB3IOS, pointed out that it was just about perfect for holding together a coil of coax.

After collecting a few of the strips I began to think of other uses for this material. It is easy to cut with a scissors or sharp knife. I find it to be ideal for keeping together the coiled cord of a wall-wart. I use a couple of strips on the coil of each radial wire on my version of Phil (AD5X) Salas's Ultimate Portable HF Vertical Antenna.[1] I also use a couple of strips to hold together the tubing of my folding J-pole antenna.[2] I rolled up several strips of this material and carry a roll in my tool box as well as a roll in my radio "travel box." See Figure 4. — *Larry Wolfgang, WR1B, ARRL Senior Assistant Technical Editor;* **wr1b@arrl.org**

[1]P. Salas, AD5X, "The Ultimate Portable HF Vertical Antenna," *QST*, Jul 2005, p 28.
[2]M. Heiler, KAØZLG, "A Backpacker's Delight — The Folding J-Pole," *QST*, Mar 2005, p 31.

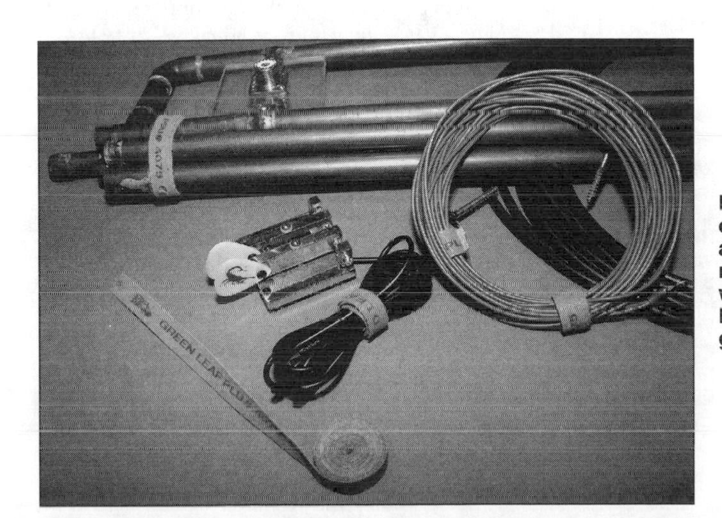

Figure 4 — A variety of uses for the hook and loop fastener material used to wrap heads of leaf lettuce by many grocery stores.

## VARIOUS WORKSHOP HINTS

### *Using Stainless Steel Hardware*

It is supposed to be common knowledge that you must use a non-seizing lubricant on stainless steel hardware. I don't believe I have ever seen this actually printed anywhere, at least not in Amateur Radio literature, however. Stainless steel nuts will gall on stainless bolts unless the threads are coated with a lubricant. I have seen ½ inch bolts break before the nut will come off. I used to think stainless steel was the end-all solution to antenna problems. To use stainless steel hardware successfully, however, you must use a non-seizing lubricant on the threads. You should be able to find such lubricants at your local nut and bolt house.

### *How to Open a Bag of Wire Ties*

The common method of opening a bag of wire ties is to tear or cut open one end of the bag. (Some bags have a zipper locking feature.) With one end of the bag open, even if it is a small corner cut off the end, the ties can spill onto the floor or workbench. I have seen hundreds of ties wasted on construction sites. The solution is to cut a slit in the middle of one side of the bag, perpendicular to the ties. Now you can easily remove the ties through the slit. Since the ends of the bag are still sealed, the ties will not fall out.

### *Inexpensive Digital Meters*

Harbor Freight Tools (**www.harbor-freight.com**) seems to have their Cen-Tech seven function digital meter on perpetual sale at about $5 (Item 92020-1VGA). Where else could you find a 750 V ac, 1000 V dc voltmeter, 10 A dc ammeter and ohmmeter at this price? For about $10 you could have a digital voltmeter and a digital ammeter simultaneously. — *Roger Gibson, K4KLK, 2709 Farnborough Rd, Raleigh, NC 27613;* **rogerk4klk@aol.com**

## SOLDER STRENGTHENS DC CONNECTOR

◊ The sleeve and the sleeve terminal of a dc coaxial connector are held together only by crimping. I've had several of these loosen up and become intermittent in use.

Now I solder it at the point indicated in Figure 5, and then use a file or copper braid to remove solder from the threads, so that the cover will still screw on. The result is a more durable connector.

To further fortify a connector for extreme conditions, I fill the body shell with epoxy after connecting the wires. — *Michael A. Covington, N4TMI, 285 Saint George Dr, Athens, GA 30606;* **mc@covingtoninnovations.com**

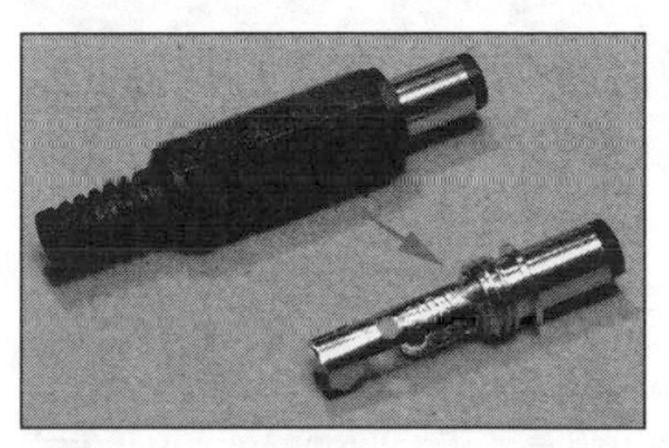

Figure 5 — The weak link for many coaxial dc power connectors is the joint where the sleeve terminal attaches to the sleeve. The arrow shows where N4TMI adds a bit of

## NEEDLE-NOSE PLIERS VISE

◊ An alternative to the May 1981 *QST*, page 45 Hints & Kinks item requires no drilling and tapping of the tool handle. Just stretch a rubber band, and wrap it tightly around the handle as shown in Figure 6. This vise is useful for holding small items during assembly, or for holding wires while you solder them. — *Jack Rosen, KA8LFX, 30811 Ridgeway Dr, Farmington Hills, MI 48334*

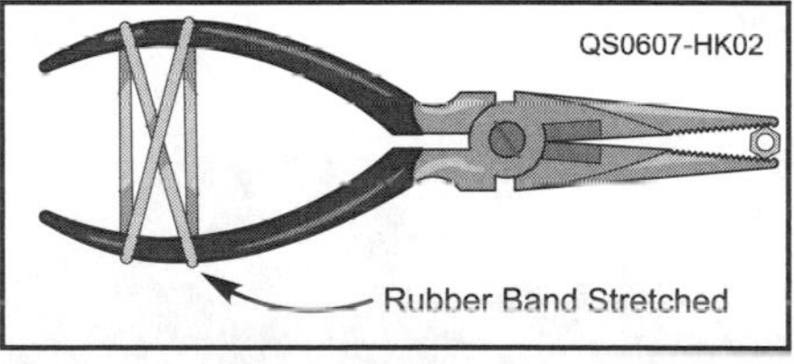


Figure 6 — Using a rubber band to make a needle-nose pliers vise.

## MORE THAN IDIOT LIGHTS

### Illuminated Barrier Strips Make Projects Interesting

◊ When I built the dc-to-dc converter module for my latest microwave transverter (24 GHz), I wanted to clearly label the output terminals so that there would be no need to "guess" at what connection does what. The chassis box was not large enough, however, to allow a nice label to be attached. See Figure 7.

Figure 7 — KH6WZ recently built this dc-to-dc converter. It has three outputs, but has little room for a nice panel label, so an alternative "marking method" was needed.

Since I use nylon insulated barrier strips to make interconnections between modules in my rigs, I thought it might be interesting and fun to insert little LEDs inside the barrier strips, to indicate an "on" condition. Then I decided to take this idea one step further, and choose certain colors to indicate voltage, since the dc converter has multiple outputs. See Figure 8. The red LED indicates 12 V dc, for the input, while green means 5 V and yellow means 24 V.

Figure 8 — LEDs illuminate and indicate "power on" and the colors show what voltages are present: Red is 12 V for the input, green is 5 V and yellow is 24 V.

This looked so "cool" that I decided to use the idea on as many of the modules inside my rig as possible. Now I know that power is going to each individual module, and also what voltage is applied. Of course, this is by no means an accurate way to actually measure the voltage. These are truly "idiot lights" that simply tell me that voltage is going there. Some of the LEDs indicate transmit or receive status (green for receive, red for transmit). I have so many lights inside the rig that I decided to name my 24 GHz transverter system "LightShow." By the way, I have also standardized wire insulation colors in my rigs, too. That simplifies tracing the wires while troubleshooting and modifying the rig.

Here's how I added LED indicators to my equipment. The barrier strips are the translucent nylon types available at most parts shops, and are available in various sizes determined by current rating. Depending on the size of your terminals, you can use 3 mm or 5 mm diameter LEDs. This trick will require an additional terminal to house the LED, so for a simple dc power connection, you need three terminal locations, one for the negative (ground) wires, one for the positive wires and one "space" for the LED. Select the location for the LED, then remove the two terminal screws and push the little metal sleeve out. Figure 9 shows the pieces.

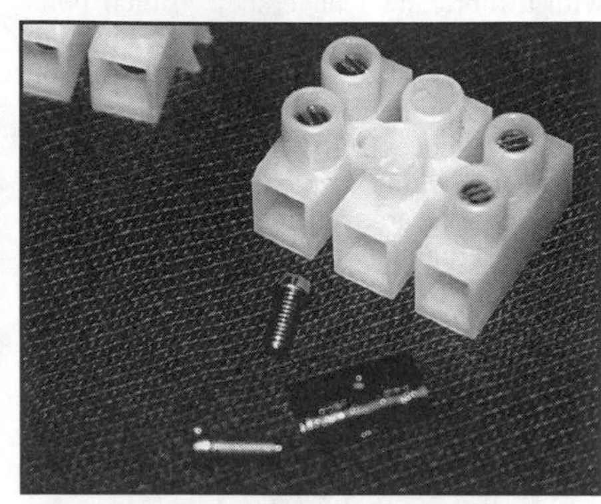

Figure 9 — Remove one of the contacts in the nylon barrier strip to make room for the LED and dropping resistor. Once the small flat-head terminal screws are removed (you may have to pry them out), the contact barrel can be pushed out.

Pre-wire an LED and appropriate dropping resistor, as shown in Figure 10. Insert the LED into the housing, and wire the LED so it is powered from the appropriate terminal.

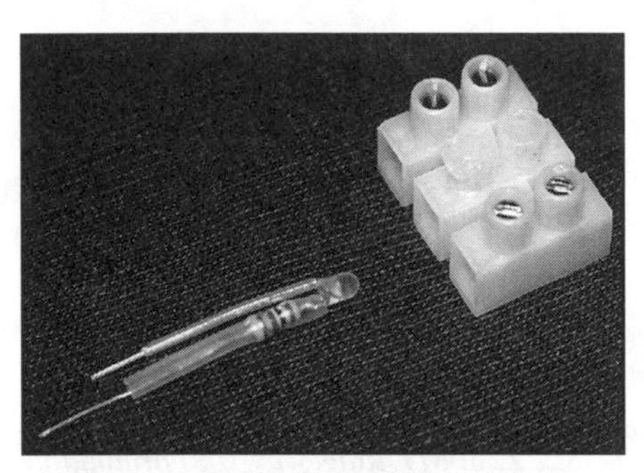

Figure 10— Pre-wire a dropping resistor on one of the LED leads, and use heat-shrink or nylon tubing to insulate the bare wires.

I use clear epoxy adhesive to hold the LED in place. Masking tape prevents the epoxy from dripping all over the place. (See Figure 11.) I used 5-minute epoxy in the prototype, but other glues are probably more safe and suitable. In fact, I now use a GE adhesive called "Special Projects Adhesive Caulk," GE stock number GE16204. This stuff does not emit an acidic smell (vinegar) when curing, so it should be okay to use with electrical things.

Figure 11 — After the LEDs are pushed into the barrier-block body, glue the LEDs in place. Here, epoxy was used, but other adhesives might be less messy. See the text for other ideas.

The completed illuminated barrier strip looks great in any project, especially in the dark. See Figure 12. In my 24 GHz rig, the LEDs indicate more than "on." The LEDs show voltage to various stages by color and relative size. For example, a 5 mm red

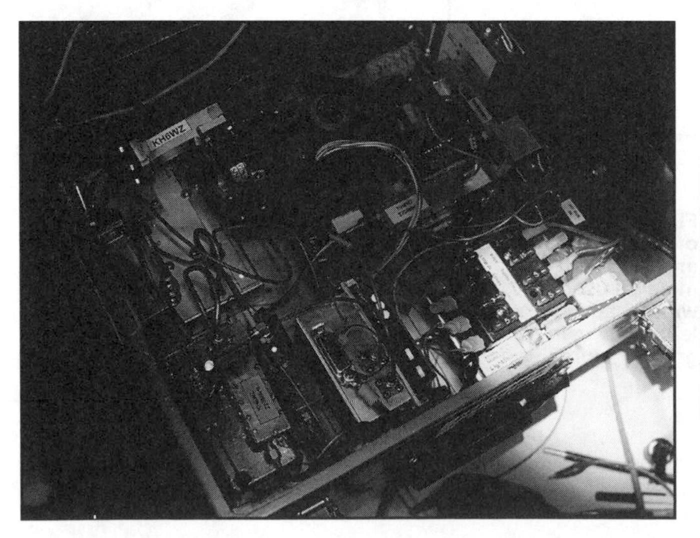


Figure 12 — The 24 GHz transverter system includes LEDs that do more than say "on." The various colors, sizes and shapes indicate specific voltages, conditions and states. It looks pretty neat in the dark, too.

LED indicates 12 V when in transmit at the amplifier power supply relay, while a 3 mm rectangular red LED shows 5 V (the transmitter bias voltage) at the transmit module. — *Wayne Yoshida, KH6WZ, 16428 Camino Canada Ln, Huntington Beach, CA 92649*; **kh6kine@earthlink.net**

## THIRD HAND FLASHLIGHT HOLDER

◊ Have you ever needed a third hand to hold a small flashlight? Well, here's what I use. My (2 AA battery size) Mini Maglite® fits into a standard microphone holder, which is attached to my microphone boom. I use a spare microphone boom on the workbench. It works for me. — *73, Paul Marsha, K4AVU, 200 Garden Trail Ln, Lexington, SC 29072;* **k4avu@yahoo.com**

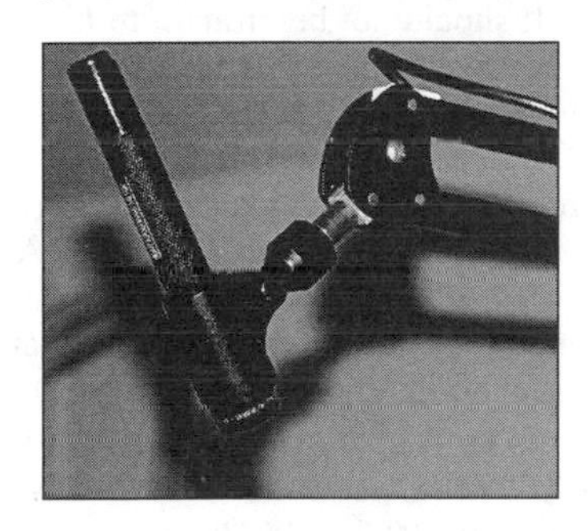

VE3OSZ PHOTO

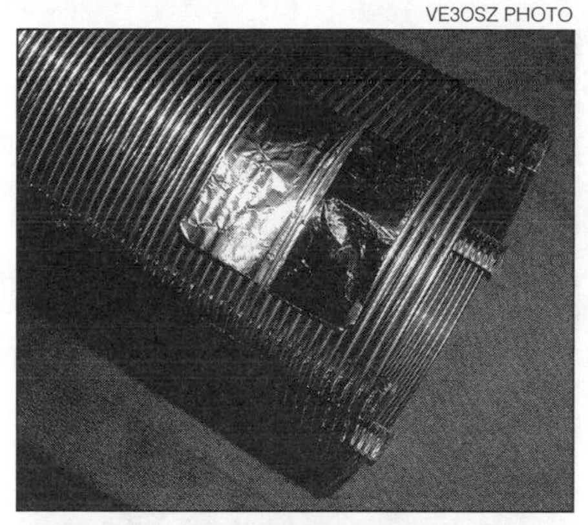

**Figure 13 — This photo shows how VE3OSZ uses a piece of aluminum foil to protect adjacent turns when soldering a tap on a piece of coil stock.**

**Figure 15 — Place a drop of Wite-Out or white paint on the top side of cables to assist in the orientation.**

## SOLDERING TAPS ONTO A COIL

[This Hint is reprinted from the May/June 2006 issue of *The Canadian Amateur* — *Ed.*]

◊ If you have ever tried to solder taps onto a close wound coil such as those manufactured by B & W, you will probably have experienced the challenge of trying to solder the tap onto the desired coil turn without also getting solder onto adjacent turns. It is quite easy to end up with two or more coil turns soldered together.

I recently had the task of soldering eight taps onto a B & W coil with a 10 turns-to-the-inch winding. Initially, I was successful in soldering some adjacent turns together, which wasn't really my objective.

I solved my problem in the following way. I cut apiece of ordinary kitchen aluminum foil to about 3 cm by 3 cm. [The size is not critical. About 1 in. by 1 in. will be fine. — *Ed.*] I folded it in half and then inserted it, from inside the coil, so that the halves of the foil protruded on each side of the turn in question. I then folded back the two halves of the foil so as to cover the coil turns on each side of the one to which I wished to solder a tap. Figure 13 shows the arrangement.

Now, it became much easier to solder the taps onto the coil. Solder will not adhere to the aluminum foil. The foil protects the adjacent coil turns and can be easily removed once the tap has been successfully soldered in place.

This technique is equally useful for the removal of taps and for removing solder from turns using some kind of solder wick. — *Bob Kavanagh, VE3OSZ, 849 Maryland Ave, Ottawa, ON K2C 0H9, Canada*; **73rjkosz@sympatico.ca**

## ORGANIZING CABLES AND CORDS WITH TWIST TIES

◊ Many hams, especially us boatanchor users, have myriad power cords, antenna lines and control lines strung around our shacks. Here's an easy way to keep them organized.

Our local "health food" emporium sells many bulk items. To facilitate checkout, they provide twist ties with "flags" attached for the customer to mark the item number chosen. See Figure 14. Use the flag to identify the wire, twist it around the connector end, and you'll always know what lead goes to what equipment.

I also found a package of cable ties in various colors at a local Dollar Store. These ties are also handy for identification. I just put a tie of the same color at each end of a lead or run of coaxial cable for quick identification. — *73, Ron Pollack, K2RP, 659 Shanas La, Encinitas, CA 92024*; **k2rp@arrl.net**

## CABLE MARKER HELPER

◊ Have you ever had one of those days when you have to crawl under your desk to plug in a USB cable or another computer cable? It seems that you can never get the cable in the right position without scrambling to find a flashlight. USB cables especially can be a real "pain" to orient in the right position.

To aid in your quest to get the plug in the correct orientation, so you can at least have a fighting chance, mark the cable with white paint or just use that bottle of Wite-Out that's probably in your desk drawer. Dab a spot on the top side of the USB/serial/printer cable and you're all set. Next time you grab the cable, you immediately know which side is "up." — *73, Tom Forrest, N4GVK, 4994 Heritage Woods Dr, Greensboro, NC 27407*; **n4gvk@bellsouth.net**

## DENTAL FLOSS DISPENSERS FOR SMALL PROJECTS

◊ I floss my teeth religiously, and end up with many little plastic floss boxes that I used to pitch in the trash. A few years ago I was building an FSK interface for my HF radio and computer. It was a simple one-transistor, few-resistor interface with several wire leads coming from it. I thought about what kind of case would be best to house such a small circuit. I settled on using the "dead bug" style of circuit assembly and an old dental floss dispenser to insulate and give support to the circuit. Figure 16 shows my completed interface.

In high-power RF environments I use shielded cable into and out of the project box. So far I have not had any problem with the dead bug circuits not being shielded. If this is a major concern one could always wrap aluminum foil around the box — or inside of it — to add RF isolation.

Using hot glue to hold the components and make a strain relief for the cables works

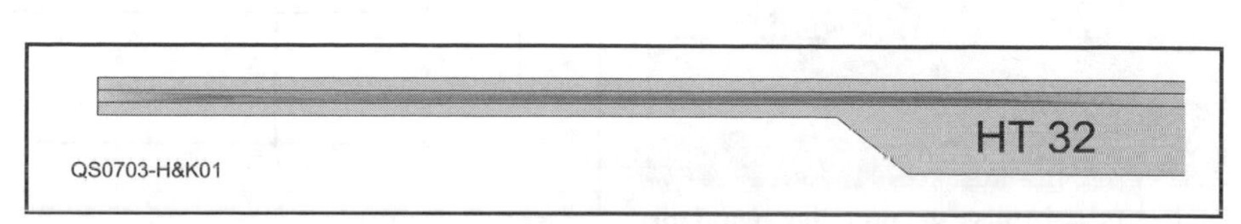


**Figure 14 — This drawing illustrates a twist tie with "flag" on one end that can be used to identify various cables and cords around a shack.**

well. In projects exposed to the weather, filling the whole box with hot glue will make it water resistant, although the plastic is not UV resistant and may break down over time if exposed to direct sunlight. Figure 17 shows a keying interface for using a computer to send CW.

NØNM

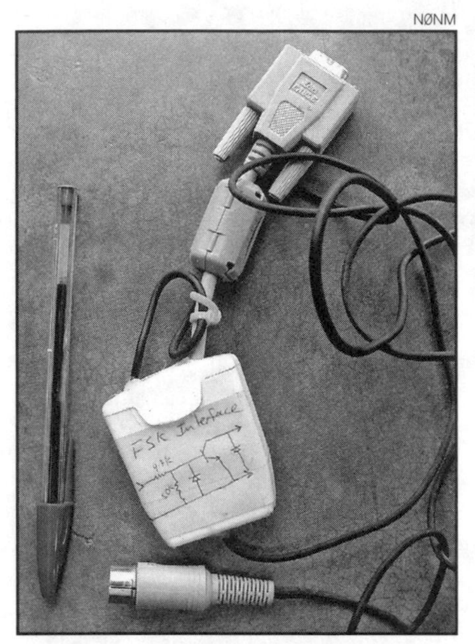


Figure 16 — This photo shows the completed FSK interface that DU9/NØNM built into a dental floss dispenser.

NØNM

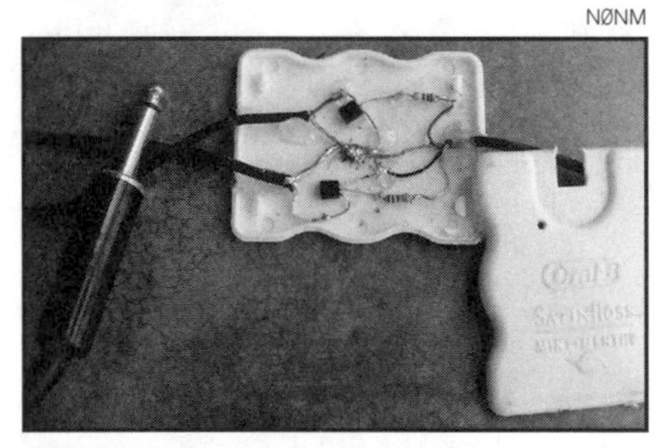

Figure 17 — This keying interface shows how hot melt glue can be used to hold components in place and to form strain relief for the cables.

NØNM

Figure 18 — DU9/NØNM built this balun for his Beverage antenna into a dental floss dispenser.

NØNM

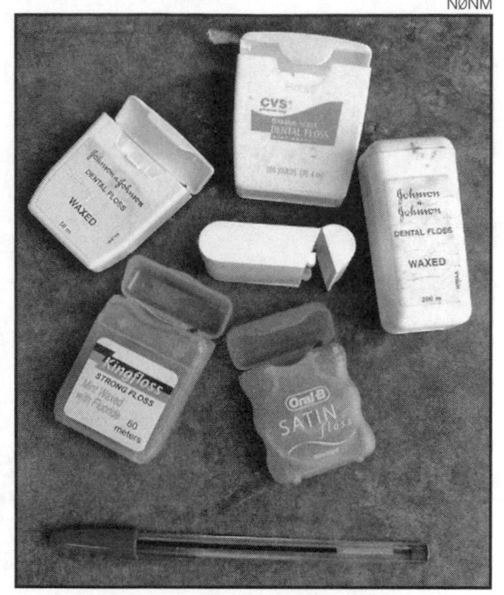


Figure 19 — Always on the lookout for a new size or shape of dental floss dispenser, DU9/NØNM has a collection of boxes waiting for his next project.

Be sure the circuit is working before sealing the box with glue! Since this first project I have used many shapes and sizes of floss dispensers to make small projects. Figure 18 is a balun I built for use with my Beverage antenna. My collection is growing and as Figure 19 shows, I am always on the lookout for interesting sizes, shapes and opening flaps for that next project! — *Jon Rudy, DU9/NØNM, Davao City, Mindanao, Philippines*; **n0nm@arrl.net**

## A FAST CIRCUIT BREAKER

◊ As experimenting radio amateurs, it some times happens that we make a short circuit, and somewhere a fuse blows. Do we have a new fuse at hand? Often not. Most transceivers have a 13.8 V output jack, useful for our many external devices. On my transceiver this outlet is fused by an *internal* 500 mA fuse. After having dismantled my transceiver for the n$^{th}$ time to replace this fuse, I decided that a quick circuit breaker could be a solution.

After some experimenting I came up with this solution, using only junk box parts. The heart of the construction is a fast reed relay. I then got the idea of supplying this relay with an extra coil, which senses the current to be protected.

The most important electrical factor for a relay is the number of ampere-turns. This is the current through the relay coil multiplied by the number of turns in that coil. I found a very small reed relay manufactured by Hamlin that operates on 5 V, with a coil resistance of 500 Ω and normally open (NO) contacts, rated to handle 0.5 A at 200 V. The closing and opening time for this relay is less than 1 ms. It should not be difficult to find reed relays with similar specifications from other manufacturers.

How many ampere-turns is necessary for this relay to close? I wound 50 turns around the original relay coil, and at 660 mA the relay closed. In other words, the relay needs 33 ampere-turns to close. I decided that the relay should close at 400 mA to protect the 500 mA fuse. Hence, a coil with 82 turns of 0.5 mm (AWG no. 24) copper wire was wound on the reed relay, and as predicted, it closed at 400 mA. The resistance in this coil is negligible.

It is important that the magnetic fields made by the two coils are in phase, so that the resulting field is not weakened. That can best be controlled by means of a compass needle. Hold a compass close to the coil and watch which way the needle deflects when you turn on the supply voltage. Then connect the new coil so the compass needle deflects in the same direction.

From the schematic diagram of Figure 20, you can see that when voltage is applied, the green LED will light. When the current through the extra relay coil, L1, exceeds 400 mA, then relay K1 will close,

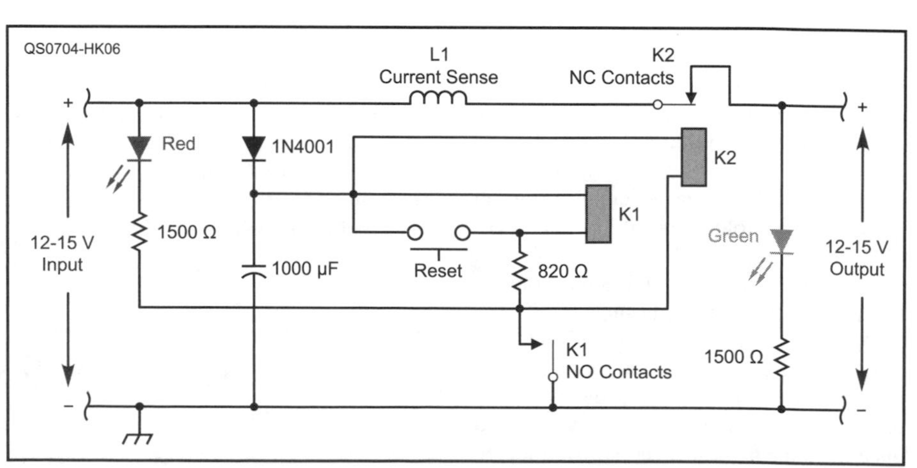


Figure 20 — This circuit uses a fast-acting reed relay with a separate current-sense coil to control a second relay as a circuit breaker.

and be latched. Relay K2, with normally closed (NC) contacts, will open. Relay K2 is a small 12 V relay with a coil resistance of 800 Ω. The K2 contacts are rated to handle 2 A. When the relay K1 closes and K2 opens, the output current will be turned off, the red LED will light, and the green LED will turn off. Once the short circuit has been removed, press the RESET button and the circuit breaker is ready to operate again.

It is not necessary to put a diode across any of the relay coils. The capacitor value of 1000 μF is not critical. It ensures that the relays always get (almost) the full voltage.

By testing this device I found to my surprise that even a 100 mA fuse could withstand a direct short circuit through this device.

This idea of an extra current sensing relay coil may also be useful in other applications where one wishes to limit the current. — *73, John Lien, LA6PB, Pettersandasem 12, 1614 Fredrikstad, Norway*; **lien.j@frisurf.no**

## ANTI-STATIC WRIST STRAP

◊ Recently, I began a construction project incorporating some static-sensitive integrated circuits. Not trusting the "touch ground before picking up the chip" method of static discharge, I opted for a wrist ground strap. Unfortunately, none was available in our rural area of the universe so I improvised using a clip lead fastened to my station ground and to the metallic wrist band of a discarded watch. A 1 MΩ resistor in series with the ground lead limits current in case of accidental contact with a voltage source. — *73, Paul Honoré, W6IAM, 100 Juan de Fuca Way, Port Angeles, WA 98362*; **psh@olypen.com**

[ARRL Lab Engineer Zack Lau, W1VT, suggests caution when using a metal wrist strap around any voltage source. Even a 12 V battery or the high-current 13.8 V dc power supply common to most shacks can produce a significant current if shorted by the metal wrist band. This Hint will help protect static-sensitive components, but a commercial wrist strap is still a good investment in safety. A quick Google search on "Anti-static grounding strap" turned up a number of such straps for less than $10, and some for less than $4 at mail-order computer parts suppliers and other dealers. Most of the commercial units seem to use a plastic wrist strap with a conductive strip inside against your skin. — *Ed.*]

## KEEPING AND LABELING A SUPPLY OF ALL SIZES OF HARDWARE

◊The following tip can be used by any "do-it-yourself" person, including hams. The next time you're in the hardware store, go to the section where they have all sorts of hardware in pull-out compartmented boxes. Take one of every size machine screw that you can lay your hands on in both SAE (American sizes like 2-56, 4-40, 6-32 and so on, up through no. 10 or so, depending on individual needs) and in the metric sizes from about M2 through maybe M5 or so, and push them into the edge of a piece of cardboard that you have brought with you. Label each screw next to where you've pushed it into the cardboard.

Buy fairly long screws. The reason for the long lengths is to have room on the screw shaft for a label listing the size, and so that you can handle them easier. Of course, you can just leave the screws in the cardboard, too. Take it one step further and you can assemble a full set of reference nuts, too, if you would like.

This will be your "reference" or "check" set of hardware, to verify the size that is needed for anything requiring additional hardware. The whole thing will only cost a few dollars, and can be indispensable for identifying all of those mystery-sized holes that we all find in ham equipment and elsewhere.

I have found this simple and inexpensive system to be quite helpful in minimizing hardware guesswork. In the field, or during another emergency, you can use that one screw in the size you need to hold something together, or you will have a sample to take to the hardware store with you to find the required replacement. If you use hardware from your "check" set, just be sure to replace it on your next trip to the hardware store. — *73, Joe Wonoski, N1KHB, ARRL CT Section Technical Coordinator, 1121 W Lake Ave, Guilford, CT 06437*; **n1khb@aol.com**

## LITTLE SPRING TOOL

◊Little springs are used inside radios for a number of devices that need tension. Some are used on relays, on the end of tuning strings to hold the string on a pulley, and they may also be used in and on other items found around the radio shack.

These little springs often have to be removed and replaced in order to service the equipment, and they tend to go flying into space while removing or putting them back on. Some work locations are surrounded by things that will hide a spring from you, and you may never see it again if it gets away from you. Little springs can jump out of your fingers and hit you in the eye, so eye protection is recommended when removing and replacing them.

So what is this miracle tool that can save your eyesight and sanity as well as keep you off your hands and knees looking for that elusive spring? It is a crochet hook! They come in a variety of sizes and are easy to use for hooking and unhooking the spring from its retainer. Simply place the end of the crochet hook into the spring and pull it off or into place. A little twist of the hook, and it can easily be removed from the spring. Keep the springs and small parts in a container so they don't escape.

No crochet hook? You could take a safety pin or paper clip and open it up, place a small hook in the end and use it the same way. But, it might be a little more difficult to handle if you have large fingers or if you need a little more "reach."

If you are using a metal crochet hook, there can be a shock hazard. Don't work inside equipment with the power on and make sure you have taken precautions so that no electrical components, like capacitors, have destructive or dangerous electrical voltages on them, just in case you accidentally touch or drop the crochet hook onto a circuit board or across energized components. — *73, Chet Chin, N1XPT, 65 Millers Falls Rd, Turners Falls, MA 01376;* **n1xpt@aol.com**

## STRENGTHENING A THIRD HAND

◊Many of us use "third hand" soldering tools with alligator clips. Figure 21 is an example of one of these devices. The shanks of the clips are easily crushed by over tightening the holders, as shown in on the left in Figure 22. In my laboratory at the University of Georgia, we prevent this by inserting a no. 6-32 machine screw into the shank and cutting off the head. The clip on the right in Figure 22 shows the result of

N4TMI

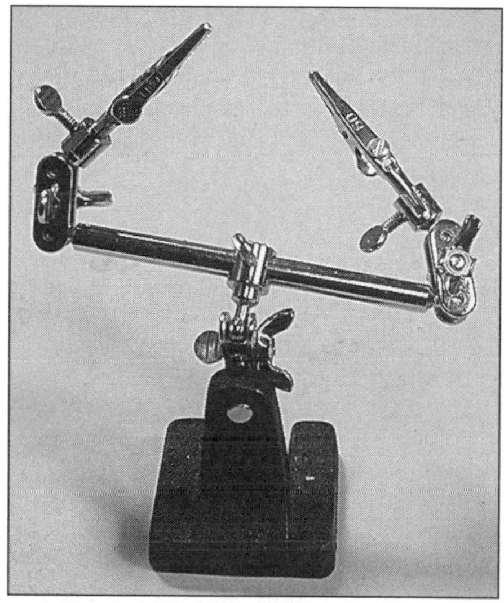

**Figure 21 — A "third hand" is useful for holding small parts and small circuit boards while soldering.**

N4TMI

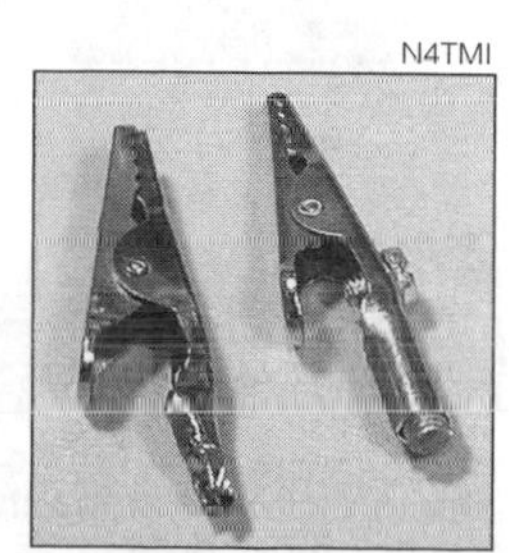

**Figure 22 —The alligator clip at the left has been crushed by over-tightening the thumbscrew to hold the clip securely in place. The clip at the right shows how a piece of no. 6-32 machine screw has been inserted into the hollow opening. This will prevent the thumbscrew from crushing the clip.**

this modification. The alligator clip is then crushproof. — *73, Michael A. Covington, N4TMI, AI Center, Univ of Georgia, Athens, GA 30602-7415*; **mc@uga.edu**

## IMPROVING ALUMINUM CHASSIS BOX RIGIDITY

◊Have you noticed that the common "store-bought" aluminum chassis boxes available at your local electronics parts stores have gotten "softer" over the past few years? I have some very old aluminum chassis boxes, and they seem to be made with thicker material than the newer versions carried at my local electronics parts shop. This did not bother me too much in the past, but recently, while testing one of my 10 GHz radio systems, I noticed that the frequency of my 2556 MHz RF synthesizer would "wobble" when I squeezed or pushed on its cabinet. Figure 23 shows the synthesizer in the original box.

I decided to stiffen the box by adding some small lengths of half-inch aluminum angle stock to the covers. I drilled holes every half-inch or so along the edges of the angle stock, and into the box. Small sheet metal screws pull the box tight, as seen in Figure 24. Figure 25 shows the 2556 MHz synthesizer installed in one of my 10 GHz rigs.

By the way, the 2556 MHz synthesizer is very handy for use as a local oscillator in 10 GHz systems. This project, and many others, appears on the San Bernardino Microwave Society Web site: **www.ham-radio.com/sbms/.**

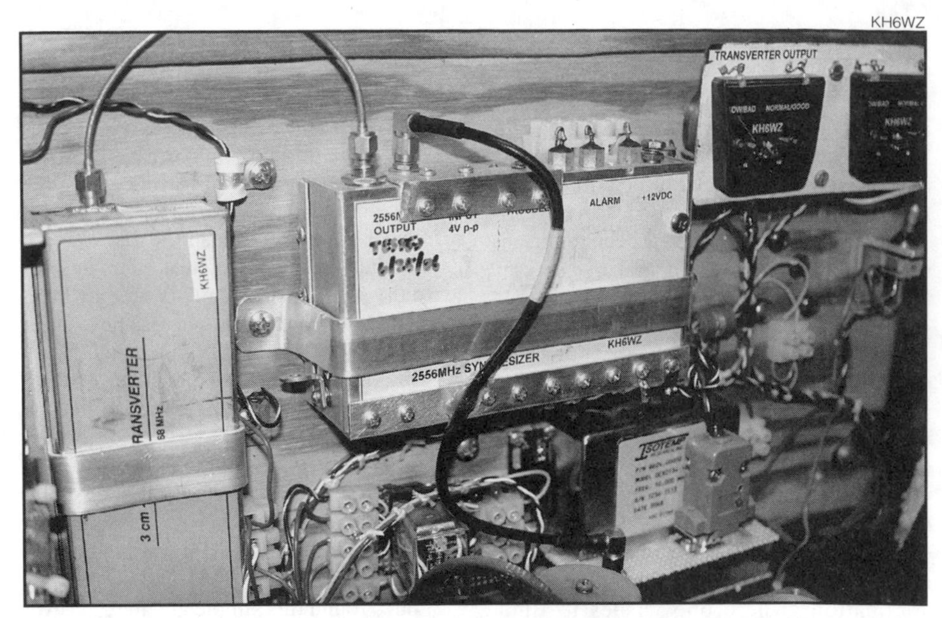

**Figure 25 — The 2556 MHz synthesizer installed in the X-band radio called "Ms June." [Many of the San Bernardino Microwave Society (SBMS) members name their rigs, much like sailors name their ships. — *Ed.*]**

For more information about my synthesizer, see Microwave Group of San Diego, "Modification of the Rectangular 3036/3236 Qualcomm Synthesizer." The Web version by Ed Munn, W6OYJ, is at: **www.ham-radio.com/sbms/sd/rplldoc1.htm**. — *73, Wayne Yoshida, KH6WZ, 16428 Camino Canada Ln, Huntington Beach, CA 92649;* **kh6wz@arrl.net**

**Figure 23 — Many "store-bought" aluminum chassis boxes these days are made of flimsy material. In the case of my 2556 MHz synthesizer, some frequency wobble was attributed to the wobbly box.**

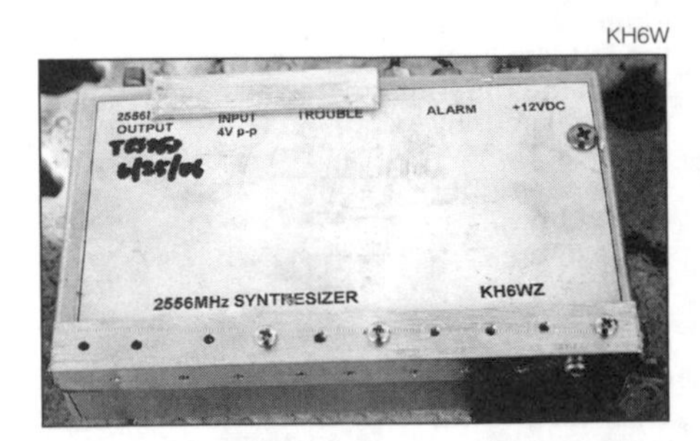

**Figure 24 — Fastening some small angle aluminum stock along the seams makes the box more rigid.**

## AN OLD TRICK — THE "GIMMICK" CAPACITOR

◊Back in 1957 I built a 6 meter converter from the *Radio Amateur's Handbook*. I needed a coupling capacitor between 2 or 3 pF. Instead of using a commercially available capacitor, the *Handbook* writer suggested twisting tightly together two lengths of solid insulated hookup wire, "pigtailing" it. Two or three twists were sufficient to provide the required capacitance and the wire insulation was sufficient to withstand the 200 to 300 V dc used in the plate circuit where the part was placed.

The bitter-ends of the wires may be clipped off and the whole thing made permanent with a liberal application of coil dope (I used nail polish). The little capacitor has behaved well and has lasted for years. — *73, Arthur M. (Mike) McAlister, KD6SF, 7570 Dartmouth Ave, Rancho Cucamonga, CA 91730*

## CRACKIN' WALL WARTS — A HOW-TO TIP

◊All of us have several wall-warts around the shack to power those hand-held radios and/or recharge battery packs. When they quit working, most often it is because of a broken wire in the dc lead. If the break is near the plug end of the wire, the repair is simple; just cut off the plug and solder on a new one. All too often, however, the break is where the lead exits the wall-wart, and there is not enough slack to make a decent splice. So, we have the choice of buying a new wall-wart, or opening the old one.

In times past I have attacked the wall-wart with a hacksaw, but that pretty well eliminates reusing the case. Somewhere I

**Figure 26 — This wall-wart has been carefully placed in a vise, and is ready to be cracked open.**

**Figure 27 — The wall-wart case has started to split open under gentle pressure from the vise jaws.**

KD5HCV

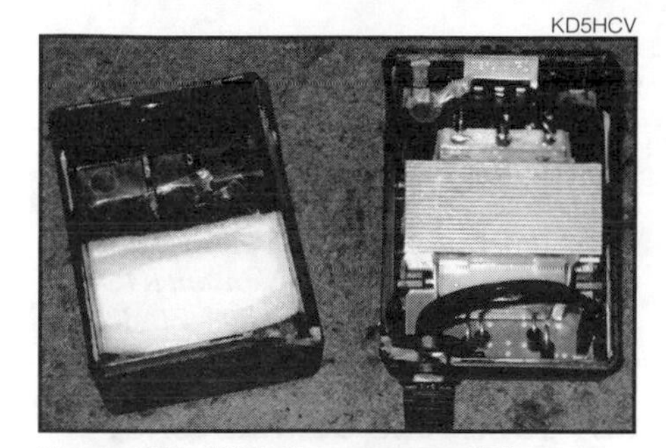

**Figure 28 — With the case carefully cracked open, you can repair a broken wire or replace a diode or capacitor.**

KD5HCV

**Figure 29 — You could glue the case back together, but heavy plastic tape will make the next repair easier.**

read a better method, and since it seems to be largely unknown, I want to pass it along.

Usually a wall-wart can be cracked open without damaging the case. Simply clamp it in a bench-top vise, corner to diagonally opposite corner, with the ac pins up and with the plastic base above the vise jaws. Figure 5 shows a wall-wart ready to be cracked open. Begin tightening the vise until you hear it crack, or pop. Loosen the vise, turn the wall-wart 90° to the other corners, and repeat the process. At this point the base and the cover should have separated. See Figure 27. If not, return to the first position and try again.

With the cover removed, it is a simple matter to replace the broken lead, or even replace a shorted diode or capacitor. Figure 28 shows the wall-wart ready for repair. After the repair, the cover can be glued back together, but I prefer to use plastic tape as shown in Figure 29, so that future repairs will be even easier. — *73, Curt Goodson, W4QBU, 12905 Watermill Cv, Austin, TX 78729*; **w4qbu@arrl.net** and *Mitch London, KD5HCV, 5601 Lewood Dr, Austin, TX 78745*; **kd5hcv@arrl.net**

## MAKING A HOMEBREW RF ENCLOSURE

◊ The usual homebrew method of making an RF enclosure involves cutting pieces of PC board to the right size and soldering them together. I had an idea for another method after a conversation with a neighbor about the siding on my house. He suggested I use copper flashing behind some siding to prevent a potential leak and told me I could get it from

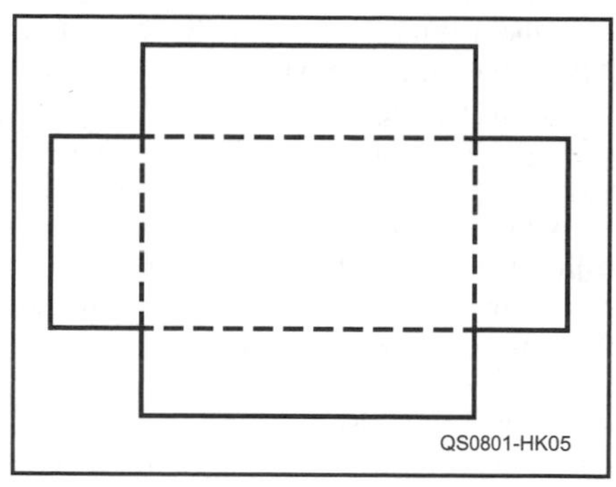


**Figure 30 — Bending template for RF enclosure made of copper flashing.**

a local heating contractor. I never bought any copper flashing but I did call the heating contractor and found that I could purchase a few square feet of 24-gage copper sheet for about $8 a square foot. An RF enclosure (minus the top) can be made from a single piece of this material cut in the shape shown in Figure 30, forming the bottom and the four sides. I have found that it is better to make the cuts with a hacksaw rather than shears. Shears may leave bends in the material that are difficult to remove.

Next, bend the sides up to form the walls and solder the corners. Copper sheet is quite malleable and easily bent. The first bend can be nicely made by placing the material in a vice and tapping the side section with a mallet or hammer. Making the remaining bends is a little more difficult. One way that works for me is to clamp a piece of hardwood in a vice and use that as an arbor to support the enclosure while tapping a side with a mallet.

Soldering the corner edges requires a large iron. I have a 40 W Weller iron that works well. Lightly clamp the enclosure in a vice with the edge to be soldered extended to one side and downward. Apply the iron to the outside of the corner and the solder along the inside seam. You could also use a propane torch.

A top piece could be made in the same way but I found another material at my local big-box hardware store that is easier to work with — 3 ounce copper flashing. This material is almost too thin for this application but it is inexpensive and available and while not very rugged, will work. A 10-foot by 8-inch roll cost about $20, which works out to be $3 per square foot. Unfortunately, this flashing is coated on one side with paper and an asphalt-like material that is supposed to make a waterproof seal around a nail driven into it. With care, you can cleanly peel this backing from the copper sheet. This material can easily be cut to size with shears or scissors, before or after the backing has been removed, and bent to form a top for the enclosure.

I'm not sure this method of making an RF enclosure is better or any less work than the conventional one. In one case you spend a lot of time messing around with PC board and in the other a similar amount of time cutting and bending the copper sheet. My method does appear to be somewhat less expensive. And I find that copper sheet is much more pleasant to work with than PC board. — *73, Gary Richardson, AA7VM, PO Box 228, Marblemount, WA 98267*; **aa7vm@arrl.net**

## REMOVING COAX-SEAL

◊ Have you ever had the frustrating experience of trying to remove old Coax-Seal from your antenna balun, vertical or beam? While working on my electrician apprenticeship many years ago I watched a journeyman weatherproof a large split-bolt connection with something like Coax-Seal and saw his solution that allowed easy removal later.

After you've tightened your coax connector, next start applying electrical tape. Start taping on the coax next to the connector. As soon as you've wound about a half turn on the coax, flip the tape over so the sticky side is now facing out. Finish wrapping the entire connector in this fashion. Cover up to, and including, the threads on the socket connector. Now take Coax-Seal and make a thin layer with your fingers and spread it over the sticky tape. Make sure you cover all the tape, extending just beyond it by at least an eighth of an inch.

Later, when you want to undo that connector, use a sharp knife to carefully cut down the length of the Coax-Seal, making sure you don't cut the coax. The black goo that has kept the connector dry will peal off like a banana with the tape stuck to it. Never again will you need to fight getting Coax-Seal off your connectors. — *73, Greg Truchinski, NVØP, 1580 Fox St, Wayzata, MN 55391*; **gregjt@mchsi.com**

## TWO SPEED SOLDERING IRON

◊ This isn't a new idea, but here's a simple way to convert your single speed soldering iron into a two speed model. It makes a 50 W iron into a 50/25 W iron, or a 100 W iron into a 100/50 W iron. This is especially useful when soldering many of the small components and wires in present-day equipment.

An added benefit is that you save soldering iron tips from burning up by leaving the iron turned on in the low heat mode until you need full heat for soldering larger items. It takes very little time to return to full heat from the low heat mode, much less time than when you used to turn your iron off and on or unplug it and plug it back in again. I've used my irons this way for over 25 years.

The modification consists of inserting a silicon diode into one side of the ac power cord, with an SPST lamp cord switch to short out the diode for high power. Figure 31 is a photo of the modification showing the 1N4004 diode mounted inside the switch.

LYLE H. NELSON, ABØDZ

**Figure 31 — Inside view of the modified in-line switch, showing series rectifier diode.**

(Any silicon diode can be used that will handle the power and line voltage of your iron. The 1N4004 is rated at 1 A at 400 V PIV, and will safely handle a 100 W iron.)

An added feature is that I have marked the visible portion of the thumbwheel on the switch with a permanent marker. This mark is visible when the iron is in the high power mode. When it is switched to the low power mode, the black markings disappear inside the switch cover. Use a good sturdy switch; there are some flimsy ones around. Mine is a Levitron 6 A, 125 V that is large enough for the diode and my iron's 3-wire ac power cord. I had to enlarge the lamp cord entry holes with a side cutter and file. I also had to cut off the sharp brass points of the prongs inside the switch that were designed to pierce the lamp cord insulation and contact the lamp cord wires. I soldered the diode across the two switch contacts and then cut one side of the line cord and soldered each end to opposite ends of the diode. Carefully reassemble the switch housing so that the diode switch and ac cord fit inside. — *73, Lyle H. Nelson, ABØDZ, 1450 201st Ave NW, New London, MN 56273*; **joannmn@tds.net**

## SECURING THIN WIRE LEADS

◊ I have been using Anderson Powerpole connectors for some time now. On occasion while installing a small diameter cable, such as a handheld transceiver's 12 V power cord, the connector would flop around and I became concerned about flexing the small wires, perhaps breaking them off.

I am a fan of hot glue and use it for many things. I cut a piece of heat shrink tubing about 2 inches long and big enough to slip over the assembled Powerpole connector. After connecting the wires and sliding the connector together, I slide the heat shrink tubing about ½ inch onto the back side of the connector. With a very hot glue gun I shoot the back end of the heat shrink tubing, filling the tube with hot glue. The tubing shrinks down, pushing the glue forward into the Powerpole creating a more resilient connector. See Figure 32.

Before I am finished I add a small shot of glue into the center locking hole on the connector, making the connection quite durable. The glue/tube provides a strain relief for the cable while retaining some flexibility. I use this with most of my portables now. — *73, Hal K. Whiting, KI2U, 955 N 2075 E, St George, UT 84770*; **galaxymap@skyviewmail.com**

## MOUSE PAD PROJECT BOX FEET

◊ I recently hit upon a simple, inexpensive solution for adding rubber feet to a project or replacing missing rubber feet on existing items. The secret? Those old, thick rubber mouse pads. I have a couple of dozen of these mouse pads in my garage, but I'm sure you could pick them up somewhere for a dollar or two. Be sure you get the thick rubber ones, not the newer thin plastic ones. The right ones are very soft, non-marking and virtually skid proof. Depending on the size you need, you can easily get dozens of feet out of a single mouse pad.

Cut the mouse pad to any size or shape you desire using a pair of scissors or a sharp knife. Then simply attach them to your project using double-back tape trimmed to fit or a good strong glue. After getting permission, I used some of my wife's nail glue. If you need extra height, the mouse pads can be stacked and glued or taped together. — *73, Michael D. Risser, KG6ECW, 42249 Clarissa Way, Murrieta, CA 92562*; **kg6ecw@gmail.com**

## FLY-TYING VISE AS "THIRD HAND"

◊ I recently needed to ground a pin on a mini-DIN connector to use to upgrade the firmware in my FT-2000. The pins are small and close together, and hams in the Internet had warned that the plastic dielectric "melts easily." The alligator clips on my "third hand" device were too big to hold the pins.

It occurred to me that the pins were about the diameter of a small dry-fly hook. I clamped one of the pins into my fly-tying vise and found that it allowed me to position the connector exactly where I needed it to be. I also found that the magnifier I use for tying flies helped my aging eyes to see the connector well enough to do the job without solder bridges to the other pins. Rotating fly tying vises are available on the Internet for under $20. Cabela's and Bass Pro Shops are two sources. — *73, Bill Hoskinson, K4CUO, 2231 Alexander Dr, Titusville, FL 32796*, **bhoskinson@cflrr.com**

## GOT CORRODED CONNECTORS?

◊ Got electronic equipment? Got cables with connectors? Of course, we all do. How do you clean the connectors when they get tarnished, dirty or corroded? It is almost impossible to find a brush small enough to get into the sockets of most female connectors.

I live on a sailboat making a three year DXpedition to the South Pacific, and the connectors on the boat really present problems. Recently I was checking all the gear on the boat prior to sailing to Johnston Island (KH3). One particular piece of equipment wouldn't work. I traced the problem down to the female connector on the cable, which was really corroded.

I called the manufacturer to see about buying one of the connectors and having it shipped to me. They wanted $59, plus shipping. There just had to be a better (spell that "cheaper") way! So I went down to my local drug store and purchased a package of regular tobacco "pipe cleaners." I sprayed a pipe cleaner with electronic contact cleaner fluid till it was soaked and then twisted the pipe cleaner into each female socket, one at a time. That fixed the problem. And it cost me considerably less than $59! — *73, Susan Meckley, W7KFI, PO Box 1210, Pahrump, NV 89041*, **wda9037a@sailmail.com**

## RIVETED CHASSIS MOUNT COAX CONNECTORS

◊ If you have an older antenna tuner, as I have, you may need to check the resistance between the ground connecting lug, and the metal shells (threads) of your SO-239 female connectors. You can easily check for problems by disconnecting all cables from your tuner and connecting an ohmmeter from the ground lug to the threads on each SO-239. (Remember to zero out your ohmmeter test leads; set on the lowest resistance scale first.) If you get anything other than zero ohms, between the ground lug and the SO-239 threads, you should check out your tuner more closely.

I had earlier discovered what I thought was an intermittent PL-259 on the transceiver connecting cable, and replaced it with a new coax cable assembly. While this seemed to alleviate the problem, later playing with the antenna connectors on the back of the tuner produced intermittent signals again. Disconnecting all cables and putting an ohmmeter across the ground lug and the SO-239 shells gave anywhere from a low

HAL K. WHITING, KI2U

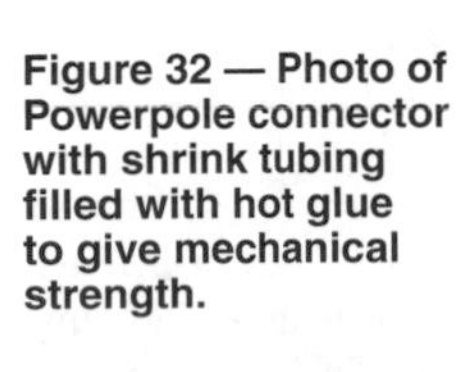

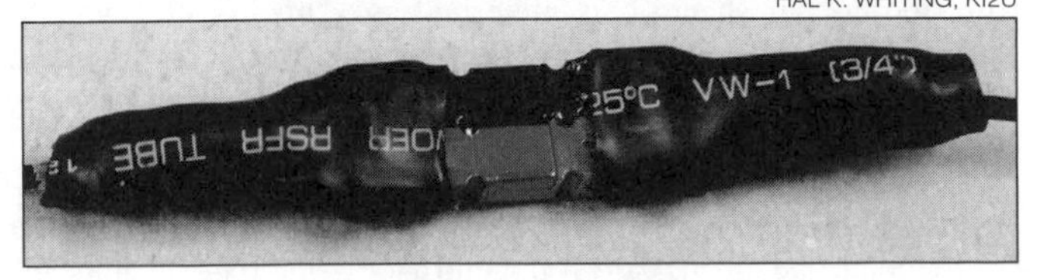


**Figure 32 — Photo of Powerpole connector with shrink tubing filled with hot glue to give mechanical strength.**

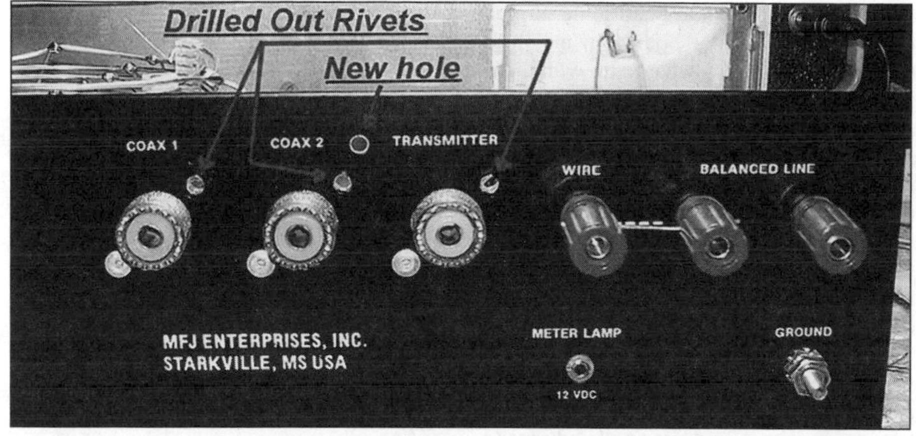


**Figure 33 — Drilling out rivets and new hole on the back of MFJ Deluxe Versa Tuner II.**

of 26 Ω, to a high of 48 Ω, instead of the expected zero ohms. I opened up my MFJ Deluxe Versa Tuner II (some 20+ years old), and found that the female bulkhead SO-239 connectors were riveted to the metal back-chassis plate. Those rivets were the only ground connections, which had now apparently become corroded.

Rather than disconnect the three coax connectors, clean, remount and reconnect them, my solution was to drill out the upper rivet in each connector (using a  inch drill bit), also drilling out a 9⁄64 inch hole near the connectors in the back-chassis plate (see Figure 33). I then inserted a ¼ inch long #6-32 flat-head machine screw, with a ground lug, split washer and nut into all four holes. Inside the box I then connected all the ground lugs with a piece of #16 copper wire (a piece of stripped #16 Romex worked well). I soldered this jumper wire at each ground lug. Voila! Received signals are now good and strong, with no intermittents. — *73, Roland Burgan, KB8XI, 2161 Jasberg St, Hancock, MI 49930,* **kb8xi@arrl.net**

## PROGRAMMABLE TEMPERATURE-MONITORING SET POINTS

◊ The ability to sense an out-of-range temperature can help prevent costly equipment failure. The Analog Devices TMP01 IC is a low power, programmable temperature sensor and controller. The TMP01's output can be used to turn on a cooling fan or a heater, turn off an overheated motor, or simply activate an alarm.

The TMP01 is available in both the DIP and SO packages and requires a single power supply between 4.5 and 13.2 V. Pin 5 provides access to the proportional-to-absolute temperature voltage ($V_{PTAT}$), which varies linearly at 5 mV per kelvin. Pins 6 and 7 supply two control signals when a temperature preset limit has been exceeded. These open collector outputs are capable of sinking 20 mA and thus can control small relays directly. A stable +2.5 V reference ($V_{REF}$) is supplied at pin 1. See the TMP01 block diagram in Figure 34.

Pin 2 sets the upper temperature set limit, and pin 3 sets the lower set limit. These set points are established by programming values for resistors R1, R2 and R3. An internal comparator compares $V_{PTAT}$ with the $V_{REF}$ values set by the resistor chain at pin 2 and 3. When a high or low set point voltage is exceeded, the transistor controlling either pin 6 ($Q_{Under}$) or 7 ($Q_{Over}$) is biased on. The transistor will conduct until a selected number of degrees over or under a set point is reached. This temperature overshoot (hysteresis) is achieved by programming a specific $V_{REF}$-to-ground current ($I_{VREF}$). To program a temperature set point, start with your desired amount of hysteresis temperature and then calculate:

$$I_{VREF} = (5\ \mu A/°C \times T_{HYS}) + 7\ \mu A$$

$$V_{SETHIGH} = (T_{SETHIGH} + 273.15) \times (5\ mV/K)$$

$$V_{SETLOW} = (T_{SETLOW} + 273.15) \times (5\ mV/K)$$

$$R1\ (\Omega) = (2.5\ V - V_{SETHIGH}) / I_{VREF}$$

$$R2\ (\Omega) = (V_{SETHIGH} - V_{VSETLOW}) / I_{VREF}$$

$$R3\ (\Omega) = V_{SETLOW} / I_{VREF}$$

Note that the factor of 273.15 in the second and third equations converts degrees Celsius to absolute temperature (kelvins). You can also convert temperature from °F to °C using the equation: Celsius = (Fahrenheit – 32) ÷ 1.8.

Let's choose, for example, a hysteresis temperature of 2°C. This would require a total current $I_{VREF} = (5\ \mu A/°C \times 2°C) + 7\ \mu A = 17\ \mu A$. Let's say that you want a high-temperature set point ($T_{SETHIGH}$) of 25°C and a low-temperature set point ($T_{SETLOW}$) of 22°C. Then $V_{SETHIGH} = 1.491$ V and $V_{SETLOW} = 1.476$ V, yielding R1 = 59.4 kΩ, R2 = 882 Ω and R3 = 86.8 kΩ.

The IC is a precision temperature sensing device and internal self heating must be avoided to maintain high accuracy. Transistors $Q_{Under}$ and $Q_{Over}$ inside the chip should not be allowed to sink more than 20 mA of current. A reed relay would be a good choice for light

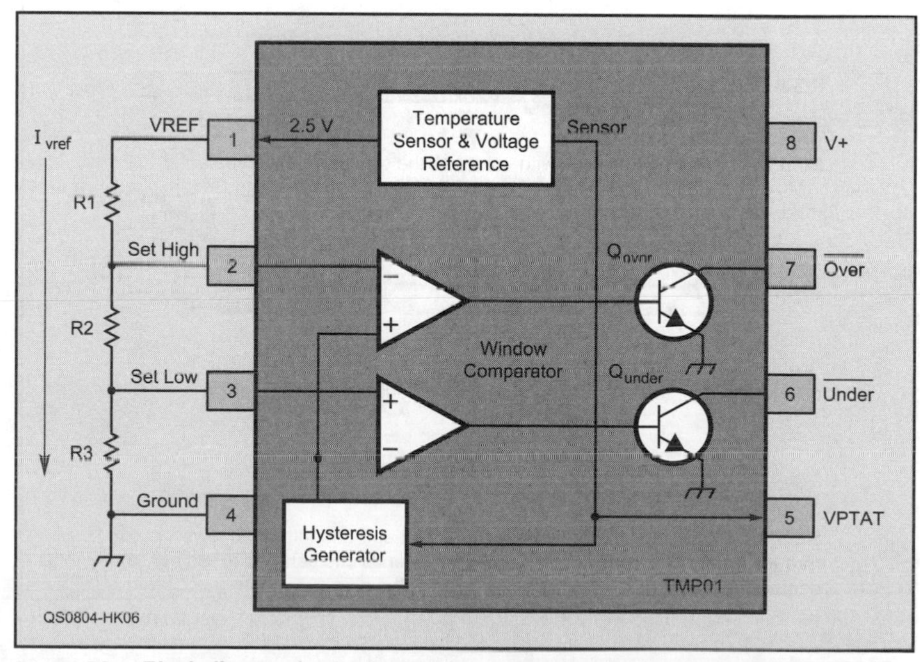


**Figure 34 — Block diagram for Analog Devices TMP01 temperature sensor/controller IC.**

loads. In addition, use a diode across a relay primary to avoid possible damage caused by inductive voltage spikes. See the Analog Devices application note at **www.analog.com/UploadedFiles/Data_Sheets/TMP01.pdf**, which provides additional details, including showing how you can read the temperature in °C or °F using a digital voltmeter.

With a little imagination you can adapt this versatile device to many combinations of cooling, heating or monitoring requirements. — *73, Robert Schlegel, N7BH, 2302 286th St E, Roy, WA 98580,* **n7bh@aol.com**

## "NAVY TECH" METHOD TO PREPARE PL-259 UHF CONNECTORS

◊ Figure 35 shows the steps needed to install a PL-259 on an RG-8 sized coax. [Note that this method is very similar to that described in late editions of *The ARRL Handbook* and *The ARRL Antenna Book*, except that WB5IZL "eyeballs" the cutting dimensions using the PL-259 body as a measuring gauge. Still, it's a good reminder. — *Ed.*] — *73, Larry Jennings, WB5IZL, 8228 E Blackwillow Cir, Apt 114, Anaheim Hills, CA 92808,* **jenninle@quik.com**

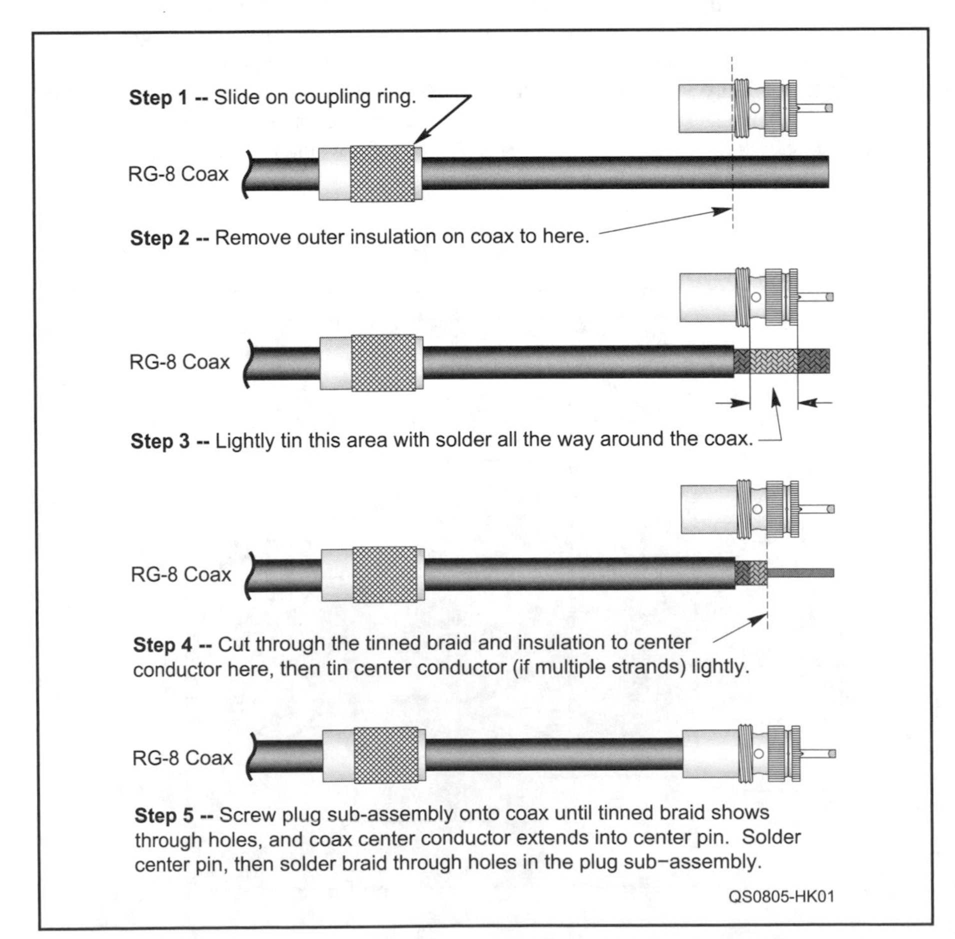


**Figure 35 — Steps for installing a PL-259 connector on an RG-8 sized coax.**

## USING SURPLUS: AVOID REINVENTING THE WHEEL

◊ While working on a frequency counter project recently, I needed a simple VHF gain stage. I remembered that I had some surplus frequency synthesizer boards in my junk box, but did not know what to do with them as a synthesizer. So I intended to harvest the components at some point. Among the "treasures" installed on the synthesizer board were printed filters, high quality surface-mount technology (SMT) RF capacitors and so on.

Just as I started to reach for my heat gun to remove some MMIC (Monolithic Microwave Integrated Circuit) devices and other components from the circuit board, I thought, "Why remove the MMICs, risking damage, when I can cut out the MMIC and its associated components, like the bias resistors and bypass capacitors, and get an already-made amplifier?"

I figured that the engineers who designed the frequency synthesizer knew more about these components than I do, and the circuit was proven to work. Conversely, if I extracted just the parts, I would have to reinvent the gain stage, with individual components, and hope that it would work.

I used a band saw to cut out a section of the circuit board containing the MMIC and a few other parts, making sure to keep enough of a border around the parts and traces to allow external connections to the circuit, and enable a way to fasten the board to a suitable support. I took off the sharp edges with a fine file and soldered wires to the board traces for RF and power. See Figure 36.

WAYNE YOSHIDA, KH6WZ

**Figure 36 — Rather than unsoldering individual components to build a VHF gain stage, a MMIC amplifier and its associated components were sawed out of a PCB assembly, making a complete, functional unit and avoiding the need to reinvent the wheel.**

Not only was the module fully functional, the little PCB looks very professional in my project! *73 — Wayne Yoshida, KH6WZ, 16428 Camino Canada Ln, Huntington Beach, CA 92649,* **Wayne.Yoshida@tycoelectronics.com**

## INEXPENSIVE PROJECT ENCLOSURES

◊ Nearly every Friday since my retirement, during the season, my wife and I hit the local garage/yard sales. They are great venues for "recycling" used merchandise. On occasion we have found some interesting treasures, including a near-mint Hallicrafters SX-88. While visiting a garage sale about a year ago, we discovered an item with great Amateur Radio potential. It was a data switch for a computer system. The switch came housed in a neat little steel box measuring approximately 2 × 6 × 4 (L × W × H) inches. A few years back, it was common for many computer setups to employ a parallel or serial data switch. The data switches are mechanical devices used to switch a computer's serial or parallel port to more than one media peripheral.

The data switch enclosures come in several different sizes and are ideal for QRP transceivers, small tuners or other accessories. With the advent of USB (Universal Serial Bus) technology, the old serial switches are now obsolete and dirt cheap on the used market.

Figure 37 is a photograph of one of the boxes I recently purchased at a garage sale for 50 cents. It is sitting atop a couple of boxes that have been stripped-out. One has an active filter built into it. The box has a switch control on the front and several serial ports on the rear panel. "Gutting it out" takes only a few minutes if you do not plan to recycle the parts. The next task is to somehow remove the lettering from the panels or to paint over the lettering. I decided to remove the lettering because of the nice quality baked-on beige finish that is original with these boxes. Rather than chance using some paint remover or solvent, I opted for "rubbing-out" the letter-

**Figure 37 — Before, during and after views of the recycling of a data switch box.**

ing with a rubbing and polishing compound known as "Mothers Polish."[1] This compound is made for polishing automotive magnesium or aluminum wheels. You can find Mothers Polish in the automotive section of any good hardware store. If you use this compound, the polish leaves a nice protective finish for the panel after the letters have been rubbed out.

I used a piece of an old terrycloth towel for the rubbing and polishing. Be careful not to get carried away with the polishing, it is possible to remove the beige finish with the lettering. By the way, Mothers Polish is good for polishing away flaws in plastic meter faces and I have used it to clean up dulled chassis and to smooth cabinet finishes on classic radios as well.

After you've removed the lettering, close up the openings on the rear panel using a thin piece of aluminum sheet metal, which can be salvaged from an old transcription disk cut to fit the inside of the rear panel with a RadioShack nibbling tool.[2] Holes for connectors and feed-through grommets can be easily drilled into the aluminum sheet metal to give the rear panel a neat appearance.

Since I started looking for these boxes at garage/yard sales, I have purchased at least eight or nine of them, each for less than a dollar apiece. With new enclosures running anywhere from 15 to 30 dollars each, recycling these old computer accessories is well worth the effort. — *73, Lawrence W. Stark, K9ARZ, 1875 Chandler Ave, St Charles, IL 60174-4601,* **k9arz@arrl.net**

## FERRITE CORE RESOURCE

◊ With today's proliferation of RF generating devices, often times it is advantageous to use ferrite cores to help eliminate interference in radio receivers. These ferrite cores are available commercially, but it is often confusing to determine the correct core mix to use. You may have to purchase an assortment to find the right one. Even then it may not solve your individual problem and you will have invested several dollars to no avail. I have found an alternate source for such cores.

Computer cables often have a ferrite core at the plug end. I have found these to be very effective. I take old cables that are not usable or have specially terminated ends and carefully remove the encapsulated ferrite core that is part of that cable. I can then use the retrieved core on my amateur cables.

Shown in Figure 38 are two different sizes taken from monitor cables. The plastic used to encapsulate these cores is tough, but with a sharp utility knife and patience you can remove the plastic covering and pull the multi-conductor cable out. The hole size inside the core is almost perfect for RG-8X sized or smaller coaxial cables, multiple wire cables or dual wire cables. In the case of the latter, several turns of the dual wire cable can be wound about the core to make it more effective.

While these cores may not meet every need, they do help a great deal and the cost is right. I find old computer cables at the local recycling center and at computer stores, which toss out unusable computer devices. There has been no cost associated with the recovery of these cables and in many cases my sources have been glad to have someone interested in taking them. — *73, Robert Brock, K9OSC, 6041 6th St NE, Fridley, MN 55432,* **k9osccw@yahoo.com**

## WEATHERPROOF SHRINK TUBING

◊ A few weeks ago, I was replacing the entire electrical system on my utility trailer. This trailer is parked outside, so all of the electrical connections needed to be weatherproof. I was going to use commercially made weatherproof crimp connectors that consist of a connector and shrink tube impregnated with hot melt glue, but they were expensive!

It occurred to me that I could duplicate this connector very cheaply by using hot melt glue and regular shrink tube. I also realized I could reduce the cost even more by soldering the wires and eliminating the crimp connector.

Here's how to do it: Solder the wires together and let them cool. Apply a bead of hot melt glue along the connection, almost the entire length of the piece of shrink tube. Let it cool before installing the shrink tubing. Next, evenly apply heat to all sides of the shrink tube — this shrinks the tubing down and remelts the glue allowing it to flow around the solder joint, rendering it weatherproof. Some glue will ooze out the end of the shrink tubing. The ooze is proof that the hot melt glue has been melted and is flowing around the joint.

To show that the glue does flow around the joint, I cut out a strip to view the inside of the connection. To my surprise, when I went to rip off the strip of heat shrink, I pulled off the insulation from the wire! The glue indeed penetrates the whole interior of the joint.

So far, I've had no electrical problems with the trailer. — *73, Benjamin Hall, KD5BYB, 102 Stoney Point Dr, Harvest, AL 35749,* **kd5byb@kd5byb.net**

## COUNTING TURNS ON VERY SMALL TOROIDS MADE EASY!

◊ I was faced with this task while building a FireFly 30 meter software defined radio kit (**www.qrpkits.com/firefly.html**). I tried to count the 56 turns on a ⅜ inch diameter toroid using a magnifying bench light. I couldn't get an accurate count. I decided to take a picture of the coil using the macro mode of my digital camera. I printed it out and was able to easily and accurately count the number of turns on the much enlarged toroid picture (see Figure 38). Also, I was able to keep track of my "place" with a pencil mark. Figure 39 shows 57 turns, I removed the extra turn. — *73, John G. Olson, W9JGO, 2707 18th Ave, Rockford, IL 61108,* **johng_olson_186@comcast.net**

JOHN OLSON, W0JGO

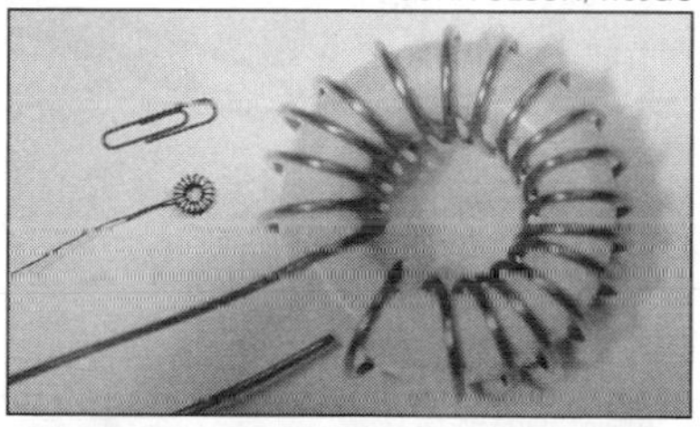

**Figure 38 — The actual toroid and its magnified view with a paper clip for perspective.**

JOHN OLSON, W9JGO

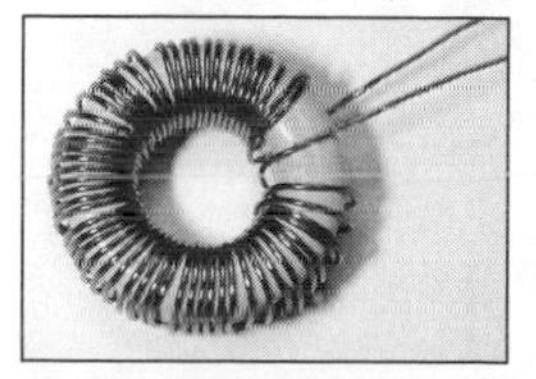

**Figure 39 — The completed toroid digitally magnified.**

[1]Mothers Mag & Aluminum Polish, 5456 Industrial Dr, Huntington Beach, CA 92649 (purchased at ACE hardware).

[2]Nibbler 29524 Metal Cutting Tool, RadioShack Stock # 55010716.

## SOLDERING RG-8X TO A PL-259

◊ I am a wire antenna experimenter. Tuning antennas that are in trees can be a challenge. At first, I was using the four hole solder method for assembling the PL-259 connector. I was always fighting broken cable sleeve connections (which can even get you going on a trimming wild goose chase). Then I remembered the method I used as a Novice. Let's call it the Threaded Sleeve Method for RG-8X coax:

1) Put coupling ring and adapter (UG-176) on the cable.

2) Tin one of the holes and surrounding outside area of the connector body; remove excess solder and let cool.

3) Strip the insulation back about 2½ inches being careful not to cut the shield (Figure 40).

4) Bend the end tight near the end of the outer insulation and open the shield without breaking any shield strands. If you break any strands, or pull any single strands loose, start over (Figure 41).

5) Pull the center conductor out through the hole in the braid (be careful not to let the center conductor spring out and cause single stray strands) (Figure 42).

6) Stretch the braid back tight around the center conductor. They should both be about 2⅜inches long (Figure 43).

7) Strip the center conductor to within ⅛ inch of the braid and twist all the strands together tight (Figure 44).

8) Remove 1 inch off the shield, cutting at an angle so the shield forms a point. Then bend the end of the shield to form a 90° angle about ⅜ inch long, positioned next to the center conductor (Figure 45).

9) Insert the center conductor into the center pin of connector. Start the shield through the hole with a pair of needle nose pliers; ensure all the shield wires come through the hole as you insert the center conductor (Figure 46).

10) Firmly seat cable into the PL-259 body and pull the shield tight through the hole in the connector (make sure cable does not twist). About 1 inch of shield should be sticking through the hole in the connector body (Figure 47).

11) Slowly tighten the adapter, keeping shield end taut. The shield should retract back through the hole a little during tightening. There should be no gap between the end of the connector and the adapter (Figure 48).

12) Check connector continuity. If open continue, if not, tear apart and start over.

13) Trim the excess shield off flush at the hole edge.

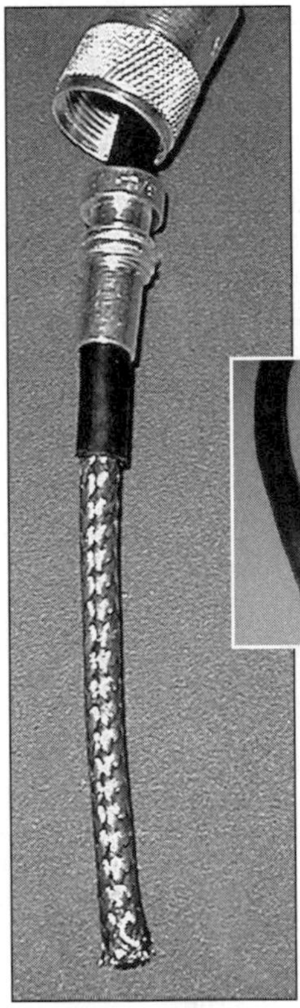

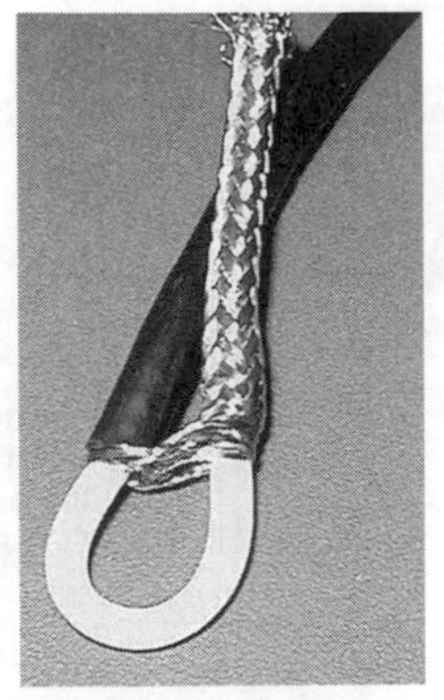

**Figure 42 — Pull center conductor out of the shield braid.**

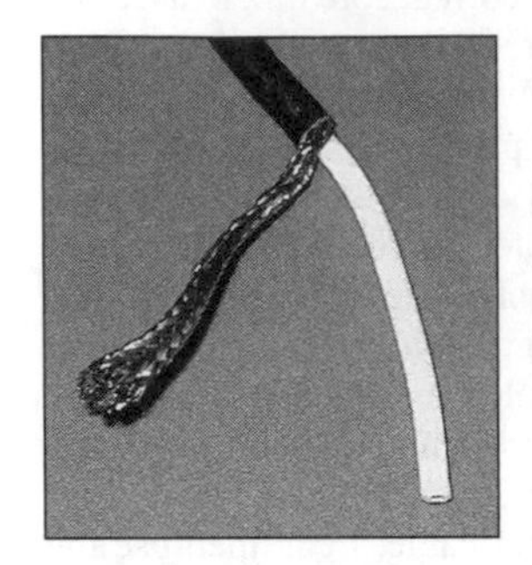

**Figure 43 — Pull the shield tight around the center conductor.**

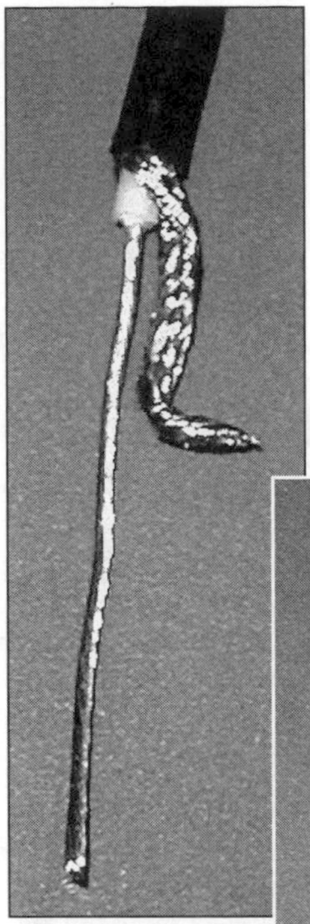

**Figure 45 — Trim and bend the shield to get it ready for insertion into the connector.**

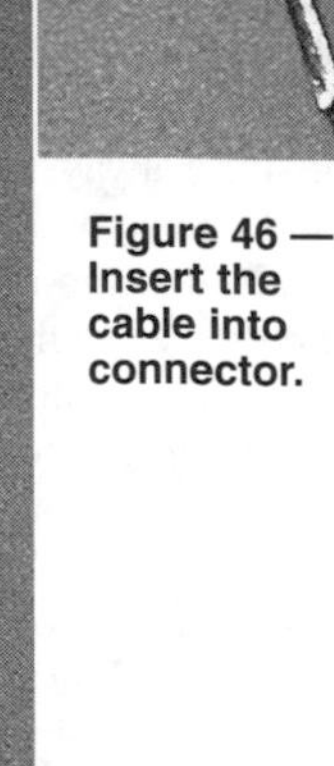

**Figure 46 — Insert the cable into connector.**

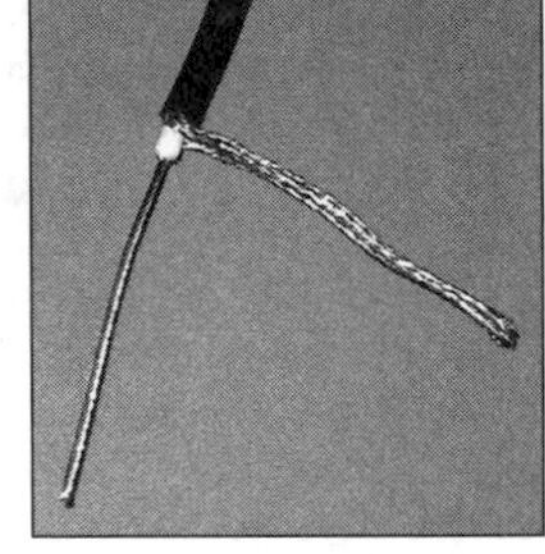

**Figure 44 — Strip center conductor down to within ⅛ inch of the braid.**

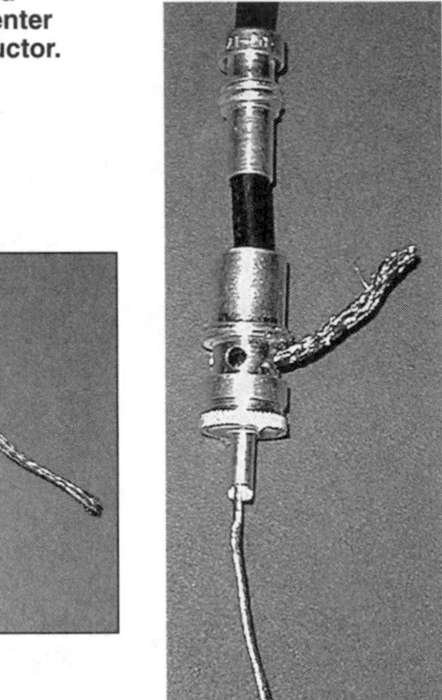

**Figure 47 — Pull shield out through the prepared hole.**

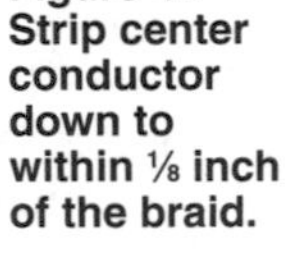

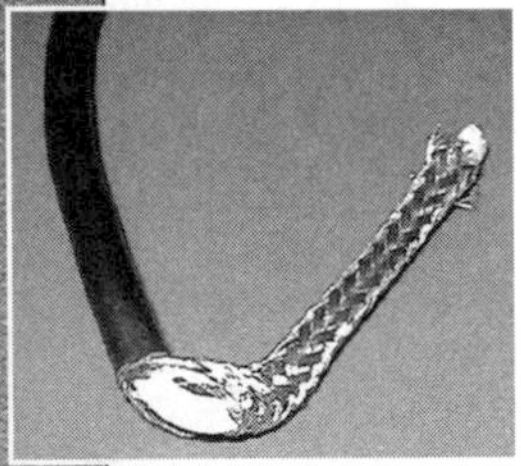

**Figure 41 — Separate shield braid to reveal center conductor.**

**Figure 40 — Strip off 2½ inches of outer insulation.**

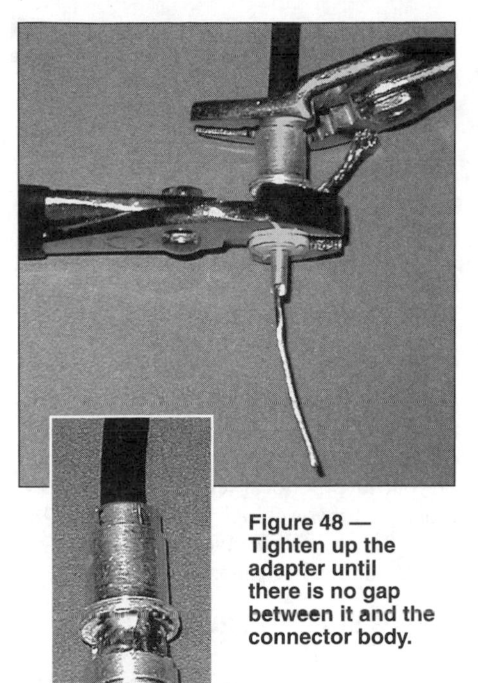

Figure 48 — Tighten up the adapter until there is no gap between it and the connector body.

Figure 49 — Solder the shield onto the connector body being careful not to overheat the coax.

14) Solder the shield in the hole using a 100-140 W soldering iron until hole is filled. Do not overheat or the coax will short. Let cool (Figure 49).

15) Check shield to center conductor continuity; if open when wiggled, trim the excess center conductor flush and solder. Let cool.

16) Check shield to center conductor continuity.

I think this method makes a much more rugged connector. Since I started using it, I have not had any broken or intermittent shields at the connectors. Also, if the antenna SWR is low, this method handles medium power just fine (my amp delivers 500-600 W output).

Network analyzer measurements (I used an Agilent 8753) of insertion loss, return loss and time domain discontinuities shows the threaded sleeve method outperforms the molded connectors on purchased jumper cables and is just as good as the four hole solder method. It looks like the molded crimped connectors have a lot more variability in them. Out of the four advertised low loss jumper cables that I bought, only one came close to performing like the ones that I soldered. The losses were still minimal (below 50 MHz).

This technique is only for the smaller diameter cables that use an adapter. It will work for RG-58 and RG-59 cables, too, but you have to strip off a ⅛ inch piece of center conductor insulation from RG-8X cable to be used as a bead on the center conductor of these cables. Otherwise, the smaller center conductor insulation will slide into the pin of the connector allowing the shield to contact the center pin and short out the connector. *— 73, Kurk Radford, K6RAD, 19471 Moon Ridge Rd, Hidden Valley Lake, CA 95467. Figures 3-12 by K6RAD. He can be reached at* **k6rad@arrl.net**.

## SIX SCREW-STARTING TIPS

◊ Starting a screw in a tight place can be difficult. If you are lacking a screw-holding, split-blade screwdriver, here are five ways to do it:

- A dab of rubber cement or superglue on a screwdriver's blade.
- A piece of paper wedged in with the screwdriver's blade.
- A rubber band twisted around the shank of a screwdriver and then a screw.
- A bit of plastic or vinyl tape around the blade and screw.
- A stiff piece of plastic tubing with the screw head pressed into the end can reach into small openings.
- [Another method: Cut a small piece of double sided tape (about ¼ inch × ¼ inch) and stick it onto the end of a piece of dowel or the eraser end of a pencil. Stick the head of the screw onto the other side of the adhesive tape. As you twirl the dowel or pencil, you can thread the screw into the fastener. This also works for starting nuts onto threaded portions of screws that are not otherwise easily accessed. *— Ed.*] *— 73, Gene Cabot, WB4ZS, 225 N Cove Blvd, Panama City, FL 32401,* **gncabot@aol.com**

## CARE OF SOLDERING IRON TIPS

◊ K4ZA mentions iron plated tips briefly in his article "Soldering — Tips for Shopping, Survival and Success" [*QST*, Mar 2006, p 51]. Here's how to maximize the lifespan of an iron plated tip. Never wipe the tip on the sponge after a soldering operation. Wipe it before soldering only. When shutting down the iron, clean the tip on the sponge, shut off the power, then tin the tip with solder and let it cool down. Never shut down an iron with an iron plated tip dry; always wet it with solder. *— 73, Dennis R. Murphy, KØGRM, 111 W Arikara Ave, Bismarck, ND 58501-2604*

## A 5 MINUTE SURFACE MOUNT DEVICE HOLDER

◊Recently, I was asked if I could repair a few Motorola Spectra radios. These radios have a known surface mount (SMD) capacitor problem and the most expedient way of repairing them is a wholesale change-out of its electrolytic surface mount capacitors.[1]

Rick Littlefield, K1BQT, published an article on building an aid for installing surface mount components.[2] There were, however, a couple of issues that precluded me from using the tool in the article. First, even though K1BQT's design is elegant and requires a minimal amount of assembly work, I needed something to work on circuit boards installed inside of a chassis. This would require some sort of modification to permit a vertical height adjustment for reaching over and into a chassis to the internal circuit boards. Secondly, I wanted to use something that would be even quicker to assemble.

### Description

Recently, I purchased a magnetic base for about $7.[3] A magnetic base is used by machinists to hold a precision measuring indicator in place and has to be adjustable in three axes to make measurements. It has a weighted base and a permanent magnet that can be turned "on or off" to aid in holding it in place. See Figure 50.

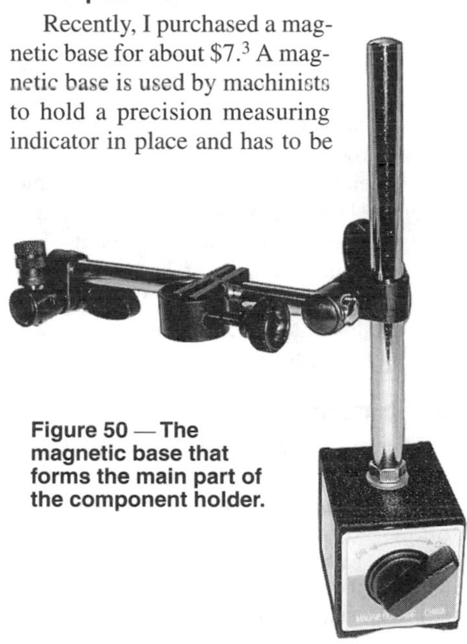

Figure 50 — The magnetic base that forms the main part of the component holder.

I wanted to build an SMD holder that could be adjusted vertically as well as horizontally. While thinking about something I might have around the shop that could do this, the magnetic base came to mind. I needed a component holder that would be adjustable in two axes and that could reach over and into the radios. The magnetic base can be adjusted in three different axes and had a vertical adjustment range of about 10 to 12 inches.

Looking at the base, it appeared it might just work, if I could find some way to modify the end to hold some sort of component hold-

[1]For more information on repairing these radios, see www.repeater-builder.com/motorola/spectra/spectra-caps.html by Robert Meister, WA1MIK, edited by Mike Morris, WA6ILQ.

[2]R. Littlefield, K1BQT, "Build a Simple SMD Workstation," *QST*, Jul 2000, pp 56-57.

[3]The magnetic base I purchased was from Harbor Freight tools (www.harborfreight.com). The catalog number is 5645.

Figure 51 — The indicator clamp with the bushing installed and the "needle" hole visible on top.

Figure 52 — Remove this part (bushing) to prepare the base for use as an SMD component holder.

down device. In the original *QST* article, a threaded rod was used.

An idea hit. I keep a large sewing needle in my tool box for making electrical measurements on conformally coated PC boards. The needle is 8 inches long, stiff, has a very sharp point and was stainless steel. Such needles are available at many craft stores and places where sewing and embroidery items are sold.

### Construction

The reason I call this the 5 Minute Surface Mount Device Holder is because that is about all the time it takes to assemble the unit. First, follow the instructions included with the magnetic base to assemble it. After assembly, on the end of the horizontal rod there is an indicator clamp. Attached to this indicator clamp is an adjustment bushing. Remove it. See Figures 51 and 52.

Rotate the indicator clamp 90° until you see a smaller hole. Push the 8 inch sewing needle through this hole and tighten the clamp to hold it in place. You're done!

### Operation

Operation is as simple as construction. Move the magnetic base and attached 8 inch needle over to the device under construction or repair. Rotate the needle point down. Adjust the clamps on the horizontal and vertical rods to position the needle over the top of the component to be held into place. See Figures 4 and 5. Tighten the clamps to hold the rods and needle in place.

If you have a non-magnetic workbench, or if you find the base not quite heavy enough to suit you, you can place a piece of sheet metal under the base and turn the magnet "on" for added stability. I found I did not need to do this, but it's another option available.

You'll spend more time purchasing the needle and magnetic base than you spend on assembly of this project. The few minutes spent assembling it will pay back many, many times over verses the tedious time spent trying to hold down a surface mount component with a toothpick or some other manual means. — *73, E. Kirk Ellis, KI4RK, 203 Edgebrook Dr, Pikeville, NC 27863,* **ki4rk@arrl.net.** *All photos for this item by E. Kirk Ellis, KI4RK*

## FOAM PCB HOLDER

◊ Here's a simple fixture that makes assembling small PC boards a snap. Figure 53 shows it in use.

A scrap piece of ¼ inch aluminum sheet forms a base for the fixture. Mount two 1 inch standoffs about ½ inch farther apart than the longest dimension of the PC board to be assembled. Now cut a piece of 1 inch thick soft foam packing material to slightly larger than the size of the board so that it fits snugly between the standoffs. Mount two crimp type insulated wire terminals to the standoffs; keep the screws slightly loose.

To use the fixture, insert several components into the board, turn both the board and the fixture vertical, press the component side of the board to the foam and return both to the horizontal position. Press down on the board to compress the foam and rotate the two wire terminals to hold the board in place.

The weight of the fixture keeps the board in place and the pressure of the foam holds the components firmly to the surface of the board. Soldering them in place is easy. — *73, Paul Jacobs, W2IOG, 6861 S Gannett Hill Rd, Naples, NY 14512,* **phjacobs@gmail.com**

## ANOTHER USE FOR A WATCH-BACK LIFTING KNIFE

◊ The watch-back lifting knife is designed for lifting the back of a watch case in a nonmarring fashion. I've found the watch-back lifting knife very useful for removing the backs from the many plastic cases that we run into everyday in ham radio and in electronics in general. It is helpful for opening everything from portable radios to cordless telephones and beyond. If carefully done, the knife will lift the backs of those devices in the least marring way (much more so than any screwdriver would).

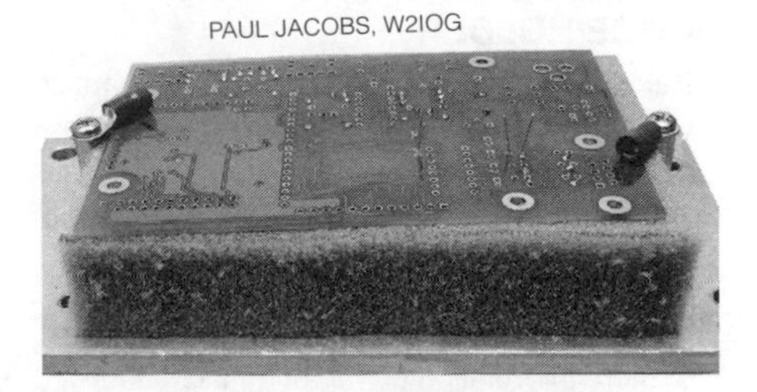


Figure 53 — The W2IOG PCB holder in action keeping components in place for soldering.

Figure 54 — This special tool designed for opening the backs of watches without damaging them can be used around the shack to open a variety of cases.

Watch-back lifting knives (see Figure 54) are available from any jewelry supply house (there are a number of them on the Internet), or your favorite local jeweler can probably order one for you. — *73, Dave Miller, NZ9E, 1216 Terry Andrae Ave, Sheboygan, WI 53081*, dmiller14@juno.com

## O-RING SEAL FOR PL-259/SO-239 CONNECTORS

◊ The popular and low-cost PL-259 UHF plug and its mating SO-239 UHF connector were never designed to be waterproof. In spite of this, many of us find ourselves having to use these connectors out-of-doors for an antenna or cable connection and then have the experience of trying to apply some form of weatherproofing.

If this is not done, and done well, then intermittent connections soon result, especially with wet freezing and thawing conditions, resulting in moisture and corrosion getting into the connectors and even into the coax itself.

One well-known solution is to use a bondable rubber tape wrapping. That often seems to work using one of the soft, moldable, stretchable tapes that are available for the purpose. In practice, I have found that the hardest area to seal effectively using this method is the area from the rim of the male PL-259 shell to the mating SO-239 socket face. This is especially true when the SO-239 socket is located partially inside the cowl of the antenna mount (such as the Solarcon/Antron vertical or other similar types) and it is impossible to get sealing tape up into this tight location.

To overcome this problem, simply slip as many tight fitting O-ring(s) over the SO-239 barrel (see Figure 55) as needed to fill the space between the base of the

Figure 55 — Place the O-ring(s) at the base of the SO-239 barrel and use the PL-259 coupling ring to compress them into a weather-tight seal.

SO-239 and the coupling ring. Lubricate the O-rings with a generous amount of silicon grease, often sold in the automotive part stores as "dielectric tune-up grease."

*Important*: Check that when the mating PL-259 plug is installed and tightened down, the outer rim of the PL-259 shell butts up against and slightly compresses the O-ring(s).

The O-rings I used were ones I had on hand, about 9⁄16 inch ID × 0.10 inch thick. I am sure anything that's a snug fit will work. Of course, you will still need to wrap the lower half of the PL-259 shell and where the coax exits the connector with tape to seal that part. — *73, Don Dorward, VA3DDN, 1363 Brands Ct, Pickering, ON L1V 2T2, Canada,* **va3ddn@arrl.net**

## MAKING FRONT PANELS — THE EASY WAY

◊ My first home brew project was a real time-consuming effort, not for the electronics involved, but rather for the final front panel lettering. Using rub-on lettering was a tedious task that took a wizard's magic to insure the letters were straight and evenly spaced. Then, shortly after, the letters would rub off in places due to normal handling. I even tried spraying the lettering with clear spray. This solved the "disappearing" problem; at least making the lettering last much longer. The bottom line was that it just was not a good method of lettering.

Sometime later, several lettering devices with tapes began showing up in stationery stores. I bought one with anticipation it would replace the rub-on lettering task. It did okay, but on larger chassis it left much to be desired. The tapes were expensive and there were few options available; colors and fonts were very limited. It was easy to see the tape and this made the front panel look "cheap."

Looking for an alternative method, I developed a quick and satisfactory way to make professional looking front panels. From start to finish the sample project shown took about 3.5 hours to complete and most of that time was involved in laying out the overlay on the computer and drilling the holes through the metal front panel. Things you will need are: (1) a computer, (2) a printer and (3) a laminating machine. For my own panels, I use the software *Visio,* which permits total versatility for fonts, colors and designs. There should be several other good programs available, which will suffice.

First, use *Visio* to lay out the design. Measure the front panel, then decide where you want the controls, switches, lights, etc. Setting up the drawing, use a 1:1 scale (see Figure 56). For larger panels, such as the 19 inch rack mount type, you may want to use a scale that will fit your full page drawing on the screen. *Visio* has grids, rulers, various unlimited shapes and colors, with many other options you may need to lay out the overlay.

If it is your design, you can lay everything out on the *Visio* drawing. If you are designing an overlay for an existing panel, you will have to use a micrometer to measure the center-to-center distance of each hole to transfer to the computer overlay. Should you decide to switch or add anything to the layout, it can be easily moved, added or even deleted. That's the beauty of the computer.

Once you are satisfied with the drawing, save it, then print out a copy. Cut the overlay to size, fit it on the panel and check the layout. If you need to make changes, just go back into the program and modify your original. When you go to save it, the program asks if you want to overwrite it. I would advise keeping your original drawing and save the new updated one by another name just so you can get back to the original drawing if needed. Once the layout is correct, you are ready to begin the final task.

Print out two copies of the overlay drawing. One will be used as a drill guide and will be destroyed. The other will be used to laminate and use for the final product. Securely tape one copy cut to size to the metal front panel and using a center punch, mark all holes by punching through the printed copy. This will dimple the metal panel under the overlay. My program gives me crosshairs, circles, or square boxes for the hole location on the overlay. You can make one copy with the crosshairs and delete them for the final copy to be laminated.

Once all the holes are center punched on the metal underneath, remove the paper and drill all the holes. Start with a small drill and increase the hole diameter by increasing the drill size in about 1⁄8 inch increments until the desired hole size is obtained. This may take as many as five or ten drills, but the center will remain fairly close to the desired location. After all the holes are drilled, use a de-burring tool or a larger size drill to remove any burrs on the hole edges. Now you can remove the old overlay and discard it.

Next, put the laminated overlay on the panel and hold it up to the light to see how close the holes were drilled to the target (see Figure 57). If it is within tolerance, with all

FRED E. BOYER, N3QK

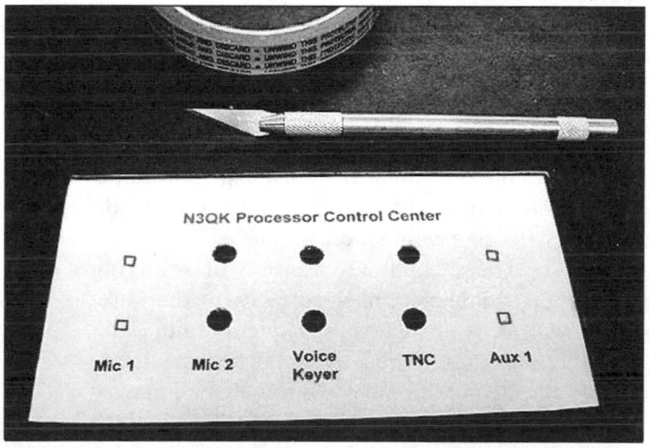

**Figure 57 — Trimming and cutting the holes with an X-acto knife using the metal front panel holes as a guide.**

FRED E. BOYER, N3QK

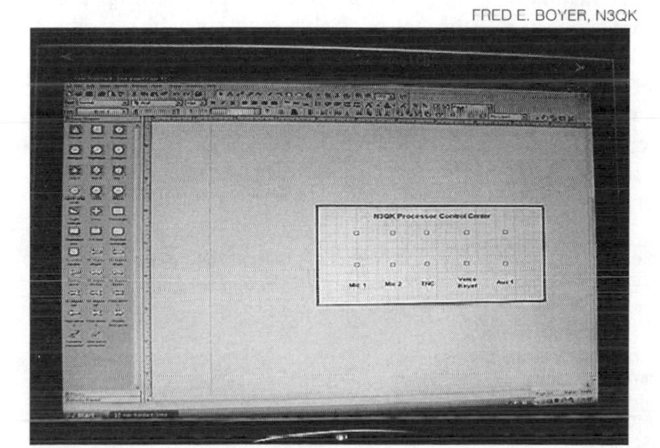

**Figure 56 — This is the *Visio* layout of the front panel as shown on the computer display.**

FRED E. BOYER, N3QK

**Figure 58 — Apply double-sided tape to the front of the drilled panel as a base for the label sheet.**

clearances looking good, apply double sided scotch tape (available at any office supply store) to completely cover the overlay side of the metal front panel. One layer of tape is sufficient and should not be doubled (see Figure 58). Carefully align the overlay to the panel, and then press down all around it to make sure the double-sided tape adheres to it. Trim any access overrun on all edges with a sharp X-acto knife. Using the front panel metal holes as a guide, take the X-acto knife and carefully cut out the overlay holes through the lamination. Keep the knife inside the metal holes as you cut away to prevent damaging the overlay outside the intended cut away areas.

Now you are ready to "stuff" the front panel with all the electronics. All work is done and the lamination will protect it from most harmful elements such as dirt, dust and wear. If an addition to the overlay is later needed, the tape lettering using the same style and size font works just fine. (This is one thing to consider when choosing your font in the overlay program.) My tape printer is an inexpensive Brother P-touch labeler. Check the fonts available on your overlay program software and use one that is likewise available on the tape printer.

Cleaning of the overlays should be done with a dry brush or cloth. Liquids should be avoided because the paper under the laminate tends to wick up the moisture. This will damage the overlay and probably discolor it. — *73, Fred E. Boyer, N3QK, 8531 Trout Ave, Palm Bay, FL 32909*, **n3qk@att.net**

## SIMPLE PROTOTYPE PCBs

◊ I know this subject has been covered more than once but here are two ways I use to make print circuit boards when I cannot justify having them made or laying them out myself on the computer.

The first is to start with a piece of copper clad board and a straight edge. Using a fine line marking pen, lay out a grid pattern of squares or rectangles the size you might need on the board; remember to allow for a ground and power bus as well. Then using a Dremel tool and the plastic or metal cutting blade, carefully cut along the lines you just laid out. Do not use too much pressure. Just enough to remove the copper or you might cut right through the board.

When I am done I have a board with a number of separate little copper pads on which to mount my parts. Depending on one's ability with the tool, one could also align the pads to match dip ICs as well as power transistors of various case styles and other parts. Even some surface mount chip resistors and capacitors, though I would not recommend this procedure for direct mounting of the smaller surface mount ICs. The best part of this procedure is that you do not need to etch the board, which saves considerable time and etchant.

The second method is to lay out the circuit on the board using paint pens. These can be purchased at any arts and crafts store. They are available from very fine lines to large and broad. The paint dries very quickly and they make a far better resist than even the resist markers that are usually sold. They also come in a number of colors so I can color code the traces if I wish. Once the board is etched, the paint can be removed with lacquer thinner. — *73, Chris Maukonen, WA4CM, 1043 Chatham Pines Cir #307, Winter Springs, FL 32708*, **chris@ucf.edu**

## MOUNTING PL-259 CONNECTORS

◊ My hint consists of two basic steps to simplify the assembly of PL-259 coax connectors onto RG-8/RG-213 coaxial cable.

The first step involves the tool used to separate the cable's outer sheathing from its shield. The procedure described in *The ARRL Antenna Book* suggests using a sharp knife.[1] That is the "old standby" method, but it is difficult to judge the requisite depth of the cut as the knife is moved around the periphery of the cable, often resulting in damage to the shield.

The procedure can be simplified somewhat, as previously described in your column, by replacing the knife with a small electrical conduit cutter.[2] The depth of the cut remains the same as the cutting tool is rotated around the outer sheathing of the coax and the depth is increased as the cutting routine progresses. Removal of the sheathing after the cut has been completed can be facilitated by using a sharp electrician's knife to make a continuous lengthwise cut into the sheathing. Then, open the sheathing along the lengthwise cut peeling it off the cable.

The second step consists of soldering the connector inner surfaces to the outer surfaces of the exposed shield braid. After the connector body is routinely rotated onto the end of the cable, the shield braid may be viewed through "solder holes," which in the parlance of the trade are called solder cups.

The problem arises that the solder cups often refuse to accept sufficient heat to bond the braid and the connector body. Oftentimes, the technician heats up the entire connector, trying to make the solder "stick."

The root cause for the problem is exposure to the atmosphere. A layer of oxidation impurities coats the connector surfaces. When heat is applied, the oxidation flakes off and creates a heat barrier between the iron and the connector.

I came across this problem early on in my various "careers" in and around electronics, and it was resolved by an Elmer's advice, which I am now passing on to Hints and Kinks readers. The gimmick involved hasn't failed me over the past 40-odd years.

Apply a few drops of lightweight machine oil (hardware stores stock it as "3-IN-ONE" oil; previously it was called "sewing machine oil") into the solder cup before applying the soldering iron and solder. The oil should cover the braid and the sides of the hole.

Dipping the nose of the hot soldering iron into the hole and holding it there for maybe 2 seconds maximum will cause the impurities to float to the top of the oil. Then, quickly apply the solder into the hole along the nose of the iron. The molten solder will "cup up," forcing some of the oil out of the solder hole. Instant removal of the solder and the hot iron will facilitate fast cooling of the braid-connector joint.

The job won't take more than 5 seconds — 10 max — and there's no need for any additional heat. If the cooled solder has risen over the top of the cup, slide the side of the soldering iron laterally across the solder to smooth it out. You may elect to wipe off any excess oil. I seldom bother, because the oil also serves as a corrosion inhibitor on the outer surface of the connector. Rotate the cable and duplicate the soldering application in as many holes as necessary (I usually stop at two).

I use a middleweight Weller iron, dual temperature, 100/240 W. It's more than enough. This "oil-n-heat" method can also be utilized when making ground connections on large chassis areas. Then it's convenient to prepare the solder spot by using a little child's (modeler's) clay to form a cup around it. That way, the oil isn't apt to migrate and a lot of impurities can be boiled off the mating surfaces. — *73, Mike McAlister, KD6SF, 7570 Dartmouth Ave, Rancho Cucamonga, CA 91730-1534*

[1]Available from your local ARRL dealer, or from the ARRL Bookstore, ARRL order no. 9876. Telephone toll-free in the US 888-277-5289, or 860-594-0355, fax 860-594-0303; www.arrl.org/shop/; pubsales@arrl.org.

[2]R. Arnold, AF8X, "Preparing Coax," *QST*, Feb 2006, p 69.

## PROJECT BOXES ON THE CHEAP

◊ How many times have you built a slick little circuit that needed just the right small project box? How many times have you dipped into the junk box only to come up with something that just plain old won't do the job?

What's worse, take the time to run down to your favorite discount electronics store (and in some cases, not so discount stores), only to find that the exact box you want is too expensive, too ugly or not there. Not that amateurs are "thrifty" or anything, but we would rather find the exact part in our junk boxes or at least find the parts rather inexpensively.

As we all know, hams are a resourceful lot. Some of us addicted to that resourcefulness, as alluded to above. This ham is very partial (I won't say addicted) to frequent trips to the local thrift stores. Visit the store of your choice —the Veterans, Goodwill, St Vincent de Paul, etc. Take a look in the "electronics" section. I find them often in the back or in the corner of the store where you will find the ubiquitous old black and white and color TVs and the equally ubiquitous VCRs (maybe even a Beta model). Occasionally you may run across some antique finds, such as a tube radio, lots and lots of AM/FM radios and recently a section for PCs and associated peripherals. Stop here.

If you were into PCs in the '80s and early '90s you will remember that printers were expensive, as were modems, and to share a video monitor, keyboard and a mouse was almost unheard of. To the rescue came a multitude of small well made switch boxes. I have found parallel switch boxes to share printers, serial switch boxes to share RS-232 devices and even early model KVM (Keyboard, Video, Monitor) switches to share one set of peripherals with several CPUs. There are even video signal switchers.

The prices on these little jewels beat the socks off of the "discount" electronics store prices. I have spent from $1 to $3 for these boxes. My best find so far (and my last) has been at the local Goodwill store where I ran into a cache of 18 switch boxes still in bubble wrap and in their cardboard boxes for $1 each; total outlay was $18. My wife will not concede that I made a good purchase, but then, I'm not sure I have 18 cute little circuits to construct and put in my boxes!

Some of these boxes come with a few pretty good components. For instance, a big multiposition switch is almost always included (sometimes push button switches); several jacks, generally DIN, RS-232 or parallel ports; DB25s, and a pretty good adult sized knob. The boxes are generally small, say 6 inches wide × 2 inches high × 4 inches deep, and fit on the operating desk nicely.

My first attempt to use one of these small boxes was when I was re-experimenting with crystal radios. As a kid I was fascinated by the "free radio" that a crystal set provided. I had built many kinds of crystal radios and housed them in things like matchboxes and other small pill boxes. They worked very well.

I had just built a high performance crystal radio using toroids and even though the sensitivity and selectivity were very good for a crystal radio, I still found that strong stations nearby could really wipe it out. Harken back to the old AM days, where we (some of us who remember AM) used the "Exalted Carrier" method to bring up the weak stations. That same technique can be used with crystal radios. I found a slick oscillator circuit on the Web that I built and rebuilt, finally building it in one of the cast-off switch boxes. (Works well!)

Need a project box? Go to your local thrift store! — *73, Dennis Merritt, WB6UHQ, PO Box 247079, Sacramento, CA 95824-7079,* **wb6uhq@arrl.net**

## SPRAY GREASE WARNING

◊ While reworking a rotator that survived Katrina, I decided to use up some spray lithium grease instead of buying a new tube of grease. Everything was fine until I reassembled it and tried it out on the bench. There was the kind of "*whump*" nobody wants to hear; it sounded like the kind of an arc that welds wires together and ruins circuit breakers.

After I reopened the case, I found nothing wrong. The fuse in the controller was intact and the gears ran fine. Then I noticed a spark in the contacts. The propellant in the spray can was petroleum-based and there was enough vapor dissolved in the grease to reach lower explosive levels inside the rotator case after I closed it up. If you use spray can grease on a rotator, give it an hour or so for the propellant to evaporate before reassembling it. — *73, Patrick Hamel, W5THT, and WD2XSH/6, 1157 East Old Pass Rd, Long Beach, MS 39560,* **pehamel@cableone.net**

## ALTERNATIVE GROUNDING MATERIAL

◊ Ed Sutton's, KD7PEI, informative article has, I'm sure, inspired many of us to take a good look at our ground systems. Whether or not one has the luxury of using 1½ inch or 3 inch wide solid copper strap, a good ground is important. Most amateurs use large-conductor wire or flat braided grounding strap. Both round-conductor wire and flat braided grounding strap can bend, sometimes sharply, and this may introduce an unacceptable amount of inductance.

A readily available alternative is to use plumbers' hanging strap, available at a hardware store. This strap is about ¾ inch wide and is usually made of steel, either plain or copper coated. It has holes at lengthwise intervals of about ¾ inch. Plumbers' strap can bend, but because it is a flat and solid conductor, it is likely to bend only as intended. — *Bob Raffaele, W2XM, 5 Gadsen Ct, Albany, NY 12205-1309,* **w2xm@arrl.net**

## CRACKING WALL WARTS

◊ I too have done battle with the recalcitrant wall warts unwilling to give up their inner secrets without a fight. I use a Dremel tool with a number 409 cutoff disc to cut along the factory glue line. This provides the precision necessary to cut the case without damaging the transformer and wiring. I close the wart with tie wraps versus tape to avoid the mess of adhesive residue. — *73, Mark Snowdon, KG4UDL, 409 Mills Ave, Pensacola, FL 32507,* **msnow20989@aol.com**

## ASSEMBLING TYPE N CONNECTORS

◊ Going through the April 2008 issue, I read the article on the proper assembly of N connectors with interest as I use them extensively at work.[1] I have a different method:

[1] J. Hallas, W1ZR, "Those Type N Coax Connectors," *QST*, Apr 2008, p 69.

EDWARD COLE, KL7UW

**Figure 59 — Use a flat headed screwdriver to press the braid onto the clamp's rim, and then trim off the excess braid that extends over the rim.**

Everything up to combing out the braid is the same, but I do not pre-trim the braid. I find I can comb the braid over the clamp, and then use a straight screwdriver or other straight blade to push the braid tight into the corner at the base of the clamp's rim (see Figure 59). This causes the extra braid length to extend outward radially where I trim it flush with very sharp side cutters.

I do not pre-tin the center conductor. It makes it difficult to slide on the center pin. I do spray the copper conductor and pin with circuit cleaner (or alcohol) to remove any machining oils that interfere with heat transfer in soldering. This is usually adequate to get a nice quick flow of solder. Be careful not to slop or glob on too much. Leave the iron on the pin long enough to see the solder flow. If necessary, you might have to remove some of the solder on the surface of the pin for it to slide into the outer connector body.

[In the April article, the author suggests turning the connector body instead of the nut. Following a number of knowledgeable comments from readers, the author reports that he is now convinced that it is preferable to turn the nut while holding the body and cable steady. — *Ed.*]

Assemble the tower-top ends of your cable inside on the bench so you are not fumbling with that task up on the tower. I place a short (2 inch) piece of heat shrink tubing over the back shell of the connector to ensure weather tightness. Be sure to double wrap the connection with electric tape/Coax-Seal. You can cut the cable to length on the end that goes to the shack and assemble that end much easier.

If you are making cables up for 144 MHz or higher I recommend you upgrade to a better coax than RG-213. I like Times Microwave LMR-400 but many prefer Belden 9913 (be sure to get the version that uses solid foam insulation as opposed to the old air dielectric with polyethylene spiral bead). RF Parts (**www.rfparts.com**) makes a good connector that fits these two cables, which are equal in RF performance. Its part number is RFN-1002-1si (male). Both LMR-400 and 9913 are larger than RG-8, so the 82-202 connector will not fit. If you insist on UHF then I recommend Amphenol (**www.amphenolrf.com**) over generic brands for better soldering. I prefer the TNC to BNC as a much better connector for RG-58. Yes, they cost a bit more ($5.50-$6) but the increased performance and reliability are well worth it. With practice, you will prefer their assembly to the old PL-259 (UHF) connector. — *73, Ed Cole, KL7UW, PO Box 8672, Nikiski, AK 99635-8672,* **ecole@cispri.org**

## ATTACHING FRONT PANELS

◊ In the March issue of *QST,* a "Hints and Kinks" hint involved using double sided tape to fasten the faceplate to a project case.[2] The picture showed exposing all of the tape before sticking the faceplate on. I would recommend only exposing the top row of tape to allow easy alignment before sticking to the exposed tape. Then just remove one adjacent tape backing at a time pressing the faceplate from the top down and from the center out to the edges this will help eliminate air pockets from forming. — *73, Dave Palmgren, N8DP, 6132 Co 420 - 21st Rd, Gladstone, MI 49837,* **n8dp@arrl.net**

## REPAIRING SLUG-TUNED COILS WITH DENTAL FLOSS

◊ Many miniature slug-tuned coils and transformers use a thin strand of rubber band-like material running the length of the tuning slug to secure it in place and keep it from vibrating out of adjustment. When working on older radios, I often find that the rubber has become brittle and disintegrates when the slug is turned.

A quick, simple and readily available replacement is waxed dental floss. It is durable and the wax provides just the right "stickiness." One strand is usually not enough, but it is easy enough to twist several stands together for the needed thickness. I find that four strands twisted tightly together are usually about right for miniature coils. Remove the slug, put the twisted strands into the center of the coil perpendicular to the threads and thread the slug back into place. Snip off the excess, adjust the slug as necessary for the circuit and you're in business. — *73, Richard Manner, KGØXO, PO Box 630724, Highlands Ranch, CO 80163,* **RLManner@yahoo.com**

## SINGLE PADDLE OPERATION WITH IAMBIC PADDLES

◊ My first electronic keyer was primitive by today's standards. It used 6SN7 tube(s) and had 300 V on both sides of the paddle. It worked well but conditioned me such that I've never been able to master the iambic feature. The availability of non-iambic paddles is very limited and they're costly. I found that it's possible to find good, inexpensive, used iambic paddles for sale.

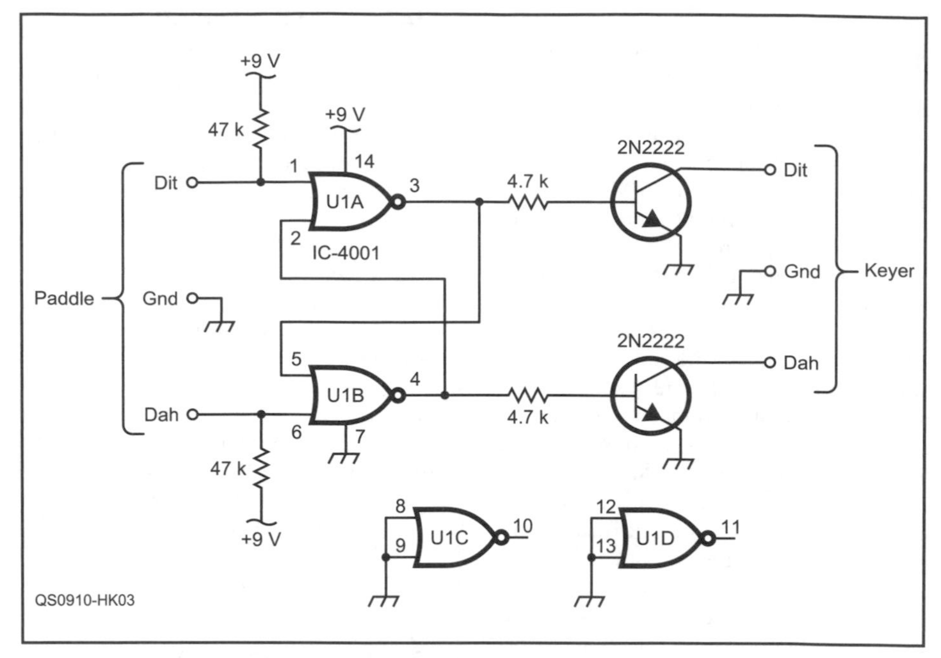


**Figure 60 — The multivibrator circuit to make an iambic paddle act like a single paddle key.**

So I thought, could I make an iambic paddle behave like a single paddle? I developed a circuit (see Figure 60) that converts an iambic paddle to non-iambic operation. It uses an SR flip-flop to prevent both keyer inputs from being active simultaneously. It's simple, the cost is minimal and the size is small. I used a CMOS CD4001 because of its low current requirements and the life of a 9 V battery should be close to its shelf life eliminating the need for a power switch. A 7402 TTL (transistor-transistor logic) chip should work just as well but would require an external 5 V power source plus the pull-up resistors on the inputs should be reduced to 4.7 kΩ and the pins will have to be adjusted.

My paddle is now comfortable to use with my CMOS-III keyer. — *73, Larry Winslow, WØNFU, 4500 Whitman Ave N #4, Seattle, WA 98103-6664,* **w0nfu@arrl.net**

## PIGEONHOLE WORK TOP

◊ Here is a method of working on projects which will enable you to keep up with small hardware and tools. It is a pigeonhole wood frame that rests on your workbench top and performs a number of functions not possible with a normal flat work top (see Figure 61).

**Figure 61 — The pigeonhole frame can hold a multitude of small parts, tools and other workshop sundries, keeping them all in easy reach during a project.**

The project you are working on is laid on the center front of the frame's dividers, convenient to your hands. You can store tools, parts, solder (I keep four rolls of solder in the bottom left holes) and even a small clamped light to illuminate the work. Note that pliers fit handily in the dividers, as do screwdrivers, knives and even the DVM when mounted on a paint stirring stick with a rubber band. Small equipment, panel meters and PC boards can be placed at an angle convenient for testing and you can temporarily store line cords and meter test leads in any convenient hole.

A few 2 × 6 inch wooden tabs with rubber bands are handy to keep parts visible while working on equipment. Pairs of alligator clips, soldered together back-to-back can be clamped on the dividers to hold wires or components.

[2]F. Boyer, N3QK, "Making Front Panels — The Easy Way," *QST,* Mar 2009, pp 75-76.

A couple of well placed nails in the top of the supporting workbench will keep it from sliding. You can place supports under the far end to tilt the surface or if you prefer a variable tilt, a pair of quarter inch bolts and angle brackets on the sides will suffice. If you need a flat surface, a clip board can be laid on top, using the clip to hold a PC board or diagram.

My pigeonhole work top was bought at a yard sale, doubtless originally designed to hold small beverage glasses when mounted on a wall of a club or meeting room. Laid flat on your workbench, it makes an ideal work surface that keeps control over tools, knives, small parts and everything but the soldering iron, which is kept on the side for safety.

After my garage sale purchase, I found a similar item in the Harriet Carter catalog. It is called a *Shot Glass Curio* and sells for under $20. If you have woodworking skills, mine is constructed from plywood and its dimensions are 20 × 27 × 2½ inches. The depth of 2½ inches seems optimum because deeper would make retrieval of parts more difficult. Each hole is 2⅜ inches square.

It will seem strange at first to work on a "holey" surface, but after a few minutes you will appreciate the greater utility of the uneven pigeonholes. — *73, John Townsend, W4RIZ, 2406 Foxcroft Rd, Wilson, NC 27896*

### 12 V SPICE JAR ADAPTER

◊I had a 12 V socket I had mined from an old portable air compressor that didn't work. My small camping power supply (a 14 A switching Jetstream) did not have a 12 V socket. I had hoped to run my handheld transceiver and another radio from the power supply, but my only option was a 12 V adapter.

I took an onion salt spice jar I had cleaned out and took the lid with all the shaking holes off to mount the 12 V socket to. I used my calipers to determine the diameter of the hole to drill. I then used a Forstner bit to slowly drill the hole in the lid. I then attached the 12 V socket to the lid and wired it up with a small jumper of #16 AWG wire (see Figure 62). (This was more than sufficient for my application. Heavier gauge wire should be used for higher amperage equipment.)

GREG LOTT, K4HOE

**Figure 64 — The adapter attached to the power supply.**

I then drilled a small hole in the side of the jar near the bottom for the wires to come out. In my case I put Powerpoles on the wires (see Figures 63 and 64). I had originally thought to tin the wires and use the banana plugs that screw down on the wire. This would allow you to push the wires all the way in the jar when packed up. Drop the banana plugs in the jar as well and put a piece of electrical tape over the hole for transport.

I do not know if the seal made when the spice jar lid is screwed down over the 12 V socket is waterproof. — *73, Greg Lott, K4HOE, 611 Hambaugh Ter, Homewood, AL 35209*, **k4hoe@arrl.net**

### KEYLESS CHUCK ADAPTER

◊ I'm not a regular tool guy but I do have a battery powered drill with a ⅜ inch chuck, which I find handy for crafts work. My drill uses a "Jacobs Keyless Chuck," which depends upon a friction fit to hold a bit. Often when working on some hard material, the bit will get stuck in the material and, when I pull the drill away, the bit stays in the material. It takes a pair of good pliers to remove it.

So I was surprised to learn that Jacobs also manufactures keyed chucks that can be installed on the drill's keyless chuck. My Jacobs keyed chuck is a Model 30247 Multi-Craft chuck and should be available in any well equipped tool store.

The replacement chucks attach to the drill using a single machine screw, which you'll find deep within the biting edges of the chuck. It took me a few minutes to realize that the screw has a left-hand thread; meaning that when loosening the screw, it must be turned clockwise. The replacement chuck comes with sufficient information to permit it to be attached to your drill. — *73, Arthur McAlister, KD6SF, 7570 Dartmouth Ave, Rancho Cucamonga, CA 91730-1534*

### HOMEBREW RF ENCLOSURES — ANOTHER APPROACH

◊ In "Hints and Kinks" of January 2008, Gary Richardson, AA7VM, described how to fabricate enclosures from copper flashing.[1] I have been making copper flashing enclosures for several years and would like to describe my method.

1. First, lay out your design on paper.
2. After you are satisfied with your layout, make a mockup of the enclosure using card stock (I use old file folders). Interior partitions can be added and you can confirm your measurements for proper form and fit at this time.
3. Transfer your design to the copper flashing using a straightedge and scriber.

I make my cuts using ordinary 3 inch tin snips. Taking time to plan and lay out the cut lines is well worth the extra effort. This helps maintain RF shielding integrity — and makes for a nicer-looking box. The end result is better than factory-made.

My design is a bit different (see Figure 65), utilizing four pieces of copper for the two sides and the two covers. This makes it far easier to form the four sides of the basic box. The "top" and "bottom" covers are made

GREG LOTT, K4HOE

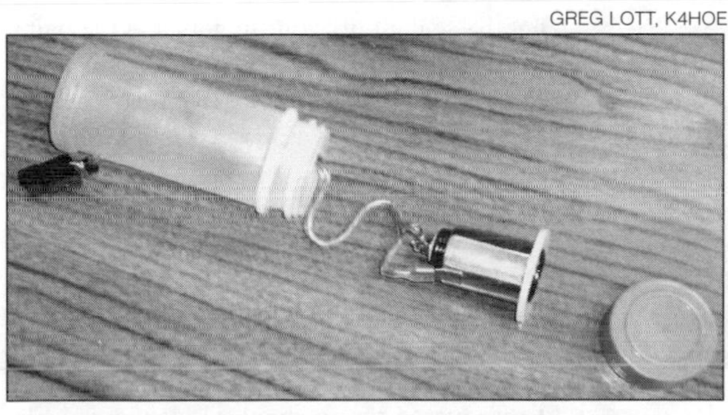

**Figure 62 — An exploded view of the spice jar power adapter.**

GREG LOTT, K4HOE

**Figure 63 — The adapter ready for action.**

[1]G. Richardson, AA7VM, "Making a Homebrew RF Enclosure," *QST*, Jan 2008, p 68.

separately. I have found it much easier to make the box this way, rather than to try to bend up the four sides. The resulting box corners are much stronger with the overlapped seams. Notice that the cover lips that mate with the sides can be inside or outside of the box — depending on project requirements. There are no tricky cuts or bends using these methods. The covers, properly made, will be RF tight.

The side pieces are clamped at the corners with a pair of hemostats during soldering. I use a 200/260 W soldering gun and rosin core solder. After the top and bottom covers are made, they can be pushed onto the box and drilled for the securing screws. This way alignment between the box and cover holes will be absolutely perfect. If you don't require access to both ends of the enclosure, you can solder one of the covers in place permanently. Prior to mounting parts inside, the enclosure can be cleaned with acetone or alcohol.

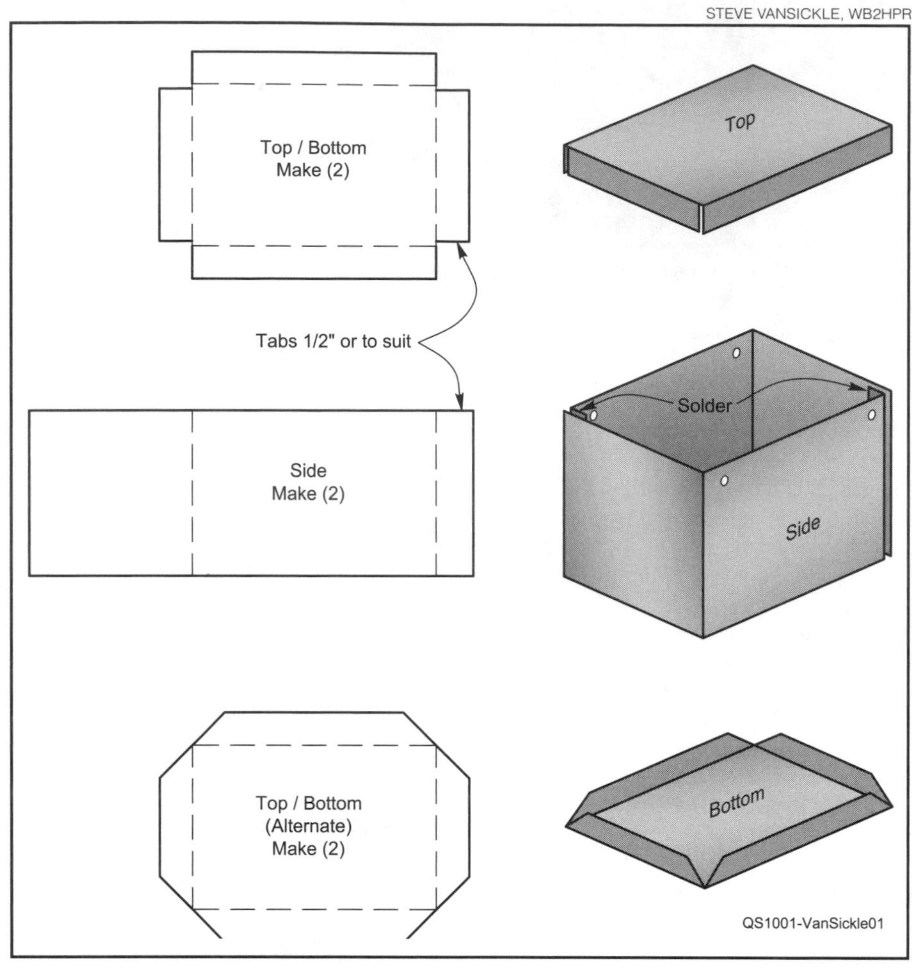


**Figure 65 — The mechanical layout of the box components showing how the cuts are made and how the box is assembled.**

**Figure 66 — A completed UHF/VHF diplexer and RF sampler showing the completed copper boxes and the two different cover construction methods.**

The examples shown (see Figure 66) include a UHF/VHF diplexer (center) for a dual-band antenna and an RF sampler for SWR measurement. The parts for each enclosure were cut from a single piece of #20 AWG step flashing using ordinary tin snips and bent into shape using a small ball peen hammer, wooden block, a 4 inch vise and pliers. A light coat of clear Krylon preserves the shine of the finished enclosures.

I have had no mechanical difficulties or electrical problems using this method for several years. Give this method a try and you will have a rugged, good-looking and functional enclosure that will last for many years. — *73, Steve VanSickle, WB2HPR, 3010 Tibbits Ave, Troy, NY 12180*, **wb2hpr@arrl.net**

## NEATER PROJECT WEATHERPROOFING

◊ Hams often use common hardware-store silicone sealant (RTV) to weatherproof their radio projects. It's waterproof, durable but *very messy*. An easy way to keep it from being lumpy and ugly on your projects is to thin it before application.

A 50-50 mix of the sealant with turpentine, paint thinner or lighter fluid works well. Squeeze a tube of sealant into a pint glass jar with a good lid, add the solvent and stir thoroughly. You now have a thick "paint" that applies easily with a brush, stays in place, covers well and flows out nicely. I use it for many projects, for example, to weatherproof and hold the turns of a home-brew 40 meter coaxial trap.

The only shortcoming is that the cure time is much longer for the thinned sealant — a day or so. But once exposed to the air the sealant becomes weatherproof in a couple of hours. Just let it finish curing and harden further in service. In the jar the mixture remains liquid and won't cure. And by the way, you can get the sealant off your fingers and the paint brush with turpentine as well. — *73, John E. Portune, W6NBC, 1095 W McCoy Ln Spc 99, Santa Maria, CA 93455-1105*, **w6nbc@arrl.net**

## "AMPLIFILER"

◊ I wanted to find a nice enclosure for my 3CX1200A7 amplifier and power supply that would have a clean look. Traditional rack enclosures are massive, ugly and expensive. I discovered that a standard legal sized file cabinet can be purchased for just over $100

JACK MORGAN, KF6T

**Figure 67 — The amplifier mounted in a drawer of the legal "amplifiler" cabinet.**

at local office supply houses and on the used market for considerably less. A legal sized file cabinet makes an attractive, inexpensive and practical enclosure for amplifiers and power supplies.

A file cabinet has the advantage of slide out drawers that allow your electronics to be serviced in place or easily removed (see Figure 67). You can optionally lock the drawers for safety reasons.

The back of the cabinet can be cut to allow easy access to cables (that may have to be unplugged in order to slide out the equipment.) Cooling can be accomplished from the back of the cabinet. I used a push-pull set of fans to keep the amplifier cool. You can also exhaust hot air through a grill on top of the cabinet.

I added an overlay layer of 14 gauge aluminum plate for front panel strength but it might not be needed in your application. The bottom drawer houses a heavy power supply. [*Tipping hazard:* In using this method be sure there is enough weight in the lower drawers to stabilize the cabinet or bolt it to the floor. — *Ed.*] The power supply is built on a piece of ¾ inch plywood and lowered into place. This is a must because the metal is too thin to support heavy transformers.

I still have two drawers left for future expansion or for storage of manuals, logs or QSLs. — *73, Jack Morgan, KF6T, 2040 Pheasant Hill Ln, Auburn, CA 95602-9673,* **kf6t@arrl.net**

## TWEEZERS TOOL FOR TESTING SMDS

◊ In looking for tools to work on Surface Mount Devices (SMD), I found several LCR (Inductance, Capacitance, Resistance meter) tweezers units on the market, costing $60 to hundreds of dollars, to hold and test SMD components. This got me thinking about making a tweezers adaptor for my meters. At first I thought of cutting some circuit board and fashioning a tweezers, but then I realized that three pieces of heat shrink, a steel tweezers and two pieces of solid hookup wire or embroidery needles could be fashioned into a simple tester.

Figure 68 shows the resulting adaptor — crude, but it works. It also works to test tiny LEDs and diodes with a power supply. This design is for ballpark accuracy and convenience rather than precision work. — *73, S. Premena, AJØJ, PO Box 1038, Boulder, CO 80306-1038,* **premzee@juno.com**

## MORE ON TUNER WEATHERPROOFING

◊ In response to the hint from N1GY in the December 2009 issue, Randy, N7CKJ, and Walter, KC2KZJ, have some additional thoughts.

■ Randy, N7CKJ, would like to point out that the Z-11Pro can be fitted with an internal AA battery pack for power. Since the Z-11Pro is also fully automatic (meaning they only need RF applied in order to tune), the need for a control cable is minimal. Once you've found a suitable weatherproof container drill holes for the antenna cables and ground only. With internal batteries installed, all one needs to do is apply a small amount of RF to the tuner (I use the AM mode on my radio) and watch the SWR indicator on your radio. When the SWR settles down, you're tuned and can start making contacts.

A similar setup was used by Ric, K5UJU, at a recent Field Day operation. He mounted his LDG tuner to a post, attached a balun and doublet antenna and simply covered the tuner with a plastic bag. With the coax run into their communications bus, he only needed to go to AM and transmit a low powered carrier until the SWR settled. With this arrangement, he was capable of operating on virtually any HF band he chose. — *73, Randy Jones, N7CKJ, PO Box 162, Colville, WA 99114,* **rjones@theofficenet.com**

■ Regarding the watertight plastic container, Walter, KC2KZJ, points out that this would be satisfactory for a daytime event where the temperature did not drop. If this is left sealed during a temperature drop, water condensation would develop inside and damage the tuner. For example, let's say that the container is sealed at 70° F with a dew point of 55° F. If left overnight as the temperature drops to 45° F, the air inside would reach saturation at 55° F and water would be condensing all the way down to the 45° F overnight temperature.

To prevent condensation a desiccant pack should be placed inside the container. This pack works by absorbing the water in the air and lowering the dew point temperature to a very low number, perhaps –10° F. These packs usually contain silica gel and should last several weeks in a sealed container. They can be restored by heating in a slightly warm oven to drive off the absorbed water. See the manufacturer's instructions for the exact recharge temperature. New electronics packed in plastic bags are often packed with a small desiccant pack inside the bag to stop water condensation as a result of temperature excursions during shipping. — *73, Walter Mellish, KC2KZJ, 13 White Oak Dr, Livingston, NJ 07039-1220,* **wmellh@aol.com**

## PL-259 CONNECTOR TOOL FOR MOST ½ INCH COAX CABLES

◊ Tired of using pliers to screw on the PL-259 when you are preparing a cable? Pliers always seem to do some damage.

I use an inexpensive ½ inch PVC female to male coupler. Simply use a step drill and ream out the female end (shown by the arrow in Figure 69). PL-259s do vary in diameter. Be sure to measure yours before you ream out the PVC adapter. It will not take a lot of ream ing for the connector to fit snugly. The outer

S. PREMENA, AJØJ

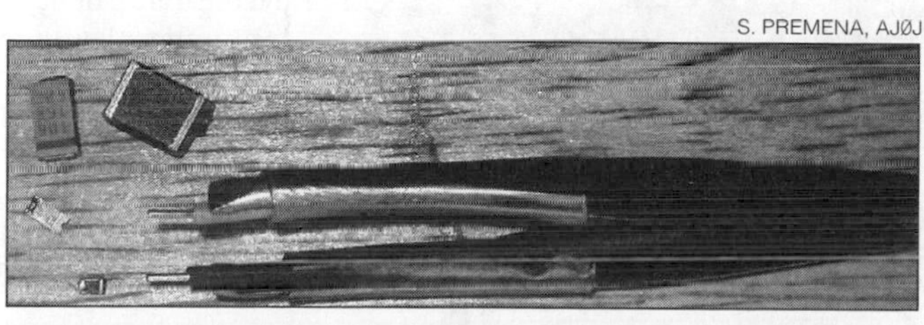

**Figure 68 — A detailed view of the tweezers tool showing how it is constructed. An assortment of SMD components is also**

PAUL MARSHA, K4AVU

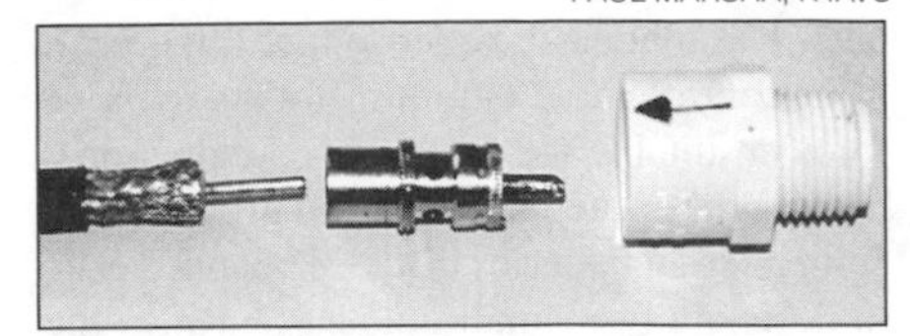

**Figure 69 — Using the PL-259 tool to help install the connector on the end of a coax cable.**

part of the PL-259 that fits into the connector is 0.55 inch diameter. This makes a secure fit. If you wear out the adapter, purchase another. They are inexpensive. — *73, Paul Marsha, K4AVU, 200 Garden Trail Ln, Lexington, SC 29072-7341*, **k4avu@yahoo.com**

## WRANGLING WALL WARTS

◊ Small low current transformers are a problem because the little transformers often take up two outlets in a conventional power strip. There are some $20 solutions, but they weren't exactly what I wanted.

For a simple and less expensive solution, I took four $1.50 extension cords, cut off the ends then soldered and wire nutted them back together into a single unit. The extension cord outlet ends would take a transformer on either side so I had a total of eight possible outlets when everything was back together. You can also run the extension cord as far as you want before you wire in the ends or use a higher rated cord and use more ends for larger loads.

There are two cautions: First, the small cords will only take low current loads and since most of my transformers were only around a half an ampere that wasn't a problem, but if there is a question check the load to be sure that it doesn't exceed what the cord is rated for. Second, when you connect everything back together, be sure that the grounds and the hot legs are connected together separately. A simple $6 solution to "what do I do with all of these little transformers." — *73, Stewart Nelson, KD5LBE, 8 Deerwood, Morrilton, AR 72110*, **kd5lbe@arrl.net**

## COTTON SWAB TIP

◊ When working with the ever popular Anderson Powerpole connectors, you might find you have run out of the factory recommended 1/4 inch long, 3/32 inch diameter metal roll pins, used to lock the housings together. Don't panic — just ask your spouse where she stashes the cotton swabs. The shaft of one cotton swab will easily make up to eight such locking devices (see Figure 70). — *73, Joe Morse, AD4W, 317 Westlawn Rd, Columbia, SC 29210-5622*, **ad4w@sc.rr.com**

JOE MORSE, AD4W

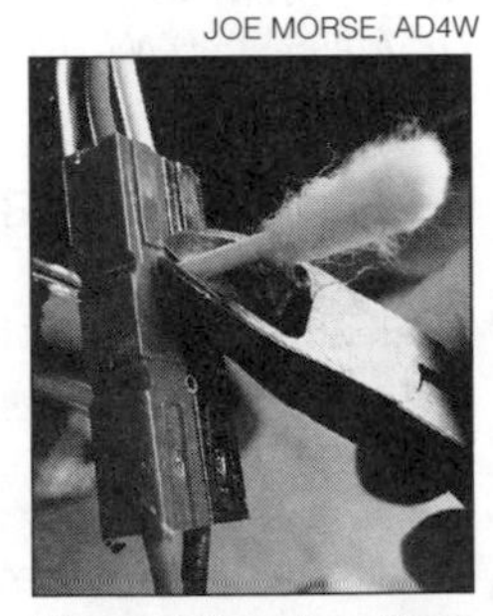

**Figure 70 — A cotton swab makes a quick and easy roll pin substitute to hold those Powerpole connectors together.**

## EASIER PERFBOARD LAYOUTS

◊ There are a number of ways to build small semiconductor circuits, ranging from dead-bug construction, Vectorbord, wired perfboards and Manhattan-style construction [small pads are glued to the board to use as attachment points for leads — *Ed.*] to printed circuit (PC) boards designed for those circuits. Some methods are easier to change, others are neater and more finished in appearance (though as *Experimental Methods in RF Design* points out, a circuit's appearance doesn't necessarily indicate its effectiveness).[1]

[1] Available from your local ARRL dealer, or from the ARRL Bookstore, ARRL order no. 9239. Telephone toll-free in the US 888-277-5289, or 860-594-0355, fax 860-594-0303; **www.arrl.org/shop; pubsales@arrl.org**.

**Figure 71 — A pattern of traces ironed onto a blank piece of perfboard.**

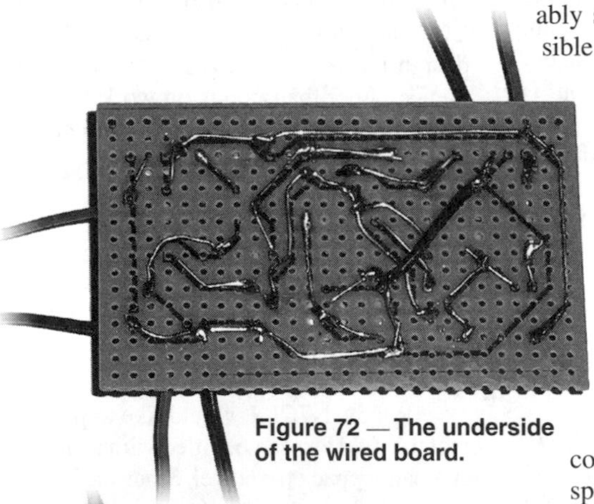

**Figure 72 — The underside of the wired board.**

**Figure 73 — The component side of the wired board.**

Wired perfboards are a useful compromise between the extremes. They can be laid out neatly, built quickly and changed easily. Plus they can look good and be fairly easy to follow. In putting one together, it is often difficult to keep track of where the connections are supposed to go on the unmarked board.

Printed circuit design and construction have been simplified in recent years by programs that help arrange components and route traces, and by ironing laser-printed patterns onto the raw boards. The following method takes advantage of these technologies to simplify wiring circuits on perfboards.

1. Design a printed-circuit layout for your project using one of the several PC design programs that are available online.
2. Print the resulting pattern of traces with a laser printer in black and white on a transparency.
3. Iron the pattern onto a piece of blank perfboard.
4. Use the image of the traces to guide the process of populating and wiring the board, using wire in place of the traces.

For example, Figure 71 shows a pattern of traces for the discrete component audio amplifier on page 1-12 of *EMRFD*, designed in this case with the freeware version of *DipTrace1* and ironed onto a piece of perfboard.[2] This pattern guided the wiring of the circuit. Figure 72 shows the wired underside of the completed board and Figure 73 shows the board from the top.

Since the traces on a board like this are laid out by hand, you needn't (and probably shouldn't) try for the greatest possible component density. Set the program to generate wide spaces and leave yourself room to work. Similarly, draw wide traces so they will be easily readable when transferred to the board. Note that even an imperfect transfer of the pattern to the board provides lots of help.

A few more hints:

■ Set the PC program to 0.1 inch spacing, so that component leads will line up with the holes in the predrilled board and use component patterns with the same pin spacing.

■ When ironing the pattern onto the board, use a piece of paper between the iron and the transparency, use pressure rather than movement and be careful not to melt the transparency with the iron.

■ Trim the perfboard after you've ironed the pattern onto it; the board is easier to handle when it's larger.

[2] **www.diptrace.com**

■ When wiring the board, you can fasten long wires to the board at intermediate points with short wire loops through two holes.

■ You are not obligated to follow the traces exactly. The program may place some long and winding roads; your wires can jump more directly, as can be seen by the black wire in Figure 2, which follows a much more direct route than the trace it replaces.

■ Don't be too concerned about traces that the program cannot place. You can use jumpers on either side of the board.

This technique can be extended in several ways. You can print a mirror image of the board's top side and iron it to the component side, in registration with the traces. Now you have a guide for placing the components. You can use this technique on undrilled blank boards. In addition to indicating the connections, the pattern will indicate where to drill holes. Also, it might work for boards with single-hole pads and for laying out Manhattan-style boards. Be sure to remove toner where leads will be soldered. *— 73, Bryant Julstrom, KCØZNG, 1945 30th St S, St Cloud, MN 56301*, **kc0zng@arrl.net**. *Photos 1, 2 and 3 by KCØZNG*

### A BETTER ELECTRET ELEMENT FOR HOMEBREW MICROPHONES

◊Many amateurs who build their own microphones use the readily available Radio Shack electret condenser microphone elements (**www.radioshack.com**). These elements, like so many similar elements on the parts market, have a flat response over a wide frequency range (for example, 20 Hz to 20 kHz). They work fine but require equalization if one desires crisp audio on SSB.

While building or modifying some microphones I was pleased to discover an inexpensive ($2) electret element manufactured by a world leader in the field (Knowles Acoustics) that has a tailored speech frequency response. It is available as Digi-Key item 423-1097-ND (**www.digikey.com**). The response, centered on 1 kHz, is down about 17 dB at 100 Hz and rises about 8 dB at 3 kHz. I've used the element in several microphones and have always gotten great audio reports on SSB. The element will work as a substitute element in almost any stock hand microphone that has a two terminal electret element. *— 73, John J. Schultz, W4FA, 302 Glasgow Ln, Greenville, NC 27858*

### REPAIRING SWITCH ARC-OVER DAMAGE

◊The band switch in my home-built amplifier (four 811-A tubes) arced between one of the grounded switch wafer support screws and adjacent switch contacts. Since the amplifier had been in service for many years, I expect that accumulated dust, etc caused the arc. Cleaning the ceramic wafer didn't help and I didn't have a replacement.

The solution was quite simple. I replaced the screws and spacers with nylon screws, nuts and washers. The amplifier is now back in service without further problem. If the switch contacts are damaged, as mine were, they can be replaced by drilling out the ends of the hollow rivets and attaching contacts (from a junk box switch) with #2-56 machine screws and nuts.

There are many sources for nylon hardware. I obtained the items I needed from Fastener-Express (**www.fastener-express.com**). For my switch, #6-32 screws of appropriate length along with nuts to fit and a number of #6 nylon washers did the job. Screws, nuts and washers were each sold in packages of 50 or 100 at very nominal cost. I ordered them online and they appeared in my mailbox in less than a week. *— 73, Dean Elkins, K4ADJ, 212 Old Orchard Ln, Henderson, KY 42420-4755*, **k4adj@arrl.net**

### CHILLING TEFLON IMPROVES FERRITE ASSEMBLY

◊I was building a 1:1 balun to use in a homebrew balanced tuner. This is the current type balun that places 50 ferrite beads on Teflon cable. The beads were a tight fit for the most part. I put the cable in the freezer for about 30 minutes and even though it was Teflon and double shielded, it shrunk enough to allow the beads to slip on much faster. There were still a couple really tight beads. I set these aside and placed them on last to minimize having to push them down the cable very far. *— 73, Charlie Liberto, W4MEC, 619 Hidaway Cv, Hendersonville, NC 28739-6915*, **w4mec@arrl.net**

### BUTTON BATTERY TABS

◊It is often difficult to solder leads to button batteries. I devised this alternative attachment method for when you lack the proper tabbed battery. You'll need a length of hookup wire, superglue or hot melt glue and heat shrink tubing.

Strip about ¾ inch of insulation off one end of the two wires. Form the stripped portion into a zigzag/curlicue. Lay the thoroughly cleaned battery on your work surface. Press and hold the shaped portion against the battery (with a piece of plastic) and apply the glue to hold the shaped portion to the battery. Make sure the glue doesn't get between the wire and the battery terminal. When set, verify the continuity of the connection, then repeat on the opposite battery face.

Place heat sink tubing over the assembly and apply heat. As the tubing shrinks, it will conform to both the battery and the formed portion of the wire leads. This will provide an electrical connection and mechanical retention of the leads at the battery. Non-formed leads would simply lack retention. Solder the exposed leads as usual.

This hint suffices for low current memory batteries. I've used this approach many times at repeater sites with no detrimental issues on NiCad battery sizes up to and including D cells. *— 73, Wayne Troutman, KC7FKW, PO Box 568, Payson, AZ 85547*, **usaf1@joimail.com**

### PERF BOARD DRILL GUIDE

◊ After etching a printed circuit board, drilling the holes accurately can be a formidable challenge. I've found that if you lay out your project using a 0.1 inch grid pattern, your hole locations will line up perfectly for using "perf" board as a drill guide. After etching the printed circuit board, I tape the perfboard on the printed circuit board and mark the hole locations with a pencil. I then use a drill press to drill the holes. Perfboard can be purchased at RadioShack, part number 276-1396. *— 73, Dave Palmgren, N8DP, 6132 Co 420 - 21st Rd, Gladstone, MI 49837*, **n8dp@arrl.net**

### PANEL LABELING

◊"Hints and Kinks" has had several good suggestions for custom label materials and procedures. In my search, I found that by using Microsoft *Word* or Corel's *Word Perfect* I can make ideal labels using any one of several available sizes of the "Clear Mailing Labels" from Avery.

After printing the label I applied it in place and then sprayed acrylic spray over the area to protect and seal the label. You can also use any clear fingernail polish. Figure 74 shows an example of this approach with the label on a wood frame. The size used in this case was 1⅓ × 4⅛ inches. It has been on my boat for three seasons so far without any changes or deterioration. *— 73, Art Bartlett, KA1RX, 3605 Britt Ter, Virginia Beach, VA 23452*, **ka1rx@verizon.net**

### TACK IN THE SHACK

◊I've found a new use for a product designed to hang items on the wall. Things like maps, charts and other items can be stuck to your wall using a low-tack putty sold under many names. This putty makes it easy to remove and reposition your item without damaging the wall. This alone makes it very useful in a ham shack.

ART BARTLETT, KA1RX

**Figure 74 — An example of an Avery clear mailing label used for equipment identification.**

STEVE SANT ANDREA, AG1YK

Figure 75 — Tacking putty is inexpensive and readily available at a variety of stores sellinghouseholdaccessories.

A small pea of this stuff under the feet of your key eliminates chasing it across the desk. When stacking equipment, a piece on the top unit's feet will keep it from sliding and prevents scratches. A small piece placed on screw heads between rigs on the desk prevents (new) scratches. I even use a gob of it to hold wires in corners to keep them neat and out of the way. Put a piece in the corner, press the wire into the putty and another small piece over it to keep it all secure.

Usually found in the stationery aisle of office stores (see Figure 75), this low-tack putty has literally hundreds of uses around the shack. I wonder if it'll waterproof a PL-259? Gotta go... — *73, Jim Philopena, KB1NXE, 265 Frost Hill Rd, Marlborough, NH 03455*, **kb1nxe@arrl.net**

## PRO FRONT PANELS

◊I read the article in *QST* on creating custom front panels with great interest.[1] I would like to suggest a very useful resource for this that many may not be aware of, but I have been using since discovering it. A company called Front Panel Express, based in Seattle, offers a free software program for designing front panels. Once designed the panel can be submitted to them to be manufactured or you can print it out for application to a panel that you make yourself. Their prices are quite reasonable for a professionally drilled, engraved or screened panel.

The free software is quick and easy to learn and use. You can go to their Web site at **www.frontpanelexpress.com** and click on DOWNLOAD to get a copy of their *Front Panel Designer* software. After creating a design you can request a quote if you want them to produce the panel for you, otherwise you can print out a paper copy for your own use. With this software I have been able to duplicate many equipment panels for modification or restoration. — *73, Scott Lichtsinn, KBØNLY, 406 E Bradley St, Tyler, MN 56178*, **kb0nly@mchsi.com**

[1]F. Boyer, N3QK, "Making Front Panels — the Easy Way," *QST*, Mar 2009, pp 75-76.

## SCREW STARTING TIPS

◊If you need help starting a screw in a difficult spot, rub the tip of any kind of screwdriver on a wax ring gasket used to seal the bottom of a toilet bowl to the floor. I bought mine about 10 years ago for $1.26 at the local hardware store. I use it any time I have to position hardware in a hard to reach location and I still have enough to last the rest of my life. It's sticky, like soft beeswax and wipes right off anything you use for an insertion probe. You'll find numerous uses for the stuff anytime you want to hold something light in position.

A second tip involves when you have to remove a screw from a location where the screw head isn't visible. What kind of tool do you use? Press a fingertip on it for a few seconds. You'll find a perfect image of your target embossed in the tip of your finger. — *73, Robert Barnes, W8SEB, 168 Belmont Ln, Whitmore Lake, MI 48189*, **w8seb@arrl.ne**

## KITCHEN HELPER

◊I needed a center support for a log periodic antenna I was building. Because the support bars are also a transmission line for the elements, I needed to find a strong insulating material that could be attached directly to the bars and could have "U" clamps attached to it for mast mounting. After some research I found that the material in plastic kitchen cutting boards is ideal. It is a quarter inch thick, is automatic dishwasher safe (will not deform with heat), is white and nonconductive, and cuts nicely with a hand or hack saw. As with most plastic, drilling is a breeze and produces nice clean holes. Depending on the project, one board would supply several insulating blocks and this material could serve for other purposes. — *73, Britt Belyea, W4GSF, 92 Waterview Dr, Newport News, VA 23608*, **w4gsf@cox.net**

## THREADING PL-259S ONTO CABLE

◊When soldering PL-259s we all know that it's a real pain to thread it onto the outer jacket of the cable. I have a method to simplify this process. My method is to start the connector on the cable and, once the thread is started, I grasp the connector with a 1 inch "Kant Twist" clamp (see Figure 76). These clamps are inexpensive and are available for a couple of bucks at any industrial supply dealer. (MSC, Travers Tool or Penn Tool) I merely clamp it onto the connector and twist it on with no damage to the connector. — *73, Ed Swiderski, KU4BP, 108 Tori Ln, Lexington, NC 27295*, **ku4bp@arrl.net**

ED SWIDERSKI, KU4BP

**Figure 76 — The "Kant Twist" clamp attached to the PL-259 connector simplifies the task of threading the connector onto the coax.**

## ALIGNING DRILL HOLES

◊I had the idea to use a two-section telescoping TV pole to increase the height of my 6 meter beam. To bond the pipes mechanically I had to drill a pass-through hole for a 4 inch bolt. In the past when I drilled a hole on the side of a pipe, the adjoining hole on the other side was always misaligned.

My solution was to drill the first hole in the pipe, but stop short of drilling through the other side. Next, take a short piece of string and cellophane tape it to the center of the hole, then run the string around the pipe until it meets the end you just taped. Cut the string to match the diameter of the pipe. Then untape the string, cut it in half, then retape one end to the first hole again and wrap it back around the pipe. It will end exactly one half way around and the end will be at the second drill hole. This method worked perfectly the first time I tried it. — *73, Randy Miller, AA5OZ, 4122 Mary Ann St, Lake Charles, LA 70605-4102*, **ka5flm@aol.com**

## MOLEX PIN REMOVAL

◊For Molex connectors, in order to remove a pin for repair you need a special pin removal tool to compress the fingers that retain the pin in the connector. When I recently could not find my pin tool, I came up with an alternative from my junk box. A broken telescoping antenna from a portable AM/FM radio can be a source of various sizes of pin tools. They can be easily cut and will fit over the pin allowing compression of the locking fingers so that the pin can be removed. Telescoping tubing sold in hobby shops can also be used. — *73, Art Carlson, WAØNJR, 2707 15th St N, St Cloud, MN 56303-1656*, **wa0njr@arrl.net**

## POWERPOLE TOOL

◊Sliding Anderson PowerPole connector housings together or apart can be difficult. To simplify this task, I used a bench grinder to offset the jaws of two pairs of inexpensive

LYNN BURLINGAME, N7CFO

**Figure 77 — Using the modified pliers simplifies the task of joining a pair of PowerPole connectors.**

LYNN BURLINGAME, N7CFO

**Figure 78 — Joining and separating PowerPole connectors requires that two pliers be modified in a mirror image of each other.**

slip-joint pliers (see Figure 77). The jaws are mirror images — one set is for joining the connectors and the other is to disassemble them (see Figure 78). Be sure to work from the bottom of the connectors as shown in the illustration — this positions the jaws of the pliers against the strongest part of the housing. — *73, Lynn Burlingame, N7CFO, 15621 SE 26th St, Bellevue, WA 98008*, **n7cfo@n7cfo.com**

## WELL WIRE

◊During these tough economic times everyone is looking for a bargain or a way to reduce costs. This suggestion may be of interest to those looking for an economical supply of radial wires for a vertical antenna.

I acquired 200 feet of three-conductor well wire (#12 AWG) with plastic insulation. This gave me twelve 50-foot lengths of radial wire for use with my antenna. All this wire was obtained at zero cost as it came out of a water well that was being replaced. I am told that the old wire is not reused but scrapped or given away.

I suggest contacting a water well maintenance company for a source of this wire and enjoy an almost endless supply of free or low cost radials. — *73, Michael Janis, KE3OQ, 149 Silo Cir, Nazareth, PA 18064*, **mjanis@ptd.net**

## L-BRACKET CABINETS

◊I, too, have joined the homebrew enclosure club (January 2010 *QST*), as the cost of commercial cabinets — even small ones — is prohibitive.[2]

While constructing a seven band CW transmitter, I found that an economical cabinet for it would cost about $65, locally or imported. So I made my own.

It consists of 0.080 inch sheet aluminum panels and ½ × ½ inch aluminum L-bracket stock (from Home Depot). I cut Vs into the bend sites and bent the L-bracket into rectangles for the front and rear cabinet openings (see Figure 79). I drilled #43 drill holes in the L-bracket, which is thick enough (⅛ inch) to sustain threads, and tapped them for 4-40 fillet-head machine screws. The panels are drilled with a #33 drill for screw clearance.

Reality must creep in here. L-bracket aluminum bends readily, but snaps off if you bend it too far then try to bend it back. In addition, the open corner of each frame must be secured. I used short pieces of L-bracket, epoxied into place and left to dry overnight (see Figure 80). Epoxy is not the strongest of glues, but it does not expand like some other types. An alternative is to machine-screw the corner pieces into place.

As I'm not the most accurate machinist in the world, some of the clearance holes had to be "adjusted" to enable the screws to grasp the threads in the frames through the panels. I used a thin rat-tail file and a lot of patience. [To improve accuracy, clamp the panel to the frame and drill the #43 holes through both and then enlarge the panel holes with the #33 drill. — *Ed.*]

The front panel of this transmitter is in two pieces to allow

[2]S. VanSickle, WB2HPR, "Homebrew RF Enclosures — Another Approach," *QST*, Jan 2010, pp 64-65.

CHARLES HOOKER, VE3CQH

**Figure 79 — A piece of aluminum L-bracket that has been cut and bent to form a rectangular frame. The transmitter breadboard is shown above.**

CHARLES HOOKER, VE3CQH

**Figure 80 — The completed enclosure showing details of the final assembly. Note the reinforcing bracket pieces in the corners.**

me to add and subtract parts as the design matures. The cabinet corners are approximately square, good enough to contain the transmitter.

A plywood board was employed as a base for prototyping; otherwise I would not have been able to determine the required cabinet size. — *73, Charles Hooker, VE3CQH (ARRL Life Member), 431068 19th Line, RR #2, Orangeville, ON L9W 2Y9, Canada,* **chuckh@sympatico.ca**

## HOT GLUE FIX-IT

◊We all have some sort of kit we keep handy for repairs. Here's an idea you'll learn to value like a Swiss Army knife. Go to a hardware or craft store and buy a small box of hot glue sticks. They come in various diameters and lengths. Also, if you don't carry one, get a couple of ordinary butane cigarette lighters. Keep the glue sticks and the lighters close at hand.

Simply rotate the stick's end in the flame of the lighter until the blob at the end is soft enough to apply and then dab it on where you want. If you're fast, you can stick a second item on top of the glue, but if it has cooled, simply reheat it with the lighter flame.

Hot glue is reusable, waterproof, works as an insulating coating, fills holes and large cracks and can be removed from most hard surfaces either hot (with a flat screwdriver blade) or cold (with fingernails, etc). You can use it as a grommet to protect wires that pass through a chassis hole. It can be used to "tack down" wires to a chassis or to attach a small project box to the outside of another plastic, wood or metal object. (It can be removed later, though expect it to leave smudges or minor heat markings.) It can be used to affix signs and labels, to seal pots from external moisture or to secure twisted wires when you don't have solder available.

To fill an unwanted hole in most chassis material, stick a piece of electrical tape over the outside of the hole. Now, using the lighter liberally, heat the glue stick until it runs in a small string into the hole, overfilling it just slightly. Once the hot glue has cooled, remove the tape from the outside and use a permanent marker to color-match the repair.

You can shape hot glue, while it's still hot, if you put a good layer of saliva on your fingertip to both shield your skin from the heat and prevent the glue from sticking to your skin. I've used this technique hundreds of times to form moldings around a repaired miniplug for headphones, patch cords, etc. You can also trim hot glue with a sharp knife blade, thread small screws into it and thumbtack into it. Small globs of hot glue make excellent shock bumpers, too.

Just use good sense and a little caution. Hot glue can catch fire and when it's hot enough to melt, it's hot enough to burn unprotected skin and delicate materials. Also, don't use it to hold any object that heats up during its normal use. Keep it *off* the soldering iron, too. — *73, Paul Schlueter III, KB3LIC, c/o DLARC, PO Box 3026, Easton, PA 18043,* **twelvevdc@aol.com**

## CRACKING WALL WARTS — A BETTER WAY

◊"Hints and Kinks" has published many methods for opening wall warts. I have been servicing wall warts for many years and offer the following method:

1. Put on some safety goggles. Place the wall wart on the work bench on top of a leather glove or similar cushioning with the seam side up.
2. Using a dull pocketknife, lay the edge of the blade lengthwise along the seam. Use the edge of the blade — not the point.
3. Give the blade a few taps with the hammer along the seam, then flip the wart over and repeat the process. The case will easily and neatly open along the seam.

After making your repairs, apply plastic model cement to the seam and clamp the case together till the glue dries. The repair is barely noticeable, since the case is undamaged.

I have used this method for several years and have not had mechanical difficulties or electrical problems since there is less chance of damage to the case and internal parts. Give this a try the next time that pesky little power supply needs repair. — *73, Steve VanSickle, WB2HPR, 3010 Tibbits Ave, Troy, NY 12180,* **wb2hpr@arrl.net**

## ASK YOUR DENTIST

◊While going to the dentist with my daughter, I struck up a conversation with him as he was cleaning my daughter's teeth. As he was using an assortment of stainless steel picks and other tools, I made the mention of how handy they would be for working on radios and soldering. He generously offered an assortment of used dental tools that have lived past their use in a dental office.

He said that these tools cannot be resharpened and he is always buying replacements, leaving him with many items that he must dispose of. As I was graciously accepting his gift and telling him of the uses that I will have with them in working on my radios, he began telling me that his uncle was a ham operator when he was a kid. So for all those people who fear the dentist, next time think of the new (to us) tools we can take home for our hobby. — *73, Jim Derra, KF6HYD, PO Box 106, Grenada, CA 96038,* **kf6hyd@arrl.net**

## FASTENING FACTS

◊I've worked in many industries over the years and a common problem I have encountered is the overtightening of fasteners and misuse of tools. This is true in the application of Dzus fasteners and shoulder screws used to secure electrical panels as well as stainless steel hose clamps.[1] Observing the difficulties with stripped fasteners the astronauts had to contend with when repairing the Hubble Space Telescope proves this out. Some assume "tight is good," not realizing that the fastener will fail, especially when tightened with a wrench, ratchet or ill-fitting screwdriver.

The popularity of the new 43 foot telescoping vertical antennas depends, in part, upon hose clamps, which simplify the raising and lowering of the antenna. Hose clamps were designed to secure a rubber hose over a metal tube fitting. The clamp is properly tightened when the hose begins to bulge up through the helical cuts in the clamp. This doesn't happen with aluminum tubing. If you use a wrench, the correct torque is finger pressure. A slug that fits the slot will never allow you to overtorque and strip the helical screw. You can file a slug for a snug fit in the screw slot and secure it with a drop of superglue to use as a knob.

### *Size Matters*

Hose clamps come in more than a dozen sizes and there is a reason for that. The anvil that contains the helical screw drive is bent for a specific diameter range. Many hams purchase larger clamps assuming they can cut off the excess band, but the anvil is too flat for the application and tightening to force it to conform stresses the housing and strips the drive screw. If you purchase clamps from a hardware or automotive store be aware of the design diameter range of the clamp and that the screw is the same color as the band (yellow means the screw is rustable steel).

It has been suggested that you use an electrical antioxidant joint compound between the antenna's sections. These compounds are made to connect soft aluminum and copper wire; that is, either aluminum to aluminum or aluminum to copper. They contain an abrasive to break oxides and an oil-based shield that can be messy. T6063 aluminum tubing is much harder and does not require it.

Another method to prevent oxidizing is to clean surfaces with a Scotch-Brite pad or fine wire brush, spray on WD-40 or CRC lubricant, wipe off any excess, then respray surfaces and reassemble. Also spray the screw drive to prevent metal to metal galling. This will make it easier to collapse the antenna.

Clean is good. I have used this process on automotive battery connections and high-current, cast iron, third rail railroad shoes wiping off the excess before assembly. Never use a lubricant that contains Teflon as it is an

[1]Quarter-turn fasteners used to secure equipment panels that allow the panels to be opened and closed quickly. — *Ed.*

insulator. You could use an antiseize paste, *but* it must be compatible with the metal and if it drifts it could make a disastrous conduction path. Any slight oil traces will be forced away under pressure and have no effect on the connection, except to occupy surfaces that do not mate, which would attract oxidation. It is good practice to wipe off any excess. — *73, Peter Murricane, WB2SGT, 200 E 63rd St, New York, NY 10065*, **wb2sgt@arrl.net**

### SO-239 CENTER PIN PROBLEMS

◊I experienced intermittent connections with the SO-239 jacks on several pieces of equipment. I traced the problem to the center contact, which had rotated and damaged an inside connection when the PL-259 was inserted. The center contact is supposed to be fixed to prevent rotation but in some inexpensive connectors it apparently isn't.

Although the PL-259 has little antirotation pips on the shell, you always have to rotate the plug to line them up with the detents before screwing the shell on. If the center contact is not pinned, this rotation will put stress on the interior connection. Some of the imported connectors only have 4 detents, making it worse, and I have found they may be more easily damaged by soldering heat. My solutions in both cases were to first repair and align the center contact and then fix it in place with a drop of epoxy. — *73, Don Dorward, VA3DDN, 1363 Brands Ct, Pickering, ON, L1V 2T2, Canada,* **va3ddn@arrl.net**

### LED MOUNT INSERTION TOOL

◊Mounting clips with press on retaining rings are convenient for mounting LEDs to a faceplate (see Figure 81), but getting the retaining ring on is not always easy. This is particularly true when there is limited access. The most common technique is to use two nut drivers, one from each side. This works well if space is available, but cannot be used with LEDs having attached leads. Another technique is to use needle-nose pliers. Because the pliers do not exert uniform pressure around the retaining ring, the typical result is scratches to the faceplate, a torn retaining clip finger or an occasional gouge to the clip face. Also, pliers are not usable for panels having lips.

I made a simple tool from two aluminum blocks that works on faceplates with or without lips, where there is limited access and with LEDs having attached leads. The tool is made from two blocks of 1$\frac{1}{8}$ inch long pieces of $\frac{1}{2} \times \frac{1}{4}$ inch bar stock. The overall size is not critical. There are two pieces, a clip block and a ring block. The clip block is made by drilling a single hole with a number 7 drill bit. Make this hole centered, about a half inch from one edge.

The retaining ring block is made in three steps:

1. Use a $\frac{25}{64}$ inch pilot point drill bit to make a flat bottomed recess slightly larger than the outer diameter of the ring approximately $\frac{12}{64}$ inch deep (not critical). Make this hole centered, near one edge of the block, positioned to mate with the hole in the clip block.

2. Using a $\frac{9}{32}$ inch bit, drill a through hole centered in the recess hole. This will leave a small lip to support the retaining ring.

3. Finally, saw and file smooth a slot between the hole and the narrow edge as a pass-through for any wire leads.

JOHN M. FRANKE, WA4WDL

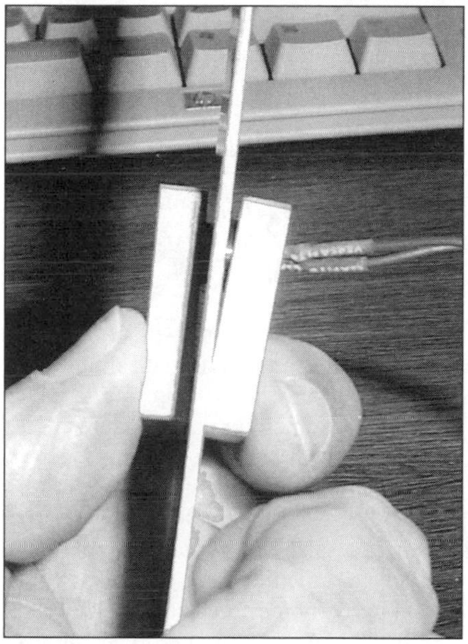

**Figure 82 — Press-on mounts make make it easy to install LEDs in a faceplate.**

The operation is simple and quick. The retaining clip is inserted into the faceplate hole with the chamfered inner edge facing the panel. The faceplate hole should be deburred but not chamfered.

Place the retaining ring, chamfered inner edge facing outward, on the LED and insert the LED into the clip. The clip block is placed against the front of the retaining clip. If the LED has leads that are soldered at each end, the wires are passed through the slot in the tool before the ring. Slide the tool along the wires until the ring is seated in the tool. Now, with both tool blocks in place gently squeeze the blocks together (see Figure 82). A snap is heard and the job is done.

I can now mount an LED almost effortlessly in almost any panel (see Figure 83). — *73, John M. Franke, WA4WDL, 4500 Ibis Ct, Portsmouth, VA 23703*, **jmfranke@cox.net**

### AUDIO INTERFACING AND GROUND LOOP SOLUTIONS

◊Having built my own version of the Pac-12 portable vertical antenna (**www.njqrp.org/pac-12**) I decided that I wanted to use my Yaesu FT-817ND transceiver and Toshiba laptop for portable digital communications, but I didn't have a spare sound card interface. I came across Ernie Mills', WM2U, excellent soundcard interfacing site (**www.qsl.net/wm2u/interface.html**) where he suggested that you try the simplest possible interface and only get more complicated if you have ground loop problems. I tried his simple interface and it worked great, except I had a terrible 60 Hz hum on transmit — the feared ground loop. So I decided to try the next simplest idea that Ernie suggested — use isolation transformers. I didn't have any in my junk box so I placed an order with RadioShack.

While waiting for the transformers I got a new radio/cd/mp3/aux player installed in my car. Driving home it hit me. I should be able to connect the audio from my Yaesu

JOHN M. FRANKE, WA4WDL

**Figure 81 — This tool makes it easy to snap an LED clip and retaining ring together even when the LED is located in a hard-to-reach location. Just position the two halves of the tool on either side of the LED retainer and press.**

JOHN M. FRANKE, WA4WDL

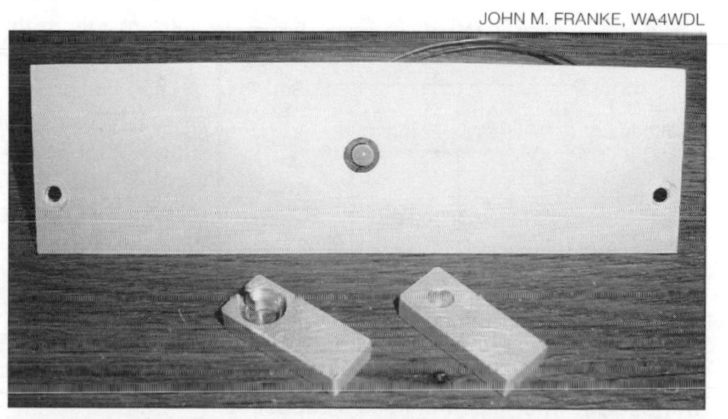

**Figure 83 — Using this tool makes for a simple installation that avoids damage to your equipment's faceplate.**

FT-857D transceiver to my new car radio's auxiliary input and get rid of that unsightly speaker stuck to my dash. So I did. The receive signal sounded wonderful. I then tuned to 7.258 MHz and checked into the MIDCARS net (**www.midcars.net**). Wow, the RF feedback on my car radio was terrible and when I quit transmitting my new car radio acted strangely. After a few minutes of cold sweat thinking about what I might have just done to my expensive new radio, it occurred to me that all microprocessor controlled stuff today has a reset button — which, thankfully, returned it to normal operation.

I went back to the Internet where I landed at Alan Applegate's, KØBG, excellent site (**www.k0bg.com**) all about mobile installations, including ground loop problems. Simply put, you have a potential ground loop whenever two pieces of equipment are interconnected with some noticeable resistance (like a car body) between their respective ground terminals. My Yaesu has a #10 AWG negative (ground) lead coming directly from the battery. I have no idea where the car radio ground lead goes. I could easily imagine that when I transmitted, a voltage would develop across the ground resistance, raising the rig's ground above chassis ground. Incidentally, according to KØBG my ground and power leads should probably be #8 AWG or even larger.

While I pondered that experience, I got tired of waiting for those isolation transformers for my portable setup. The thought occurred to me that I might try modifying Ernie's interface to isolate the dc ground between the FT-817 and my laptop by putting a capacitor in the ground side of the audio cables. I modified the interface by cutting the cables and putting a 0.1 μF capacitor in the ground lead between the transceiver and laptop (see Figure 84). No more hum and my digital signal is perfectly clean on the receiving end. To compensate for the approximately 1.5 kΩ reactance added to the audio path by the 0.1μF capacitor I replaced the 10 kΩ series resistor with 5.6 kΩ. I've had a number of contacts with it and it seems to be working well. [This interface may work for PSK31 but results may vary for other digital modes, especially if a reduced frequency response or roll-off causes problems. Also, before going on the air check that the audio level isn't overdriving the transmitter. Add a gain control if needed. — *Ed.*]

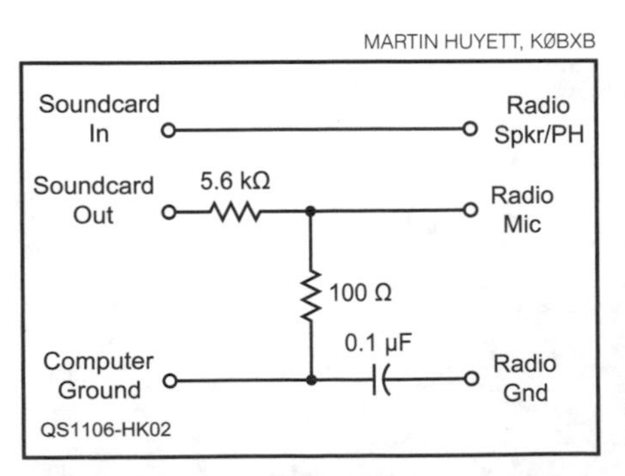


**Figure 84 — Schematic of the laptop to transceiver interface that uses a capacitor in the ground line to avoid ground loop problems.**

That got me thinking about my car radio problem again. Digging around in my capacitor junk box I found a 1 μF nonpolarized capacitor that I had removed from an old oscilloscope. So I cut my patch cord and inserted the capacitor in series with the ground lead. Not even a little RF was noticeable in the car speakers on any band at 100 W and my new car radio's microprocessor doesn't even know there is a transceiver in the neighborhood.

I realize this is probably considered an unconventional way to isolate the gear, but it is so simple and worked so well for me I wanted to pass it on to others for consideration. — *73, Martin Huyett, KØBXB, 7735 Big Pine Ln, Burlington, WI 53105*, **huyettmeh@tds.net**

## RECYCLED SPEAKER HOUSING

◊Need a quick speaker cabinet? Try using a 25 pack CD-ROM case. It does a nice job and was simple to build. For grille cloth I used porous-nonadhesive shelf liner. The audio wire is brought out through the center to simplify screwing the case together. I have not put any sound deadening material inside since the sound quality is pretty good without it. — *73, Thomas Hart, AD1B, 54 Hermaine Ave, Dedham, MA 02026*, **tom.hart@verizon.net**

## BAYONET BULB GRIPPER

◊Have you ever had a difficult time trying to remove a burned-out lamp from a bayonet socket? Some pilot lamp fixtures don't provide enough space to get a good grip on the bulb to enable you to "press-in, turn counterclockwise and remove the defective bulb." It may also be as difficult to replace the bulb in the same fixture or get the bulb into a tight location in a crowded radio cabinet.

I have found an easy way to accomplish the removal or replacement of those bayonet socket lamps. The rubber grippers that are found on common ballpoint or gel pens are just the right size to snugly fit the commonly used numbers 44, 47, 53, 1814 and 1820 lamps.

I recovered rubber grips from Uni-Ball and Pilot brand pens. I'm sure there are other brands that are similar in size and would work just as well. For smaller sized lamps I frequently use rubber fuel-line tubing designed for lawn-mower engines that I purchase from a local lawn equipment repair shop. The soft rubber of the grips really does make lamp removal and replacement an easy task. — *73, Lawrence Stark, K9ARZ, 1875 Chandler Ave, St Charles, IL 60174-4601*, **k9arz@arrl.net**

## FAUX PHILLIPS TOOLING

◊The cross-head recess screws in Japanese radios are *not* Phillips head screws; they are JIS (Japan Industrial Standard). JIS screwdrivers are available from a number of sources online and they are not particularly expensive. The JIS recess is dimensionally different from the Phillips standard and it can be stripped very easily, especially if your Phillips screwdriver is worn. This is easily shown by placing a JIS screwdriver vertically in the recess and it will remain there when your hand is removed. This is something to keep in mind on any product manufactured in Japan. I learned this while working on my Japanese motorcycle after I had stripped the heads on a couple of screws. It was an expensive lesson. — *73, Chuck Percy, N7FZ, 3343 Laurel Mountain Rd, Murfreesboro, TN 37129-2548*, **n7fz@arrl.net**

## COLOR CODING POWERPOLES

◊Mal Eiselman's, NC4L, article on Anderson Powerpole connectors is very apt, due to their ever-increasing popularity among amateur operators.[3] If there is a caveat, it is the fact that the same housings are used for 15, 30 and 45 A contact assemblies. Further, a good portion of these connectors are used with multioutlet power centers like West Mountain Radio's RIGrunners (**www.westmountainradio.com**). Most of these come preinstalled with ATO fuses in various ratings. As a result, one could inadvertently plug a power cable into an outlet with the incorrect fuse rating. While this might not cause a problem, there is a very simple solution.

Anderson Powerpole connector housings are available in black, gray, violet, pink, brown, red, blue, yellow, green, orange and white. These correlate very nicely with ATO fuse ratings 1, 2, 3, 4, 7.5, 10, 15, 20, 30 and 40 A, respectively. The 5 A ATO fuse is tan colored and the remaining white housing could be substituted, if required.

The only remaining caveat is assembling the connectors as Mal suggests. If you do, it's impossible to cross connect them, even if both housings are the same color. Purists might insist on always using black as the negative, but that still leaves plenty of leeway for the most popular ATO fuse ratings of 3, 10, 15, 20 and 30 A. — *73, Alan Applegate, KØBG, 3202 Notting Hill Ave, Roswell, NM 88201-0403*, **k0bg@arrl.net**

## PUT A CORD ON THAT CORDLESS TOOL

◊If you are like me you have at least one cordless tool lying around your shack that has outlived its battery pack. Consider doing what I did; put a cord on that old dead pack.

[3]M. Eiselman, NC4L, "One Ham's Power Connector Preference," *QST*, Sep 2010, pp 36-37.

JOE MORSE, AD4W

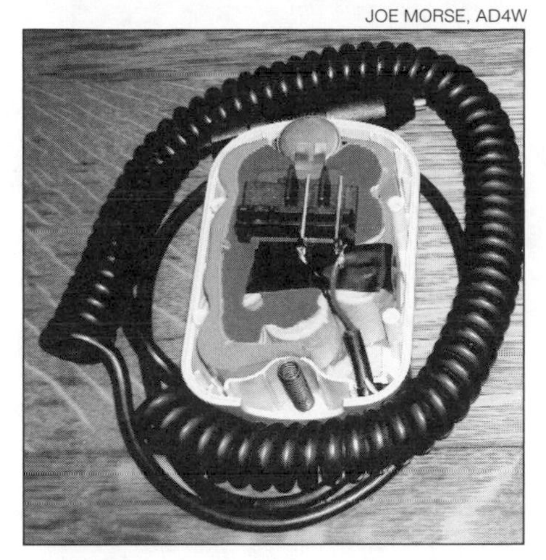

**Figure 85 — The 12 V cord is attached to the connections removed from the dead battery pack.**

If you don't have a suitable cord and cigarette lighter plug in your junk box, get one that can handle the current from an electronics house like All Electronics (**www.allelectronics.com**) or your local auto supply store. Just ask for a "cigarette lighter extension cord."

Clip off the socket end and save it for another project. Open up a dead battery pack but don't remove the dead cells. The contacts are attached to and supported by the cells in the center of the pack so they need to be left in place. You might have to remove one or more of the outer cells to make room for your cord where it enters the pack.

Use a cable tie to secure the cord to the pack case (see Figure 85). Usually one of the contacts is spot welded directly to a cell and the other is wired. Cut the wired connection and solder the cut tool wire to the cord (You should disconnect at least one of the tool's contacts from the cells so the old battery pack isn't in the power circuit). You do not have to disconnect the welded contact, just solder it to the other conductor of your cord (Of course, make sure you have maintained proper polarity). I have successfully done this modification on 12 and 14.4 V tools. The 14.4 V tool might turn a bit slower but will still be quite useable. — *73, Joe Morse, AD4W, 317 Westlawn Rd, Columbia, SC 29210-5622*, **ad4w@sc.rr.com**

## GROUND CLAMP BATTERY EXTENDER

◊I had to connect a heavy ground cable to a 12 V auto battery with top posts. There wasn't enough room to get a wrench in, but there was room for a screwdriver. I solved my problem using a two-piece bronze ground clamp that I maneuvered around the post and then screwed on tightly. The ½ - 1 inch size easily accommodated both the 00 wire and the binding post. There has been no sign of any corrosion after 6 months. — *73, Howard Burkhart, KA6EMT, PO Box 91021, Los Angeles, CA 90009*, **hburkhart98@yahoo.com**

## A SIMPLE TAP GUIDE

◊When tapping threads into metals or other materials it is important to control the angle the tap makes with the material surface. If the angle isn't correct the result could be a broken tap or a screw, if it has to pass through a thick cover piece, might not engage the threads without stripping or binding. In the vast majority of cases, the tap should enter the material perpendicular to the surface. I find doing this by eye can sometimes result in a large angular error. [A drill press, operated manually, can be used as a tap guide if the work space permits.—*Ed.*] So, I made a simple handheld guide to align the tap while starting the threads.

I drilled a series of five equally spaced holes in a bar of aluminum 3⅜ inches long (see Figure 86). The aluminum bar has a cross-section measuring ½ inch wide by ¼ inch thick. The bar should be wide enough to fit flat against the surface to be tapped, yet narrow enough to fit in relatively close spaces. A thicker bar would prevent the tap from entering the tap hole very far before the expanded portion of the tap shank is blocked by the guide hole. A thinner bar would allow a nonperpendicular entry angle.

Although I used aluminum, any other metal, including brass, iron or stainless steel, could be used. The guide holes are spaced approximately ⅝ inch apart. The spacing is not critical. The hole diameters *are* critical. The holes should be close fit clearance holes for the screw sizes you most often use. I primarily use 2-56, 4-40, 6-32, 8-32 and 10-32 hardware, so the holes were drilled with

ELLEN FRANKE

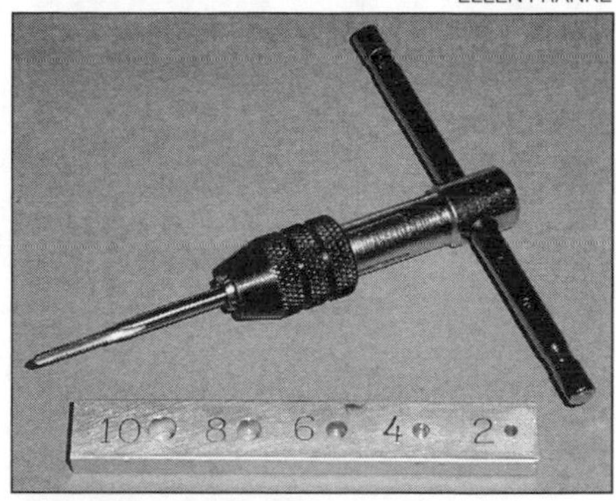


**Figure 86 — The tap guide helps to keep the tap tool perpendicular to the work project's surface.**

ELLEN FRANKE

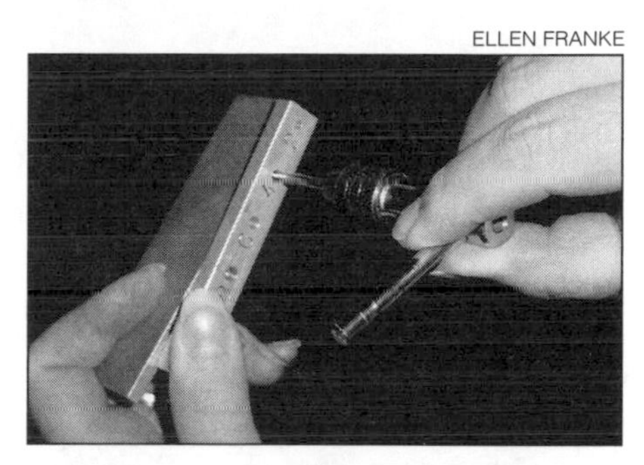

**Figure 87 — Holding the tap guide against the work while starting to tap the hole will prevent the tap from entering the work at an angle, distorting the threads.**

#43, 32, 27, 17 and 9 drill bits, respectively. I used number stamps to label the holes and filled the stampings with black paint to improve contrast. Decals or even hand lettering could be used.

In use, the guide is held against the surface, cutting fluid is applied, the tap is inserted into the appropriate hole and the tapping effort goes forward (see Figure 87). For screw sizes with expanded shank taps, the threads are cut until the expanded portion of the tap shank gets close to the guide. Then, the tap is backed out, the guide removed and the tap is reinserted in the hole. Then tapping continues to the desired depth. This tool helps reduce the number of broken taps and gives a more professional look to the job. — *73, John Franke, WA4WDL, 4500 Ibis Ct, Portsmouth, VA 23703*, **jmfranke@cox.net**

# Chapter 8 Antenna Systems

## AN ANDERSON POWERPOLE-BASED PORTABLE DIPOLE

◊ I have been using this system to attach the center feed point of dipoles for ARES use. It allows easy connection or removal of single or multiple wires. Figure 1 shows a Van Gorden Engineering center insulator, although any similar feed point/balun could be used. I have soldered a length of #14 AWG THHN wire to the lug on each side, with a 30-A Anderson PowerPole connector on the end of each wire. There are two snap hooks (carabiners), one on each eyebolt. Each element wire also has a PowerPole soldered to the end, and a loop formed to accept the snap hook. This system lets you coil or spool the element wires without fussing with the feed point. Two or more elements can be attached at once, but each wire requires another connector at the feed point. The PowerPoles handle more current than RG-8 coax.—*Robert C. Bell Jr, N1OCM, 8 Hilltop Dr, New Fairfield, CT 06812-2328;* **n1ocm@arrl.net**

## CAC-6 (ALUMINUM SHIELD) VERSUS COAXIAL CONNECTORS

◊ *QST* Technical Editor Stu Cohen, N1SC, has spent years around cable TV equipment, so when we met, it wasn't long before he began telling me about the virtues of "CAC-6." That's the generic name for the cable used by CATV companies for "drops" to the subscribers' homes.

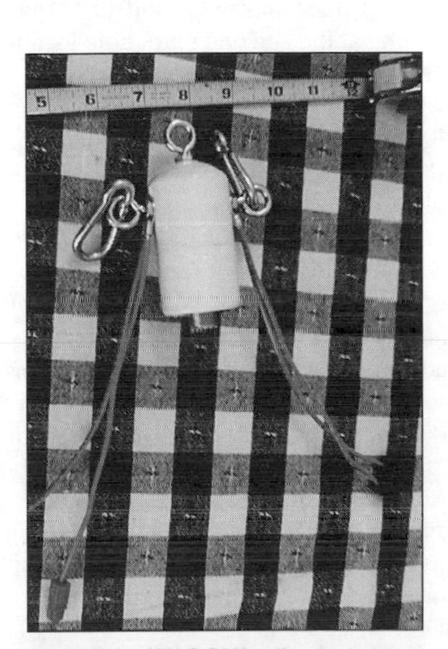

**Figure 1—W1OCM's dipole feed point. Wire elements attach to the spring clips for physical support and to the PowerPole connectors for electrical connection.**

Cable for this purpose must be tough, strong and lightweight. Its losses must be low—well into the gigahertz range. Finally, CATV companies use it by the mile; it must be inexpensive (cheap!). I soon bought a spool end of 200 feet or so for $10 at the October 2003 "Hosstraders" Amateur Radio fleamarket at Hopkinton, New Hampshire.

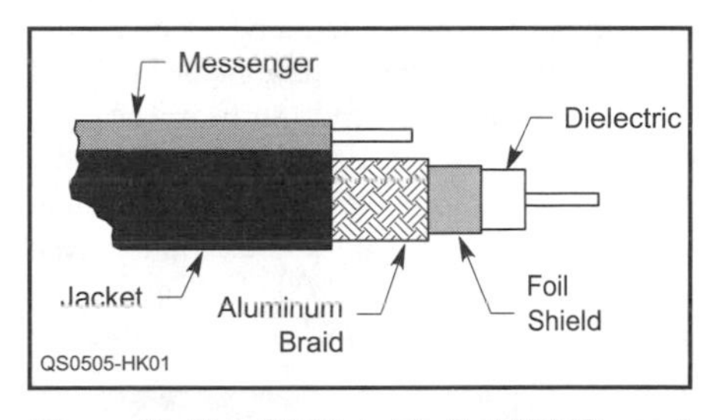


**Figure 2—The CAC6 cable that KU7G used looks like this. Other RG-6 type cables with aluminum shields may differ.**

It is nice feed line. It's light, yet the jacket is stiff and tough.

The stuff I bought is labeled "Belden 9117." It is an RG-6 type cable with "duo-bond" shield. This shield has both foil and braid shields (see Figure 2). It has a jacket-covered steel messenger wire attached to the outside of the jacket. It's handy to secure the messenger at the balun and the house cable entrance, thus removing the strain of cable support from the connectors. Connectors—oh yes.

So there I was, standing in the backyard, in the dark, save for the light of a floodlight up on the second story. It's about 30°F, and the first snow of the season is coming tomorrow. I need to get the new feed line installed before the snow flies.

I have F connectors and a cheap crimping tool in the basement, but I'm not very

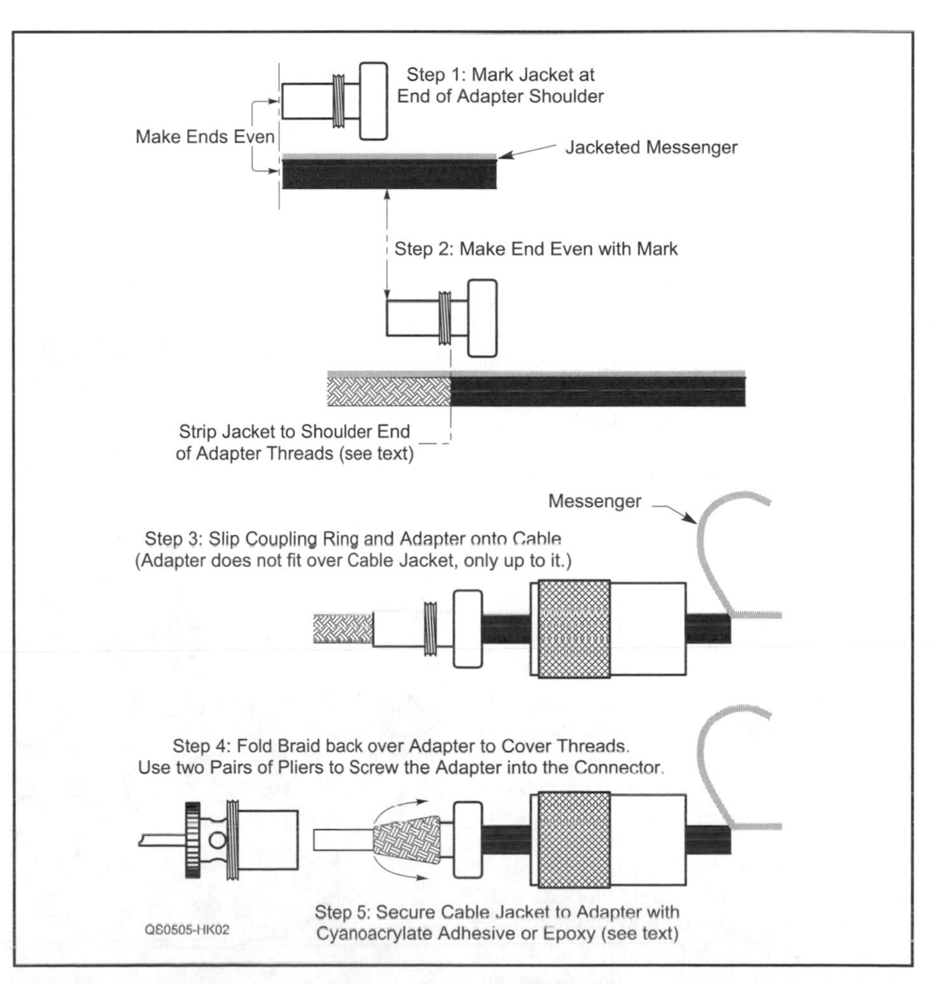


**Figure 3—The procedure for installing a PL-259 with a UG-175 or UG-176 adapter on cable with an aluminum shield braid.**

confident in my coax-crimping skills. I did have some PL-259s and adapters for RG-6 (UG-176, but I couldn't solder the aluminum braid to the adapter, especially in that cold, hostile environment. What I needed was a PL-259 crimp connector. Such a thing may exist, but I didn't have one—yet I did have the adapter. What if I used the adapter to crush (as in crimp) the braid against the PL-259 body? I did. It works, and you can do it, too. Here's how:

I'm right-handed, so that's how these instructions are written. I hold the cable in my right and, with the cut end to my left (refer to Figure 3).Trim the cable end so that it is square.

Hold the adapter alongside (the adapter does not fit over the cable jacket) of the cut cable end so that the ends of the cable and adapter are even. Slide your (right) thumbnail against the adapter shoulder to hold the place, then bring your left thumb up to hold the place. Now reposition the adapter with its small end against your left thumbnail and mark the jacket just past the shoulder end of the threads.

This procedure sounds complicated, but it is not. The point is to mark an appropriate length for stripping the cable jacket. Because the adapter ID doesn't fit over the cable jacket, we must remove the jacket to expose enough braid. The exposed braid must be long enough to pass through the adapter once and then fold back over the adapter to cover its threads. By using the adapter to make the measurement, we account for any length differences in adapter length.

The mark tells us where the cable jacket will meet the adapter, so we can separate the messenger from the main cable a small distance beyond the mark and know it won't interfere with the connector installation. Now remove the jacket up to the mark.

Now slide the coupling ring and adapter onto the end of the cable. (Make sure the adapter goes on with the shoulder *away* from the cable end.)

Position the adapter so that its end is even with the trimmed cable jacket (see Figure 3). Work a little looseness into the braid weave by sliding the cut end of the braid along the dielectric, and fold it back over the adapter (Step 4). Using two pairs of pliers, turn the adapter into the connector until its shoulder seats against the back of the connector body. (If it is too difficult to screw the adapter into the connector body, back it out and trim away some of the braid until it fits.)

Solder the center pin to the connector and seal the cable jacket to the adapter with your favorite adhesive. (I use cyanoacrylate glues ("hot" glues), but some silicone sealant or quick-setting epoxy might work as well.

These connectors are now in the middle of their second New England winter, and they're working fine.—*73, Bob Schetgen, KU7G, ARRL Staff;* **rschetgen@arrl.org**

## INSTALLING RADIALS THE EASY WAY

◊ Recently, while I was talking to a fellow amateur about his plans for putting up a vertical antenna, the subject of installing radial wires came up. He was trying to come up with the easiest and least damaging way to dig shallow trenches, for the wires in his established yard was out of the question. One low-impact method often proposed is to first lay the wires out on the surface and then pin them in place with short "bobby pin" shaped "pins" of stiff wire. However, there is always the risk that some of the "pins" will work their way loose and get caught in the lawnmower. This is especially true in those areas where there is a lot of freezing and thawing over the winter. An effective alternative is to simply "weave" the radial wires through the blades of grass—right on the ground's surface. The grass will keep the wires in place for years to come. This method also makes it easy to add more radials each year as you find the time.—*Ken Samuelson, WB8IEA, 1485 Edenwood Dr, Beavercreek, OH 45434-6838;* **wb8iea@arrl.net**

## HOSE CLAMPS TUNE VHF ANTENNAS

◊ I was having trouble getting my homebrew VHF antennas to come out on frequency. These were loop antennas with no exposed ends for trimming. I then realized that a hose clamp could provide some easily moveable reactance to slightly change the resonant frequency. Figure 4 shows some of the ways I've applied this technique. Larger clamps may work below VHF, or the small clamps could be used to secure wires for trimming to resonance. A pair of clamps could fasten two tubes side-by-side, while providing for easy element-length adjustment.—*Roger Wagner, K6LMN, 2022 Thayer Ave, Los Angeles, CA 90025-5927;* **k6lmn@arrl.net**

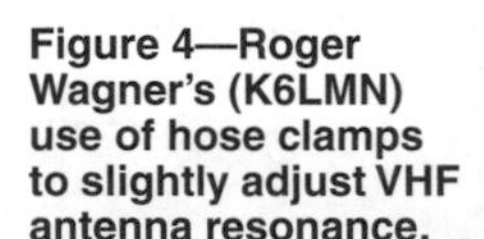

**Figure 4—Roger Wagner's (K6LMN) use of hose clamps to slightly adjust VHF antenna resonance.**

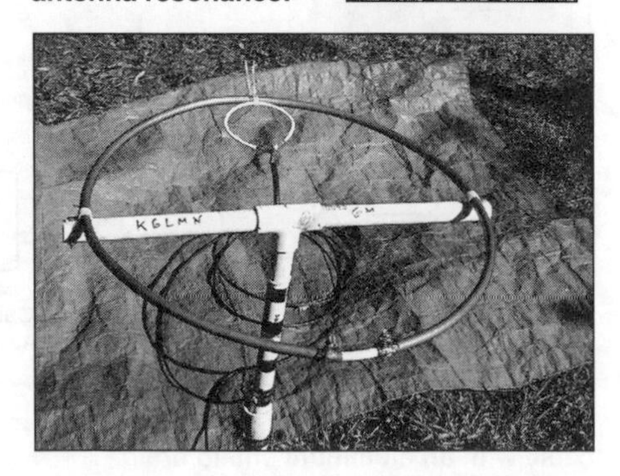

## MORE ON FEED-LINE LOSS MEASUREMENTS

◊ In the March 2004 column, Dan Wanchic, WA8VZQ, suggested that line-loss measurements made with SWR Analyzers could be improved by adding a 4-dB attenuator to the test setup. Frank Witt, AI1H, challenges the accuracy of that method and presents some alternative measurement methods in "Measuring Cable Loss," which appears in the current (May/Jun 2005) issue of *QEX*.—*Ed.*

## HEAVY-DUTY STANDOFF INSULATORS FOR LADDER LINE

◊ 450 Ω ladder line offers a broad surface to windy weather conditions and requires strong tie points to resist wind buffeting and eventual breakage. So I decided to design a strong standoff insulator. Figure 5 shows four different styles, for attaching the base of the insulator to a wall or pipe. The other end of the insulator is fabricated to secure the ladder line.

Figure 6 is a photo of ladder line secured with a wire tie. Table 1 is a Bill of Materials for the various styles.

The fabrication of the one-inch PVC tube is performed with ordinary shop tools. Here is a step-by-step procedure.

1. Lay out a 3½ × 3½-inch card-stock template (shown in Figure 7) for accurate hole locations, then punch a 1/16-inch diameter hole at each location.

2. Wrap the template around the cut length of tubing at the end and mark hole locations with a black marker or pencil.

3. Center punch each hole location and drill 7/32-inch diameter holes at locations A and B. Drill these holes through both sides of the tube.

4. Saw a 1/16-inch-wide kerf around the long path from hole X to hole Y as shown in the photos and Figure 5. To get a saw kerf this wide, mount two 18 teeth per inch hacksaw blades in a hacksaw frame, simultaneously. Use a couple of wraps of electric tape near the mounting holes to keep the blades together.

5. Deburr the holes and the saw kerfs, both inside and outside of the tube, then check the clamping action with a piece of ladder line. This completes the fabrication of the insulator end that holds the ladder line.

For styles 1 and 2, some of the cap of the T must be removed so that the remaining cap walls will present a stable surface for mounting (see Figure 5). Also drill 7/32-inch holes as needed for mounting.

**Figure 5—Four styles of PVC ladderline standoff insulators. From left to right are styles 1 through 4. Styles 1 and 2 both have the top half of their T caps removed. Style 1 is secured by long screws through the remaining T cap to a wall. Style 2 has tie wraps through holes in the remaining T cap securing it to a pipe. Style 3 has no T, but its tube is secured to a surface with pipe straps. In style 4, the T is secured to a wall with pipe straps.**

**Table 1**
**Bill of Materials**

| *Standoff Style* | *1* | *2* | *3* | *4* |
|---|---|---|---|---|
| 1" schedule 40 PVC pipe (inches) | 6 | 12 | 12 | 12 |
| PVC T | 1 | 1 | | 1 |
| #12 × 5/8 sheet metal screws | 1 | 1 | | 1 |
| Tie Wrap (1/8 ×1/16×6 inches) | 1 | 1 | 1 | 1 |
| 1-inch two-hole pipe straps | | | 2 | |
| 1¼-inch two-hole pipe strap | | | | 2 |
| Tie wrap (7/32 ×1/16×11 inches) | | 2 | | |

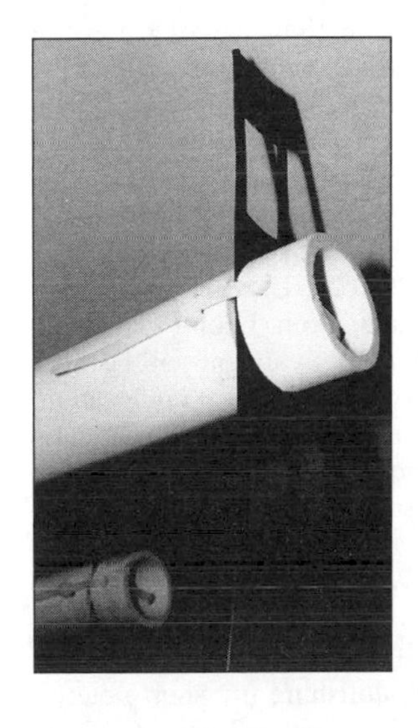

**Figure 6—A tie wrap clamps a kerf in the tube on the ladder line, securing it in the standoff.**

Styles 1, 2 and 4 have the tube secured to the T. After the tube is in the proper position in the T, drill a 11/64-inch hole through the T and tube, and install a #12× 5/8-inch sheet-metal screw.

I hope my design helps ladderline users achieve neater, more secure installations. If you have any questions, please contact me by US Mail (please enclose an SASE) —*Charles Simon, NØMS, 12440 Glennoble Dr, Maryland Heights, MO 63043*

## SOMETIMES, YOU FEEL LIKE A WIRE NUT!

◊ Hams faithfully obey Ohm's Law and take the path of least resistance. This is the fundamental reason for using wire nuts to fabricate wire antennas, but over the years, there have been added benefits. Many of us returned to ladder line or have recently discovered the lower costs and reduced losses it provides. Several articles have mentioned that ladder-line is difficult to use because the constant flexing from wind causes work hardening, which leads to broken connections and wires. Blind luck and laziness may have saved me some grief.

Making solder connections on wire antennas has always been difficult to do in an outdoor environment. A light breeze takes all the heat away, and most electric irons are way too small. My propane torch always seems to have an empty tank, and finding the soldering tip requires a search through a box of parts. Open-flame soldering oxidizes the wires too fast and makes a good joint impossible. Even if the whole process is performed indoors with proper equipment, the wires for an inch or two either side of the joint are usually overheated so that they lose their original properties. Often they will become the sites of future failures.

Why not use wire nuts? "Nuts are fine for Field Day," you say, "but certainly not a permanent installation!"

Well, consider this. My dipoles have been up for seven years with the same hardware and never a failure. So just what is "permanent"?

Common sense and good engineering practice should prevail here. Notice in Figure 8 that the ladder-line conductors pass up through the holes in the center insulator to provide some stress relief, and the dipole wires are twisted to provide relief for each of them. There's not much pull on the joint made by the nut. It's easy to accommodate stranded, solid, copper-clad and size variations with the large selection of wire nuts available. The dipole photographed is made with #10 AWG stranded copper wire and the ladder-line is #16 AWG copper-clad steel.

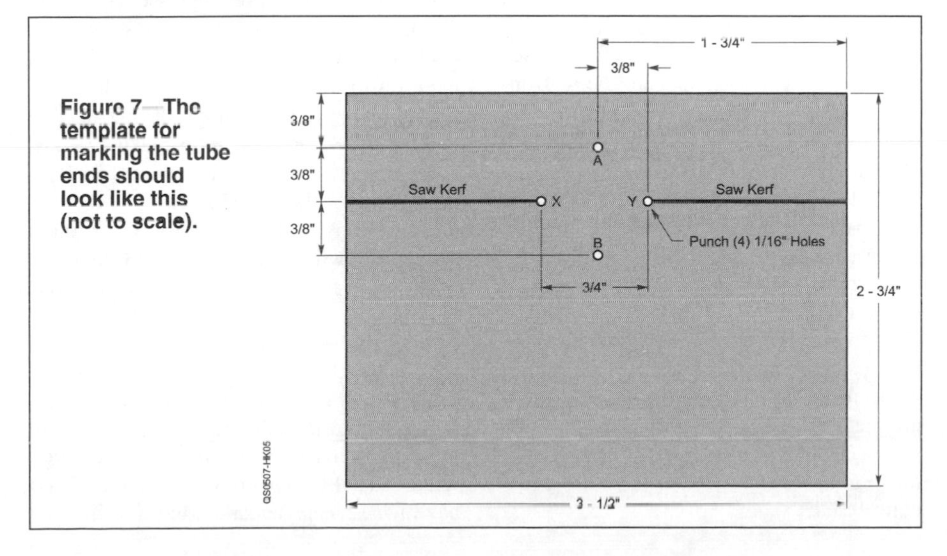


**Figure 7—The template for marking the tube ends should look like this (not to scale).**

**Figure 8—A Delrin insulator used at the center of a dipole and showing how wire routing can provide strain relief so that wire nuts are adequate for connections to the feed line.**

These variations were not a problem.

It's a nasty environment out there. Wind, rain, ice, heat and cold take a toll on anything up in the air. The theory is that the metal-to-metal contact forced by the compression of the wire nut is a gas-tight seal. The weather in Kansas hasn't caused any failures over the years. My antenna near Galveston, Texas, however, has a good coating of no-oxide grease applied to the wires as an additional barrier against the extreme humidity. Wire nuts are intended for use in a closed box, so the plastic is not likely to be UV resistant, and the sun may cause deterioration. When my dipole was lowered for a photo shoot, however, the plastic still looked nearly new. Perhaps my UV worry is unfounded. Some wire nuts are made from ceramic material and are intended to survive high temperatures, and I expect they would be very UV resistant.

Ladder line does present some mechanical challenges. You can't simply roll up the excess like coax so it has to be cut to length.

Providing stress relief to the line coming down seems to be the worst problem. I have made stress relief blocks from plastic and pull them with shock cord to keep tension on the line. The tension block in Figure 1 is made from Delrin but nylon or other plastic should work as well. The flat line does catch the wind and the only thing that seems to help is a few twists per yard. The number of twists isn't critical but don't overdo it either. When it twists in the wind and catches the sunlight there's the appearance of some sort of animated pop art that's probably only beautiful to hams.

The dipole needs to remain in the same place all the time so the feedline doesn't become slack or snap too tight. A healthy spring from an overhead garage door provides tension for my 150-foot dipole. It has survived a major ice storm with over an inch of ice buildup. Springs seem to perform better than weights.

Keep your cool. Put the irons and torches away and have a nut or two. No more burnt fingers, oxidized connections, cold solder joints and brittle wires. At least give this idea a try on Field Day.—*Ken Shubert, KØKS, 1308 N Leeview Dr, Olathe, KS 66061-5028;* **k0ks@arrl.net**

## A WEATHERPROOF N CONNECTOR CAP

◊ I'm installing an HF station in my car. The antenna is on a receiver hitch mount, so it can be conveniently removed when it's not needed. This means that the coax cable from the radio to the antenna won't be connected all the time. I used an N connector because it's weatherproof when connected, but I needed a way to protect it from road ice and dirt when it's not connected.

The usual solution to this problem is a plastic cap that pushes into the coupling ring or over it and seals the end off. This didn't appeal to me. I drive a Lexus RX300, and it just seemed like that would look a little cheesy—and besides, it could easily be knocked off without me noticing it. I asked some friends, and they mentioned that Amphenol makes a part that would work, for both N and UHF connectors (the coupling rings use the same thread). I wasn't able to find the Amphenol part locally.

**Figure 9—This shows where to cut the PL-259 plug. An intact plug is shown for comparison.**

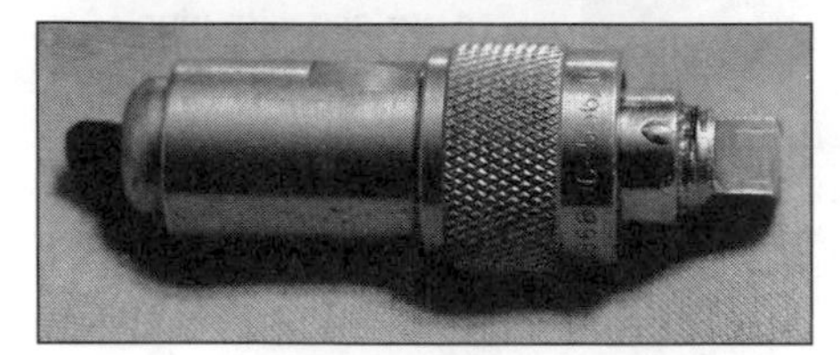

**Figure 10—The finished cap, mated to an N connector.**

The part about both connectors using the same thread gave me an idea. The body of a PL-259 has a part that has that same thread. I realized that—with a bit of surgery—the connector body itself would serve. The result can be seen in Figures 9 and 10.

Cut the plug body (be sure to wear eye protection!) just below the threads, as seen in Figure 1. A Dremel rotary tool with a cutoff wheel makes short work of this job. Grind away any sharp edges. Then, insert a ⅛ inch flare pipe plug. The threads are different, but since we never intend to remove the plug, this doesn't matter. Use a bit of brute force to seat the plug firmly in the connector body. Don't twist it too far in, or you won't have enough room for the center pin. Solder the plug in place to make sure it's watertight.

The cap will seat against the rubber ring inside the connector, and the coupling ring will hold it on, making a watertight seal. This will work for PL-259s, too, but it won't be weather tight by itself. It will protect the center pin from damage. If you use it for this purpose, you might want to consider grinding a couple of notches to receive the plug body's teeth. —*Jay Maynard, K5ZC, 1831 Oakwood Dr, Fairmont, MN 56031-3225;* **jmaynard@conmicro.cx**

## USE A "STRAP HOIST" TO LIFT ANTENNAS

◊ If you've ever been atop your tower trying to wrestle a large HF beam into place, it's not easy, is it? Here's an idea you might try. Visit your local auto- or building-supply store and get one of those large cargo tie-down straps, the ones with the heavy ratcheting mechanism to draw it tight. Make sure the strap is rated for *several times* the antenna's weight.

Back at the tower, secure one end of the strap to the tower, near the antenna destination. When your beam comes up the tramline, fasten the strap's other end to it and use the ratchet mechanism to hoist the antenna into position. Let the strap support the antenna's weight while you bolt the antenna into place.—*Dave Hough, W7GK, PO Box 2072, Elko, NV 89803-2072;* **dhough@chilton-inc.com**

## STEALTH ANTENNA TACTICS

◊ While we can make no antenna (like a dipole or inverted V) invisible, some tricks can go a long way toward that goal. First, consider colors: dark is good, white is bad. Unfortunately, almost all baluns and many insulators are made from stark white plastic. Even dim ambient light makes the white plastic very visible. A can of good quality flat black [or gray—*Ed.*] enamel will work well to make those parts less visible. Get a can of spray paint that says it is plastic friendly. This is very convenient and the second coat can usually be applied in a few minutes. Tape over the connections. Use black insulated wire for the antenna run. I don't know how long the black paint treatment will last, but I can attest to two plus years at my location.

Bare copper wire is fine for the antenna; a few months of oxidation will darken it to yield good results.—*Sam L. Grider, KJ8K, 2810 Blackberry Tr, Cincinnati, OH 45233-1722*

## SWR AND LINE LOSS—AN UPDATE

◊ Thanks for publishing my short note on using SWR to measure line losses (*QST*, Jan 2003, p 62). Here are two updates:

1. I noticed recently that for low- and medium-loss lines, you don't need the table. To a high degree of accuracy, the line decibel loss for any SWR reading is equal to 9/SWR. If the SWR is 18, the line loss is 0.5 dB. If the SWR is 9, the loss is 1 dB. If the SWR is 4.5, the loss is 2 dB, and so forth. This is quite accurate down to an SWR of 3 (3 dB loss), and it's still fairly accurate down to an SWR of 2 (simple formula gives 4.5 dB, while the actual value 4.77 dB). Neat, eh?

2. Thanks to W1VT for pointing out that the table, as given, assumes a matched system. I think the technique will still work in most cases if you have a mismatch, but you will need a different conversion table. If you have a 75 Ω meter and a shorted 50 Ω line, the

**Table 1**
**Line Loss versus SWR for Various Line/Meter Mismatches**

| | | *SWR* | |
|---|---|---|---|
| *Loss (dB/100 ft, matched line and meter)* | | *(50 Ω line, 75 Ω meter)* | *(75 Ω line meter, 50 Ω meter)* |
| 0.2 | 43 | 65 | 29 |
| 0.4 | 22 | 33 | 14 |
| 0.6 | 15 | 22 | 9.7 |
| 0.8 | 11 | 16 | 7.3 |
| 1 | 8.7 | 13 | 5.8 |
| 1.2 | 7.3 | 11 | 4.9 |
| 1.4 | 6.3 | 9.4 | 4.2 |
| 1.6 | 5.5 | 8.2 | 3.7 |
| 1.8 | 4.9 | 7.3 | 3.3 |
| 2 | 4.4 | 6.6 | 2.95 |
| 2.5 | 3.6 | 5.4 | 2.38 |
| 3 | 3.0 | 4.5 | 2.01 |
| 3.5 | 2.6 | 3.9 | 1.74 |
| 4 | 2.3 | 3.5 | 1.55 |
| 5 | 1.93 | 2.9 | 1.28 |
| 6 | 1.67 | 2.5 | 1.11 |
| 7 | 1.50 | 2.25 | 1.00 |
| 8 | 1.38 | 2.07 | 1.09 |
| 9 | 1.29 | 1.93 | 1.16 |
| 10 | 1.22 | 1.83 | 1.23 |
| 20 | 1.02 | 1.53 | 1.47 |

effect is to simply multiply each SWR by 1.5.

If you have a 50 Ω meter and a shorted 75 Ω line, things get a bit messy because for line losses above about 4 dB there are two values of loss for each SWR, one giving an impedance less than 50 Ω and one giving an impedance greater than 50 Ω. For lines with losses less than 4 dB/100 ft, it's pretty straightforward as shown in Table 1. The loss is approximately 13.5/SWR for the second column and 6/SWR for the third column.—*Ned Conklin, KH7JJ (ex-K1HMU), 2969 Kalakaua Ave,Honolulu, HI 96815;* **ekc@forth.com**

## A BACKPACKER'S DIPOLE

### *An HF Antenna to Take Along When You Can Only Take One*

◊ Whether you are backpacking, vacationing or assembling a "go kit" for emergency service, the desire can arise for a low-cost, easily transportable, reusable, versatile HF antenna. This article describes one antenna that fits those criteria. The antenna is a simple half-wave dipole. What makes it noteworthy is that its length is adjustable so it can be used on any frequency higher than the resonant frequency at the antenna's maximum length. The key to this almost unlimited adjustability is the use of bare, stranded Flexweave wire and open reels made from wooden dowels. When the wire is accumulated on the reels, it ceases to act as part of the length of the dipole. Therefore, for example, if the dipole is made long enough for 40 meters, but about half of it is collected on the reels the antenna will resonate at 20 meters. Later, the wire can be unrolled from the reels until the antenna is long enough to be used on 40 meters—or shortened to work on 15 meters.

As shown in Figure 11, the reels are constructed of hardwood dowels. Each reel consists of two 9-inch long ⅝-inch diameter dowels joined together, about 4 inches apart, by two 9-inch long ⅜-inch dowels about 4 inches apart. The ⅜ inch dowels pass through the larger dowels and extend outward to serve as crank handles for rolling the wire on or off of the reels. This really speeds the work of laying out the antenna and avoids kinking of the antenna wire. The feed-point is a manufactured center insulator with an integrated female UHF connector. Alternatively, a regular center insulator can be used and the coax soldered directly to the antenna wire.

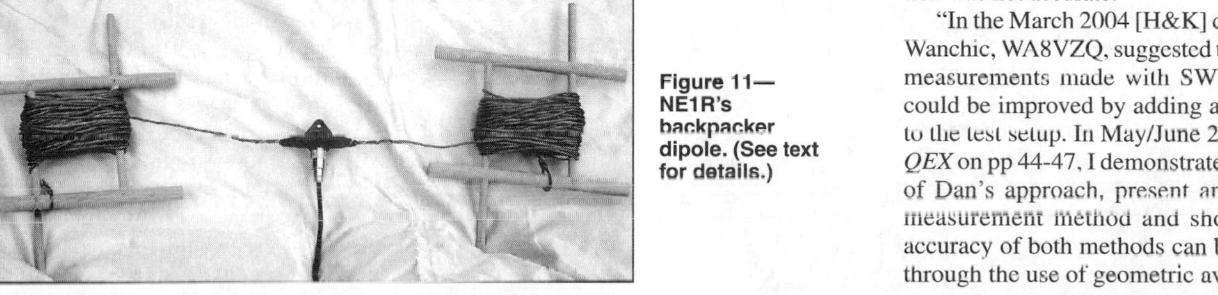

**Figure 11—NE1R's backpacker dipole. (See text for details.)**

The proper length of the antenna for each desired band can be determined by experimentation before leaving home. Once the proper lengths are determined, the wire on each side can be marked with paint or tape to speed adjustments in the future. Notice that capacitance formed by the bundles of wire on the spools will shorten the over-all length of the wire as compared with the well-known formula (468/f), particularly on the higher frequency bands, when more wire is on the spools. This does not seem to affect adversely the performance of the antenna to any noticeable degree, however. Proximity to the ground will also affect the resonant frequency. Therefore, if you are going to install the antenna 10 feet off the ground in the field, experiments at home should simulate that height.

To deploy the antenna, string the support lines over tree branches or some other supports. Unroll the wire from each reel to achieve the desired lengths, which should be approximately equal. Gather the remaining wire on each spool in a bunch and tie it with one end of the support line. Then raise the antenna by pulling the support lines.

The sample shown in the photo contains approximately 100 feet of bare wire on both reels and resonates from approximately 5 MHz to 54 MHz. The antenna, 30 feet of coax and some nylon support line, weighs less than five pounds. The weight of the assembly could be reduced by using thinner antenna wire. Happily, there is no weight of an antenna tuner because no tuner is needed once the antenna is adjusted.

I would enjoy corresponding with others experimenting with this antenna.—*Thomas C. Carrigan, NE1R;* **ne1r@arrl.net**

## STILL MORE ON FEED-LINE LOSS MEASUREMENTS

◊ The June 2005 installment of "Hints & Kinks" ["More on Feed-Line Loss Measurements," page 63] described a *QEX* article by Frank Witt, AI1H, as "challenging the accuracy of" a method of measuring feed-line loss that was presented earlier in Hints & Kinks. Frank, AI1H, points out that that characterization was not accurate:

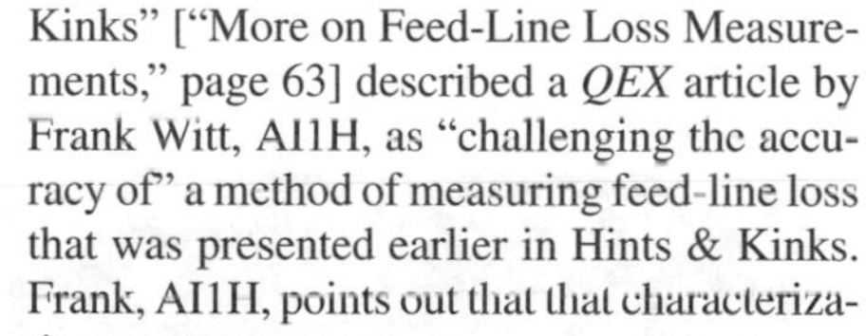

"In the March 2004 [H&K] column, Dan Wanchic, WA8VZQ, suggested that line-loss measurements made with SWR analyzers could be improved by adding an attenuator to the test setup. In May/June 2005 issue of *QEX* on pp 44-47, I demonstrate the validity of Dan's approach, present an alternative measurement method and show how the accuracy of both methods can be improved through the use of geometric averaging."

## MAKE A PORTABLE WIRE ANTENNA WITH PIPE COMPRESSION FITTINGS

◊ Many hams build antennas with connectors to accept wire segments of different lengths for operating on multiple bands. Some operators use alligator clips, phono plugs or other connectors. They work, but those that are strong may not provide good connections, and those with good connections may not be very strong. Ease of assembly and disassembly adds another requirement to the search.

One day Jerry Hessenflow, WBØTWT, stopped by ARRL headquarters with a truly elegant solution.

Jerry brought a brass compression union fitting for very small (¼ inch and smaller) tubing. Jerry's sample and details of the installation are shown in Figures 12 through 14.

I was initially concerned that it would be difficult to find such small fittings, but I found a selection at my local building-supply store, and a Web search for "compression fitting brass ¼" led to several mail-order sources with fittings sized down to ⅛ inch tubing.

Thanks for the hint, Jerry!—*Bob Schetgen, KU7G, Hints and Kinks Editor;* **rschetgen@arrl.org**

KU7G

**Figure 12— Reverse the fitting nut and slide it onto the wire (just like a coax connector), then fold the wire back over the compression ring. I suggest rounding the sharp edge of the compression ring with sandpaper or a file to reduce the possibility of the ring edge shearing through the wire.**

KU7G

**Figure 13—Solder the wire to the compression ring.**

KU7G

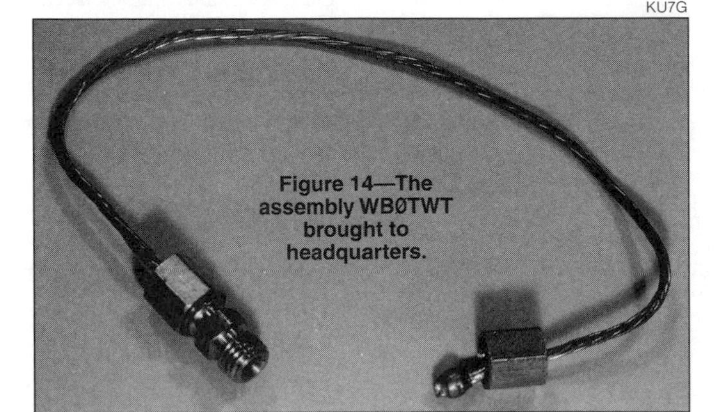

**Figure 14—The assembly WBØTWT brought to headquarters.**

## PREPARING COAX

◊ Many times while stripping the insulation off of coax for the installation of a connector, I have cut into the copper braid or otherwise made a circular cut that did not meet at the ends. I found that using a small tubing cutter made a much neater cut. The cutter I use is called a Little Inch, however, there are a number of small cutters available. Another one is the Mini Tubing Cutter from Harbor Freight (**www.harborfreight.com**) for $4.99.

Figures 15 through 18 provide the details of how to use this approach.—*Richard Arnold, AF8X, 22901 E Schafer St, Clinton Twp, MI 48035;* **af8x@arrl.net**

**Figure 15—The relative size of cutter compared to the end of the coax.**

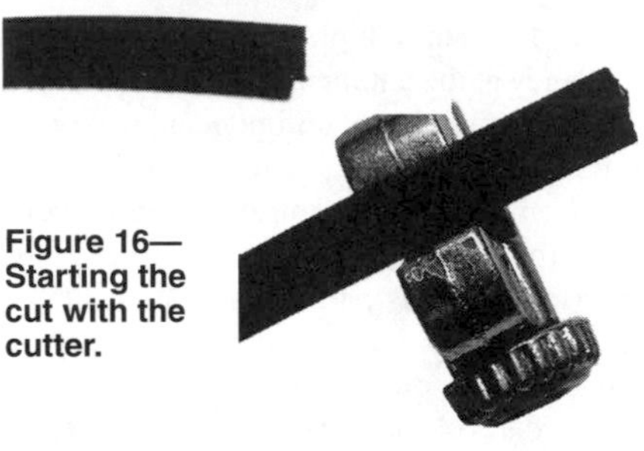

**Figure 16—Starting the cut with the cutter.**

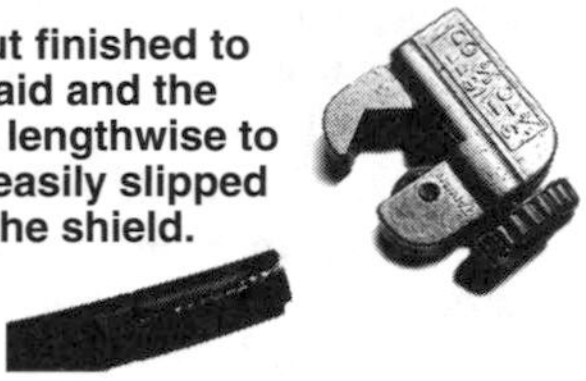

**Figure 17—Cut finished to the copper braid and the insulation slit lengthwise to allow it to be easily slipped off exposing the shield.**

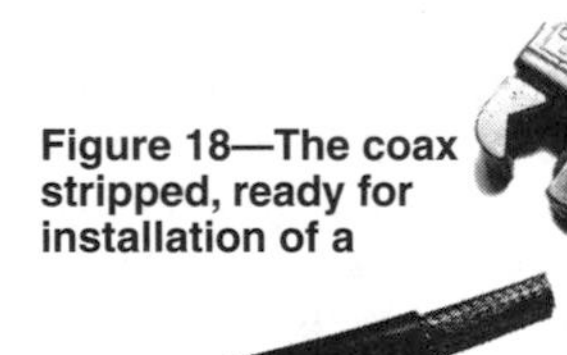

**Figure 18—The coax stripped, ready for installation of a**

## STEALTH ANTENNA BASE

◊ If you've ever looked at the great projects in *QST* every month and wished you had the time, tools and tenacity to complete one, here's one you can do—and it's guaranteed! It's a really simple solution to the problem of living in an area populated by antenna paranoids.

If the local antenna mafia won't allow you to have a *permanent structure* or a mast *attached or affixed to a dwelling or other building*, then don't! Just attach the thing to this temporary base made from stuff you can get in 20 minutes at The Home Depot. (That's the reason for the guarantee—they take the parts back if you don't want 'em!)

There's nothing electromagnetic, rectifiable or modulational about this, it's pure gravity. The basic form is a portable bucket of cured concrete that supports a mast to which you attach whatever you want to stick up in the air. That's it. Construction time is less than 30 minutes plus the time you spend wandering around with all those folks in orange aprons. The most difficult part of this project is moving an 80 pound bag of concrete from the store to your place. As for technique, if you've ever stirred really thick oatmeal, you've mastered the process.

If you have trouble selling the idea to others who share your estate, explain how well it can support a patio umbrella too. Remember, it's not permanent. Even if you want to add guy wires temporarily, it does not make it more permanent than a pup tent. The version shown has a handle to make it easier to move around and a lid to make it look more "finished." If you really need a finished look to match the décor of your grounds and gardens, you could select a white 5 gallon bucket instead of the one I chose, and use some rather chic *faux* paint to allow a more aesthetically appealing or stealthy unit.

### *The Process*

Take this shopping list (Table 1) to The Home Depot (or fax it to their Pro Desk ahead of time and they will have the stuff all ready for you when you get there). The mast section(s) are in the electrical department, the handle hardware parts are in hardware, the bucket and lid are in the paint department (orange ones are all over the store!) and the cement is in the building materials department. The set of parts is shown in Figure 19.

**Table 1**
**Parts for Stealth Antenna Base**

| *Quantity* | *Item* | *SKU#* | *Price ($)* |
|---|---|---|---|
| 1 | 5 gallon bucket. | 131-227 | 3.98 |
| 1 | 4½ foot antenna mast. | 114-572 | 7.95 |
| 1 | 80 pound bag general purpose cement. | 420-584 | 2.67 |
| 1 | Bucket lid. | 756-849 | 1.37 |
| 1 | Stanley chest handle. | 240-214 | 3.79 |
| 4 | ¼×6 inch galvanized carriage bolts. | 188-198 | @ .78 |
| 8 | ¼ inch galvanized nuts. | 538-841 | @ .09 |
| 4 | ¼ inch galvanized cut washers. | 538-914 | @ .08 |

At the construction site (your patio), assemble the parts for the handle, Figure 20. The carriage bolts will be pushed into the wet concrete at the end of the process, but it is easier if they are ready to go ahead of time. Spin one nut on each of the carriage bolts far enough so that the handle, a washer and another nut can be attached. Then insert each into the bottom side of the handle's four holes, add a washer and tighten a nut to complete the job. Set the assembly aside for now. If you will use a lid, cut the 1¼ inch hole in the center now.

You mix the concrete right in the bucket you will use as the base. The directions call for about 3½ quarts of water for the 80 pound bag. I used a little more to make it easier to work. A pail with the pre-measured water makes it easy to add to the mix. Set the cement bag, top side up, next to the bucket. Use a dry, 16 ounce tin can as a scoop to add the dry mix to the bucket without making a mess (important). I found that four scoops of cement followed by enough water to keep it mushy worked well. Mix with a stick long enough to do the job and one that you will probably throw away. When the bag is empty, the wet cement will be about ²⁄₁₂ inches from the top of the bucket as shown in Figure 21.

Figure 19—This is all you need.

Figure 21—The 80 pounds of concrete fits nicely in a 5 gallon bucket. The handle, below the mast, makes moving much easier. It folds flat under the lid.

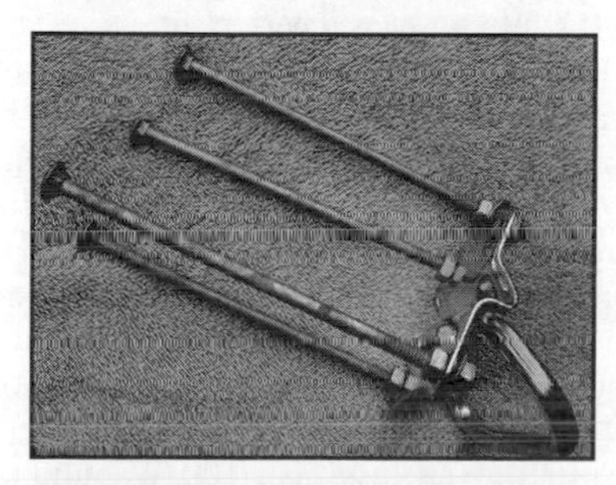

Figure 20—A handle makes moving the base much easier. The carriage bolts are pushed into the

Figure 22—Ready to attach whatever antenna you choose—normally with U-bolts. Additional antenna mast pieces may be added to the top.

You're almost done! Insert the 4½ foot antenna mast into the mix all the way to the bottom with the warning label at the top. [You may want to set the bucket on a level surface and use a level to make sure the mast is vertical. Some temporary supports to make sure it hardens in that position is a good idea, too—*Ed.*] The label reminds you that even temporary antennas should be away from overhead wires. You can add more mast sections to the top later if you want. Now push your handle assembly into the mix with the hinge toward the near side of the bucket, so it is level with the top of the cement. Slide the top down to use as a support to make sure that the mast is not tilting. The cement will cure in a couple days depending on the weather, so if you want to add your initials don't wait too long!

Whenever you want to enjoy your hobby with a little more range than you get with an indoor setup, remember what your mom told you: "Take that thing outside, now!" And don't let the bad guys mess with your temporary umbrella holder. *—Don Eaton, KNØTME, 1700 Tulip La SW #31-102, Tumwater, WA 98512;* **Nooz@aol.com**

## A SHEPHERD'S CROOK FOR ANTENNA WORK

◊Antenna wire or tubing can snag on tree limbs, parts of buildings and even other antennas while the antenna is on its way up. The "shepherd's crook," shown in Figure 23, used with an extension pole, can either push or pull the wire or tubing around obstacles. A Yagi element that has slipped (due to loose hardware) and is no longer parallel with others in an array can be nudged back into its proper place. It's also helpful for pulling dangling coax into a window when the overhang of the roof holds it more than an arm's length away.

The crook is made from a discarded paint roller. To use it with an extension pole, the handle must be the common threaded type. A used roller with paint on it works fine, so this can be a no cost project. If you are one of those who always cleans up your tools when finished, you may have to buy a cheap roller at a discount store.

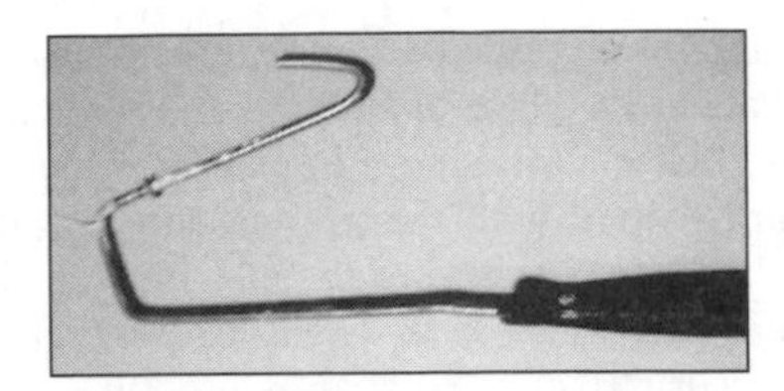

**Figure 23 — A "shepherd's crook" for clearing antenna snarls.**

Break the plastic end that holds the roller by striking it with a hammer. Remove and discard it and all other parts except the plastic handle and metal rod. Form the rod to make two hooks using a vise and a hammer. One hook holds the wire while pushing up or away and one is used to pull down or toward you. Remember to be very careful when working with antennas around power lines or any exposed electrical circuits. For safety use this crook with either a wooden or fiberglass extension pole.

I have a fiberglass painters' extension pole that telescopes out to 24 feet. It works well in this application and does double duty as a mast for portable wire antennas. Some brooms are made with a screw-on head that has the same threads as a roller. Its handle is a good length for fishing a length of coax into a window. — *Larry G. Ledford, KA4J, 553 4th St SE, Cleveland, TN 37311*

## WEATHERPROOF THOSE OUTDOOR COAX CONNECTORS

◊To weatherproof your coax connections, you may not have to travel any farther than your local bicycle shop or supply store. A bicycle inner tube will provide five or six rubber sleeves that can cover and protect a coax connection, such as two PL-259 coax cable terminations surrounding a barrel adaptor. An 8 inch length covers the entire connection and, with cable ties at the ends, provides a neat, moisture resistant barrier.

While the more expensive N and BNC connectors are waterproof if properly assembled, the usual amateur UHF type connectors are not unless external covering is applied.

I used a Bell (Bell Sports, Inc) 27 inch inner tube (no. 2445001013), designed for 1 inch tires. The diameter of the inner tube is ¾ inch, which makes for a nice fit over PL-259 coax connectors. Begin by cutting the tube on either side of the stem. Since the interior of an inner tube is covered with a talc-like powder to prevent the rubber from self-adhering during storage, you'll want to flush out the interior with water. Cut the tube into 8 inch lengths. The residual powder can be removed from each sleeve by running a damp paper towel through it.

Slide a coax sleeve over the connection and secure each end with a cable tie, as follows. At one end, wrap the sleeve around the coax and secure the folds with a cable tie about an inch from the end. Pull the cable tie tight by hand (not with pliers). Repeat at the other end. Trim the cable ties and you're done, as shown in Figure 24. Remember to create a drip loop at the coax connection, as well.

A nice feature of the coax sleeve is the ease with which it can be removed and then reinstalled. Be careful when snipping the cable ties so as not to nick the rubber sleeve.

How well does the inner tube coax sleeve waterproof the connection? I've used it successfully for more than a year. Of course, your mileage may vary. — *Al Bregman, W4GHQ, 2009 Antler Dr, Catawba, SC 29704-9483*

**Figure 24 — Completed coax connection waterproof system.**

## NOTES ON LIGHTNING PROTECTION FOR OPEN-WIRE FEED LINES

◊ Reading about lightning, and its disastrous results, caused me to look into ways of improving my ground system around the house. Because various publications already adequately explain the grounding of towers, coax fed antenna systems and radio room equipment, I won't cover it here, but I will write about protection for open-wire or twin-lead feeders.

*The ARRL Antenna Book* has a couple of ideas for feeder protection, but I had one of my own I wanted to experiment with. Since the automobile spark plug is nothing more than a spark gap, why not use them for feeder protection? They're cheap, and the spark gap is adjustable.

Out of all the spark plugs available, how does one make the correct choice? Since this was my experiment, I set the selection criteria based entirely on intuition. (Little technical theory was actually used to implement this idea.)

First, the spark plug had to be the non-resistor type. (Yes, most plugs have a 5 kΩ internal resistance in series with the main conductor.) Finding a non-resistor plug wasn't easy. Some clearly indicate in the part number (with an R) that it's a resistive plug. Others do not. Just remember that non-resistor types are generally for older model vehicles.

Second, the spark plug needed to have a threaded tip, so the feeder could be secured with a nut and lock washer. Without a threaded tip, some other way of connecting the feeder would have to be devised, and it probably would not be as solid a connection.

I actually found the plugs I needed by accident. (Salesmen tried to help, but didn't seem to be very knowledgeable about their stock. Their first question was, "What kind of car is it for?" Once I told them it was for an antenna system, the conversation quickly went down hill.) I happened to look up over the counter and saw a clearly labeled box. Turns out it contained exactly what I needed.

The plugs were Japanese AC-Cel high performance, non-resistor 274s. The box indicated that they're a replacement for the following: AC Delco 43TS or 42TS, Old Autolite AF32B or AF22B, New Autolite 14 or 13 and Champion BL9Y or BL11. Presumably any of these original types should work as well. To be safe, bring a multimeter and check for a resistor. Also, the top brass looking connector should unscrew to reveal the small 4 mm threads. The spark plug base should be 14 mm.

My next step was to find a nut to fit the 14 mm base of the spark plug. Not an easy task. The local nut and bolt company came close, but couldn't match the fine threads. I even went back to the auto parts store, but no nuts were found. However, I stumbled across something called a "spark plug non-fouler." Not being an auto mechanic, I didn't have a clue as to its use, but the device fit perfectly on the base of the plug. They come in different depths, so make sure you get the correct one. These little devices not only secure the spark plug to the ground buss, they also cover the exposed spark gap, and protect it. The details are shown in Figure 25.

Next, I took a spark plug to the local hardware store and matched a metric nut for securing the feeder to the top of the plug. A 4 × 0.7 mm nut was a perfect fit.

Before mounting the spark plug to the ground buss, I needed to adjust the gap for best performance. (I didn't want RF jumping the gap when the transmitter was keyed.) I came across an article on the Web written by a broadcast engineer. His research on spark gaps for AM broadcast towers indicated a need for 0.029 inch spacing for a kW station, and 0.045 inch spacing for 2.5 kW. Since I normally run less than 100 W output, I opted to set the spark plug gap to 0.025 inch (.635mm). [Note that the voltage increases

Figure 25 — Lightning protector rear view. Note "spark plug non-fouler" attachment mechanism.

Figure 26 — Completed balanced line lightning protector; front view showing feeder

with the square root of the SWR. Since many balanced line systems are operated at an SWR of 10:1 or greater, keep that in mind as you calculate gap size — *Ed.*]

To complete the project, I drilled 9⁄16 inch holes in the ¼ inch ground buss and mounted the spark plugs using a liberal amount of OX-Gard, an anti-oxidant compound for wire connections. I then grounded the buss to the system ground. The completed system is shown in Figure 26.

Running 100 W output has not resulted in any arcing or intermittent changes in reflected power. No attempt was made to check impedance changes in the line resulting from the addition of the spark gaps. A very minor adjustment of the antenna tuner was required to re-match the antenna system, likely caused by reducing the overall length of the feeder by 3 feet. — *Joe Hutchens, WJ5MH, 4712 San Simeon Dr, Austin, TX 78749,* **wj5mh@arrl.net**

## ADD SOME GREASE TO THE RECIPE

◊In the August 2005 Hints & Kinks item, "Sometimes, you feel like a wire nut," Ken Shubert, KØKS, suggested that wire nuts could be used to fasten ladder line to wire antennas. As Ken wrote, these nuts are not specifically intended to be exposed to the elements and may deteriorate over time, but there may be other alternatives. Why not use the "grease caps" made to fasten lawn sprinkler control wires to the control valves? Designed to be used outdoors, even for direct burial, products such as Orbit's WaterMaster grease caps typically are sold in blister-pack kits at lawn and garden sections of "big box" stores here in the South and may be available generally throughout the country. The typical kit, which sells for less than $3, contains four medium size wire nuts, four grease caps (plastic sleeves filled with waterproof silicon grease) and, as a bonus, four insulated spade lugs. After the wire nut is installed on the wires, the grease cap is pushed on over the wire nut resulting in a moisture-proof connection.

Several manufacturers make grease cap kits, including a new model cap from TORO that doesn't require a wire nut; just open the split cap, insert the wires, and snap the two halves together to effect a water-tight connection. A light coating of brush-on electrical tape can be applied to the caps if desired for added UV protection.

For those who may find the grease caps too bulky (they actually are smaller than some center insulators), another alternative is to use the special waterproof wire nuts such as Dryconn #62110, now available in the electrical department of most hardware stores. A bit more expensive than the standard versions (#12 to 22 gauge starting at about 50 cents each), they contain silicon grease, are self-sealing, and are touted as "corrosion proof" and suitable for damp and wet locations. I've used the latter type to attach ribbon line to coax on a half-size G5RV here in Florida (I also sealed any exposed breaches in the insulation with silicon sealant). A layer of electrical tape will provide added mechanical strength to the junction if needed. — *William Wornham, NZ1D, 1819 W Schwartz Blvd, Lady Lake, FL 32159,* **nz1d@arrl.net**

## ANOTHER METHOD OF EASY RADIAL INSTALLATION

◊I wanted to be sure my new radials were deep enough to be out of harm's way, from not just the lawn mower, but also rakes and other things that might uproot them. I also wanted to do as little lawn damage as possible. My solution was to rent a gasoline-powered walk-behind sod cutter.

After laying out the pattern with stakes, I simply ran the machine the length of each radial, beginning a little way out from the center. I then just lifted the sod and slipped the wire under, just as if the sod were strips of carpet. I then kept the sod strips watered until they re-rooted. —*Bill Wheaton, K4DER, 101 Carriage Ln, Concord, VA 24538,* **bwheaton@moreinformation.net**

## THE POCKET MAG MOUNT

◊ Looking for a small, efficient magnteic-mount antenna that complements the convenience and portability of your handheld radio? If so, this inexpensive project may be for you.

You will need a magnetic microphone holder, a chassis-mount female BNC connector and a solder lug to fit over the threads of the BNC connector. You will also need a length of RG-174 or RG-58 coaxial cable and a BNC male connector or other mating connector for your radio. You may want a second BNC male connector to hold a piece of wire for a ¼ λ antenna to attach to the magnetic mount.

Use needle-nose pliers to bend the entire "finger" part of the mic holder up at a 45° angle. Next, bend the last half of this section so it is parallel with the magnet. See Figure 27A. The last bend is made to bring the ends of the "fingers" a little closer together to form a partial loop. See Figure 27B. Place the

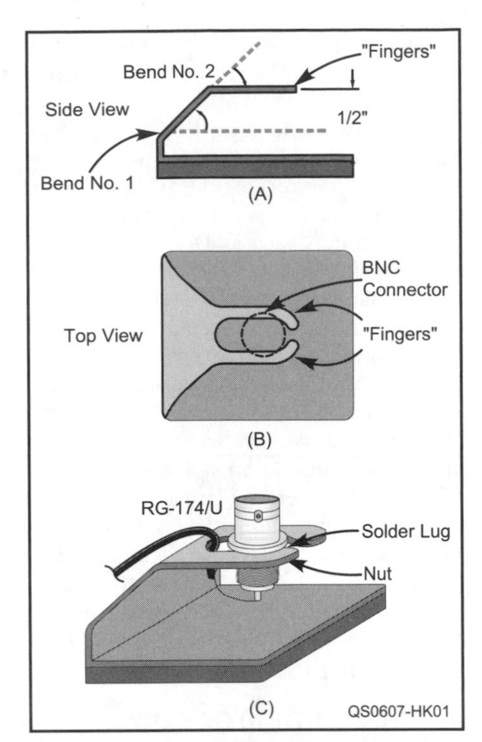


**Figure 27 — Part A shows the details of how the mic holder is bent to form the base of the pocket mag mount. B is a top view, showing the mic-holder fingers bent to hold the BNC chassis-mount connector. Part C shows the final assembly of the pocket mag mount.**

BNC chassis connector through the opening, and tighten the nut to hold the solder lug and connector in place. You may have to squeeze the loop to ensure a better fit. Prepare the end of the coaxial cable, and solder the shield to the solder lug. Next, solder the center conductor to the connector pin. To provide strain relief, thread the other end of the cable through the space between the mic holder and the connector. See Figure 27C. Seal the exposed connections for weather protection, and install the male BNC connector on the other end of the coaxial cable.

You can now use a rubber duckie antenna on the mag mount, or you can make a ¼-λ antenna using another male BNC connector and 19 inches of wire or length of brass brazing rod. The mount serves nicely as a highly flexible patch cord when using the handheld radio with additional equipment or antennas. Results have been gratifying. When not in use, the mag mount can be stored in the glove compartment of your car or carried in your pocket or a pack. — *Steve Brtis, WA6FGW, 23801 Friar St, Woodland Hills, CA 91367-1235;* **brtis@juno.com**

[The original article referenced RadioShack part numbers for the microphone holder and BNC connectors. While RadioShack still lists the BNC connectors, they no longer list a magnetic microphone holder. Quick Silver Radio Products lists all the required parts on their Web site (**www.qsradio.com/**), and is one mail-order source for the pieces you would need to build this antenna mount. — *Ed.*]

## REPAIRING A DAMAGED BOOM/ MAKE YOUR BEAM ANTENNA STRONGER

◊ One of the most common reasons for having to take your beam down is that elements have shifted or rotated on the boom. One of the most common problems you'll encounter in refurbishing a beam is crushed, kinked, or severely dented boom ends.

Usually, someone, (maybe you!) trying to ensure that the elements would not shift or rotate on the boom, kept tightening until the U-bolt began to crush the boom's aluminum tubing. If the dents are deep, they will seriously affect the strength of the boom and make it difficult for the U-bolts to ever properly grip. Here's what I've found to be an effective fix.

Buy or borrow an automotive tubing straightener, used when a muffler or exhaust pipe is to be reused. They're available in nearly any store carrying automotive tools. Most will straighten 1.5 to 3.0-inch-diameter tubing, which exactly covers the range of common boom sizes! The straightener's bolt and spreader cones are lightly greased, the tool inserted in the damaged boom, and the bolt turned until the outside of the tubing is restored to its original size. See Figure 28. For 2 inch booms (as used by the Mosley Classic 33 and many other beams) cut an 8-inch-long section of common 2-inch-diameter ABS plumbing pipe. It is very slightly too large to fit inside the Mosley boom, so a lengthwise slot is cut with a saw, just wide enough so that when the pipe is driven into the end of the boom the slot will be closed. You may find it easier to first cut the lengthwise slit before cutting off the 8-inch length of pipe: this gives you something better to grip in the vise! While 2-inch ABS pipe has lower stiffness than the 2-inch aluminum boom, it is very incompressible, and adds greatly to the strength of the repaired boom end when slipped inside. Now the element's U-bolt can be safely tightened considerably more than previously, giving the elements a much better grip on the boom. For smaller booms (1.25 to 1.75 inches or so) hardwood dowels (or broom handles) of suitable size to snugly fill the straightened boom tubing can be obtained at most larger hardware stores and lumber yards.

I have successfully restored two TA-33 tribanders with bad boom kinks at the center (driven) element by slightly tapering one end of an 18-inch length of hardwood dowel, then driving it into the boom with a smaller diameter rod until it forces the central kink straight and rests across the kink.

If you use hardwood dowels, you should waterproof them with linseed oil, thinned oil paint or urethane, because if they absorb water they could swell and possibly split the boom tubing. Of course a damaged boom isn't required: these ABS or hardwood boom strengtheners will make any beam stronger and more wind and ice-resistant. — *Fred Archibald, VE1FA, 25 Canard St, RR#1, Port Williams, Nova Scotia, B0P 1T0, Canada,* **hfarchibald@ns.sympatico.ca**

## MAKE YOUR OWN 300 Ω OPEN WIRE FEED LINE

◊ I read "The Doctor is IN" in the October 2004 issue of *QST* about "Stealthy Ladder Line." Open-wire line is always a problem for Japanese hams because that type of feed line does not seem to be sold outside of the US. I often make an open-wire line from 300 Ω twin lead by using a paper punch and scissors. First, make a series of holes along the center of the twin lead, say every 2½ inches. Then cut out the vinyl part between

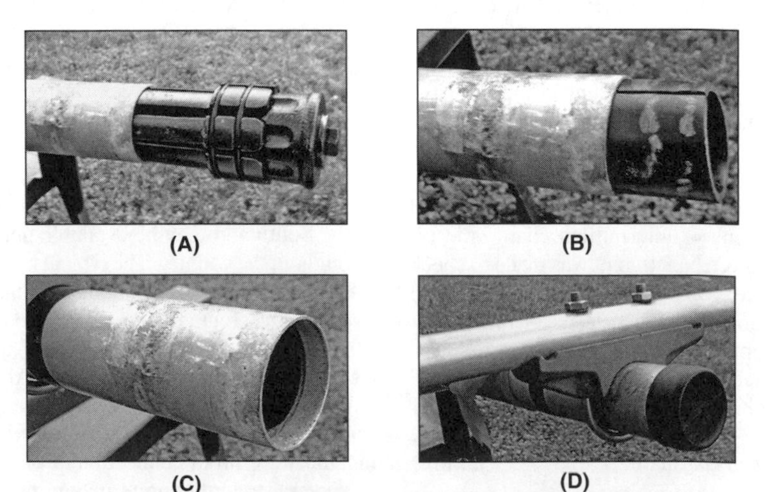


**Figure 28 — Photo A illustrates the use of a tubing straightener to restore the end of a Yagi boom. Photo B shows a length of ABS pipe started into the boom, and C shows the pipe set a short distance inside the open end of the boom. Part D shows the assembled Yagi, with the element tightened in position and the end cap in place.**

every other pair of holes with scissors. This is a very simple way to make an open-wire line. I recommend using Japanese or European stationery punches because those seem to make a smaller diameter hole than a US three-hole punch. You may need to draw a guide line on the surface of the punch to help you align the center of the twin lead in the punch. Do not punch out the copper leads. Carefully check from the side to make sure the punch is properly aligned before you punch the hole. — *Hiroshi Horiike, JA3JNA, 12-22 Mihoga-oka, Ibaraki City, Japan*; **cacos400@hcn.zaq.ne.jp**

**Figure 30 — Piling caps used as the manufacturer intended.**

## COAXIAL CABLE TO WINDOW LINE ADAPTER

◊I use a Budwig Hye-Que connector as an interface between a length of 450 Ω window line feed line to coaxial cable. Intended to be used as dipole center insulators, these connectors are available from Ham Radio Outlet, and may be available from other Amateur Radio equipment dealers as well. They sell for $7.95 each. Figure 29 shows how I secure the window line to the connector with a cable tie. — *Kermit Raaen, W7BG, 15509 SE Mill Plain #42, Vancouver, WA 98684*; **kerm@e-z.net**

## GPS ANTENNA RADOME

◊ This project began with the purchase of a GPS timing receiver. These receivers should be left running continuously for best time and frequency stability. I built the antenna described by Mark Resauer, N7KKQ, intending to mount it permanently outside.[1] Mark protected his antenna from the weather by enclosing it in a cream cheese container. Since mine would be a permanent installation in an area with frequent rain, wind, and occasional snow, I wanted a rugged radome that would keep out rain, prevent the buildup of snow and prevent birds from roosting on and otherwise messing with the antenna. There followed a strange interest in all sorts of oddly shaped plastic objects. After several weeks of staring at traffic cones, kitchen utensils, various pieces of plumbing hardware and other odd objects, the solution turned out to be, in the words of Otis Redding, "Sittin' on the Dock of the Bay."[2]

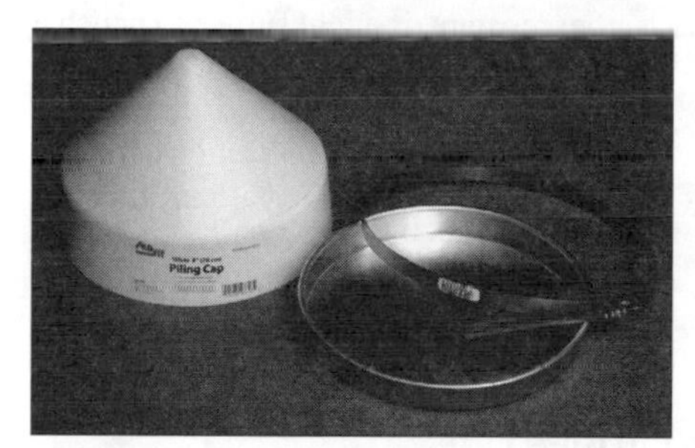

**Figure 31 — An 8 inch pizza pan just fits in the bottom of the 8 inch piling cap.**

**Figure 32 — The finished radome, side-mounted on a tower.**

Marine supply houses sell "piling caps" — plastic cones used to protect the tops of pilings at docks. See Figure 30. Mine came from West Marine (**www.westmarine.com**). It is 8 inches in diameter at the base. (Other sizes are available.) It is made of polypropylene, about ⅛ inch thick, and comes in black or white. Except for the label proclaiming it to be a pile cap, it could have been built to be a radome! I chose white to reflect heat during summer months, something to consider if you need to install a preamp at your antenna. (Placing both black and white caps over a GPS receiver showed both to be about equally transparent to RF.) West Marine also stocks large stainless steel hose clamps. I bought one big enough to fit around the cap at its base.

Next came a visit to a restaurant supply house: an 8 inch diameter aluminum pizza pan fits snugly into the bottom of the piling cap. Figure 31 shows the parts, ready for assembly.

The pan has a rolled edge, which prevents it from going all the way into the piling cap. I cut off the rolled part, leaving a pan about a ½ inch deep. (Don't forget to file the resulting edge down smooth!) When the pan is turned upside down, it fits snugly inside the cap, allowing the lower edge of the cap to protrude below the edge of the pizza pan. This way the joint between cap and pan is protected from rain.

There is enough room inside to mount the GPS antenna, a preamp if needed, or probably an entire GPS receiver module. With the hose clamp holding the cap in place, the whole package is air tight. You can devise a mounting arrangement to suit your situation. Mine is mounted on the end of a piece of PVC conduit, side mounted to a tower. Two 90° bends and one straight section produce the

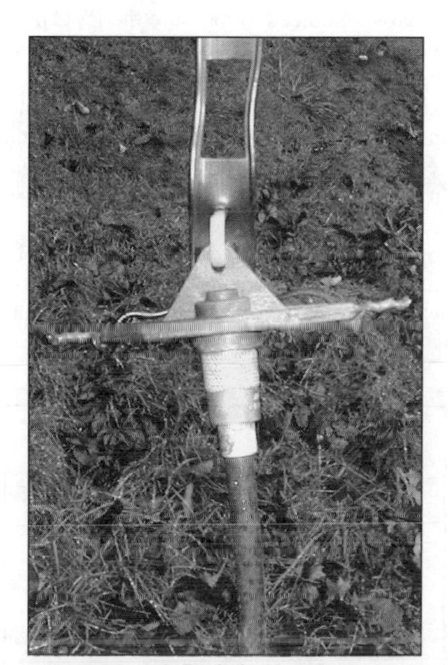
**Figure 29 — A Budwig Hye-Que dipole center insulator provides a simple way to transition from 450 Ω window ladder line to coaxial cable.**

[1]M. Resauer, N7KKQ, An Inexpensive External GPS Antenna, *QST*, Oct 2002, p 36.

[2]Otis Redding and Steve Cropper, copyright East/Memphis Music.

Figure 33 — The assembled pizza-pan mount for the GPS antenna. The mounting hole for the antenna is drilled and tapped into the center of the pipe cap, and a corner of the cap has been cut away to allow the coaxial cable to pass through the conduit and enter the radome.

Figure 34 — Here the GPS antenna is fastened to the pipe cap.

arrangement you see in Figure 32.

One inch conduit is big enough to pass RG-8 with an N connector. Use of PVC instead of metal eliminates the problem of corrosive electrolytic action between dissimilar metals exposed to moisture. A PVC conduit fitting called a "terminal adapter" glued to the end of the conduit terminates it in a male thread. A hole in the pizza pan allows you to attach it to the terminal adapter, just as you would attach an electrical box to conduit. Use a gasket under the pan. Gaskets and the terminal adapter are available where you buy the conduit. Consider routing of the coax and mounting of the antenna when you decide where to drill the hole. For my arrangement, off center was best.

To avoid drilling unnecessary holes, I mounted the antenna directly on the end of the conduit. A 1 inch pipe cap does the trick. I couldn't find caps in the electrical department, but PVC caps for water pipe fit. You may have to search a bit to find a cap with a flat end. Most are slightly convex. I found a flat one in the higher pressure "Schedule 80" water pipe. Drill and tap a hole in the end of the cap, then cut a large opening in the side to allow the coax to pass through, as shown in Figure 33. The pipe cap screws on the end of the conduit, acting like a nut to hold the pizza pan in place. A screw through the antenna ground plane secures the antenna to the cap. (If you don't have a tap, just drill it, and use a nut underneath. I used a nylon screw to avoid electrolysis problems.) See Figure 34.

When the antenna is mounted and coax laced through the conduit, press the pile cap in place. The hose clamp goes around the bottom part of the cap and cinches it firmly against the pizza pan.

Just how weatherproof you make this depends on your local conditions, and what kind of equipment you have inside. For maximum protection, the entire assembly can be slightly pressurized to guard against any entry of moisture. (If you choose to do this, use very low pressure. Conduit and normal PVC pipe are not rated for use with compressed air. If they rupture, flying sharp fragments may result.) A "cord grip" fitting at the other end of the conduit can provide an airtight seal where the coax exits. An alternate approach might be to place a bag of desiccant crystals inside and seal all the joints with suitable calking compound. In drier climates, or if you are using an antenna that is already weather resistant, it might suffice to punch a hole in the pan, venting the enclosure to the atmosphere, to prevent condensation buildup.

In this enclosure, the GPS antenna should give good performance for a long time, in any kind of weather. —*Mike Piper, WA7QPC, 205 W 36th Ave, Eugene, OR 97403;* **wa7qpc@arrl.net**

## MAKE YOUR OWN 600 Ω LADDER LINE

◊ I've been a ham for about 8 years, and I've never had much money to spend on radios and accessories. As the saying goes, "Necessity is the mother of invention." I've always built my own antennas and have always thought about building my own feed line. I've tried 300 Ω twin lead, but it always broke if there was a strong breeze. About a year ago I bought some 450 Ω ladder line to use with a 40 meter horizontal loop that I built.

After a few months of fun with my new loop antenna, it all came crashing down in a storm. We have some really bad storms in Nebraska. The day after the storm passed I went outside to check my antenna, just to make sure everything was okay. When I got outside all I saw was a big mess of tree limbs and wire. As I was cleaning up the yard, I kept asking myself "What am I going to do now?" I decided I would order some more 450 Ω line. When I purchased the ladder line the first time it was about $10 dollars for 100 feet, but now I found that it was $20 dollars for 100 feet. That's a 100% increase. I had heard that the price of copper went up but that's just crazy.

Figure 35 — This photo shows the completed 600 Ω feed line attached to the 40 meter loop.

Being a financially strapped ham operator, and also being the new daddy of a beautiful baby girl, I decided that I could build my own feed line and save quite a bit of money in the process. After some trial and error I came up with the feed line described in this article. It works great with my new 40 m horizontal loop. See Figure 35. It should work with any balanced antenna, when used with a tuner. I use an automatic tuner and a 4 to 1 balun.

Here is a list of everything that I used to build my ladder line. With all of this you can make 40 feet of 600 Ω ladder line:

- 10 feet of ½ inch Schedule 40 PVC pipe.
- 84 feet of AWG no. 14 wire. (I'm using THNN-insulated wire because I have a big spool of it. I build all of my antennas with it also.)
- Hot glue gun and glue sticks.
- Drill with ⅛ inch bit or the size to fit your wire.
- Tape measure.
- Pipe cutter and hacksaw. (You will need something to cut the PVC pipe. I just happened to have a pipe cutter, so I used that.)

I had all the parts except the PVC pipe, so my total cost for this project was less than $3.

Now you need to measure the PVC pipe into 6 inch sections and cut them apart. These will be the spreaders to hold the feed-line wires apart. After you have them all cut, measure ½ inch from both ends of the 6 inch sections and mark them for drilling, as shown in Figure 36. Try to make it uniform, because that will help when it comes time to put it all together. Now, using a drill with a ⅛-inch bit, make a hole through the pipe on each side where you marked it. I found that the insu-

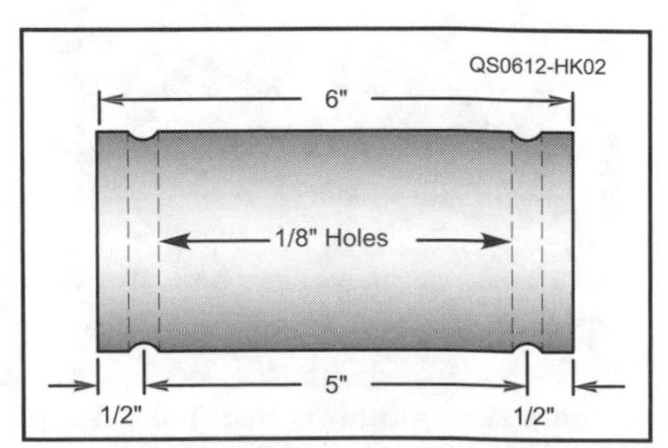


Figure 36 — Diagram of the spreaders (with size and dimensions for drilling).

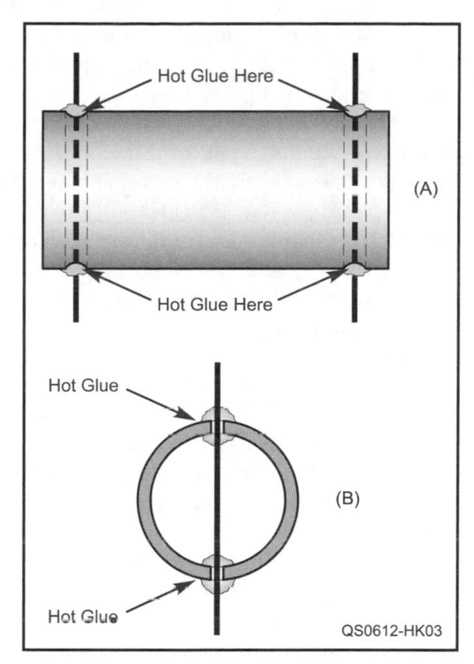


**Figure 37 — Part A is an outside view, showing where to glue the wire to the spreader. Part B is a side view.**

lated wire I used would fit through the ⅛-inch holes. It's a little snug, but not too tight.

Now comes the fun part; putting it all together! I first tried a spacing of 3 feet between the PVC spreaders but I think that's a bit too wide, especially if you're using stranded wire like I did, as the wire had a tendency to wrap around itself. I used a spacing of 2 feet between spacers.

Cut your 84 feet of wire in half so you have two wires measuring 42 feet each. Feed one wire through one of the holes on the PVC spreaders and slide it to the center of the wire. Now hold the wire and spreader in place while you hot glue it. Glue the wire and spreader on both sides and put a dab inside the spreader on both sides of the wire to hold everything in place. See Figure 37. It will take a few moments for the glue to cool down and harden up, try not to move the wire or spreader too much. It is also possible to use some epoxy to hold it all together too but I found the hot glue to be convenient.

Now measure about 2 feet from the first spreader and hot glue the wire and spreader again, just like you did with the first one. Pull the wire tight, as this will help to keep it from wrapping around. Repeat measuring and gluing the spreaders till you get to the end of the wire, and then add spreaders from the other end of the wire, measuring 2 feet between spaces as you go. Thread the second wire through the holes in the other end of the spreaders. When the second wire is in place, glue the spreaders in place.

You now have 42 feet of 600 Ω ladder line. You can make the ladder line as long as you want but I only needed 42 feet. Remember it's the spacing between the wires that's important. You can use fewer spreaders if the wire you're using is more rigid than what I am using, such as solid conductor Copperweld wire.

Since I've been using this feed line I've noticed that my field strength meter indicates less RF in the shack. Also, I can now use my loop on all bands from 40 m through 10 m, which I could not do with the 450 Ω line I used previously.

I hope you find this project helpful! I know I enjoyed making it and presenting it to you. — *Hal Schlotfeld Jr, ACØAX, 432 Duke St, Octavia, NE 68632*; **ac0ax@arrl.net**

[For longer lengths of ladder line, you might try cutting a slot into the ends of the spreaders, and then gluing the wires into those slots, rather than threading the wire all the way through the spreaders. There are other insulating materials that would make suitable spreaders, such as Plexiglas and even wood, such as dowels. If you use wood, be sure to weatherproof the wood. An old-timer's trick is to boil the spreaders in paraffin. (Be very careful with the hot paraffin, though!) — *Ed.*]

## USING A COAXIAL STUB TO SHIFT ANTENNA-SYSTEM RESONANCE

◊ A simple dipole will not present a good load when used at a frequency that is far from where it is resonant. For operation on different frequencies within a band, most of us resort to using either a compromise length or an impedance-matching device called an antenna tuner. The versatility of a tuner is impressive, and I wouldn't be without one! There is, however, a limited but very useful and interesting alternative solution for those who feed their antennas with coaxial cable — coaxial stub tuning. "Stub matching" and "stub tuning" are names also given to a method that was used for tweaking at the antenna, usually on 20, 10, 6, or 2 meters. My system is well suited for the low bands; impedance adjustment is accomplished near the transmitter.

For each band on which you want to use widely separated segments, I propose using a half-wave dipole cut for the lower-frequency portion of the band and fed with a specific length of coaxial cable (an integral number of half-wavelengths for the higher-frequency portion of that band). Near the resonant frequency, there is nothing remarkable about the configuration. In a higher-frequency portion of the band, you can achieve a good SWR by attaching an open-ended stub of the right length at the beginning of the feed line. See Figure 38.

Our example, shown in the drawing, is a dipole cut for 3650 kHz, where its feed point impedance is assumed to be 50 Ω resistive. At 3850 kHz, the configuration has a radiation resistance close to 50 Ω, but also a significant feed point reactance (known to

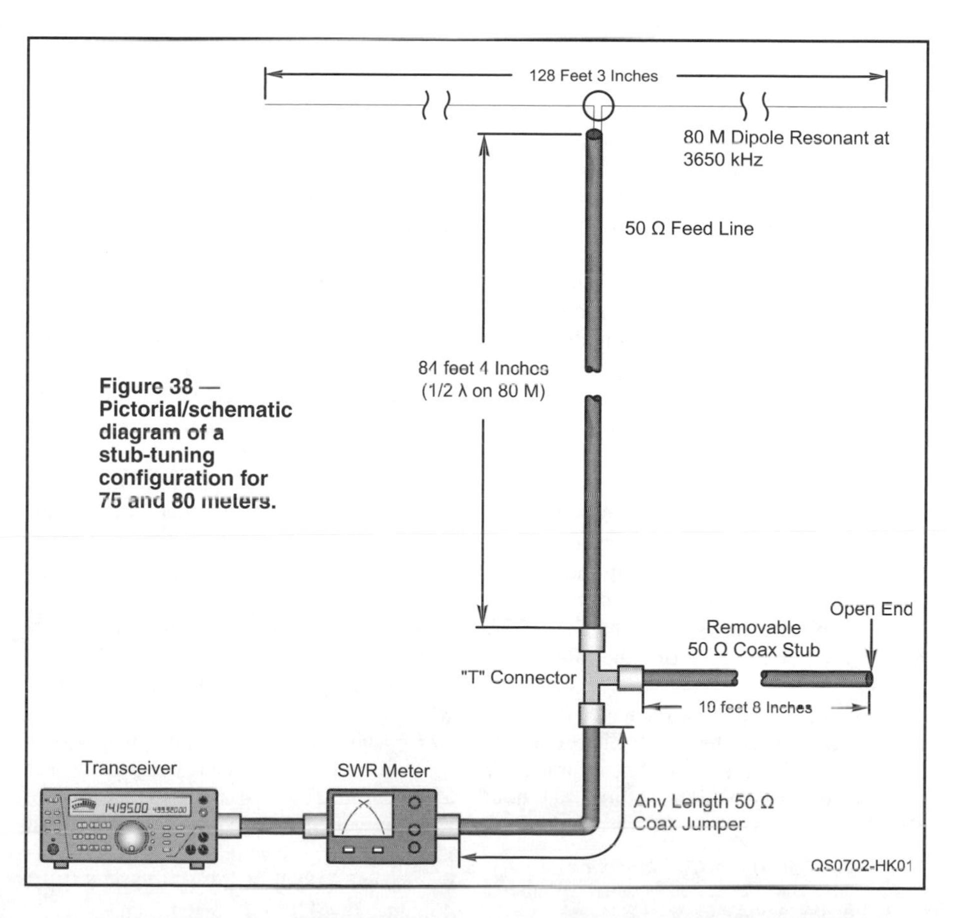


**Figure 38 — Pictorial/schematic diagram of a stub-tuning configuration for 75 and 80 meters.**

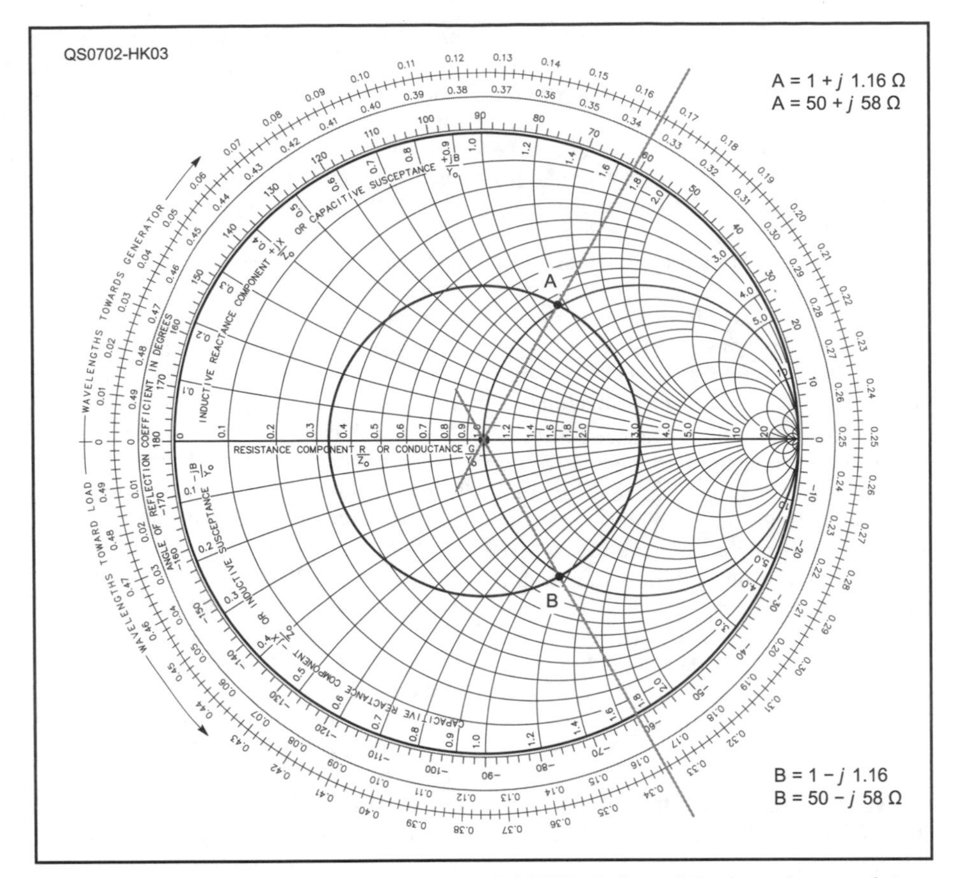


**Figure 39 — This Smith Chart graph shows a 3:1 SWR circle and the impedance points used to design the coaxial cable stub for use with an 80 m dipole.**

be inductive because the dipole is longer than a resonant length of wire). Impedance on a simple transmission line repeats every half-wavelength, so, back at the transmitter, one half-wavelength away, each component of impedance is the same as at the antenna end of the cable. The reactance can be determined and then canceled, giving the transmitter a resistive 50 Ω load.

A half-wavelength in space (in feet) is equal to 492 / f, where f is the frequency in megahertz. To find the length of our dipole, we must multiply the free-space half-wavelength by 0.95 to account for end effect; the result is the familiar half-wavelength dipole equation:

$$\text{Length (in feet)} = 468 / f(\text{in MHz}) \qquad \text{(Eq 1)}$$

To find the physical length of this same electrical length of coax, we multiply the free-space value by the appropriate velocity factor. For many types of coax used by hams a typical value is 0.66; see *The ARRL Handbook*.[1] Figure 38 shows the calculation results.

To find the amount of reactance that is to be canceled, I used the Smith Chart normalized to 50 Ω. Beginners, don't be intimidated; I'll walk you through this and you will see how I determined the reactance. First, establish the uncorrected SWR (the SWR measured without a stub) at 3850 kHz, which we'll now assume to be 3:1. Then, draw the appropriate SWR circle on the chart; that would be a 3:1 SWR circle in the example case, as shown in Figure 3. Finally, find the value of inductive reactance that sits on the 3:1 SWR circle along the 50 Ω resistance line. The answer is 58 Ω (Point A on Figure 39, or 1.16 on the normalized Smith Chart).

A stub with 58 Ω capacitive reactance cancels the inductive reactance at the beginning of the feed line. The Transmission Lines chapter of *The ARRL Handbook* contains a section called "Lines as Stubs," which gives an expression that approximates the reactance of a short (less than ¼ λ), open-ended stub.

$$X_C = Z_0 \cot \ell \qquad \text{(Eq 2)}$$

where $X_C$ = the stub's capacitive reactance;

$Z_0$ = characteristic impedance of cable of which stub is made;

$\ell$ = stub's electrical length in degrees.

Rearranging the formula, we calculate that a piece of 50 Ω coax, 40.8° in length can serve as a stub with 58 Ω capacitive reactance. That is exactly what we want. At 3850 kHz, a half-wavelength (180°) of my RG-58 is 84 feet 4 inches, so 40.8° is 19 feet 1 inch.

With the stub in place, the transmitter sees an impedance equal to the radiation resistance at the dipole. Radiation resistance does not change much with frequency excursions within a band, so if the original SWR at 3650 kHz is 1.2:1, the apparent stub-corrected SWR at 3850 kHz will also be about 1.2:1. Getting that even lower, which is unnecessary, can be accomplished by an adjustment of feed line length (followed by stub readjustment).

Shifting system resonance with stubs is not limited to 75 and 80 meters. Suppose that you have a 40 m dipole with an SWR of about 1:1 at 7050 kHz, and an SWR of about 2:1 at 7250 kHz. The Smith Chart tells us that the inductive reactance to be canceled is 35 Ω. Equation 2 shows that an open-ended stub of coaxial cable with a 50-Ω characteristic impedance and a length of 55° can accomplish this. Using coax with a velocity factor of 0.66, an open-ended stub of cable 13 feet 9 inches long placed at the beginning of a 44 foot 9 inch (½ λ) or 89 feet 7 inch (1 λ) piece of feed line, will allow the transmitter to see a good SWR at 7250 kHz. Let's not forget about the 160 meter band; this type of stub tuning would work very well there.

When using a stub as a tuner in the described configuration, a significant SWR exists on the feed line, and there will be some line loss. This same warning, of course, applies to any "antenna tuner" connected at the transmitter. Also, note that there may be a high voltage at the open end of the stub; insulate it, and keep it out of reach. For operator convenience and shack neatness, several switch-selectable stubs can be kept coiled and out of the way.

I operate both 75 and 80 meters regularly; it is a pleasure to tune the antenna system simply by connecting or disconnecting some coax, rather than readjusting an antenna tuner. The ability to hop between subbands so easily may be useful on Field Day. — *73, Bob Raffaele, W2XM, 5 Gadsen Ct, Albany, NY 12205*; **w2xm@arrl.net**

## MORE ON USING A COAXIAL STUB TO SHIFT ANTENNA-SYSTEM RESONANCE

◊ In the February 2007 Hints & Kinks column I (Bob Raffaele, W2XM) presented an article about using a coaxial stub to shift the antenna system resonance. The idea of tweaking an antenna system by attaching a stub at the transmitter end of a particular length of feed line is sound. My technique for establishing both the reactance to be canceled and its location were, however, incorrect. In the previous article, I made no attempt to improve upon an acceptable SWR; the SWR at the "new" resonant frequency can be made to be 1:1 with some careful adjustments.

[1]M. Wilson, Ed., *The ARRL Handbook for Radio Communications*, 2007 edition, Chapter 21, Transmission Lines, p 21.2.

Several readers commented that my calculations for the earlier article were based on placing a reactance in series with the antenna system impedance, when the coaxial stub was actually being placed in parallel with that system impedance. I had incorrectly calculated stub length and location; however, estimates, inaccurate measurements as well as a radiation resistance that was not 50 Ω made me think I was correct. ARRL Technical Advisor John Stanley, K4ERO, and I corresponded extensively about the correct calculations. This article presents a correct method of calculating the stub length and placement for an antenna system. Further, we'll respond to the recent changes in FCC Rules by talking about shifting the antenna resonant frequency from 3550 kHz (CW) to 3850 kHz (SSB) on the 80/75 meter band.

Our technique for achieving two dipole "resonances" within a band requires that the antenna be cut for the lower frequency. Use the well-known formula given in Equation 1.

length (in feet) = 468 / f (in MHz), [Eq 1]
length (in feet) = 468 / 3.550 = 131.8 ft

Cut the wire a bit longer, and then build your dipole to be 131.8 ft long. The feed point impedance at resonance will be near the theoretical free-space value of 73 Ω. At 3550 kHz, the SWR is 1.3:1. With any length of feed line, the installation works well at 3550 kHz.

Making no changes to our system, we might observe a 5:1 SWR at 3850 kHz. An open-ended stub, cut to a particular length and connected at the right place along the feed line, will act as our antenna tuner, and will reduce the SWR at the transmitter to 1:1. See Figure 40. Evaluating the parameters is more complicated than was stated in the previous article. We'll explain the procedure now, in the hope that readers will find it useful, educational, and interesting. Those who do not care to learn how to make these calculations now may wish to "experiment," starting with a 35-foot stub and with 70 feet of transmission line between the stub and the feed point.

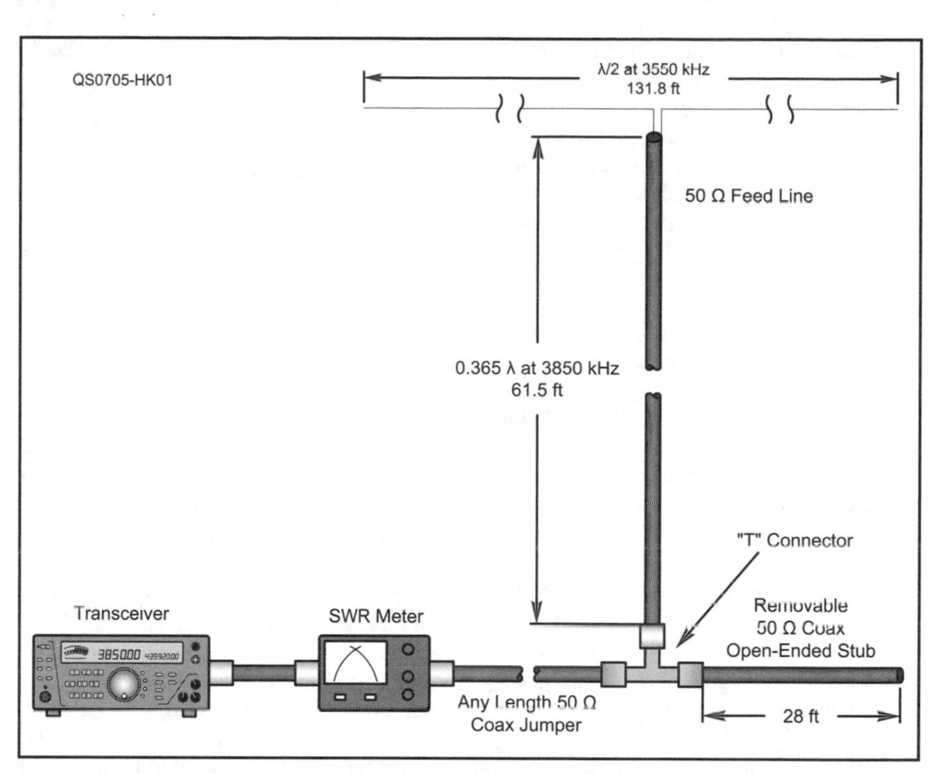


**Figure 40 — This diagram shows an 80 m dipole resonant at 3550 kHz connected to a transceiver through an SWR meter. For operation at 3850 kHz the open-ended stub is added at the T connector.**

Figure 41 shows a Smith Chart with two auxiliary circles drawn on it. The 5:1 SWR circle is the one that is centered; it indicates the complex impedance at all points along a transmission line characterized by a 5:1 SWR. The other circle is our 50-Ω matching circle, from which can be determined every complex value of impedance that could include an equivalent parallel resistance of 50 Ω. Our 50-Ω matching circle is the same as the 20-millisiemens (mS) conductance circle; it is symmetrical to the 50-Ω resistance circle.

Our educated guess for resonant feed point resistance on my antenna system is 65 Ω. A 300-kHz increase in frequency on 75 meters will cause this resistance to increase slightly, perhaps to about 80 Ω. Point A on Figure 41 is where the 80-Ω (1.6 on the normalized chart) resistance curve crosses the 5:1 SWR circle.

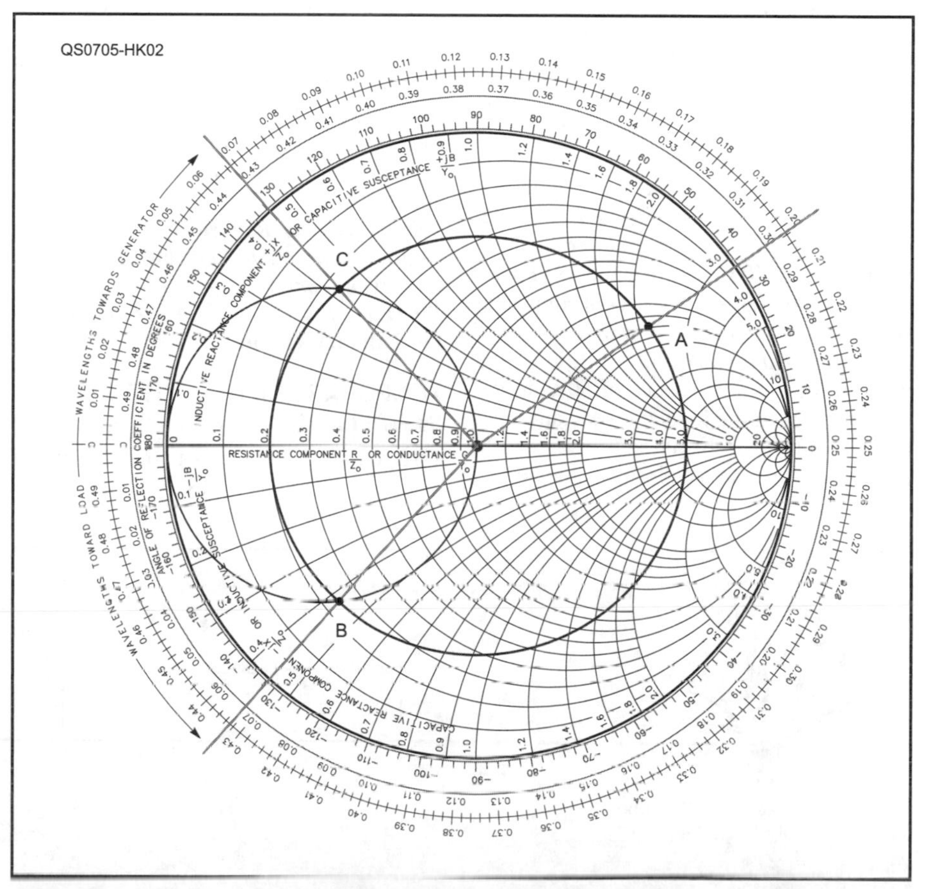


**Figure 41 — This Smith Chart graph shows a 5:1 SWR circle and a 50 Ω matching circle. Point A represents the antenna impedance when operating at 3850 kHz. Point C represents our desired matching point. Point B represents an alternative possible matching point.**

Feed point reactance can be read directly at this point as 110 Ω (2.2 normalized). The feed point impedance appears to be an 80-Ω resistance in series with a 110-Ω reactance.

Next we must find a point on the transmission line where the impedance is equivalent to a 50-Ω resistance in parallel with some reactance. If we can determine this location and the amount of parallel reactance, we can cancel it. There and then, the impedance of the line would be a resistive 50 Ω. There are two places on the feed line that we could use; these are indicated by the two points where the 5:1 SWR circle intersects the 50-Ω matching circle. These are Points B and C on Figure 2. Because of the length of line that was already in place, Point C is being used. Point A is at the feed point; Point C is 0.365 λ nearer the transmitter than Point A, as can be seen by the calibrations along the chart's circumference. Since λ is 168.5 feet in coax with a velocity factor (VF) = 0.66, then 0.365 λ is 61.5 feet.

[For readers who are not familiar with reading values from a Smith Chart, look at the two scales along the outside edge of the Smith Chart circle. One scale is labeled "Wavelengths Toward Generator" and the other is labeled "Wavelengths Toward Load." One trip around the Smith Chart in either direction represents a distance of 0.5 λ. To find the distance from the generator (transmitter) to the load (antenna), start at the radial line through Point A. That line crosses the outside scale at 0.202 λ. Proceed clockwise to the zero point at the left edge of the scale. Here you will have to subtract 0.202 from 0.5 to find the distance from the antenna toward the generator. This result is 0.298 λ. Now continue clockwise from the zero point to the radial line through Point C, which crosses the same scale at 0.067 λ. Adding 0.067 and 0.298 gives a result of 0.365 λ. — *Ed.*]

The series impedance values at Point C are read directly from the Smith Chart as 50 (0.24 + *j*0.43) Ω; that is, (12 + *j*21.5) Ω. This impedance has a parallel equivalent, with values of 50 Ω resistive and 28 Ω reactive. [For more information about equivalent series and parallel circuits see page 4.45 of *The ARRL Handbook*, 2007 Edition.] We want to present a resistance of 50 Ω to the transmitter, and the inductive reactance can be canceled with a coaxial stub. The formula for the reactance of a capacitive stub is:

$$X_c = Z_0 \cot \ell \quad \text{[Eq 2]}$$

where $X_c$ = stub's capacitive reactance;

$Z_0$ = characteristic impedance of cable of which stub is made;

$\ell$ = stub's electrical length in degrees.

Rearranging Equation 2,

$\ell = \text{inv cot } X_C / Z_0$.

Converting electrical length to physical length,

length of stub = $(\ell / 360)\ \lambda$

For this particular case,

length of stub = 28 feet.

Therefore, a 28-foot-long open stub will provide the capacitive reactance which, when connected so that it is in parallel with the transmission line at Point C, will cancel the inductive reactance. Looking into that junction, a 3850-kHz signal sees a 50-Ω resistance.

The capacitive stub acts as an antenna tuner, providing the transmitter with a 50-Ω resistive load. The feed line between the stub and the antenna is characterized by a high SWR; the SWR-related losses may be considered significant. If you wouldn't mind using a tuner, you may want to try a coaxial stub. It is okay to leave the stub coiled up while in use.

The high-voltage warning included in the original Hints & Kinks article is worth repeating here: There may be a high voltage at the open end of the stub; insulate it and keep it out of reach.

We have suggested using a feed line length of 61.5 feet. This is likely too short for many installations. You should feel free to use cable that is one half-wavelength longer. A more appropriate length might be 145.8 feet.

*Submitted by Bob Raffaele, W2XM, 5 Gadsen Court, Albany, NY 12205;* **raffaele@nycap.rr.com** *and John Stanley, K4ERO, 524 White Pine Lane, Rising Fawn, GA 30738;* **jnrstanley@alum.MIT.edu**

## TEMPORARY DIPOLE ANTENNA INSTALLATION

◊If you are putting up a temporary dipole antenna you might try using bungee cords to maintain tension and provide some strain relief. See Figure 42. — *Richard Mollentine, WAØKKC, 7139 Hardy St, Overland Park, KS 66204*

## LIGHTNING PROTECTION FOR OPEN WIRE FEED LINES REVISITED

◊The June 2006 Hint & Kink from Joe Hutchens, WJ5MH, on grounding for open wire feed lines caught my eye and I read it with interest. I recognized the ground bus bar as close to a type we use for our telecommunications systems, and the use of spark plugs was pretty clever. What also caught my eye, however, was the apparently under-sized Earth ground lead.

In our Telecomm rooms the lead to building ground is a minimum AWG no. 000 cable. Yes, about ½ inch diameter of stranded copper wire! In this case, since it appears that the ground bus is mounted on a block of wood outside, I would sink a ground rod right there and tie it directly to the bus. The original article is not clear about where the ground connection is made. The "system ground" mentioned should be tied back to the bus or ground rod, using at least an AWG no. 6 stranded ground wire (another Telecomm spec).

In any case, it gave me a good idea for future lighting protection, even though in my home area we have few lighting storms and no history of any strikes on a house, tree or building. (I've seen no lighting rods on rooftops anywhere in southern California.) Right now I have a separate ground rod for my two J poles, tied together with no. 6 wire. The damp location next to the house makes for a pretty good ground point. (Now I'm going to have to check for corrosion!) — *73, John Powell, KF6EOJ, 8325 Otto St, Downey, CA 90240;* **jpowell@csulb.edu**

◊A few years ago I tried the same type of setup as WJ5MH described. Mud Dauber wasps fouled the "spark plug non-fouler" by daubing their mud into the holes clear back to the spark gaps. Needless to say this "fouled" up my antenna system.

I suggest daubing a little bit of silicon in

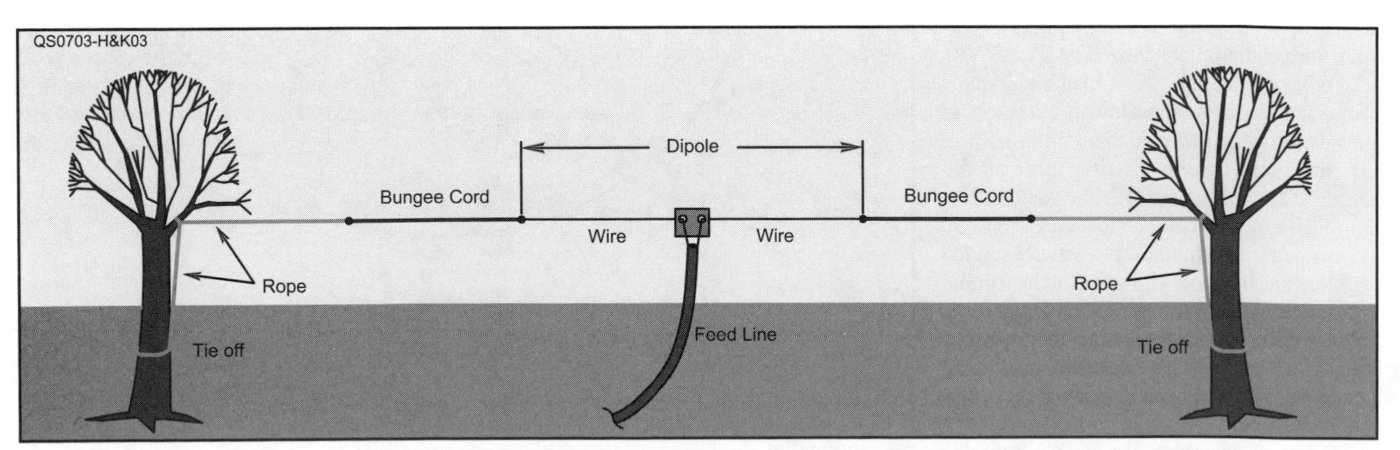


**Figure 42 — For a temporary installation, bungee cords can provide strain relief while also holding a dipole wire under tension.**

the holes to keep the wasps out. Wasps also filled my banana plug socket system, which I used as a disconnect switch for my open-wire feed line. I have had to clean the mud out of the sockets several times each year. — *73, Bill Hall, K3CQ, 747 Rodney Dr, Nashville, TN 37205*; **k3cq@arrl.net**

## THE KA3IXF APARTMENT DWELLERS ANTENNA

◊Having moved from a house — where I had a 5-band vertical and a Kenwood TS-940AT transceiver — after many years of being active, to an apartment, it is very easy to get discouraged. I went several years of being inactive, with little or no hope of putting out a good signal. Recently, with the birth of Elecraft and the K2/100, my interest was peaking to a high level, similar to when I was a Novice. It was like Heathkit was back in business. That warm fuzzy feeling was back, when I used to sit in front of my Galaxy on a cold winter's night and get warm from the heat coming off of my finals.

I had looked through many issues of *QST* and other publications, trying to come up with a "ready-to-buy solution." There were many indoor antennas and an antenna that connected to the windowsill, but it protruded straight out horizontally. None of these would do. I personally went to the rental office of my apartment complex, and tried to explain Amateur Radio and the need for an antenna, but they were not buying it.

The September 1996 issue of *QST* carried a Strays item on page 84 about an antenna I designed for an apartment and had some luck with it. At that time, however, I had a balcony. The apartment that I now live in has no balcony, and although I live on the third (top) floor, the possibility of having an outdoor antenna seemed hopeless. I thought I might have to settle for an indoor antenna.

I purchased a K2/100 kit, and had many hours of fun building this incredible radio. (I had a little help from Alan Wilcox, at Wilcox Engineering — **alan@wilcoxengeneering.com** — in Williamsport, Pennsylvania.) Then it was time to decide on an antenna.

After spending $1000 on the K2/100, and not having a whole lot of money left over for an antenna, I thought about the idea that I had back in 1996. I took two cardboard tubes from a couple of rolls of Christmas wrapping paper, and placed them end-to-end. Then I wrapped them completely in black electrical tape to give them some strength and a little weather proofing. Figure 43 shows the beginning of this process.

I went to our rear bedroom in the apartment, because it has a double window, similar to a picture window, with a little width to it. I set the cardboard tube on the windowsill and cut it to fit snug in between the cut-out for the window.

I then took some wire and some feed line, and created a limited-space antenna for 10 through 40 meters. I placed the feed line in the middle of the tube, and secured it to the tube with black electrical tape. I then began to wrap the antenna wire around the tube, being careful not to allow the turns to touch each other. I used heavy-duty insulated copper wire for each side of the antenna. I left about a half inch of space between each turn, as shown in Figure 44. It felt as though I was wrapping one of those Elecraft toroids again.

After making what appeared to be a big wire-wound resistor with a coax feed line coming out of the middle, I again wrapped it all up in heavy duty tape, color optional, weather proofing it and adding a little bit of camouflage. You might choose different colored coatings, depending on the color of your apartment building.

I securely placed it on my windowsill(see Figure 45) and connected the feed line to my MFJ Tuner. I was able to tune up on 10 through 40 meters on the CW portions of the bands, with little or no reflected power at the transmitter; the SWR was just what I wanted. Being a CW-only operator, I am unaware how this antenna will do on the phone portions of the bands.

KA3IXF

Figure 43 — This photo shows how KA3IXF wrapped Christmas-wrapping-paper tubes with electrical tape to form a core for his helically wound antenna.

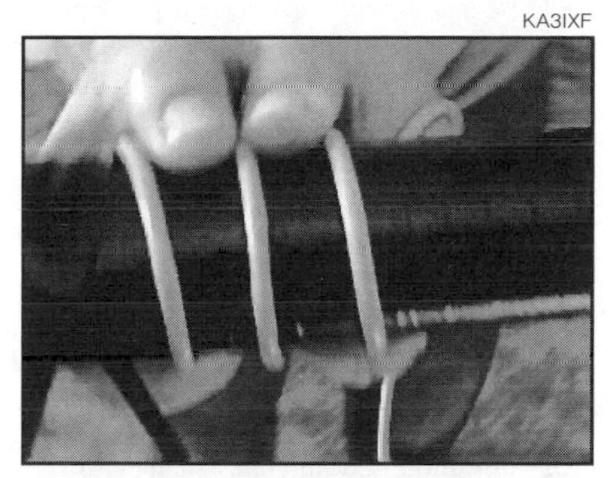
KA3IXF

Figure 44 — After securing the coaxial cable feed line to the center of the cardboard tubes, KA3IXF wound the antenna wire around the tube to form a helically wound antenna.

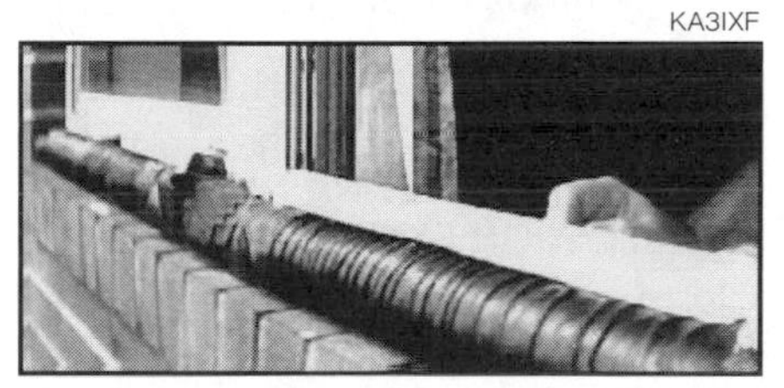
KA3IXF

Figure 45 — This photo shows the completed KA3IXF Apartment Dweller's Antenna on the windowsill outside his apartment.

I began working the 40 meter CW band, and having some luck with the K2 at 15 W, working up and down the east coast. Not being satisfied, I added the KPA 100 to my K2. My contacts increased 100%, and to my surprise I began to work DX again. I worked Cyprus Island on 14.027 MHz. I was running approximately 50 to 75 W out, because I did not want any TVI issues in the apartment complex.

To my amazement, this antenna has performed flawlessly. The overall cost of The KA3IXF Apartment Dwellers Antenna, is a hefty $20. Of course, the longer the tube can be, and the larger the tube diameter can be, without attracting too much attention, the better the performance. Individual sizes will vary on the band portions you wish to work. I am not claiming this antenna to be the next big thing in Amateur Radio. For me, however, it was this antenna or nothing, and to my surprise, along with using the excellent overall features of the Elecraft K2/100, this antenna keeps KA3IXF on the air, answering your CQs on 40 and 20 meters from a third floor apartment in "rare" Delaware. — *73, Bill Parker, KA3IXF, 3314 Old Capitol Tr K-12, Wilmington, DE 19808*; **wwjp123@verizon.net**

## EASY MAINTENANCE OF A LOW VISUAL PROFILE ANTENNA

◊As a newly licensed ham, I found myself facing the same dilemma I've read about so often — getting an antenna up into the air where it could do some good, and not irritating the neighbors. After studying antennas, talking with the neighbors, and surveying the types of things the neighborhood has accepted, I settled in on the idea of a flagpole mast. In the process of finding a workable solution, I came up with an antenna mount system that extends to raise the antennas so that they clear my roof, yet is easy to lower for maintenance.

The flagpole I finally settled on is a Sunsetter 20 foot telescoping flagpole (**www.sunsetterflagpole.com**). The flagpole comes complete with a special sleeve that is set in concrete so the pole is easily removed for repair or replacement. I ordered the bronze anodized pole to reduce visibility and "shine," and had a concrete base poured that was a bit more than what the spec sheet required. Figure 46 shows my concrete base. I figure that the wind load of the antennas I was planning to use would actually be a lot less than the wind load of two 3 × 5 foot flags, which is what the flagpole was designed for.

Once the base was poured, set and cured, erecting the pole is easy. It just slides into

KD7TOG

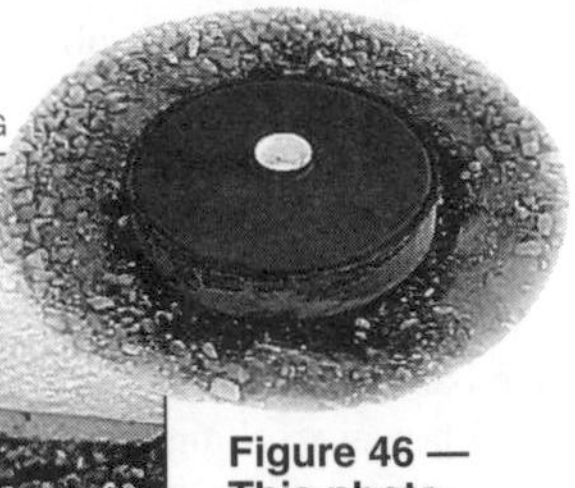

Figure 46 — This photo shows the concrete base, ready for the flagpole to be

Figure 47 — The bottom of the flagpole slips into the sleeve in the concrete base.

KD7TOG

Figure 48 — It is a simple matter to lower the flagpole to work on the antennas.

KD7TOG

Figure 49 — With the flagpole fully extended, the three antennas are above the top of my house roof.

the concrete sleeve, and is locked in place with a setscrew to stop it from twisting. See Figure 2. The top section of the four-section pole is 1.5 inches in diameter, perfect for most standard antenna mounting hardware. The greatest feature of using the telescoping pole is that it can be collapsed down to a comfortable working level, as shown in Figure 48, and then extended back into position rather easily.

For my main antenna, I decided on an Arrow dual band J-pole (**www.arrowantennas.com**). The antenna pattern has proven to be very effective for working the complete Las Vegas Valley, as well as easily making packet contacts via the International Space Station. One benefit of the Arrow J-pole design is that the longest/tallest element of the antenna is actually a grounded element. This will help drain to ground any ionized charges in the air around the antenna.

Over time, I have added a cross arm to the top pole section, and added two receive antennas for my ICOM R-75 and PCR-100 radios. The J-pole is used with a Yaesu FT-8800 2 m and 70 cm FM transceiver. The profile is a little less stealthy with the cross-arm-mounted antennas than with the J-pole alone. Figure 49 shows the flagpole and antennas in the fully raised position. I hope this idea helps give other hams another tool for getting in the air, and on the air! — *73, Tony Messina, KD7TOG, 5452 Painted Gorge Dr, Las Vegas, NV 89149;* **kd7tog@arrl.net**

## HOMEBREW ANTENNA INSTALLATION EXTENDER POLE

◊It seems a good portion of our setup time during Field Day is spent trying to get a rope over a branch on a tree for a wire antenna. For Field Day 2005, we were fortunate to have a ham with a strong and accurate throwing arm. The first rope went up nice and easy. The second one, however, proved to be problematic. After several failed attempts, the thrower's arm gave out and I wound up having to climb the tree. No doubt this is a common problem for hams trying to put up antennas when they do not have the luxury of a cherry picker, crane, or large extension ladder. So, how best to tackle this problem?

Starting with a light line and a bow and arrow or slingshot, or surf casting are all possibilities but I discounted those ideas because of the inherent danger and unpredictable results. An extender pole of some sort seemed the best way to go. Unfortunately the commercially available fiberglass poles have such a thin top section that they wave around in even the slightest breeze. Invariably the weight at the top gets tangled around the pole, so it takes a lot of wiggling and jiggling to get it to drop.

I was determined to come up with a simple, quick and reliable solution to the problem. In addition, my solution should

K2RFP

Figure 50 — The top of the pole in its collapsed state, showing the cross arm, hose clamps and weights.

K2RFP

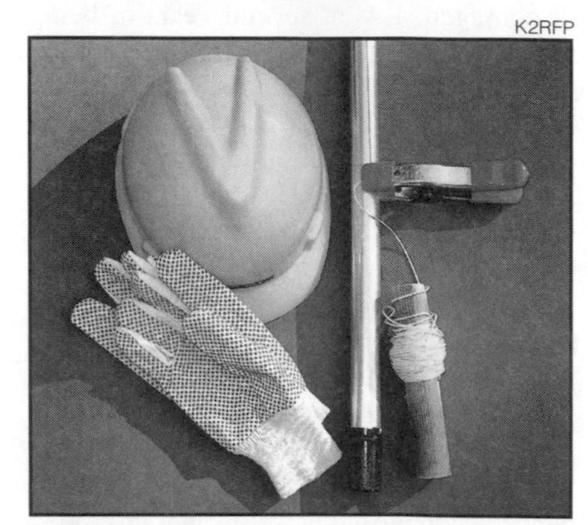

Figure 51 — The lower end of the pole, showing the string being held taut and the suggested safety equipment.

K2RFP

Figure 52 — Rope and pulley arrangement to be hauled up.

K2RFP

Figure 53 — The extender pole in use.

make use of materials I already had or that would be readily available. Finally, it had to be easy to put together and use. Here is what I came up with.

I assembled an extender pole made with three sections of EMT (electrical metal tubing) type steel electrical conduit and a wooden cross arm at the top. Figure 50 shows the top of the pole in its collapsed state with the cross arm attached.

The cross arm was made from a piece of ⁵⁄₄ pine I found in my scrap wood box. The dimensions are not critical. A measurement after construction showed the distance from the pole to the front screw eye as about ten inches. I tapered the front of the cross arm to reduce weight. The top section of conduit is ½ inch conduit with an outside diameter a little shy of ¾ inch so it was a loose fit to the ¾ inch hole in the cross arm. A few wraps with duct tape made a snug fit. Hose clamps above and below the cross arm keep it in place.

The center section of conduit is ¾ inch and the lower section is 1 inch. These tubes make for a very loose fit when telescoped together, so again I used a few wraps of duct tape to make a snug fit. I cut some slits in the upper ends of the lower and center tubes with a hacksaw, so I could squeeze down tightly on the inner tube with a hose clamp. The upper and center sections extend two feet into the section below. With this arrangement the total length of the pole is about 25 feet. I taped a 35 mm film canister to the bottom of the lower section to keep dirt out when standing the pole up on the ground.

To use the pole, lay it on the ground and extend it to the length you want, then secure it with the hose clamps. The golf ball weight works fine with low friction nylon string, but the extra weight provided by the rock does better with the thicker mason line. Pull the string taut and hold it temporarily at the bottom end of the pole with a spring clamp or some tape. See Figure 51. As with any situation when raising something overhead, it is a good idea to wear a hard hat. Sure grip gloves are also helpful.

Putting the bottom end against the tree or having someone hold it down with a foot makes raising the pole real easy. With this pole you can easily reach branches at 25 feet. With the aid of a ladder I was able to reach 32 feet.

Once you have the light line over the branch, use it to haul up the rope and pulley arrangement shown in Figure 52. Make sure you have enough rope to accommodate your particular situation. Figure 53 shows the extender pole in use.

If you use this pole for setting up a temporary antenna for field day or a special event station, you should consider using the pulley rope to haul back a light line to leave there when you take the antenna down. Choose a durable line that will make it through the weather until the next time you want to use it. For a more permanent situation you can periodically lower the rope and pulley arrangement for inspection and replacement if needed. Finally, you should be reminded that this is a *metal pole* and it should not be used anywhere it is possible to come in contact with overhead *power lines*. — *73, Richard Pav, K2RFP, 85 Radio Ave, Miller Place, NY 11764*; **k2rfp@arrl.net**

### DIGGING A TOWER FOUNDATION IN SANDY SOIL

◊After Hurricane Katrina blew over my tower, I decided that I had to find a way to make the base stronger in this sandy soil. It is difficult to safely dig a deep enough hole for a tower foundation. A construction contractor recommended a Sonotube form. It turns out these forms are inexpensive and served my purpose well. I hope this article may save someone else from a lot of extra work and expense.

The treated cardboard forms used for pouring bridge supports are light enough for this old man to manipulate. There may be other brands of these forms, but Sonotube seems to be the most common. A long-handled posthole digger and a length of this tube allowed me to dig and form the hole for my replacement tower in one afternoon.

Sonotubes come in various diameters, in 10 foot or longer lengths, but half that length is sufficient for my guyed tower. The 16 inch diameter size I used cost $2.30 per foot at a local contractor supply house. You may be able to buy a piece of the tubing cut to the length you need. If you have to buy the entire 10 foot length, you can easily cut a piece the length you need using a circular saw or other method.

As the wet sand layer about 3 feet down tried to cave in, pushing the 5 foot long piece of tubing down into the hole allowed me to safely dig to the selected 5 foot depth. Be sure the tube is plumb as you dig the hole. Eleven bags of cement filled it up.

I used a drilled steel plate and ¾ inch threaded leveling anchor rods. See Figure 54. A tower section could be embedded in a form easily, but holding it perfectly vertical while the concrete sets would be more work. — *73, Patrick Hamel, W5THT, 1157 E Old Pass Rd, Long Beach, MS 39560*; **w5tht@arrl.net**

### AN IMPROVED COAX STANDOFF ARM FOR CRANK-UP TOWERS

◊I have never been happy about the way coax and control cables are supported on the top of typical crank-up towers. The tower manufacturer will sell you an expensive clamp-on metal strut that causes the cables to bend at sharp angles. The weight of the cable and the constant flexing in the wind may lead to premature failure. The coax may also deform and disturb the impedance match, especially at high frequencies.

KF6T

**Figure 55 — This photo shows a cable support arm attached to the top of a crank-up tower.**

W5THT

**Figure 54 — A 5 foot length of Sonotube provided a safe way to dig a tower-mounting hole in very sandy soil. W5THT used ¾ inch threaded rod leveling anchors to secure a tower base to the concrete support pad. As shown, the bottom tower section is in place, but the washers and nuts have not been added to secure it. The aluminum flashing is for grounding the tower.**

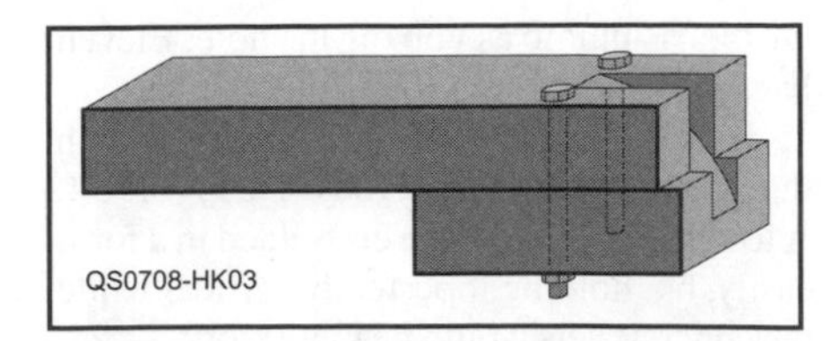


**Figure 56 — This drawing shows the construction details of the wood-polymer support arm that KF6T built.**

**Figure 57 — In this photo, you can see the completed wood-polymer support arm attached to the top of a crank-up tower, with coax cables and rotator control cable clamped in place. The tower is tilted over in this view.**

A trip to my local building supply house provided the necessary material to construct a "better" arm. Wood-polymer decking is designed to survive rain and sun. It is easy to cut, drill and shape, and is quite strong. I got a two foot sample display piece from the store for free. Also, there are so many decks being built with this material; you could probably find a scrap cutoff piece for free in your neighborhood.

The material that I selected is 1 × 6 decking (15⁄16 inch thick × 5½ inches wide actual). As you can see from Figure 55, I doubled the thickness at the end of the support. A short 3 inch section is held on with two stainless bolts, lock washers and nuts. The short section is offset by about ¼ inch as shown in Figure 56 to reduce the amount of grinding required. The cable channel is ground using a bench grinder followed with a flat file. The resulting coax bend radius is about 2 inches, and that is a lot more coax friendly than my old metal support arm. The channel should be wide enough to fit your coax bundle. The channel also contains the coax from side to side.

Figure 57 shows the three coax cables and rotator control cable attached to the arm. Two or more cable clamps secure the bundle to the horizontal surface. The clamps are held on with stainless steel screws. Add electrical tape to make a tight fit under the clamps. The arm is secured to the top of the tower with another set of stainless steel hardware. You may have to cut some of the deck material away to clear a thrust bearing. In this application, the arm holds the coax away from the corner of the tower by about 8 inches. About 35 vertical feet of cable is supported by this arm.

While installing a support arm, check the coax service loop to make sure there is ample slack and clearance for full rotation. Remember, some rotators turn more than 360°.

So far, my inexpensive support arm has survived a very wet California winter with plenty of wind at my location. The support shows no sign of "sag" and the material still looks new. — *73, Jack Morgan, KF6T, 2040 Pheasant Hill Ln, Auburn, CA 95602*; **kf6t@arrl.net**

## QUICK AND EASY PORTABLE ANTENNA MOUNT

◊When you need a quick and easy portable antenna mount for Field Day or other operations consider a metal paint-roller pan. They are available for about $3 at the local home supply store. I drilled a hole for the antenna and three smaller holes for radials. Number 8 or 10 machine screws and wing nuts in the smaller holes work well for attaching radial wires. The mount works well for my mobile 20 m Hamstik type antenna, as Figure 58 shows. This is certainly the most affordable mounting method I have tried. — *73, Ed deBuvitz, W5TTE, PO Box 13372, Albuquerque, NM 87192*; **w5tte@arrl.net**

**Figure 58 — An aluminum paint-roller pan works well as a portable antenna base.**

## PROTECT THE BUTTERNUT HF-6 ANTENNA 30 METER DOORKNOB CAPACITOR

◊My Butternut HF-6 has served me very well for the past 20 years. The only problem I've had is that the 67 pF doorknob capacitor on the 30 m coil has failed three times . This problem is characterized by loss of sensitivity and high SWR on 20 and 30 m — usually intermittently at first, which makes it difficult and frustrating to determine where the problem is. I believe this problem is due to stresses placed on the capacitor when the antenna sways in the wind. If you examine the way in which this doorknob capacitor is mounted compared to the other doorknob capacitors, you can see why this could occur (the capacitor is mounted with aluminum brackets to a possibly flexible piece of dielectric material). To resolve this problem, I cut out a small 1 × ½ × 0.037 inch aluminum bracket (aluminum sheet available from hardware stores, such as ACE Hardware). I drilled ⅛ inch holes on either end of this bracket, and mounted the 67 pF capacitor to one side of the existing bracket on the 30 meter coil using a no. 4 machine screw, lock washer and nut. (I used all stainless steel hardware.) Next, I made up a 1 inch piece of wire with no. 4 solder lugs on each end, and used this to connect the other end of the capacitor to the 30 m mounting bracket, again with no. 4 hardware. This wire eliminates any potential stress on the capacitor. Figure 59 shows the details.

I am confident that the mechanical stress on this capacitor was causing the failure. I made this modification to the antenna more than two years ago, and have had no further problems. — *73, Phil Salas, AD5X, 1517 Creekside Dr, Richardson, TX 75081*; **ad5x@arrl.net**

## SECURING RADIAL WIRES AND OTHER LINES TO THE GROUND

◊Anyone who has ever laid ground radials in the woods for a vertical antenna knows that every little twig along the way prevents the wire from lying on the ground. I recently decided to modify my 80 m wire Four Square antenna to use ground radials instead of elevated radials. I had to lay 160 radials, each

**Figure 59 — AD5X added an aluminum bracket and a length of wire to reduce mechanical stress on the 30 m doorknob capacitor on his Butternut HF-6 vertical antenna.**

66 feet long. Not only did twigs and brush make it difficult, but it was easy to catch one with my foot and pull it out of place.

I tried to come up with an easy way to stake them to the ground. I bought a box of a hundred 24-inch insulation supports, sold in home improvement stores for holding ceiling insulation between joists on 24 inch centers. I cut the stiff 16 gauge support wires into three eight inch lengths, and bent the rod at about an inch from the end so it was almost bent back on itself. The rods were easy to push into the ground, and they hold the wires in place where I want them. Three hundred stakes cost about $10! — *73, Jack Schuster, W1WEF, 408 Thompson St, Glastonbury, CT 06033*; **w1wef@arrl.net**

[That's a great idea, Jack. I have a box of "Garden Staples" that I carry with my portable equipment/Boy Scout demonstration equipment. Those stakes are about 4 inches long and bent in a U shape, designed to hold landscape fabric secure to the ground. A box of 40 cost several dollars "a few" years ago. Your insulation-rod stakes would be even cheaper, and could double as emergency tent pegs. — *Ed.*]

## COPPER TAPE ON GLASS, A NEW WAY TO MAKE ANTENNAS

◊There are very few really new antenna designs as far as the actual theory or geometry are concerned. Most novel antennas are just new ways to build an existing design. This is no exception.

I have been experimenting with antennas made by using sticky copper tape on a dielectric supporting medium, mostly glass. I believe this approach has several advantages. Copper tape in various widths from less than ⅛ inch to ½ inch wide is available cheaply from hobby sources. It is provided for persons who make stained glass art. The tape is bright copper on one side with a sticky tape back side. The tape is meant to be applied to the edges of the colored glass pieces so that they can be soldered to each other. There are a number of sources on line.

K4ERO

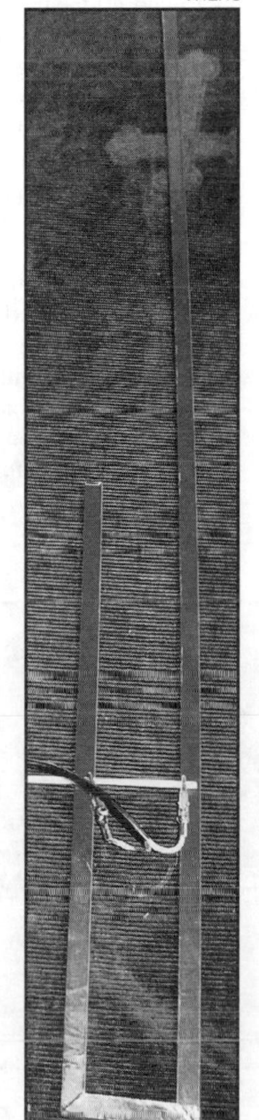

Figure 60 — This photo shows a 2 meter J-pole antenna made with copper tape applied to a piece of glass.

This tape sticks to the flat surface of glass as well as to the edges. It is quite inexpensive and works well for antennas. The approach is to tape the copper to the glass in the desired shape and then solder leads to it as needed for the interconnections and the feed point.

Most modern homes have large expanses of glass. Even windows that seem to be broken up into separate panes are often two large sheets of plate glass with nonconductive plastic separators inside the dead air space. The inside and outside surface of the glass is flat and continuous. Many high rise apartments with limited antenna possibilities have a number of large glass windows. It makes no difference whether the antenna is laid out on the inside or the outside of the glass. If done neatly it can be family and landlord friendly.

Taping up a half wavelength dipole and connecting a lightweight coax such as RG-174 at the center makes a very useful and quick 2 meter antenna. A tape on glass vertical dipole is the default antenna for my 2 meter handheld radio, when at home. My experiments have all been at VHF but an HF loop would fit on a larger picture window. UHF bands would allow use of multi-element arrays. Of course, rotation is probably not in the cards, so high gain would be in only one direction. Perhaps a hinged glass door would make a rotating beam, but a sliding one would not!

There are a few things to learn. One is that glass has a rather high dielectric constant and even though it is on only one side of the copper, it will change the velocity factor. I have found that on the glass I use, 0.80 or 80% is close to the correct value. Thus, on 2 meters, a half wavelength would be closer to 80 cm than to 1 m in length. You may have to make a test dipole on your glass and measure its resonance with a bridge of some kind to determine the value for your windows. Be careful of "tinted" glass that has a film on it to reduce sun light. The film may be conductive. Don't overlook this technique for a no holes car antenna for the VHF/UHF bands.

The copper tape is easy to work with. If you accidentally cut it too short, you can easily solder on an extension. On the other hand, it is so cheap, you can easily replace it. If you get the tape too hot and/or put a lot of force on it, the glue may fail. You will soon get a feel for this. If you pull a piece off, a new piece is easily spliced in. If you remove an antenna, some sticky residue will be on the glass. This is easily removed with a single edged razor blade. I suppose you could break a window if you heated it too much in one spot, although this has never happened to me. Just use the minimum heat necessary and have the feed wires tinned beforehand. Since the tape has no thermal mass, solder flows very quickly and easily.

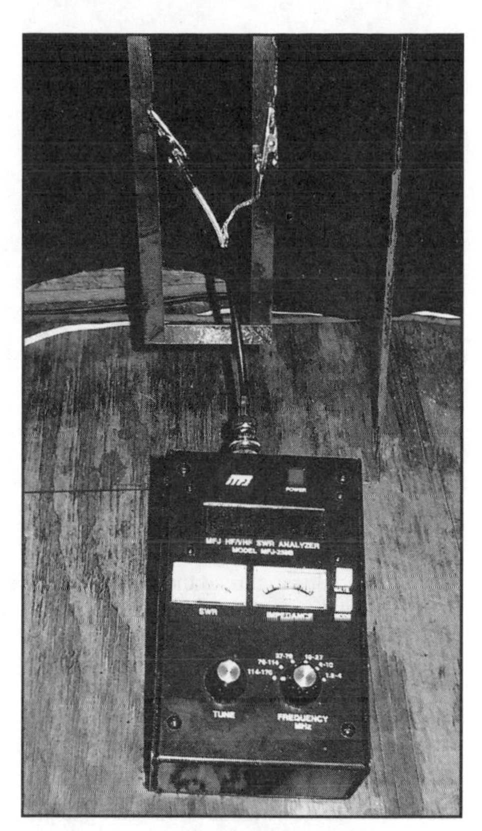


Figure 61 — A short length of coaxial cable with alligator clips on one end makes a convenient test fixture for finding the feed point on the copper tape J-pole

The J-pole is a popular antenna, so I will use it as an example. I laid out a J-pole with ½ inch tape. The total length was 49½ inches and the folded back section was 17 inches. The two tapes were separated by 2 inches. Figure 60 shows the antenna. At the corners, I folded the tape back on itself so that no solder joint would be needed, even though the tape solders easily. I simply moved the input coax up and down the doubled line until a good match was found at about 6 inches from the bottom. To make it easier to move the tap points evenly, I clipped the alligator clips on the end of a short piece of coax to a plastic insulated rod. The other end of the coax connects to my antenna analyzer. It was a simple matter to slide the insulated rod up and down along the J-pole base, with the back of the alligator clips touching the copper tape for electrical contact. It took 5 minutes to lay out the antenna and 2 minutes to find the match. See Figure 61.

The same technique would work on many dielectric surfaces, such as vinyl siding or fiberglass roofs. Copper tape can also be applied to indoor walls to make a stealth dipole for HF. This could be painted over and be almost totally invisible. Of course, RF safety considerations might limit this to receive only or QRP operation. To avoid noise keep away from hidden wiring in the wall. I have run copper tape up a fiberglass pole to make a vertical antenna. The tape can also be used to repair PC boards and even lay them out if you are very careful not to overheat. It also can be used to seal corners to make an RF tight box. When you get tired of all that, you can always make stained glass art. Hey, who will be the first to lay out a stained glass window that has gain on 2 meters? — *73, John Stanley, K4ERO, 524 White Pine Ln, Rising Fawn, GA 30738*; **k4ero@arrl.net**

## TIP ON INSTALLING A PL-259 CONNECTOR

◊ The cable jacket of LMR-400 (and some varieties of RG-8) is a very tight fit inside PL-259s and the newer style 2-piece N-connectors (made by Andros). Although a common work-around is to apply lubricant to the jacket, an alternative that makes them even easier to install is to thread the jacket using a 7⁄16-14 thread die. This is especially helpful if you need to install a lot of connectors at one time.— *73, Michael Tracy, KC1SX, ARRL Lab Test Engineer;* **kc1sx@arrl.org**

## FEED LINE ENTRANCE TIP

◊Several years ago my family moved to a newer home. Knowing that I was going to put up several antennas, I had to come up with a way to get the coax outside. At that time I ordered glass-block windows to replace the single pane originals. I asked the salesman if he could have the installers leave one glass block out of one of the windows and cement three 2¼ inch PVC pipes in its place. He agreed to do this and since this involved no drilling in the wall my wife, Debbie, N8FPA, gave her full blessing.

I purchased a 4 foot length of 2¼ inch PVC pipe and cut three pieces about 8 inches long. Then I took the caps and drilled five holes in each and covered them with Coax Seal. I made each hole large enough to feed coax a little larger than RG-8 through it. This would allow me to have 15 runs of coax or other cables to the outside. I drilled the caps ahead of time because once you have coax in one hole you cannot remove the cap. It is a lot easier to drill them in a vise rather than outside.

When I gave the pipes to the installers I had put the cap ends on each pipe since you cannot touch the pipes while the cement dries and because I did not want any vermin coming in. I made sure to have them leave about 3 inches of pipe sticking out of the cement. The caps need over 1 inch to attach to the pipe. I let the cement dry for several days. See Figure 62.

Currently I have a lot of wires running through this system. There is *Buryflex* coax for my remote SGC tuner, a run of direct-bury ac cable I use to provide dc power to the SGC tuner, a coax for a dual band vhf/uhf antenna, coaxes for a 6 meter and a 2 meter loop, RG-6 to a UHF corner reflector for HDTV, a coax for an FM broadcast omni and a heavy duty ground wire.

This system works great for my purposes, especially since my shack is in the basement. Another benefit of glass block windows is that they are very energy efficient. — *73, Zack Schindler, N8FNR, 1103 Butternut, Royal Oak, MI 48073*, **vtnn43e@comcast.net**

## NN8R "MINITENNA" SYSTEM

◊ I built a simple two band portable antenna system using two "Hamstick" type of whips mounted onto a wooden support structure. This mount is secured with 10 inch spikes pounded into the ground through mounting holes in a 30 × 30 inch sheet of aluminum flashing that acts as a ground counterpoise. See Figure 63. I chose this size of aluminum sheet because it just lies down flat in the trunk of my car. I use alligator clips on one end of a 40 foot long RG-174 subminiature coax for connection to the aluminum sheet and to the antenna feed point.

Mounting the two Hamsticks side-by-side doesn't affect the operation of either, and without a tuner I get nearly a 1:1 SWR on both bands. On Field Day I operated for 5 hours running 3 W output from a battery-powered KX-1 and I worked 24 states. — *73, Ray Grob, NN8R, 26 Shaker Ct, Fremont, OH 43420*, **kc8ayj@juno.com**

## INDOOR ANTENNAS AND FIRE SAFETY

◊ I just escaped burning down my house and I want to pass along what happened. When I bought my home in Florida, I knew CC&R restrictions would require me to be innovative in my antenna designs. The roofline of my house is sufficiently long for me to install a four band trap NVIS antenna about 20 feet above ground in the attic. This would be a compromise antenna, but the use of hardline coax should at least minimize some of the loss.

I used a trap described by N3GO in the 17th Edition of *The ARRL Antenna Book*. I made the traps from RG-58 and PVC pipe. While running only 90 W input power to my old EICO 720 transmitter during Straight Key Night I noted that my antenna tuner was abnormally sharp in tuning, and was using different capacitance and inductance values on 40 meters. Since I normally operate on 20 meters, it was unusual for me to try loading on 40 meters, so I don't know how long it had been this way. But I decided to shut down the transmitter and climb up in the attic and take a look up there.

Am I glad I did! One trap was fine, but the other one's connection between the center conductor and the shield had burned through, leaving the antenna tuned for 20 but unusable on 15 and 40 meters. Judging by the smoke stains, damaged insulation and discolored PVC pipe, the arcing that had killed my trap must have been substantial. Because I had suspended the antenna about a foot below the rafters I avoided starting a fire, but I think I am very fortunate. Apparently less than 50 W can generate quite a lot of high voltage and current in a trap.

Antennas are normally used out of doors, and for good reason. I am presently figuring out whether the neighborhood architectural committee will permit me to put up a flagpole. At the very least you should install a fire detector if you install an antenna in your attic. — *73, Jim Crutchfield, AC4WR, 1020 Pinehurst Ct, Oviedo, FL 32765*, **ac4wr@arrl.net**

ZACK SCHINDLER, N8FNR

**Figure 62 — PVC pipe through one open frame in glass-block window in the basement allows cables easy, water-tight (and vermin-tight) entry to the house.**

RAY GROB, NN8R

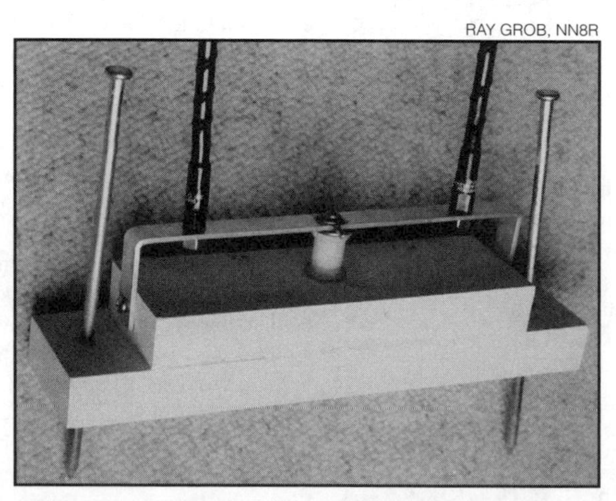

**Figure 63 — Mounting system for a pair of Hamstick whips. Long spikes go through the ground plate and are pounded into the ground underneath.**

## WATER VAPOR CONDENSATION IN AIR DIELECTRIC 9913 COAX

◊ Information about the problem of water condensation in air dielectric types of coax is well-known in the UHF community, but not in the HF community, where air dielectric types of coax may be used to drive several different antennas. This has been an intermittent problem that seems to affect some installations and not others, and I wondered why. After having some coax up for only two years at my new shack and running into this problem, I began to wonder what was actually causing the water to get into the line.

I have been a ham for almost 30 years and have homebrewed my share of complete antenna systems. Sealing out moisture to prevent corrosion of the coax connectors has always been a priority, at least since I first saw what corrosion could do to VSWR and my signal! So I pride myself in doing a first rate job in sealing the coax, yet I was still getting water in my 9913 cable.

The problem seems to stem from the fact that 9913 uses air dielectric to keep losses low. This system works only when the air inside the coax is dry. If the coax was assembled in high humidity, as is the case most of the time in Austin, Texas, then chances are good that water vapor is trapped in the assembled and sealed coax. As temperatures drop with the seasons, the interior cools to the dew point and water vapor condenses to liquid water. Vapor condensation results in a decrease in internal atmospheric pressure as temperatures are lowered and the gas is converted to liquid. This negative pressure then pulls more moisture into the coax until finally I had a wet noodle for a transmission line. Preventative measures include drying out the interior of the coax periodically, assembling on a dry day (commercial installations will pressurize with dry gas), sealing it well and crossing your fingers or scrapping the air dielectric coax for something without an air space to collect moisture.

I salvaged a long line of 9913 years ago by using compressed air from a scuba tank, since the air they are filled with is dry and the pressure can be regulated easily. This kept me going for a number of years after that until I finally sold the house and took everything down. My most recent episode of wet-noodle coax occurred just 2 years after the initial installation. This is a 60 foot length of coax that runs from the ham shack, through the ceiling near an outside wall, down through the soffit, then up a short 25 foot tower beside the house to a dish for cell phone use. I dried out the coax using a small compressor with a crude Type N connector adapter to screw directly to my coax. I wrapped enough splicing tape around the ⅜ inch brass male adapter on the pressure hose to allow me to screw a female Type N connector shell on to the wrapped adapter. I then taped the whole thing with splicing tape and went over that with stiff electrical tape to prevent air blow by.

After attaching this to my coax (60 feet of 9913 with Type N connectors) I adjusted the pressure for about 20 PSI and blew it out for 2 hours. You don't need high pressure, just air flow to keep the water moving and evaporate it. At first, water dripped constantly out the end of the coax (I removed the opposite end Type N connector). After an hour, the dripping stopped but the end was still moist. Another hour of compressed air blew it dry and VSWR looked normal again. The proof of the pudding was putting the cable back on the dish antenna and seeing signal strength go back to normal levels. To be specific, the inside of the coax is not dry, just drier than it was. Eventually, I'll get this cable swapped out for some that doesn't condense moisture. — *73, Stephen Bosbach, KW5V, 9122 Circle Dr # C, Austin, TX 78736-8013,* **sbosbach@austin.rr.com**

## A SOLID, WELL-GROUNDED HF ANTENNA MOUNT

◊ I recently lost my Yaesu ATAS-120 (**www.yaesu.com**) due to metal fatigue in a portion of the antenna mount.

I decided to follow several of my ham friends in purchasing a High Sierra Sidekick antenna (**www.cq73.com/index.php**). Naturally I wanted to mount it more securely. The owner's manual of the Sidekick emphasizes the need for a very solid ground connection to the car body, so my design had to keep that in mind. I discarded the idea of strengthening and modifying my original mount. In studying the rear of my 1998 Camry, I noticed that the gap between the trunk lid and the trunk opening was wide enough (just over ⅛ inch) to accommodate a piece of sheet metal without interfering with the opening and closing of the trunk. The construction of the car's body provides a convenient flat, strong area at that point to attach a mounting bracket.

I determined the shape and dimensions of the bracket and used a tool called an "Angle Devisor" to measure the offset from vertical to match the angle of the edge of the trunk opening. With dimensions and angle determined, I made a full size mock-up of the bracket out of thin cardboard.

The next step was finding a willing welder who would do the welding for a reasonable fee. Luckily, a fellow employee stepped up after hearing me talk about trying to find a welder. I found a piece of ⅛ inch cold rolled steel and gave him the mock-up. Several days later he had the bracket tacked together. We held it in place to check the angle and agreed a slight tweak would get it just right. Final welding included the two angle braces. My cost was a case of beer.

The next steps were mine. I bought three ¼-20 × 1 inch stainless steel oval-head screws with washers and lock washers. I drilled three ¼ inch holes and one ½ inch hole in the bracket. Then I marked and center punched the hole locations on the car. To assure a good ground, I drilled and tapped for the ¼-20 screws. You need a number 7 drill bit and a ¼-20 tap. This provides a very solid attachment and a very good ground.

JIM AUGUSTEIJN, K9LDX

Figure 64 — View of the completed mount.

The antenna itself is mounted on an SO-239 with ⅜-24 thread coax connector. The ½ inch hole accommodates the nylon step washer that is positioned between the bottom of the Sidekick and the top of the bracket. Apply a bead of Permatex Clear RTV Silicone Adhesive Sealant to the base of the antenna and the mounting plate. When the antenna is mounted to the base, the two beads combine to form a firm waterproof seal for the antenna. Figure 64 shows a view of the completed mount.

The mounting bracket is so solid that one can grasp it and move the whole car. The ground is highly effective, in that I've been able to tune the antenna to less than 1.5:1 on all bands between 6 meters and 40 meters. — *73, Jim Augusteijn, K9LDX, 1542 Mellow Ln, Simi Valley, CA 93065,* **k9ldx@arrl.net**

## PROTECTING LADDER LINE FROM ABRASION

◊ I like to use 450 Ω ladder line for wire antennas for its light weight and low loss, but a continuing problem has been abrasion of the plastic insulation where the feed line runs over the edge of the roof and moves up and down in the wind. This eventually bares the Copperweld conductor, abrades the copper coating and the steel core rusts. I have had to replace my feed line on two occasions due to this problem.

I have tried lengths of split garden hose

JOHN WEATHERLEY, AB4ET

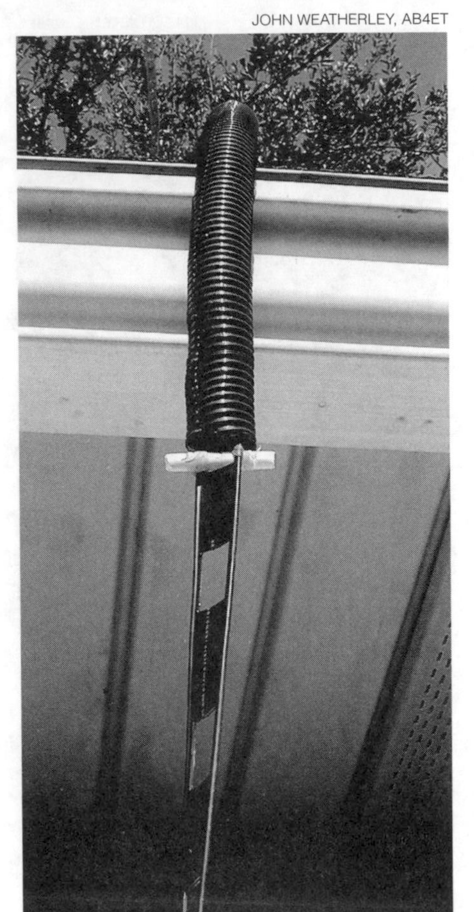

Figure 65 — The split tubing installed on the ladder-line with the plastic pen stop in place.

with limited success. It is relatively heavy and two split lengths are needed, which are difficult to keep in place. Cable ties used to secure it tended to catch and hang up on projections.

After several attempts to find a simple and inexpensive solution I finally opted for split flexible tubing sold to contain wire harnesses. It may not last quite as long as garden hose, but it is neat, flexible and inexpensive, and can be replaced in a few minutes should the need arise.

I found the tubing at a local Lowes hardware store, but it should be available at most electrical and hardware outlets. The diameter is 1 inch and it comes in 5 foot lengths. The name is 1 inch Split Flex Tubing and it costs under $4.

It is installed by inserting the ladder line through the longitudinal split. The tubing should then be rotated to bring the split perpendicular to the plane of the ladder line to prevent it from exiting the tubing.

To stop the tubing from sliding down the ladder line, I installed a piece of non-conductive material through the ladder line. In my case this was a piece of the barrel of a discarded ballpoint pen. It is held in place with waxed nylon string used for securing wire bundles. A couple of small black cable ties or weather resistant twine would work just as well. This has proven to be a satisfactory solution to the problem and has no effect on the performance of the feed line. On one occasion, the ladder line found its way out of the split tubing. I fixed this by adding three black cable ties around the tubing, one near each end and one in the middle. I also used another black cable tie around the tube and anchored it to a gutter bracket. This stopped an annoying rasping noise during the night when the wind caused the tubing to move up and down against the edge of the gutter. *— John Weatherley, AB4ET, 1575 Harlock Rd, Melbourne, FL 32934,* **ab4et@arrl.net**

## USING POWERPOLE CONNECTORS TO GROUND LADDER LINE

◊ In looking for a simple way to ground the ladder line running between my shack and tower I have found Powerpole connectors to be an effective solution. Basically, I attached Powerpole connectors to the ladder line and have the ladder line coming down the tower to a grounded Powerpole connector. The line from the shack is stored about 6 feet from the tower. When I want to operate, I disconnect the ladder line from the grounded connector, unroll the shack side of the line and plug it into the antenna side of the ladder line. This setup allows me to disconnect the shack from the antenna when not in use, as well as ground the antenna. The Powerpole connectors do not seem to cause any appreciable issues with the feed line. *— Ron Wagner, WD8SBB, 5065 Kessler-Frederick Rd, Troy, OH 45373,* **wd8sbb@arrl.net**

## GO PORTABLE WITH AN MFJ-1775 ROTATABLE MINI-DIPOLE

◊ When MFJ introduced its rotatable, half-wave horizontal mini-dipole for 40, 20, 15, 10, 6 and 2 meters designed for home use, I immediately thought, "RV and portable operation?" So I purchased one and was entirely pleased. It's rugged and lightweight, has lower ground loss than quarter-wavelength mobile antennas, exhibits lower noise than a vertical, has moderate directivity, is capable of full legal power and is small enough to use on a recreational vehicle (RV). These are big advantages compared to what many hams employ for RV and portable operation.

### Basic Concept of the Antenna

The fundamental design of the antenna is technically sound and offers better performance than most ham antennas seen on RVs. For the HF bands it consists of four shortened end-loaded dipoles on a common boom. Six and 2 meters are included as a fan dipole.

The MFJ-1775 loading coils are stacked end-on-end on the same boom and top whips are replaced by capacitive hats made up of four short radials. A metal strap on the side connects the bases of all the loading coils together. Now, when RF is applied to the antenna, the currents "ignore" all but the one pair of loading coils and capacitive hats that brings the antenna to resonance on that frequency. The others appear as high impedances; that is, essentially switched out of circuit. My own experience bears this out. There is only modest interaction between bands.

There was, however, a problem. For RV use, the antenna is awkward to transport fully assembled. I had to perform constant assembly and disassembly on my first trip. It wasn't because the antenna was too large; it was because the capacitive-hat radials just "stuck" out inconveniently in all directions. But the solution was easy: I bent the radials at right angles near the boom. Now I can easily rotate them in against the boom for transport or storage (Figures 66 and 67). This configuration also makes fine tuning easier. Rotate the radials on any band toward the others and the tuned frequencies will decrease.

JOHN PORTUNE, W6NBC

Figure 66 — View of radials ready for stowing.

I wasn't sure if bent radials would disturb the tuning too much, so I first assembled the antenna with the unmodified parts as supplied and then compared the resonant frequencies, before and after bending, with an MFJ-259 Standing Wave Ratio (SWR) analyzer. I mounted the antenna on a 6 foot wooden ladder over average

JOHN PORTUNE, W6NBC

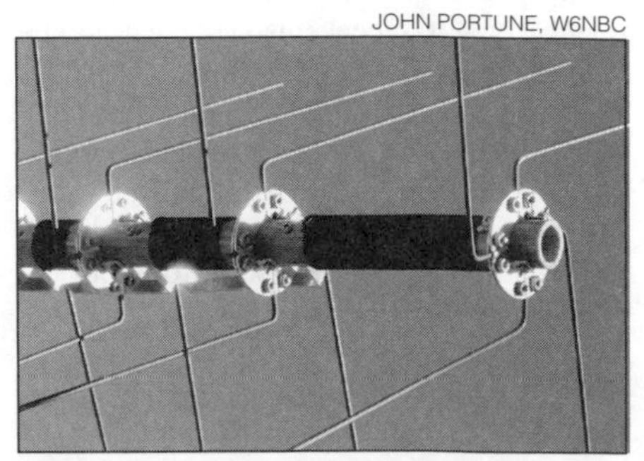

Figure 67 — Put a right angle bend in the radial about a ½ inch from the mounting ring.

soil out in the clear. In the worst case, 10 meters, the resonant frequency increased by 600 kHz, easily within the tuning range afforded by trimming radials as supplied.

To bend the radials, use pliers to grasp each radial while installed on the antenna and make a right angle bend roughly ½ inch from the outer edge of the center mounting ring (Figure 66). The exact distance isn't critical, but do not bend right at the ring. This little extra space is necessary to let the radials remain clear of the aluminum strap that connects the loading coils together. Then trim the now-bent radials to bring each band to resonance, following the guidelines in the brochure. These work well and my bent-radial version performs as advertised on my RV.

As modified, I have found it to be the best of many RV antennas I have ever used. It is also a good choice for Field Day and portable situations. As a home antenna, it is not much larger than a TV antenna. The neighbors may not even notice it. — *73, John Portune, W6NBC, 1095 W McCoy Ln #99, Santa Maria, CA 93455,* **jportune@aol.com**

## POOR MAN'S WIND SOCK

◊ If you have a beam antenna (or several) on a tower, it is prudent to face the antenna elements perpendicular to the wind direction rather than broadside to it — particularly in high winds — to reduce the chance that the elements will be damaged. A simple wind direction finder can be made from a large plastic trash bag, available in various colors to "blend in." Cut a few small holes in the bottom to release any air or water and mount the bag near the top of the tower. It will indicate the direction and relative strength of the wind. — *73, Richard Mollentine, WAØKCC, 7139 Hardy, Overland Park, KS 66204-1710*

## PORTABLE LIGHTWEIGHT ANTENNA SUPPORT

◊ Having recently participated in several bike tours as a communicator, I realized that my heavy tree-pruning tool was not the optimum antenna support. I noticed that one ARES®/RACES ham used an aluminum 15 foot pool-cleaning pole as his antenna support. This pole collapses down to about 6 feet, so it easily fits in a car. I was able to purchase the Kem-Tek pool-cleaning pole at Orchard Supply Hardware (OSH) for $25 recently, but since it was designed to be handheld, it would not stand up by itself. At one event I found a parking sign and attached the pole with bungee cords. While that made a secure antenna mount, how would I set up operations without the benefit of parking signs? I needed an inexpensive, lightweight, collapsible tripod that could mount anywhere.

While at OSH recently, I noticed a sale flyer that advertised a Craftsman 500 W work light (#73826) with tripod stand for $10 (Figure 69). The normal price is $20.

**Figure 68 — The worklight tripod with the PVC legs attached.**

RICH STIEBEL, W6APZ

**Figure 69 — The worklight tripod before the addition of the PVC leg extensions.**

I had seen this on an earlier trip to OSH, but did not want to pay $20 just to get the tripod. The price was now right, so I purchased the light just to get the tripod for my 15 foot pool-cleaning pole.

When I unpacked the tripod, I became concerned about the diameter of the tripod's base. Was it large enough to keep the pole from falling over when the pole was extended to its full 15 foot height? Setting up the tripod, I noticed the quick-release pin at the top of the tripod post. I was able to unscrew and remove this with a crescent wrench to permit the bottom of the pool pole to fit over the top of this tripod post. Upon extending the pole to its full 15 feet in my back yard, I immediately realized that the tripod base diameter was not sufficiently large to support the fully extended pole with antenna *if* there were much of a wind.

I needed to increase the tripod diameter somehow. Each foot of the tripod had a plastic cap on the end to prevent the metal tube from digging into a floor. Those plastic caps pry out very easily. The inside diameter of the legs is just big enough to accept a piece of one half inch schedule 40 PVC pipe. I happened to have several 18 inch lengths of this pipe available, so I put a length of PVC pipe into each leg (Figure 68). This more than doubled the tripod diameter and greatly increased the stability of the tripod, even with the 15 foot pool-cleaning pole at its maximum length. One could use longer pieces of PVC pipe to provide even more stability if needed, or one could place weights over the ends of the PVC pipe to anchor the tripod in a windy environment. Another alternative, if the tripod has been set up on earth, would be to use a tent stake driven into the ground near each foot of the tripod and secured to the PVC pipe.

This tripod is an inexpensive way to mount an antenna for emergency work or when helping out at a public service event. One could also use this support system to hold a Buddipole or a dipole antenna made with two Hamsticks when working HF. This tripod is very lightweight, so it is easy to carry. I look forward to using this setup at future public service events and RACES drills. — *73, Rich Stiebel, W6APZ, 840 Talisman Dr, Palo Alto, CA 94303-4435,* **w6apz@sbcglobal.net**

## FEED LINE STRETCHER

◊ I am currently using one antenna for all of my HF work. During three hurricanes with winds of 100 mi/h each, I sustained serious damage to my tower and antenna.

I replaced the antenna with a 130 foot wire dipole fed with a homebrewed 4-inch-spaced open wire feed system. I tried my MFJ 989D Versa Tuner with the 75 feet of feeder on all of the HF bands. I was able to obtain good

SWR measurements on most of them, but not all. So, going back to my earliest days (I was licensed as W2ITD, in 1935), I began shortening and lengthening the 75 foot feed line to the shack. Yes, it worked on some bands, beautifully — SWR to be happy with, but some required more feed length changes. After a few days of changing the feed line with the band, I conceived a way to vary the length to suit the band desired — thus my "Feed Line Stretcher" idea.

This is by no means "Rocket Science," but it does the trick beautifully and with minimum cost. I wanted a means to support about the same length of open wire line I was using but with 12 gauge ladder-line.

I cut two circular pieces of ¼ inch plywood, 23.5 inches in diameter for the top and bottom of the support. Each had eight ⅞ inch diameter holes drilled, equally spaced (45°) around the circumference and centered in from the edge, approximately 1¾ inch, affording roughly a 20 inch diameter circle of ⅞ inch holes in which the eight, ½ inch ID PVC pipe supports are placed holding approximately 15 turns of ladder-line. The eight pipes were cut 30 inches in length and cemented in place, flush top and bottom, with Goop cement. The holes can be filed, slightly, for a snug fit before gluing (see Figure 70).

Winding the 15 turns of line is simple and each turn is held in place with plastic cable ties keeping the top wire of each turn about 1 inch below the bottom wire of the turn above it. With 15 turns about 75 additional feet is available for the whole line. I used an MFJ SWR bridge for my tests.

In order to get the best flexibility in band usage, I cut turns and added 2 inch alligator clips (RadioShack No 270-346B) to one side of the cut portion and large solder lugs to the other. Restoring the cut portion is easier when clipping to a flat solder lug than to another alligator clip. It is also easier to unclip when reopening is desired. In this way, any length of line is possible without any part of the entire coil assembly remaining in contact with any other.

STEVE TABER, W4ITD

Figure 70 — The Feed Line Stretcher in action at W4ITD.

Since each turn is approximately 5 feet in length, breaking open the line four turns from the top, and attaching the tuner balanced line output, would then add eleven turns, or about 55 feet to your open wire length. In my case, with 75 feet outside the shack, I would now have 75 feet plus the additional 11 remaining turns (at 5 feet per turn) or 55 more feet, so the tuner now sees a new length of 130 feet.

The input from the tuner or transceiver can be placed anywhere along the coil, depending on how many turns you need. The unused turns are completely divorced from the others. If you need only 10 feet of coil, you connect two turns from the bottom or top end, as you wish. The rest of the coil will not be in the circuit. I must point out that the choice of 75 feet of ladder line (15 turns) was arbitrary for my situation only. Just keep in mind that the feed line length to the antenna should be in odd quarter wave length units placing the end of the existing feed line at the lowest impedance. So the addition or subtraction of overall length must be considered to provide the best match to your tuner. Also, changing clips is not necessary for every band change as there may be five or six bands you can work on one setting.

For those users of the MFJ 989, pages 8 and 9 of the manual give great examples of what to avoid in trying to match balanced line antennas.

Looking back through my log, I can find many good DX results on these bands with the antenna feed point only 28 feet above ground: UAØ, Sakhalin Is PSK31 25 W 20 meters, EU stations 80 meters, VK6, P29 40 meters. — *73, Steve Taber, W4ITD, 25 Cunningham Dr, New Smyrna Beach, FL 32168,* **stevew4itd@gmail.com**

## ANTENNA COUNTERWEIGHT DEVICE

◊ As do many other hams, I use a tree to support one end of my long wire antenna. As we all know (or soon find out) trees move around in the wind. Most of us use a pulley and counterweight at the tree end to keep the antenna wire from breaking. To help stabilize my counterweight and to keep it from blowing around in the wind, I came up with the following solution.

I purchased a 4 foot piece of galvanized water pipe; placing a pipe flange on one end and a pipe cap on the other (the cap was drilled to accept an eye-bolt). I placed the pipe flange end in a plastic pail (making sure the flange was against the bottom of the pail). I then poured some ready-mix concrete into the pail. After the concrete set, I removed the pail and inserted a piece of ground rod up into the pipe and buried the bottom end slightly into the ground.

A rope halyard was attached with a removable "quick-link" allowing me to remove it from the counterweight and attach additional rope when I want to lower the antenna. The device is adjusted so as the wind blows, the counterweight moves up and down, guided by the rod (see Figure 71).

My counterweight was 4 feet long and weighed 53 pounds, but you should adjust yours to your situation. — *73, John B. Kruk, K3KR, 408 Irwin St, Lock Haven, PA 17745,* **jbkruk@kcnet.org**

JOHN B. KRUK, K3KR

Figure 71 — The counterweight in use.

## COAX CRITTER DAMAGE

◊ In response to the "Hints and Kinks" item submitted by Thomas Lally, W1NSS, concerning coaxial cable being destroyed by deer,[1] I thought I should report another danger. I recently moved from western North Carolina where I had a similar problem with squirrels. I had a multiband inverted V dipole in some trees behind my house. The coax ran along the house and was suspended from my back deck to the tree, which acted as a center support.

When preparing to move I recovered the antenna from the tree. To my amazement, the suspended section of coax had many small cuts and punctures that I suspect were caused by squirrels that I had seen using the cable as an easy access method to my deck. Additionally, there were several small sections of cable that had the insulation completely removed (down to the braid) and showed signs of gnawing. Next time I'll put a circular "squirrel guard" made of non-conductive material around the coax, similar to those seen on bird feeders. — *73, Tighe Kuykendall, NK4I, 211 Scotch Range Rd, Summerville, SC 29483,* **tighe@buoy.net**

[1] "Watch Where You Run That Coaxial Cable," *QST*, Oct 2007, p 64.

## LOOP AND HOOK SKYHOOK

◊ An emergency flexible antenna can greatly extend the performance of a handheld transceiver. The best attachment point to hang the antenna always seems just out of reach. This simple wire hook made from a coat hanger makes it easy to hang and retrieve the antenna.

Using about 18 inches of wire, bend a hook to match the attachment point. Below the hook, form a loop to which the antenna line is tied. Below the loop, bend a downward pointing V-shaped pin with a little elbow on the side to prevent the pin from slipping completely inside the lifting tube (see Figure 72).

RANDY QUINTON, WQ7Q

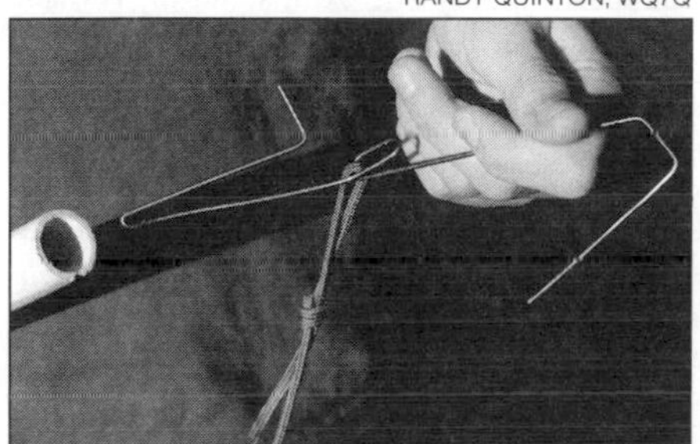

**Figure 72 — A wire coat hanger bent into this loop and hook shape makes raising that Emcomm or Field Day antenna much easier.**

RANDY QUINTON, WQ7Q

**Figure 73— Keeping tension on the support rope holds the hook in place.**

The hook is lifted by a tube or tube section attached to a pole. Tension on the antenna wire keeps the pin in the tube until the antenna has been placed (see Figure 73). The antenna is retrieved by reversing the process.

In my emergency kit, I keep a loop and a short section of PVC pipe with tape around it that can be used to turn a broom or branch into a lifting pole. As always, don't hang an antenna on electrical wires or anyplace in the vicinity of power lines. — *73, Randy Quinton, WQ7Q, 1003 221 Ave SE, Sammamish, WA 98075,* **wq7q@arrl.net**

## GROUND MOUNTED LIGHTNING ARRESTOR

◊ When changing my HF antenna to a ground mounted vertical, I realized my lightning/static discharge arrestor was located too far away from the antenna to provide good protection.

My vertical is mounted on a 5 foot long, 1.5 inch diameter galvanized water pipe with 3 feet in the ground. I use an Industrial Communications Engineers (ICE) Model 300 HF Arrestor (**www.iceradioproducts.com**). This is a small unit with SO-239 connections. I found a Bud Industries (**www.budind.com**) cast aluminum enclosure in my junk box to mount the ICE unit in. This keeps the weather out as the arrestor is not rated for unprotected exposure (see Figure 74).

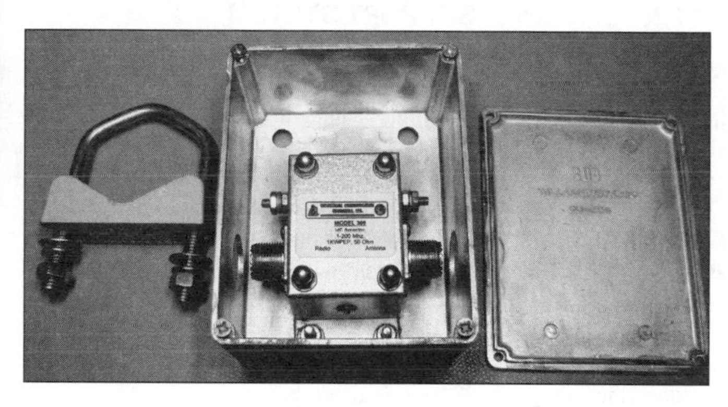

**Figure 74 — A view of the entire antenna base assembly.**

The holes in the sides of the enclosure were made with a ⅞ inch flat paddle drill bit, which is the correct size for a PL-259 outer shell. I used a hand drill and went slowly. The aluminum is very easy to cut. Wherever two metal joints occurred, I cleaned both surfaces and applied a coating of anti-oxidant joint compound. This is available at any home center or bigger hardware store. The joint compound prevents oxidation and ensures a good electrical connection.

I used stainless hardware for everything. The enclosure is mounted directly to the ground pipe, giving a very good ground connection (see Figure 75). I agree a copper clad ground connection would be better, but the galvanized pipe does a pretty good job.

**Figure 75 — The U-bolt alongside the arrestor mounted in the enclosure.**

The U-bolt I used is a very heavy duty, stainless steel model from DX Engineering (**DXEngineering.com**). I don't like the muffler clamp type; they don't hold up very well and will corrode.

I taped my PL-259s with Scotch 33 electrical tape and ran a bead of silicone sealant around the cable openings. I also drilled two small weep holes in the bottom of the enclosure to allow any condensation to drain. My enclosure cover did not have a gasket, so I applied silicone sealant around that as well.

This turned out to be a very easy and necessary project. You can use whatever you have on hand to fit your situation. Plus, I think it looks good. *All photos courtesy of the author. — 73, Phillip J. Mikula, WU8P, 10648 Aquarius Dr NE, Rockford, MI 49341,* **wu8p@charter.net**

## LADDER LINE CENTER INSULATOR

◊ I have had this center insulator up for over 6 years and it's still holding. It is made from a 0.75 inch T and an underground sprinkler riser. I drilled the T as shown in Figure 76, keeping the spacing for the ladder-line equal. Loop the ladder-line around and use cable ties to hold it on and take up the strain. I think the total cost was around $2. — *73, Len Keppler, KC7XH, 6050 Shady Pines Rd, Helena, MT 59601*, **keppler@wildblue.net**

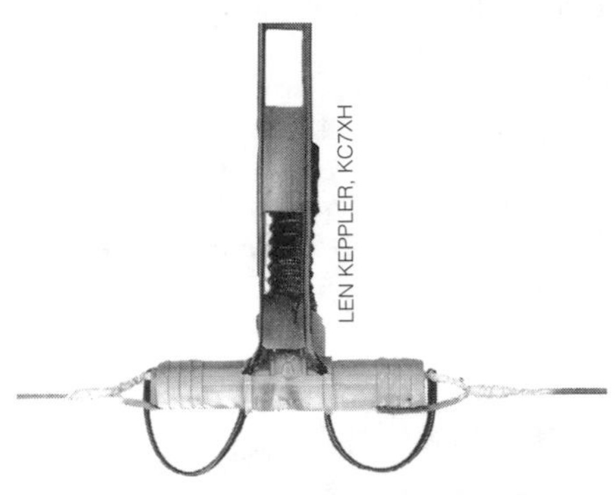


**Figure 76 — The sprinkler T repurposed to be a dipole center insulator. The addition**

## INCREASING RELAY VOLTAGE HANDLING CAPACITY

◊ Using relays for switching antenna feed lines, tuning and matching circuits, loading coils and so on requires that the relay contact spacing be sufficient to withstand the voltages that may be present. The voltages can be quite high and adequately rated relays are expensive. Luckily, garden-variety relays can be made to work well with a little ingenuity. The trick is to increase the voltage handling capability of the relay by modifying the mechanical travel limit when the contacts are open. Open-frame DPST relays, such as the one pictured here, are the easiest to work with. These relays and similar models are sold by McMaster-Carr and many other electrical supply houses.

Remove the NC (normally closed) contact assembly from the rear of the relay, if present. Next, bend the "tail" of the armature outward to allow the armature (the movable contact assembly) to move much farther, increasing the contact gap (see Figure 77).

The resulting contact gap is approximately 0.5 inch and should be able to withstand any voltages the amateur is likely to generate. Make sure any other gaps are of comparable size. The relay still pulls in with voltages as low as 12 V (it is a 24 V dc coil) taking somewhat longer than the unmodified version. The armature spring can also be weakened, if faster pull-in is required.

The result is a very serviceable RF switching relay at a fraction of the cost of a commercially manufactured one. — *73, Rick Karlquist, N6RK, PO Box 2010, Cupertino, CA 95015-2010,* **n6rk@arrl.net**

**Figure 77 — The modified Potter-Brumfield PRD7AYO-24 relay showing the rear area with the tail bent to permit an increased gap between the relay contacts.**

## PVC ANTENNA SUPPORT SYSTEM

◊ I use this arrangement to support the center of my multiband HF inverted V while also providing an integral center insulator (see Figure 78). It holds the 400 Ω feed line away from the mast and provides strain relief. The support is made from a 10 foot length of Schedule 40 PVC pipe, six T fittings, two PVC end caps and two ⅜ inch eye bolts. The end caps, drilled at the center, fitted with the eye bolts and mated to one of the T fittings with two short lengths of PVC pipe forms the center insulator. The other T fittings are spaced down the mast and fitted with short lengths of pipe to provide the feed line stand-offs. Each of the stand-off pipes is split vertically with a saw about an inch deep on the open end. The whole assembly is glued together with PVC cement. The feed line is slid into the stand-off slots and secured with ultraviolet light resistant electrical tape (see Figure 79).

The PVC assembly slides down over the top of a telescoping mast. The size of the PVC pipe is chosen so it just fits over the mast. A saddle clamp on the metal mast allows the PVC support to slide down about half way, which serves to extend the height of the mast. Finally one extra stand-off is made from another piece of PVC pipe and a right angle fitting and clamped to the mast with saddle clamps just above the roof line. Be sure the center insulator assembly is rotated to the angle you need relative to the stand-offs before gluing it in place. It's inexpensive, no center insulator to buy, no more broken feed line from wind and no problems from the feed line hanging too close to the metal mast. — *73, Tex K. Monroe, KJ4DU, 1226 Algoma St, Deltona, FL 32725,* **tmon2@bellsouth.net** or **kj4du@yahoo.com**

**Figure 78 — The complete PVC support mounted on the mast.**

**Figure 79 — A close-up view of the mast head and the stand-offs.**

## TRAFFIC CONE ANTENNA STAND

◊I was preparing for our local Prince Georges County ARES/RACES go-kit "show-n-tell" and I needed a lightweight, cheap (about $10), easily disassembled and stored antenna support for my newly completed WB6IQN DBJ-1 dual-band J-pole antenna.[1]

At my local "big box" hardware store I noticed bright orange, 12 inch "plumbers' cones" (see Figure 80) for sale (about $8) and realized that I could use one to make a perfect antenna stand for my DBJ-1!

To make my cheap and quick antenna stand, you will need the following parts:

- Five #8-32, ¾ inch machine bolts and nuts
- Five #8 washers
- One 12 inch traffic cone
- Two 10 inch lengths of galvanized steel hanger strap
- One 1 inch or ¾ inch PVC pipe cap (see text)

I made my version of the DBJ-1 from scrap 1 inch PVC pipe instead of the ¾ inch pipe Edison Fong, WB6IQN, calls for in his article. I also cut the pipe 1 foot longer than the antenna to allow mounting from the bottom, below the shorted matching stub of antenna element. I figured that mounting below this "zero impedance" point would reduce the

[1]E. Fong, WB6IQN, "The DBJ-1: A VHF-UHF Dual-Band J-Pole." *QST*, Feb 2003, pp 38-40.

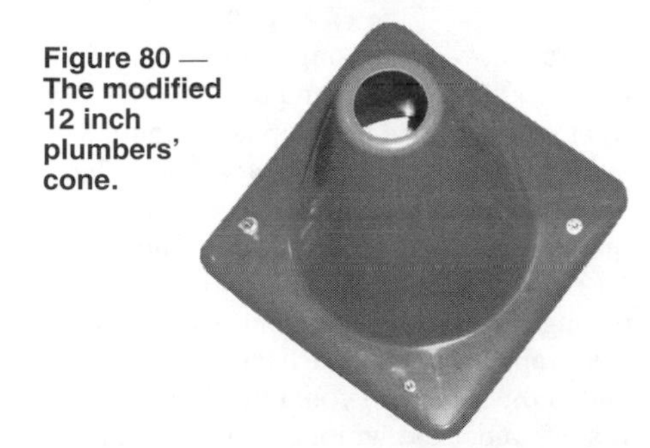

Figure 80 — The modified 12 inch plumbers' cone.

Figure 81 — The construction of the antenna base support.

Figure 82 — The antenna base support bolted to the bottom of the cone.

effects of the hardware on the performance of the antenna.

As it turned out, the 1 inch pipe fits snugly through the hole at the top of the cone. If you are planning to use ¾ inch pipe for your DBJ-1, the fit through the cone's top hole will probably not be as snug as mine is but a tight fit is not critical because the antenna is attached to the stand at the bottom support.

I made the bottom support by drilling a hole through the center of a 1 inch PVC pipe cap (flat-top caps work best) and bolting it to the center holes of the two 10 inch, galvanized steel hanger straps with a machine bolt and washer (see Figure 81). (If you made your DBJ-1 from ¾ inch PVC pipe, just substitute a ¾ inch PVC pipe cap.)

Next, I centered and bolted the completed support to the bottom of the cone with four machine bolts and washers (see Figure 82). I placed washers over the ends of the bolts on the top of the cone so that tightening the nuts would not "cut" through the cone's surface.

Figure 83 — The completed antenna stand.

Figure 83 displays the completed antenna stand with the antenna installed. Assembly is easy; just push the antenna through the top of the cone and into the support's PVC cap. Just don't push the antenna into the cap too far or you may have to use a hammer to remove the antenna from the stand later!

To remove the antenna from the stand, I found that it's easiest to lay the antenna on its side, grasp the PVC cap through the support straps and twist while pulling the antenna out of the stand.

If you plan to install the antenna and stand on a roof or other windy area, you may want to consider placing cinder blocks or bricks on the corners of the cone to keep the antenna stable.

Well, my antenna stand was a hit at our show-and-tell and it's now a permanent part of my go-kit. — *73, Jesse N. Alexander, WB2IFS/3, 7804 Westover Ln, Clinton, MD 20735,* **wb2ifs@arrl.net**. *Photos 3-6 by WB2IFS/3*

## TEMPORARY OPEN-WIRE FEED-LINE CONNECTORS

◊ Over the years we have used 450 Ω open-wire feed line to connect our dipole antennas for events like Field Day. Each time the tent or the antenna is at a slightly different location or height causing the feed line that is permanently attached to the antenna to be the wrong length. Each time we have had to cut, which is easy, or start looking for the soldering iron to add a piece of feed line. After our last Field Day I decided there had to be a better way.

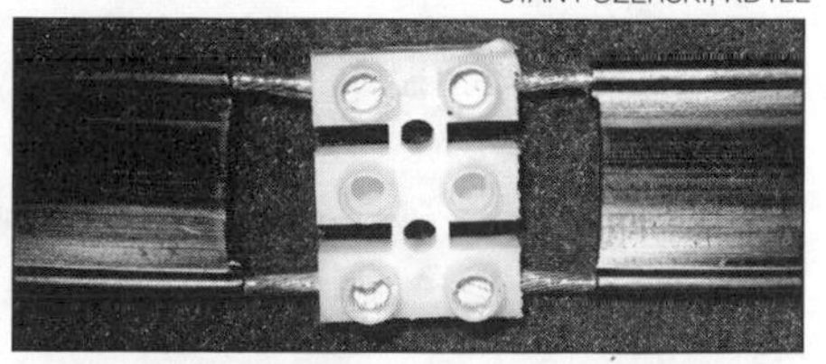

Figure 84 — A three connector segment of a terminal strip being used to join two sections of open-wire line.

The first part of the solution is to have a quick connect system. I chose European style terminal strips in 10 mm pitch because in three terminal increments it will match the spacing of the open-wire conductors with the center terminal unused. Prepare the connector by cutting off three terminal sections. Remove the screws from the center section (see Figure 84). The screws don't just fall out, so you either have to tap the terminal strip on an object or use a dental pick. When the screws are removed the center metal insert will come out, leaving only plastic between the two outside sections. Save the extra inserts and screws in a small plastic bag with your connectors.

The second part of the solution is to have pre-measured pieces of open-wire feed line to add to a run that is too short. I have pieces of 60, 30, 20 and 10 feet, which I marked for length. I put one of these connectors on each piece of feed line that is attached to an antenna and the pieces I regularly use as add-ons so they will be ready.

We have tested the connector in an RF field (the microwave test) and at full legal power on 10 meters with the connector at the voltage node and a SWR of 9:1. We have tested the mechanical connection under strain to over 40 pounds pull without separation. These terminal strips are available in 12 connection strips from Jameco (**www.jameco.com**) P/N 215029 and Digi-Key (**www.digikey.com**) P/N WM59151-ND and you can make four connectors per strip. This connector can make the use of open-wire feed line easier even in ad hoc situations. Thanks to Bob Reif, W1XP, for the strain and RF testing. — *73, Stan Pozerski, KD1LE, PO Box 527, Pepperell, MA 01463-0527,* **spozerski5090@charter.net**

## ANCHORING COAXIAL FEED LINE

◊ Here's the way I anchored the coaxial feedline from my windom antenna to my house. It is free to swing in the wind with the antenna, without becoming crimped or kinked (see Figure 85). I use ⅛ inch nylon rope and a variation of the knot used to lace

Figure 85 — RG-8X coax anchored to the author's house.

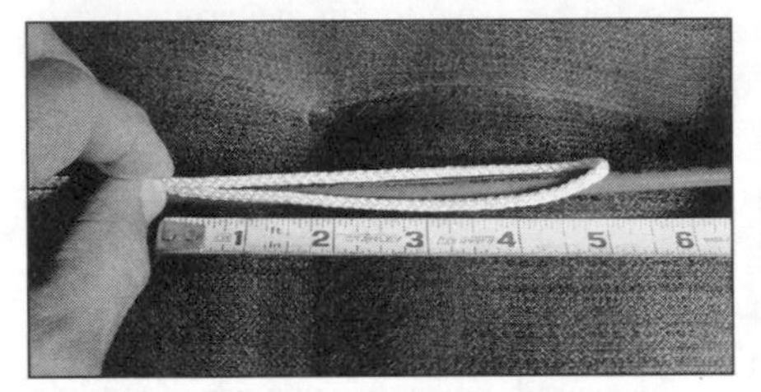

Figure 86 —First, form a 5 inch loop.

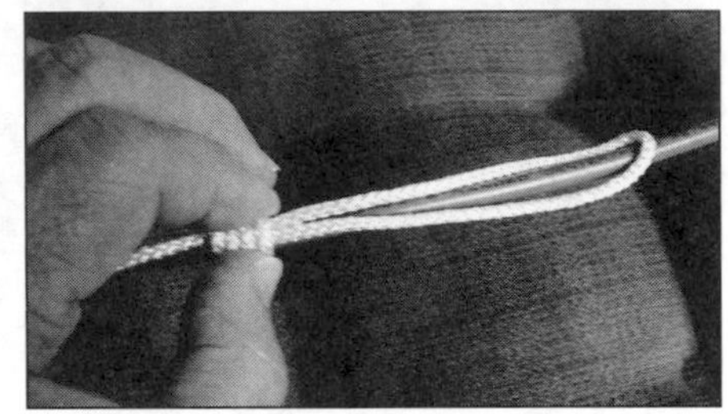

Figure 87 — Wind the end of the rope around the 5 inch loop in tight turns.

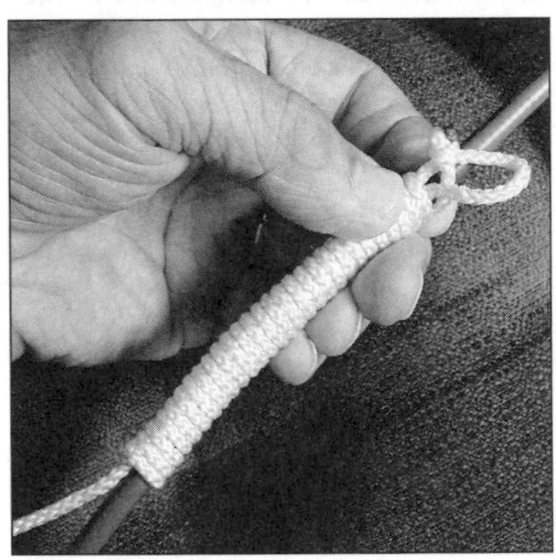

Figure 88 — When done, you will have about 3.5 inches of rope wrapped around the coax.

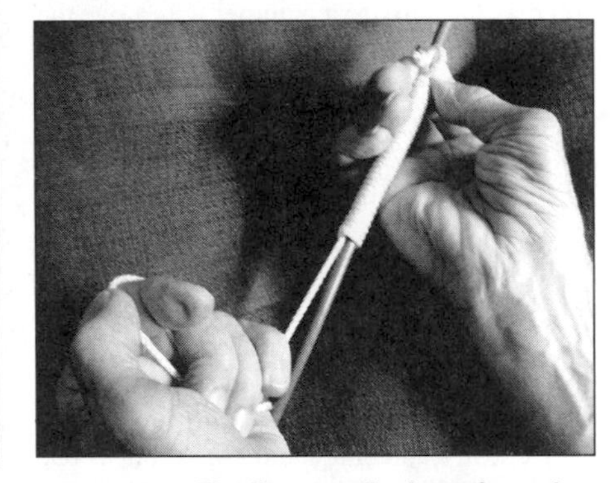

Figure 89 — Finally, pass the knot through the remaining loop and pull the end of the rope to close the loop below the knot.

telephone cables. For RG-8X coax, I start with a 5 foot piece of rope in which I tie simple overhand knots on each end.[1]

Next, I melt the end with a lighted match. When the melting nylon forms a ball, I put the fire out and press the melted ball against the flat side of a screwdriver blade or something similar, to blunt the end of the rope with the now hardening ball. If you do this just right you will have a hard glob on the end that is larger than the diameter of the rope and it will not pull through a simple overhand knot tied at that end of the rope.

Then measure 40 inches from one end of the rope and fold the rope back on itself. Place this loop on and parallel to the coax at the point where you want to secure it. Starting 5 inches from the end of the loop (see Figure 86), wind the longest loose end of the rope around this loop and the coax in close tight turns, toward the loop (see Figure 87). When you run out of rope, you should have about 3.5 inches of wrapping, ending about 1 inch away from the loop end (see Figure 88).

Now pull the knot at the end of the wrap through the 1 inch loop and while holding the wrapped coax in one hand, pull the other end of the rope with your other hand, so that it slips under the wrapped turns and closes the loop around the knot on the other end (see Figure 89). A little practice may be necessary here to get everything tight and neat.

After tying the rope to the screw eye with a couple of half hitches, the remaining rope is wrapped around the screw eye to make it neat and keep it from dangling in the wind. *All photos by the author. — 73, Lyle H. Nelson, ABØDZ, 1450 201st Ave NW, New London, MN 56273,* **lylenel@tds.net**

## FIXING AND REPLACING FIBERGLASS-TUBE ANTENNA INSULATORS

◊The two fiberglass tube insulators on my Titan DX 80/20 vertical antenna were failing. Zeke Zeanon, KJ4ASG, an engineer with expertise in fiberglass pipe, built the replacements and gave me additional advice on future failure prevention. The first of two failures I experienced was sun, rain and snow had caused the fiberglass to "bloom" in which the glass fibers separate resulting in a drop in strength. The solution is periodic application of latex paint, reflective white preferred, to seal the fiberglass from the weather.

The second failure was the lip on the two aluminum bushings that fit into each end of the tube was smaller than the outside diameter of the tube. That's a design error by the manufacturer. It eventually crushes the ends of the tube loosening them and causing the antenna to wobble in the wind. The solution was a washer with an inside diameter that fit over the bushing and an outside diameter larger than that of the fiberglass tube. I measured the outer diameter of the smaller end of the bushing and then looked for a washer with an inside diameter of that size. In my case a "USS 1⅛" would work with some burring to enlarge the inside diameter. The fastenings expert at a local hardware store took my bushing and found a washer with a perfect fit, no burring required.

There was no way to repair the original tubes so they fit into the cast aluminum clamps on the Titan DX. The fix was a two layer schedule 40 PVC pipe replacement. A piece of PVC with a slightly larger outside diameter than the original fiberglass was split down its length. A second piece of PVC small enough to fit inside the larger split piece was forced inside and the outer, split piece "welded" together with PVC rod. The result is seen in Figure 90; the new double-layered PVC insulator, the original through-bolt and

[1]R. Collins, WX3A, "The Knots of Ham Radio," *QST*, Jun 2006, pp 57-58.

BRUCE MACALISTER, W4BRU

**Figure 90 — The assembled components ready to be tightened onto the insulator.**

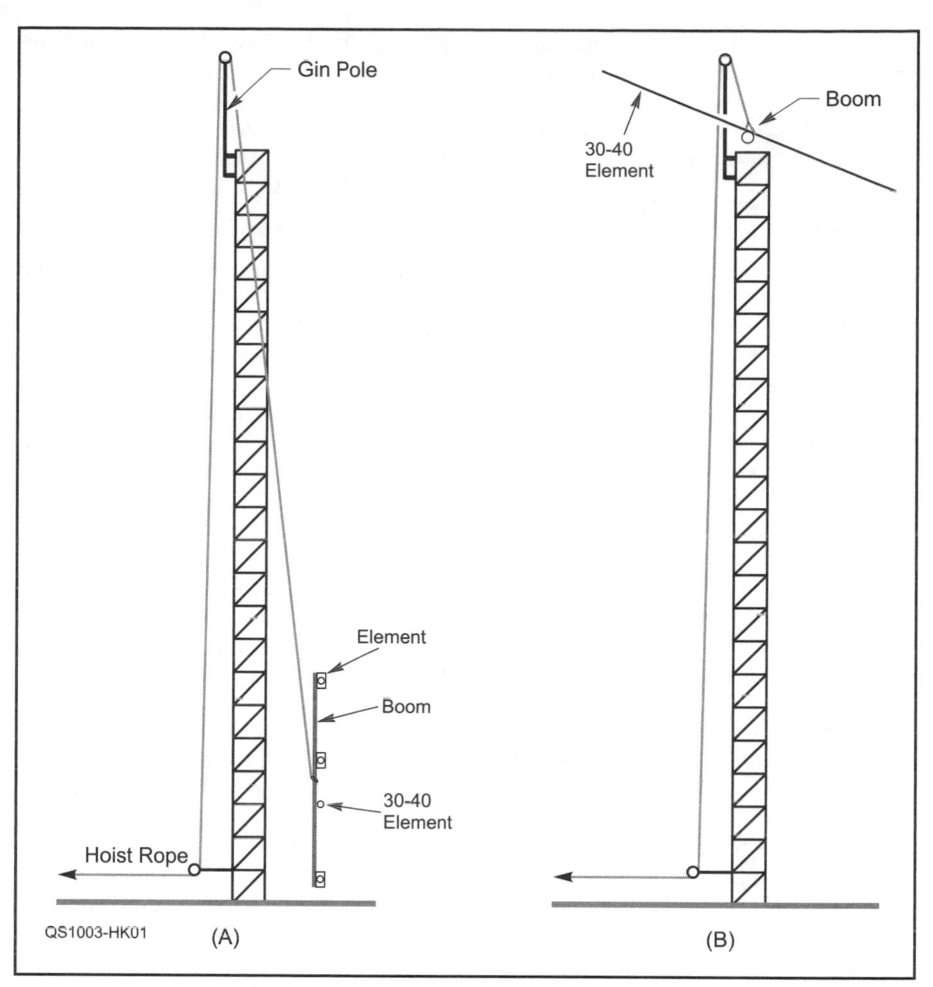


**Figure 91 — A diagram showing how to raise the SteppIR with the 30-40 meter upgrade. At (A) is the position of the antenna on the ground. At (B) is the position of the antenna at the top having been rotated with the 30-40 upgrade loop element passed over the mast.**

bushing and the new larger washer.

PVC pipe seems to be a good RF insulator, but there is a way to check the piece you buy. Put a small piece of the PVC and a cup full of water in the microwave oven. Run the microwave for a minute. The water should be hot and the PVC not, indicating there is no metallic content in the PVC that might conduct RF. — *73, Bruce MacAlister, W4BRU, 1805 Grove Ave, Richmond, VA 23220*, **w4bru@arrl.net**

## STEPPIR MOUNTING — A BETTER WAY

◊[Before starting this procedure you should put on a hard hat and review tower safety procedures. — *Ed.*] I have found a simple way to mount a SteppIR Yagi antenna with the 30-40 meter upgrade. Place the antenna leaning up against the tower with the boom vertical. Lower the rope from the gin pole so that the rope is between the tower and the antenna (see Figure 91A). Tie the rope close to the center of the boom. Begin raising the antenna while holding onto one side of the lower fiberglass rod. This will cause the antenna to flip (windmill). Allow it to rotate till the fiberglass rods are vertical and the boom is horizontal. Continue raising the antenna. At the top, flip the 30-40 meter trombone element up over the gin pole and down on the mast — like throwing a loop up over a stake (see Figure 91B).

Another thing about the 30-40 meter element is that the element tapes are flat and will bend from side to side but not up and down. Therefore the tape has a tendency to twist or bind because of sag near the ends of the fiberglass rods where they fold back.

To correct this, I made a 4 foot removable extension with two double pulleys, which I added to my mast above the SteppIR. I fastened two lengths of ¼ inch diameter antenna rope ⅔ of the way out from the boom to both sides of the 30-40 meter element. I then pulled the rope ends back to the boom before I raised the antenna. After I got the antenna up, I put the ropes through the pulleys and installed the extension and pulled the fiberglass rods up horizontal. The unit now operates a lot more smoothly. [Note that SteppIR has a Truss Kit available for the DB36. — *Ed.*] — *73, Pete Peters, W7OW, 7520 N Whitehouse Dr, Spokane, WA 99208-6144*, **w7ow@hotmail.com**

## PASSING CABLES THROUGH STUCCO WALLS

◊I've seen several ways to run coaxial cables through walls to outdoor antennas, but none seemed as flexible, easy, or unobtrusive as my method. I used a standard outdoor four outlet receptacle housing backed up by an ⅛ inch aluminum plate cut to size and drilled for four threaded UHF female-female bulkhead mount adapters. I added a piece of ¼ inch closed cell foam between the stucco siding and the aluminum plate. The receptacle box came with a foam gasket already in place on its back side.

By using 90° UHF connectors I can easily route as many as four RG-8 or RG-58 lines out to the antennas, although my installation has only two UHF adapters for now (see Figure 92). These adapters project through the stucco and wood sheathing into the basement close to my radios. The receptacle box comes with a frosted plastic snap-on cover and knock-outs that provide holes for the wires. To prevent tampering, I can lock the box. When I disconnect the

JON TITUS, KZ1G

**Figure 92 — The assembled and mounted outlet box containing two UHF connectors. A neat and tidy installation.**

outdoor coax, the box looks like just another power outlet. I protect the connectors with silicone grease. — *73, Jon Titus, KZ1G, 14253 S Trailview Way, Herriman, UT 84096*, **kz1g@arrl.net**

### STOPPING WIND DAMAGE TO LADDER LINE

◊I just finished reading Tomm's, W2BFE, question in the April 2008 issue of *QST*, about replacing his window line with open wire line to stop wind damage.[1] I have had the same problem with the ladder line feeding my G5RV. I solved the problem after noticing trucks on the interstate using straps to secure their loads. The straps that were straight and flat would move or vibrate in the wind but those that had a twist did not move. I questioned one truck driver and he says the twist acts like an air foil and stabilizes the strap in the wind. I thought this might be a solution to my problem with the wind causing my ladder line to whip like a jump rope and eventually come apart.

I secured my ladder line to the antenna support rope with zip ties about every 3 to 4 feet. After tying off the rope and before attaching the coax I give the ladder line about 3 or 4 twists (see Figure 93). This keeps the ladder line in place in moderate winds (5-10 mi/h) and I have had it withstand winds of 20 mi/h and above. I don't notice any change in performance and I haven't had to replace my ladder line because of wind damage since. — *73, Louis Pauly, N8KXM, 8610 Penny Rd, Pleasant Hill, OH 45359-9784*, **n8kxm@arrl.net**

[1]J. Hallas, "The Doctor is IN," *QST*, Apr 2008, p 67.

LOUIS PAULY, N8KXM

**Figure 93 — A view of the ladder line tied to the support rope and twisted to minimize its interaction with the wind.**

### UNISTRUT ANTENNA SUPPORT

◊ When considering a center support for the installation of an inverted V antenna I decided on the following requirements: a strong material, easy assembly, corrosion resistant and with tilt-over capability. The initial design was to use several 10-12 foot sections of antenna mast or 2 inch galvanized pipe coupled together. Due to the cost of antenna mast and galvanized pipe being more than anticipated, an alternative, 10 foot sections of 1⅝ inch Unistrut channel was selected.

Unistrut is a hams Tinkertoy. It is available in 13⁄16, 1¼ and 1⅝ inch widths, 10 and 20 foot lengths, in 12, 14 and 16 gauge sizes. It is made with dipped surface protective treatments from electro-deposition acrylic green to "hot dipped" galvanized coatings and multiple channel designs (including telescoping sections). A vast selection of brackets and fasteners is available and it is continuously slotted for easy guying. The design and choices for project use are limited only by the creativity of the user. For more information go to **www.unistrut.com**.

The tilt over base support consisted of a treated 4 × 4 inch landscape post anchored in concrete. (The wooden post was moisture sealed at the top using the dipped coating used for hand tools and along the length with several coats of wood sealer.) The Unistrut mast pivot point on the support base was a 3 inch lag bolt about 50 inches above ground. In retrospect, a section of Unistrut could have been set in the concrete as the antenna base support.

The Unistrut mast preparation consisted of black paint (except for areas where the sections were bolted together) and the addition of a hoisting pulley mounted at the end of a 2 foot PVC pipe (see Figure 94). One problem encountered was in bolting the Unistrut together. Bolting back-to-back against the base of the "U" was not possible because I couldn't get a socket on the bolt head inside the Unistrut channel. To get around this the Unistrut was bolted with the open "U" ends face to face. Square Unistrut channel would have eliminated this issue. The three Unistrut sections were bolted together with about a 12 inch overlap.

An antenna hoisting line was fed through the pulley and mast section carried to the base support for mounting. After attaching the mast to the pivot point it was tilted up into position, plumbed and secured in place with three additional lag bolts.

The V antenna was spread out, transmission cable attached and hoisted to the top of the mast. The ends of the V antenna were attached to a 12 foot landscape treated 4 × 4 that was secured to the property fence. Hmm, I wonder how the Unistrut would tune up. — *73, Louis Kobet, WB3DZD, 2223 NW Douglas Loop, Camas, WA 98607-8085*, **kobetls@comcast.net**

#### More on the Unistrut

◊Tim Fuller, KC5TCF, is also an aficionado of Unistrut [Hints & Kinks, *QST*, Apr 2010, p 66] and wants to add that it is available in stainless steel and fiberglass. These are excellent for weather and corrosion resistance with the non metallic fiberglass providing obviously desirable characteristics for antenna support.

### TROUBLESHOOTING STEPPIR ANTENNAS

◊I recently purchased a new SteppIR antenna and as with all new products, I had some problems getting it up and running. Here are a couple of tips on troubleshooting the antenna.

Make sure that the interconnecting cable is wired properly. SteppIR has a good checkout list and common faults listed in the manual. I used CAT 5 cable for the long antenna run and needed to install connector blocks. Make sure that you don't mix Telco connector blocks with network connector blocks. Some Telco blocks have an unwanted cross over in the connection between the terminals and the RJ-45 jack. The unwanted crossover prevents communication to the antenna.

When doing the checkout of the antenna at ground level, it is difficult to make RF measurements that are meaningful. After much thought, I decided to try a "stud finder" to detect the metal tape inside the fiberglass tube. It is not possible to "see" the tape, but the stud finder will easily detect where the end of the tape is located. With the stud finder, you can verify that the motors are moving the tape and each half of the element has the same amount of tape extended. — *73, Mel Farrer, K6KBE, PO Box 4005, Ione, CA 95640*, **farrerfolks@yahoo.com**

LOUIS KOBET, WB3DZD

**Figure 94 — The PVC pipe and hoist pulley device mounted to the top of the Unistrut for hoisting the antenna.**

## QUICK AND DIRTY ANTENNA SUPPORT

◊Aesthetic concerns placed many limits on the size and configuration of the antenna farm allowed at my station. In my case, the "design tolerances" allowed limited me to vertical antennas that would blend in as much as possible with the color scheme of the house. As such, I was required to use white for antennas, coax cable and mounting hardware to blend in with the trim color and the color of the PVC plumbing pipes that pierce the roof in several locations.

I find the boating industry has already solved many of these same problems faced by amateur operators. Shakespeare Electronic Products Group sells the Style 408-R Upper Bracket (**www.shakespeare-marine.com/mountshow.asp?findmount=408-R**) that can support antennas (or antenna extension masts) of 1 inch (using the insert provided) to 1½ inches.

I've used this bracket in two ways: as the mount for short no-ground plane antennas and as the support bracket for heavy duty TV masts used to support longer vertical antennas (see Figures 95 and 96). The brackets are mounted to the fascia using ¼ inch lag bolts and lock washers. I try and make sure that the lag bolt on one side of the bracket is screwed into a roof joist inside of the fascia. Shakespeare is also a source for white RG-213 coax cable and gold plated PL-259 connectors that I used to build a spouse approved cable.

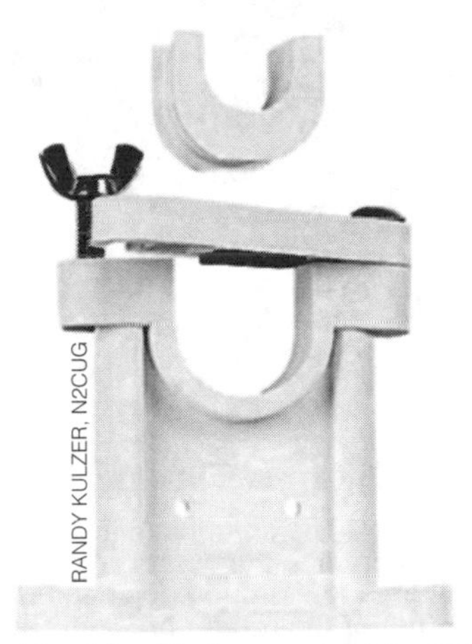


**Figure 95 — The Shakespeare style 408-R upper bracket.**

**Figure 96 — The 408-R used as a support bracket.**

Both uses of this bracket have survived several fairly windy storms here at my station. Of course, mounting the antennas so they do not extend above the roofline also helped. As with anything else, your mileage may vary when using this bracket.

I'd like to thank Syed Ali, WB2AFA, my Amateur Radio mentor and friend for encouraging me to publish this article. — *73, Randy Kulzer, N2CUG, 2235 Allenwood Rd, Wall, NJ 07719*, **rkulzer@usa.net**

## ANTENNA QUICK CONNECT

◊Like most of you, I unscrew the antennas coming into my shack upon the threat of a lightning storm. Living in the Tampa Bay area of Florida has brought with it a more intense challenge, as this is the lightning capital of the nation.

Screwing and unscrewing PL-259s after almost every use of my equipment has worn my fingers to the nub. I therefore sought a better alternative system for disconnecting my antennas. At first my mind was blank, then I remembered back 40 years ago when I was a General Radio Sales Engineer (peddler). General Radio, or GR, invented a connector during WWII that was and still is unique. This is the GR874 series of connectors.

The GR874 was a sexless connector (hermaphrodite) where one just pushed into its mate. This connector still has extraordinary characteristics as it covers dc to 9 GHz with minimal loss and can handle 1500 V peak. This permits it to handle full legal limit power plus into a 50 Ω load without any sweat. This connector comes in a full range of adapters such as PL-259 female to GR874 and PL-259 male to GR874, which is what I am using. It also comes in N series adapters that have far superior electrical properties than an N connector.

I do not believe anyone is manufacturing the GR874 anymore, but they are available at many outlets. All one has to do is an Internet search on "GR874 Connector" and a host of suppliers will pop up.

So gentle persons, save your fingers and a lot of time as this ancient connector from the 1940s will save your fingers. — *73, Burt Yellin, K2STV, 1813 Columbine Pl, Sun City Center, FL 33573*, **k2stv@copper.net**

## ARROW SAFETY TIP

◊Many areas of the country, such as my area, Long Island, New York, consider the "Sling-Shot" a "Fire-Arm." I know it's crazy but that's how politicians keep messing with us. So how do you get that antenna up in the top of that *big* tree without getting arrested? Well here, at least not yet, a kids bow and arrow set is *not* considered a lethal weapon.

Sports Authority sells a neat kids archery set. It comes with a bow, 2 target arrows, target, arm guard and a quiver. Now you need to put a safety tip on those arrows. When installed it is easy and relatively safe, to put some fishing line up in the tree. Then pull up the supporting lanyard for the antenna. Locate some Acetal/Delrin [a type of plastic with a low coefficient of friction giving it less of a tendency to get hung up in branches or leaves — *Ed.*] and make a couple of tips for the arrows. — *73, Charles Rankin, WA2HMM, 165 Hickory Ln, Smithtown, NY, 11787*, **crankin@dialup4less.com**

## ADD 70 CM WITH A COAT HANGER

◊On a recent road trip through California I was faced with having just a 2 meter quarter-wave mag-mount antenna for the car, yet many of the interesting repeaters were in the 70 cm band. Fortunately, my handheld transceiver covered both of those bands. Not willing to sacrifice a perfectly good 2 meter whip, I opted for an ad hoc conversion of the antenna to a dual-band 2 meter /70 cm antenna. With only scissors and side-cutters for tools, and a wire coat hanger, painter's blue masking tape and steel wool for materials, I came up with the following workable solution.

I bent a straight piece of the wire coat hanger into a "hairpin" so that it resembles a section of parallel transmission line that is shorted at one end. The total length of coat hanger wire was estimated to be 95% of a quarter wavelength at 440 MHz or 162 mm (see Figures 97 and 98).

This hairpin decoupler was positioned open end down 162 mm up from the base of the 2 meter whip. With no soldering equipment available, I polished the whip as well as one leg of the bent wire coat hanger hairpin with some steel wool. Then I laid the two elements, with the polished sections in contact along their entire length, on the sticky side of some painter's blue masking tape. The tape was then folded over along the length of the joint, sticky side to sticky side, and firmly pressed together to form a temporary watertight connection. The excess blue tape was then trimmed with scissors.

A quick test using the relative SWR feature on a friend's ICOM IC-817 transceiver indicated that the antenna was quite serviceable on both bands. I was able to communicate with 2 meter and 70 cm repeaters in the San Francisco area using my multiband handheld radio.

KAI SIWIAK, KE4PT

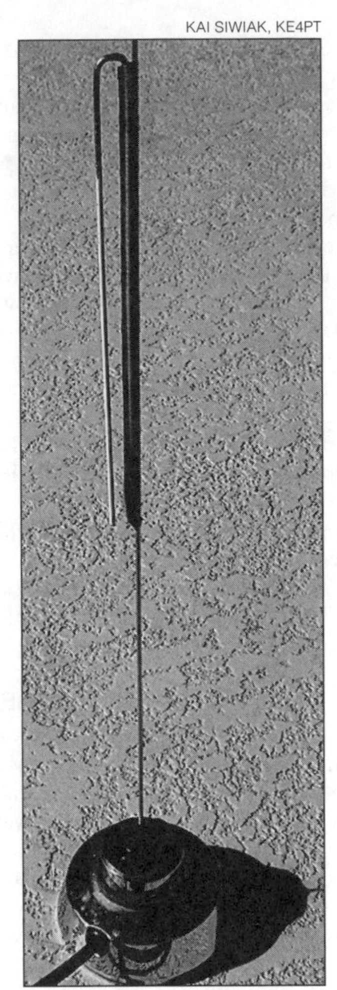

**Figure 97 — The modified 2 meter whip with the hairpin taped in place on the antenna element.**

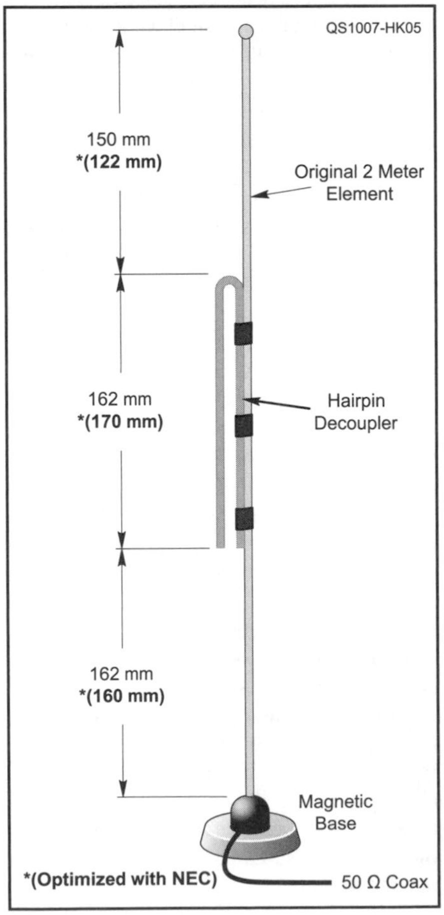

DONALD A. CRISP, W7ZNN

**Figure 99 — All the components needed to insure your beam stays where you aim it. Note the aluminum angle on the right. The corner has been filed down to better fit the rotator bracket.**

A subsequent check, later in the road trip when some test equipment became available, revealed that the antenna VSWR was indeed less than 2:1 on both bands. In fact, resonance was actually about 5 or 6% low indicating that the antenna could be trimmed for a better match. Nevertheless, the antenna functioned beautifully during the trip.

In theory, on the 2 meter band the antenna continues to function like a quarter-wave whip but the addition of the hairpin stub adds some inductance, lowering the resonant frequency, as was observed. In the 70 cm band, the lower section of the whip functions like a quarter-wave section at 440 MHz. The open circuited end of the hairpin element functions like a quarter-wave decoupler at 440 MHz and together with the remaining top element of the whip forms a half-wave radiator, which is in phase with the lower quarter-wave section. Thus, the dual-band conversion yields a ¾ wave antenna in the 70 cm band and a ¼ wavelength antenna in the 2 meter band. — *73, Kai Siwiak, KE4PT, 10988 NW 14th St, Coral Springs, FL 33071-8222,* **ke4pt@amsat.org**

QS1007-HK05

150 mm
***(122 mm)**
Original 2 Meter Element
162 mm
***(170 mm)**
Hairpin Decoupler
162 mm
***(160 mm)**
Magnetic Base
***(Optimized with NEC)**
50 Ω Coax

**Figure 98 — A diagram of the modified antenna showing the component spacing used by the author and also the optimized values generated by *NEC* computer simulation.**

## A SIMPLE CURE FOR A SLIPPING ROTATOR MAST

◊A couple of weeks after I had installed a new three element SteppIR (**www.steppir.com**) beam antenna, I noticed that the beam heading had shifted about 25°. I had very carefully leveled and aimed the antenna, so I was surprised to see that the actual beam heading did not coincide with the rotator control heading. It was apparent that a recent wind storm with gusts of up to 60 mi/h had shifted the beam direction by causing the mast to slip in the Tailtwister rotator's bracket. This occurred even though I had tightened the two rotator bracket U-bolts down securely.

In order to prevent the problem from reccurring, I made a shim out of 1¼ inch aluminum angle, as shown in Figure 99. I placed the angle between the mast and the V of the rotator bracket to lock the mast in place. The inside angle of the mast bracket on the Tailtwister is more than 90°, so it was necessary to file the outside corner of the aluminum angle in order to spread it to conform to the angle of the rotator bracket.

DONALD A. CRISP, W7ZNN

**Figure 100— The completed mounting arrangement has the corner of the aluminum angle ground down to fit into the corner of the rotator bracket. The three stainless steel hose clamps secure the aluminum angle to the mast.**

I cut the aluminum angle to 8 inches in length, which left 3 inches to protrude above the rotator bracket. Three stainless steel hose clamps, in conjunction with the rotator bracket and its U-bolts, were used to clamp the aluminum angle securely to the mast and rotator as shown in Figure 100.

The aluminum angle also serves as a rotator to mast shim, which centers the mast on the exact rotational center of the rotator. For the Tailtwister and other similar Hy-Gain and CDE rotators, the aluminum angle shim thickness should be half of the difference between the outside diameter of the mast and 2¹⁄₁₆ inches. This assures that the rotator will not try to turn the mast eccentrically with respect to the tower top bushing or bearing.

In spite of some recent gusty winds, the beam heading has now remained rock solid. I suspect that many other hams have had the same problem and may benefit by this simple, inexpensive and easy fix. — *73, Donald A. Crisp, W7ZNN, 2907 North Rambo Rd, Spokane, WA 99224-9164,* **w7znnqrz@peoplepc.com**

## PLASTIC LID INSULATORS

◊One of my ham radio passions is to experiment with homebrew antennas and it seems as though I am always missing one or more of the components to build an antenna. Necessity and not wanting to wait for insulators and other parts to come in the mail has driven this project and it has worked very well. I now have plenty of readily available raw materials to build all the parts needed. This project is also simple, fast and durable.

I save the plastic lids from various products used regularly around the house. These lids can be trimmed in either a square or circle completing the first step of the project. Trim all lids to a circle and store them until needed in one of the cans (see Figure 101). This makes plenty of raw material available when you have an antenna project.

If your project requires end insulators, trim the circle to a square and then cut the square in half. Fold it in half and in half again. This makes a 4 layer end insulator after punching holes in it with a hole punch for notebook paper (see Figure 102). If you need a really heavy duty insulator, double it again.

For the center insulator I take a round piece and quarter it into four triangles, lay these on top of each other and once again using the hole punch, punch three holes in the appropriate areas and you have a supportable center insulator for your dipole. If you feel you need a stronger center just add more layers.

Here is all you need to build your dipole, tough, fast and cost efficient. Just calculate the length of your wire. Assemble it and you are in business. — *73, Roger Odorizzi, W7CH, 195 Ivan Morse Rd, Manson, WA 98831,* **w7ch@arrl.net**

ROGER ODORIZZI, W7CH

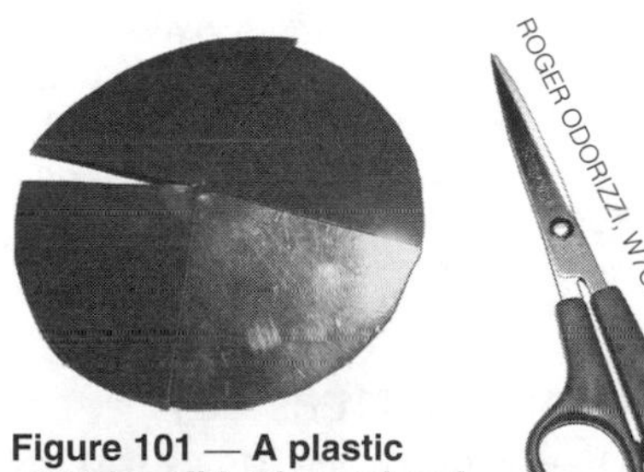

**Figure 101 — A plastic container lid, trimmed and cut into quarters, can be insulators for wire antennas.**

ROGER ODORIZZI, W7CH

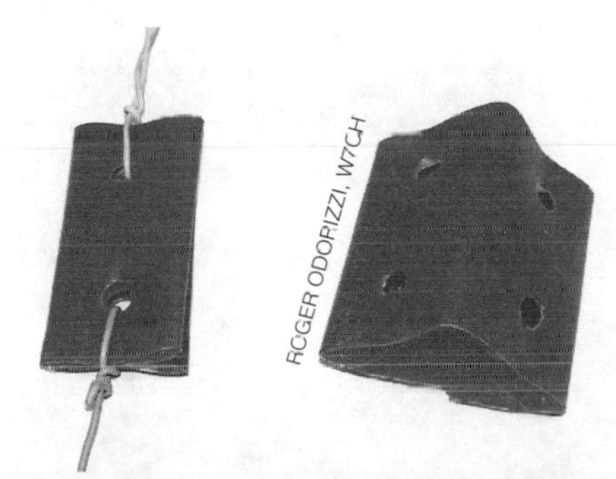

**Figure 102 — An end insulator made from a plastic container top. The top is cut into an appropriately sized square and then folded over itself to form a multilayer insulator. A paper punch is used to make the holes.**

## COAX CABLE END SEALING

◊I prefer to retain the "ole-way" of doing things, so none of these newfangled dipole center connectors for me. I found a very neat way to weatherproof and seal the end of the coax at the feed point. For that matter, it can also seal your connectors. Two coats sealed the end very nicely and it is flexible too.

This product is called "Liquid Tape Electrical Insulation." I purchased a small can from a local hardware store. To quote the manufacturer, Plasti Dip International, "Performix Liquid Tape is a rubber insulation coating that exhibits excellent acid, alkaline and abrasion protection, and seals out moisture and salt permanently. The 4 oz brush-in-cap container colors come in Black and Red." Give it a try — it's a simple cure for all those "water in the coax" problems. — *73, Charlie Rankin, WA2HMM, 165 Hickory Ln, Smithtown, NY 11787,* **crankin@dialup4less.com**

## HOSE REEL ROPE CARRIER

◊After a long day of working on a tower, one of the hardest things to do is rolling up the 200-400 feet of rope. To make this job easier I store my long ropes on an old hose reel, with wheels to help move it. I took a leaking hose reel that was to go out to the trash and cut off the hose connector. I attached the rope to the reel and start turning the handle to roll up the rope. I can get about 300 feet of ½ inch rope on the reel. — *73, Charles Stokes, WB4PVT, 494 Pamela Dr, Newport News, VA 23601-1723,* **wb4pvt@arrl.net**

## EASY LADDER-LINE STANDOFF

◊A trip to the local hardware store turned up a nice ladder-line standoff. The 1.5 inch plastic pipe hanger matches my ladder-line without modifying either. Each hanger comes with a pair of slots for a zip tie. — *73, Dave VonDielingen, AD8B, 17701 E Willow Rd, Garber, OK 73738-1027,* **ad8b@arrl.net**

## TARHEEL ANTENNA PROTECTOR

◊During the first winter using my Tarheel screwdriver antenna (**www.tarheelantennas.com**) I had wondered how it would hold up against the ice and snow. As I found out, a lot of ice and snow can collect on the mounting bracket and antenna base. I had to find something to shield the area around the mount without interfering with the tuning of the antenna.

STEVE BENELL, WA9JNM

**Figure 103 — The fence post modified for mounting on the Tarheel antenna mount. The picture shows hurried cuts that are not straight, but they do the job until I can make better ones.**

STEVE BENELL, WA9JNM

**Figure 104 — The finished cover, painted and mounted on the Tarheel.**

I found a plastic fence post at a local hardware store that was 4 × 4 inches square and 3 feet long for less than $10. I cut a length of the plastic post to 11⅞ inches to fit the MT-1 mount. In the center of one side of plastic post I cut two vertical lines about 1 inch apart (see Figure 103). This 1 inch gap is used to pry open the post with your hands. Once the cut is made, paint the post an appropriate color and then slip the post around the antenna mount protecting the lower portion of the antenna and shunt coil (see Figure 104). — *73, Steve Benell, WA9JNM, 1160 Shoreline Cir, Cicero, IN 46034-9426,* **kjbenell@comcast.net**

## COAX CABLE STANDOFF

◊ After erecting my vertical antenna, I wanted to take the strain off the coax hanging on the PL-259 connector. The typical TV standoff doesn't work very well. I did not want to tape the coax to the aluminum mast. Not finding any standoffs advertised in the various ham radio catalogs, I decided to make my own.

Figure 105 shows a hardwood dowel, these are maple, 6 inches long and 1 inch in diameter with a right angle bracket mounted on one end and a plastic cable clamp mounted on the other end. Select a clamp to hold the coax firmly. A light coat of clear acrylic spray keeps the dowels from absorbing any water and the right angle brackets from rusting. Use hex head, self-tapping screws to secure the right angle bracket to the aluminum antenna support mast.

ERNEST KAMPE, KBØLSX

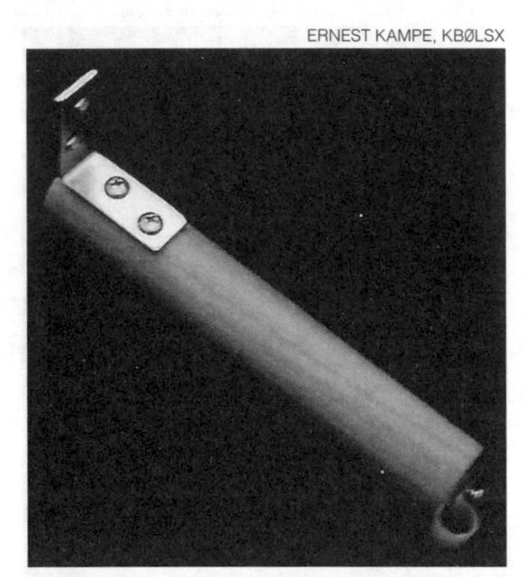

**Figure 105 — The homebrewed standoff made from a wooden dowel, angle bracket and cable clamp.**

I built four of these standoffs and they did the job very nicely and look very professional. In addition to being inexpensive, mine have been up several years without any deterioration. — *73, Ernest Kampe, KBØLSX, 417 Cheyenne Ave, Eaton, CO 80615,* **memakampe@earthlink.net**

## MOUSE TRAP GRABBER

◊I recently needed to retrieve one end of a rope that was hanging from a mast at the top of my tower so I decided to use a mousetrap in an unconventional way.

Using a typical spring-loaded trap, I removed the holding bar and the catch but left the holding bar staple (afterward I realized I could have left the bar and catch in place). I drilled two holes, each about $\frac{3}{4}$ inch from the trap edges and continued drilling through a length of $\frac{5}{8}$ inch OD aluminum tubing that I saved from an old antenna. Then I bolted the mousetrap to the aluminum tube (see Figure 106).

I bent the head of a tenpenny finishing nail into a circle and tied a suitable length of string

BILL KIRK, NJ1X

**Figure 106 — A mousetrap gimmick to catch the "tail" of a rope hanging out of reach.**

through the opening. The next step was to pull down the trap hammer and, at the same time, insert the finishing nail through the holding bar staple and over the hammer, holding the hammer in the "armed" position.

After assembling the necessary length of tubing needed to reach the rope at the top of the tower, I positioned the mousetrap next to the rope, pulled the string and, with some luck and a few tries, caught the rope. When the rope is caught, twist the mousetrap around it several times to be sure it is secured. — *73, Bill Kirk, NJ1X, 17 Bellevue Ave, Winchester, MA 01890,* **wk9879@verizon.net**

## TRI-BAND LAMP ANTENNA

◊Need to get a little more range from your handheld transceiver for the weekly net? This is a simple arrangement that can turn any household lamp into an antenna to help those 5 W along (see Figure 107). To build the "lamptenna" you will need a $\frac{1}{16} \times 1\frac{1}{2}$ inch aluminum strip, an SO-239 with threaded stud and a telescopic rod antenna with 3 mm female thread. The SO-239 and antenna are available from the author. To build the antenna:

1. Saw the aluminum strip to a 3 inch length to form the support bracket
2. Layout and drill two $\frac{1}{4}$ inch diameter holes, $\frac{1}{2}$ inch from each end.
3. Punch a $\frac{5}{8}$ inch diameter hole at one end
4. Place the SO-239 in the hole. Mark the four mounting holes, drill $\frac{1}{8}$ inch holes and mount the connector.
5. Install the telescopic rod antenna onto the SO-239.

BOB EVANS, WBØSVS

**Figure 107 — The business end of the "lamptenna" connected and ready to get on the air.**

To use the antenna, find a medium size table lamp, remove the finial and shade. Next, place the bracket onto the lamp's hoop stud and replace shade and finial. Route the coax alongside the hoop using a wire bag tie. Note that conical shaped lamp shades seem to work the best. Extend telescopic rod to 19$\frac{1}{4}$ inches for 2 meters, 12$\frac{1}{2}$ inches for 1.25 meters or 6$\frac{1}{4}$ inches for 70 cm.

For a single band version use a common SO-239 and solder a $\frac{5}{32}$ inch OD brass tube to the center conductor. Use a 6$\frac{1}{4}$ inch length of tubing for 70 cm. For lower frequencies use a 12 inch length of $\frac{5}{32}$ inch tube with a $\frac{1}{8}$ inch tube telescoped inside. For 2 meters extend the tube to 19$\frac{1}{4}$ inches or 12$\frac{1}{2}$ inches for 1.25 meters. Add a brass round head machine screw at top to finish it off. [The "lamptenna" should only be used for low power — 5 W or less — transceivers and it should be used with coax long enough to allow yourself to sit a few feet away from the antenna when operating. — *Ed.*] — *73, Bob Evans, WBØSVS, 2253 Norwegian Dr, Apt 25, Clearwater, FL 33763-2904,* **beach_bob@msn.com**

## KEEPING THE CRITTERS OUT

◊I have been putting a new M² KT34M triband beam together and wanted to make sure my neighborhood birds did not build a nest

RON TOYNE, WAØAJF

**Figure 108 — The downspout strainer secured in the end of the boom will keep the critters out.**

in the boom. I looked around for a solution and found that a rain gutter downspout strainer, found in any lumber store, worked perfectly. I drilled a small hole in the boom and used a stainless steel key ring to make sure the screen stayed in place (see Figure 108). I also sprayed the gutter screen with a clear spray just to add some extra weather protection. *— 73, Ron Toyne, WAØAJF, 1220 Hertz Dr SE, Cedar Rapids, IA 52403*, **wa0ajf@aol.com**

## ROTATOR BREAK-OUT BOX

◊Like many hams, I have an antenna rotator on the top of my tower, with the control box inside the house, and in my case, in the basement "shack."

Any form of troubleshooting of the rotator operation is very difficult, usually requiring someone inside to operate the control box and someone outside to verify operation of the rotator. Otherwise, a lot of running back and forth is required.

DON DORWARD, VA3DDN

**Figure 109 — The rotator break-out box mounted to the side of the tower. Note the modified wine bottle cork to seal the bottom cable entrance and the weatherproof seal between the cover and the box.**

A simple solution is to install a rotator "break-out box" at the base of the antenna (see Figure 109) so it's easy to access the rotator connections from the outside. This is a good thing to do at the time of installation. It allows you to easily disconnect your inside rotator control box and substitute another rotator control box outside, where you can see the antenna and communicate with the guy up on the tower.

I used a commonly available rigid PVC ½ inch conduit access fitting, a Kraloy type LB05, that has a removable cover with gasket. A plastic wine bottle cork cut in half served as a plug for the unused ½ inch port. The cork, with two ¼ inch holes drilled into the center also acts as a weatherproof lead-in plug for the rotator wire, as shown.

I used the simple "bullet" style connectors for the controller wires. Note that the controller wires can have 30 V ac on them. To use, simply disconnect the inside rotator control box at the break-out box by pulling the bullet connectors apart and plug in the rotator control box to be used outside, which can be the one from inside the house, if you don't have a spare. Obviously, any weatherproof plastic box could be used and with whatever type of rotator you have (three or four wire, etc). *— 73, Don Dorward, VA3DDN, 1363 Brands Ct, Pickering, ON L1V 2T2, Canada*, **ddorward@sympatico.ca**

## BIG SLING ANTENNA RAISER

◊Sling shot and spinning reel, bow and arrow, and crossbow and bolt all seemed like good ways to get a line up into a tree to pull an antenna into the nirvana above 45 feet.

Ha!

Monofilament and a weight can be projected by a wrist rocket to adequate heights but the weights get themselves wrapped around any handy branch and you can't readily tell where the they end up; they are nearly invisible above 25 feet.

I found crossbow bolts much too light to tow a nylon mason's string any useful height. More than a few yards of both yellow and pink string dangle from the other of my target maples. A bow and arrow is just dangerous. In desperation I even contemplated purchasing a little RC helicopter to fly the cord over the target trees but I couldn't figure out how to release the cord onto the tree top.

Then genius set in — water balloon launchers (WBL). Their ads claim a range of 25 yards to a ¼ mile and one with a modest range is cheaper than any of those other machines I mentioned. I bought a 150 yard WBL for $15 from Think Fast Toys (**www.thinkfasttoys.com**). **Amazon.com** and **www.sillytown.com** also have WBLs. I suspect that any retail toy store with an adult section would have them. They are very popular at the beach. The added cost would have been worth it since, with a softball and mason's twine, the 150 yard model cleared a maple tree about 60 feet high and several other trees behind it.

DUNCAN MORRILL, KB1THS

**Figure 110 — Duncan's first attempt to get that yellow softball over the oak. Note his use of a launcher frame to support the WBL.**

The oak at the other end is taller and I never could get even the baseball over the top of it. After five or six tries I settled for a high branch 10-15 feet below the top at about 70 feet. So I gauge that the 150 yard launcher can reach about 60 feet with a softball, maybe 70 with a baseball. It'll launch a grapefruit with bright pink cord attached over a 60 foot maple tree with ease and a couple of the trees behind it as well. An orange is easily launched over an 80 foot maple in a grove of interfering maple, pines and birches. For my purpose, a fake baseball was adequate. Brightly colored projectiles and line were invaluable. Figure 110 shows my first attempt to get that yellow softball over the oak.

In using the WBL you do need to be careful. On my second launch I was seated on the ground and pulled until I was almost flat on my back. My Super Golden Slingshot backfired — the two stakes holding the launcher's restraining ropes pulled out of the damp ground — and it slammed back at me almost breaking a couple of my ribs. [Careful! As Duncan notes this is an *adult* toy and can cause injury if misused. — *Ed.*] Take note that a water balloon launcher with any appreciable range requires a launch team of two men and a boy (a strong boy, that is), or absent these, build a launcher frame of 1½ inch PVC, as I did.

I've finally gotten the wire, balun, etc, rigged up. Right now the antenna rig is all in rope, essentially a prototype so I can work out the details, like feed line positioning, before committing to wire. Oh, yes, my antenna will be a Len Carson 120 foot Windom (**www.hamuniverse.com/k4iwlnewwindom.html**) running from 55-70 feet. Watch out ether, I'll be on the air momentarily. *— 73, Duncan Morrill, WV1J, Old Kings Rd, Merrimack, NH 03054*, **wv1j@qsl.net**

## CLEANING CORRODED ANTENNA WIRE

◊One day my lawn mowing teenaged neighbor showed up at the door with a puzzled expression on his face and a hank of bare copper wire in his hand. Well, normally I don't need much excuse for some antenna work but this was perfect. I gathered up the remnants and found only a couple of feet that had made too many trips through the mower. I could splice the rest back together but of course it had been in the air for several years and was very corroded. I needed an easier method than the sanding and scraping I was used to, so I was off to the kitchen for some experiments.

I wanted a method that involved no scraping and as little elbow grease as possible. Also, since I'm not a chemist, I avoided mixing ingredients that could cause dangerous results — and I suggest you do the same! Caution — you'll find there are no scientific methods ahead; I just recorded go-no-go results.

Bleach: nada

Ammonia: big effect on me but none on the wire

Mr Clean Magic Eraser: not even a dent

Lemon juice and baking soda: don't bother

Then my spouse suggested lemon juice and salt — bingo!

I used about a half-cup of juice and 20 seconds of shaking the salt into a butter tub. Stir until dissolved. Spread the wire strands out and soak the ends for 20 minutes, swishing occasionally (it helps to tape the wire ends to the tub). Maybe half way through you can help it along by rubbing a little with a paper towel soaked in the juice. You're done when the copper has a slight reddish tinge. It will tin up just fine. — *73, Mark Albert, KB9VKE, 1760 Riverwood Dr, Algonquin, IL 60102,* **kb9vke@arrl.net**

## FORMING ANTENNA INSULATORS FROM PVC CONDUIT

I was experimenting with various wire antennas and found I did not have enough insulators to complete the project. I needed at least 12 insulators for my project and rather than purchase them new, I decided to experiment with making insulators from blue plastic corrugated conduit that I had left over in the shack.

The flexible PVC conduit was first cut to the desired length of about 3½ inches with a ratchet tubing cutter. The ends were then heated with a commercial heat gun for about 1 minute until soft. Next, I used sheet metal locking pliers to crimp the ends, holding the clamp in place until the ends cooled. Finally, the end holes were completed with a ⅜ inch drill. After a short time, I was able to fabricate the insulators for a few cents each. Aside from the ease of fabrication, the corrugated conduit is strong and light weight. Views of the before and after insulator are shown in Figure 111.

MARK LANDRESS, WB5ANN

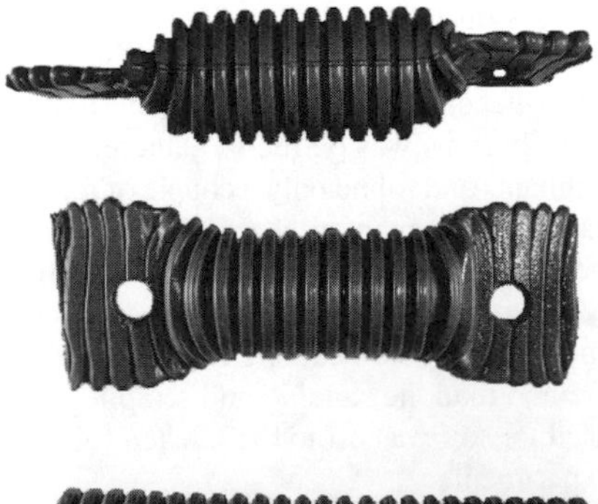

**Figure 111 — Both side and top views of a finished insulator and a piece of the conduit used for their construction.**

Corrugated conduit has many potential uses and over the years I have used the material to construct standoff insulators, ladder line separators, non-conductive bushings and other antenna parts. With a length of corrugated conduit in the shack, you will never be without insulators again. — *73, Mark Landress, WB5ANN, 1011 W31st St, Houston, TX 77018,* **wb5ann@arrl.net**

## WHIP ANTENNA WEATHER GUARD

◊The second week of December of 2010 found most of my part of Tennessee under a freezing rain watch. "No big deal," I thought to myself, as this was going to make for an excellent excuse to sit in a nice warm shack and cruise 40 and 30 meters. Unfortunately, my plans were cut drastically short when I did a quick SWR check before hitting the airwaves and found that there was no combination of transmatch settings that was going to overcome a suddenly high SWR.

A glance out the window confirmed that the antenna was still erect but the weather was bleak. A quick trip to the backyard confirmed my suspicions. Ice had already formed on the antenna bracket, effectively shorting the loaded whip to the grounded base plate. A couple of whacks with the back of a crescent wrench broke off the existing coating of ice. A quick re-check of the SWR proved that the ice had been the culprit. But the precipitation was increasing and unless I wanted to make a trek to the antenna once an hour, I had to do something.

Sitting back down at my desk and pondering my options, I took a long draw from a plastic bottle of my favorite soft drink and sat back for a moment. As I gazed at the bottle, it struck me that the top of the bottle somewhat resembled an umbrella. It was then that I had my weather guard epiphany. Put the umbrella *on* the antenna.

STEVEN ROBESON, K4YZ

**Figure 112 — The soda bottle antenna shroud over the end of the 40 meter resonator.**

I started by taking a pair of paramedic shears and cutting the top off of a two liter soda jug, being careful to make as even a cut as I could. I then removed the loaded whip from my ground-mounted mast and brought it into the shack. My initial thought was to drill a hole in the regular cap of the bottle that was just a hair smaller than the diameter of the antenna, and then force it over the rod, allowing me to "screw" the shroud I'd made from the bottle top onto it.

Unfortunately, no matter how slowly or carefully I tried to do it, the caps kept splitting. [A step drill might do better here — *Ed.*] My next thought was to use electrical tape. I set the antenna through the neck of the bottle top and marked where the top of the shroud sat. From that point, I began to wrap the rod with the tape, gradually building up a "knob" of tape at the point where I wanted the shroud to sit.

I would occasionally reseat the shroud over the end of the whip in order to see how much more tape might be needed in order to obtain a tight fit. It actually took less tape than I thought. I wound up having to unwind a couple of layers to get a snug fit without bunching up the tape as I pushed the shroud over the end of the whip. Once I obtained the desired fit, I taped the outside edge of the shroud to the antenna.

The result was a very tight fitting shroud that actually covered the entire top of the antenna bracket and keeps all precipitation off of the plate (see Figure 112). I am sure this could be done with a smaller bottle (1 liter) for mobile uses and might even be sealable along the edges for mobile installations where the antenna rides close to the ground. — *73, Steven Robeson, K4YZ, 151 12th Ave NW, Winchester, TN 37398-1061,* **k4yz@arrl.net**

## SLINGSHOT TIPS

◊A slingshot is one of the easiest ways to get a line over a tree limb but they don't come with instructions books. After hours of trial and error, I came up with an easy, surefire way to use a slingshot and get the line over a tree limb. Any good adult type slingshot will work; most have pretty much the same pull. These are our three easy suggestions.

First, use at least a 1 oz fishing weight and you can use a 3 oz; it will not hang and it will come down. Use what is commonly called "builder or mason line" from a hardware store; it is a fairly stout line. Be sure you have enough line to go all the way up and down with some extra.

Second, loosely unwind the line into a box; we used a simple plastic fishing box. Anchor the other end of the line so the whole thing doesn't go over the tree limb.

Third, the real secret is not to hold the slingshot right side up but *upside down*. Position yourself over the box and slightly back, this allows the line to hang down and come straight out of the box. When you pull the slingshot all the way back with the fork pointing up the line tends to tangle in the slingshot as it comes out. With the fork upside down it will never hang up.

Finally, do wear safety glasses and watch where you shoot. A 3 oz fishing weight launched from a slingshot could do serious harm. — *73, Stewart Nelson, KD5LBE, 8 Deerwood Dr, Morrilton, AR 72110-4416,* **kd5lbe@arrl.net**

## TRAPPED RADIAL MULTIBAND ANTENNA

◊One of my favorite things to do is head for a campground and do some hamming while communing with nature. The problem was finding the right antenna.

I wanted 80-10 meter capability and something that would handle an amplifier, as you can't expect too much from portable antennas, especially if you're in mountainous terrain. I tried a long wire with a tuner but for high power it was too much equipment to haul around and required a campsite with a couple of hundred feet of space and trees in the right places. I also tried a multiband dipole, but I had wire all over the place, the trees weren't high enough and the SWR was high.

After reading as many reviews as I could find, especially on the eHam website, as well as the antenna "shoot-out" data from the HFpak website (**hfpack.com/antennas**), I settled on a Hustler 5-BTV antenna (**www.new-tronics.com**). I was lucky and found one on Craigslist for $50. I then tackled putting together a pair of radials for each band, which I thought was the minimum setup for getting out a reasonable signal. While the radials worked well, it quickly became obvious that all those radials were just too much to deal with for a portable setup.

It then occurred to me that using just a pair of trapped radials might be the ticket. I found John's, NU3E, article on a 5 band trap dipole (**degood.org/coaxtrap**) and built that for about $35. I raised it and tuned it as a dipole and then laid it on the ground attaching it to the vertical. Voila! It took a little retuning of the vertical but I now have a 5 band antenna that will handle my SB-200 without a tuner for all but a portion of 75 and 40 meters. The 75 and 40 meter bandwidths are a little narrow but if I want to use more of these bands I can run 100 W and use a tuner. The trap dipole/radials are each only 38 feet long and pose no problem at a campsite.

Not having to hang wire in trees is a plus but the bonus here is that if the trees are high enough and in the right place, I can put up just the trap dipole and not bother with the vertical. I attach the trap "radials" to the base of the vertical with a screw and wing nut, and attach them the same way to a piece of Plexiglas fitted with an SO-239 connector for dipole operation. All of this sets up and disassembles easily, another requirement for comfy portable operation.

The trap radials could likely be used with most vertical antennas, but not many could handle the high power and cover five bands while needing no guys and being lightweight. The 5-BTV also disassembles quickly with thumbscrew pipe clamps. For the vertical ground mount, I coupled the 18 inch long, 1 inch diameter antenna mounting pipe to a ¾ inch steel pipe (also 18 inches long) so it would go in the ground a little easier. — *73, Paul Voorhees, W7PV, 10090 Misery Point Rd NW, Seabeck, WA 98380-9784,* **w7pv@arrl.net**

## ADDING 6 METERS TO THE R7000

◊The Cushcraft R7000 vertical antenna has four radial rods just below the bottom trap to enhance 10 meter operation. I decided to try adding another set of rods below them to see if 6 meter operation would be possible — and it was. The SWR is 1.5:1 and the center frequency is about 50.3 MHz with very broadband performance. I made the modification as follows:

I obtained a 4 foot long, ¼ inch diameter aluminum rod from my local hardware store. After cutting it in half, I made two right-angle bends about a ½ inch on either side of each rod's center, giving the rods the look of an elongated "Z" (see Figure 113). I used a hose clamp to attach the two Zs to the antenna about 82 inches above the base radials so as to resemble the 10 meter rods, which are located above the new 6 meter rods. — *73, Neil Klagge, WØYSE, 1102 E 3325 N, Layton, UT 84040,* **w0yse@arrl.net**

## IN-TOWER MOUNTING OF HY-GAIN/CDE ROTATORS

◊The disappearance from the market of 2.0625 inch outer diameter (OD) tubular steel mast has made in-tower mounting of the HAM-V and CD-45 rotators problematic. Other models using a fixed mast clamp on top of the bell housing also have difficulties. These old designs were intended to clamp and rotate a 2.0625 inch mast, perfectly centered in the tower. Even with the 2 inch OD masts sold by some amateur dealers, shimming will still be required for the mast to turn on the rotational center of the rotator.

One popular mast choice for HF beams mounted a few feet above the tower top is common 1½ inch galvanized steel water pipe, in either Schedule 40 or the thicker walled Schedule 80. Nominally called "inch-and-a half," it's actually 1.9 inches OD. Again, these will require a shim placed into the rotator's V-clamp; otherwise, the rotation will be an ellipse rather than a circle and cause binding of any thrust bearing or tower-top sleeve bearing.

[Caution, steel water pipe is not designed for mast applications and may not provide sufficient bending strength for your application. For more information refer to the "Mast Strength" discussion in the *Antenna Book*.][2]

To calculate shim thickness, subtract the actual mast OD from 2.0625 inches then divide the result by two. For 1½ inch galvanized steel water pipe: 2.0625 – 1.9000 = 0.1625/2 = 0.08125 inch shim. Then the question is where to obtain the shim stock economically in a small quantity?

Home Depot and similar home improvement stores stock small galvanized steel reinforcing plates used in framing wood homes. Simpson Strong-Ties is one brand available at Home Depot in the lumber section. One each of the Strong-Ties LSTA12 (20 gauge) and LSTA36 (18 gauge) will provide enough 1.25 inch wide stock to cut two shims of each gauge stacked, the 20 and 18 gauge shims will measure 0.083 inches, less than 0.002 inches from optimum.

Fit both stacks into the V-clamp; bosses cast into the clamp where the U-bolts pass through will not allow the shims to lie flat, so mark the shims to later notch them with a nibbling tool or grinder. When the shim

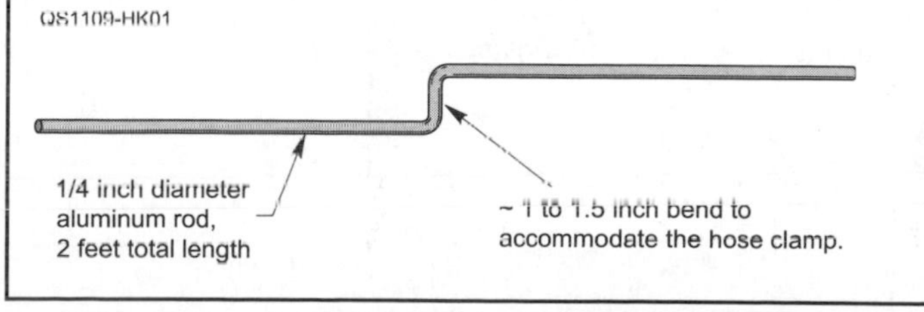

QS1109-HK01

1/4 inch diameter aluminum rod, 2 feet total length

~ 1 to 1.5 inch bend to accommodate the hose clamp.

**Figure 113 — A mechanical drawing showing how to prepare the 6 meter radial rods.**

[2]D. Straw, N6BV, Ed., *The ARRL Antenna Book* (Newington: 2007), pp 22-22 — 22-23. Available from your ARRL dealer or from the ARRL Store, ARRL order no. 9876. Telephone toll-free in the US 888-277-5289, or 860-594-0355, fax 860-594-0303; **www.arrl.org/shop/**; **pubsales@arrl.org**.

stacks lie flat in the "V," insert the mast, assemble the clamp with the shims in place and tighten the U-bolts. Total cost for material is under $4. — *73, Robert Wheaton, W5XW, 16015 White Fawn Dr, San Antonio, TX 78255-1042*

## SUPER SIMPLE GROUND PLANE

◊ The trick to making the Super Simple Ground Plane is to use the inner metallic tube and screws of a terminal strip. The only skill needed is the use of a screwdriver — no solder, no flux. Before I came up with this idea, I soldered the wires to the SO-239, but this approach makes the antenna difficult to store or disassemble.

The advantages of this technique are that you can assemble and disassemble it with just a screwdriver, in less than a minute. Several sets of wires for your favorite bands can be prepared so you can get on the air quickly. Experimenting with the length and number of radials is simple. The whole antenna can be stored in a PVC pipe in the trunk of the car ready for use at a moment's notice. Even a hiker can put it in their backpack and have fun with a handheld transceiver by hanging the antenna from a branch. Finally, it's cheap!

PEDRO MOTILLA, EA5BFT

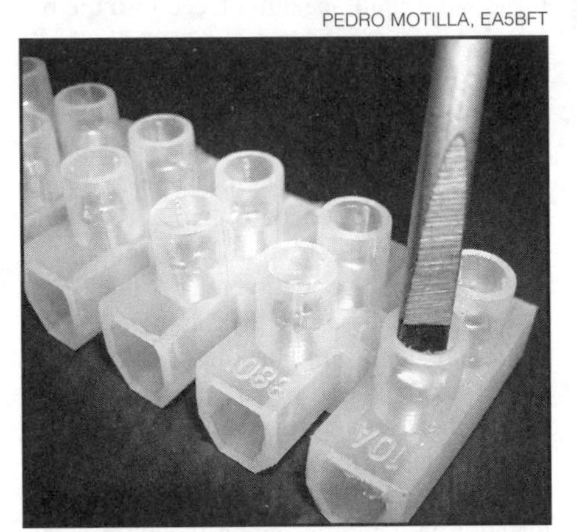

**Figure 114 — Remove the screws from the connector tube to free it from the plastic housing.**

PEDRO MOTILLA, EA5BFT

**Figure 115 — The connecting tube and screws removed from the terminal strip.**

PEDRO MOTILLA, EA5BFT

**Figure 116 — Use the screws to attach the tubes to the base and center post of the SO-239.**

PEDRO MOTILLA, EA5BFT

**Figure 117 — Bend and trim the radials until you get a match and your antenna is ready to go.**

### *Construction*

1. Cut the radial and vertical element wires to the lengths for the band you want to use. Leave an extra 1 or 2 centimeters to trim for best SWR. In my experience for VHF the element lengths should be about 49 cm and for UHF about 16.3 cm.

2. Remove the metallic tubes from a five position terminal strip by taking out the two screws (see Figure 114) and sliding the tube free of the plastic housing (see Figure 115).

3. Using one screw, attach one tube to each mounting hole of the SO-239 (see Figure 116) and one to the center pin.

4. Insert the radials into the tubes at the base and tighten the other screw. Do the same for the vertical element. For the radials, bend them to an angle of approximately 120° with the vertical element (see Figure 117).

5. Finally, connect your transceiver and SWR meter, trim the radials for best match and you are ready to go.

I should note that this approach is good for portable or temporary use but long exposure to moisture and water will create corrosion problems degrading the antenna's performance. — *73, Pedro Motilla, EA5BFT, Urb Pedralvilla, Buzón 107, 46169 Olocau, Spain*, **pedrolo2@pedrolo.com**

## ANTENNA SUPPORT FOR MILITARY MASTS

◊Setting up for Field Day I ran into a small problem when I realized that all my mast rings for the guy ropes were at home — about 40 miles away. So after a quick run to the local Home Depot I was able to create a solution for guying the masts.

I used the following parts: a 1¼ inch PVC coupler (use a PVC coupler that has a ridge on the inside to hold the coupler on the male end of the pole), a three pack of spring links and an appropriate size metal hose clamp.

To prepare the guy ring, open the hose clamp fully and slip the three spring links onto it making sure they are facing in the same direction. Then hook the hose clamp back together again. Slip this over the PVC coupler and start to tighten down. Arrange the spring links at 120° intervals around the coupler and finish tightening the clamp (see Figure 118). Then slip the coupler over the male end of the pole seating the female end of the next section on top.

I have used this both as a replacement for guy rings and to attach a pulley at the top of some joined masts. I do have to stress two things though: First, some hose clamps can have *very* sharp edges, so wear gloves. Second, this is a *temporary* means to support these masts. Replace it with proper hardware as soon as you can. — *73, James French, W8ISS, 1811 Horger St, Lincoln Park, MI 48146-1424*, **w8iss@wideopenwest.com**

JAMES FRENCH, W8ISS

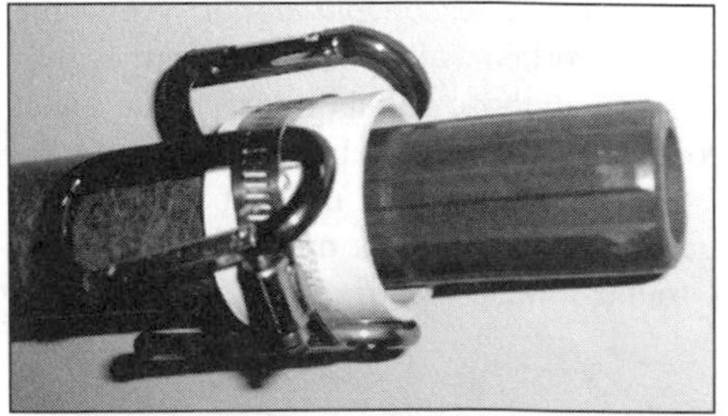

**Figure 118 — Here is the temporary guy ring installed on the male end of a fiberglass mast section.**

# Chapter 9 Operating

## AN ALTERNATIVE HOMEBREW IAMBIC KEY

◊ Sometimes, when you see a person—or an article for that matter—something "clicks inside you." If it's a person, you might end up married. If it's a ham-radio article, you hit the workbench and start building. That's exactly what happened to me when I got last February's *QST*. "Homebrew Your Iambic Keyer,"[1] gave me the final push, and I headed for the workbench, fast!

The results you see in Figure 1 probably represent my personal record: from reading to "paddling" code in less than two days! This is a delightful project requiring simple parts and simple tools. The result was well worth the modest effort. I built only the paddles, because my IC-821H has its own internal keyer.

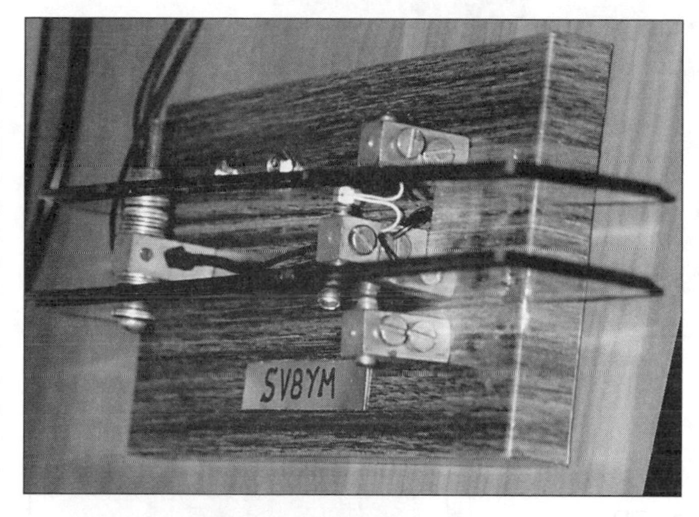


**Figure 1—Construction details of the paddles.**

Every man keeps his own treasure box. I must confess that I had always wanted to build a set of paddles, so I had put aside some odd bits and pieces that were patiently waiting to be transformed into a prince—err—"paddles." I had even found the near-perfect base. It came from an old desktop calendar, and it was just the thing for the paddles, heavy (it is filled with concrete!) and nicely finished.

If you haven't saved a similar calendar, you may use whatever material you can work with. Wood, marble or metal would make a nice base, provided you could cut it to the dimensions required and drill several holes in it.[2] I must admit, if I could work with a copper or bronze block of the proper size, I would have done so!

Whatever you choose for a base, it must be heavy enough to stay put. Four self-adhesive anti-slip pads at the corners are very helpful! Those pads can be easily procured at a supermarket.

But the thing that got me going was … the *rulers*. VU2PGB used two plastic rulers that served as the paddles, and I thought it was a great idea, simplifying the project greatly (in place of the metal bending I was getting ready to do). I selected two sturdy-looking clear plastic rulers (3 mm—about ⅛-inch) thick at the local supermarket. More advanced builders could use clear plastic sheet, cutting the material to the required dimensions—even shaping it to a desired form. That's a bit more trouble, though. You may even get two plastic isosceles triangles, to build the first pyramid or club-sandwich-like paddles in history! Okay, I am joking.

**Figure 2—Back-post details.**

Those who have read the original article know that VU2PGB used a spring between the rulers, to get the required mechanical tension against the end posts. After a little experimentation (and a *lot* of looking for the spring that tended to "piiiiing" out of sight during my tests), I found that I could do without it, because the spring's action could be created in another way. The round washers at the back post accomplish that end by separating the rulers more at the back than at the front. After tightening the screw and nut (see Figure 2), the rulers tend to press against the end-stops at the front. By adding more round washers between each ruler and the back post, more tension is created. You can create the exact "feel" that you like! If you don't want to go this way, you can always use the spring, per VU2PGB's instructions.

The nice metal posts you see serving as contact supports and adjustable end stops are parts of special switches used in electromechanically controlled lift (elevator) installations. I was very lucky to find them! Each switch has a movable contact and two adjustable fixed contacts. I used the fixed contacts for my project and returned the rest of the switch assembly to my ever-hungry junk box. I am sure you can locate similar switches without too much trouble, although building the required four posts shouldn't actually be very difficult if you start with square aluminum rods. The close up photo (Figure 3) should give you some ideas. You can clearly see how one of those posts could be fabricated. Drill and tap them for the hardware you want to use. Their general dimensions appear in Figure 4. The contacts themselves can be fabricated by cutting short lengths (about 15 mm) of round silicon bronze brazing rods. These are The same ones we build antennas with! The required diameter is 5 mm. Remember to file their ends smooth and then round them a little with a fine file or sandpaper. The "*perfectionistas*" can even polish them by rubbing them many times on a piece of thick paper, until the ends look shiny (I skipped that part).

[1]N. G. Praveena, VU2PGB, "*Homebrew Your Iambic Keyer*," *QST*, Feb 2005, p 40.

[2]Marble can be worked (very slowly) with carbide-tipped drill bits intended for masonry, carbide-abrasive jig-saw bits and power sanders (disc and belt). Gloss-finished surfaces will require lengthy sanding and polishing with a progression of grits.

**Figure 3—Front-side details.**

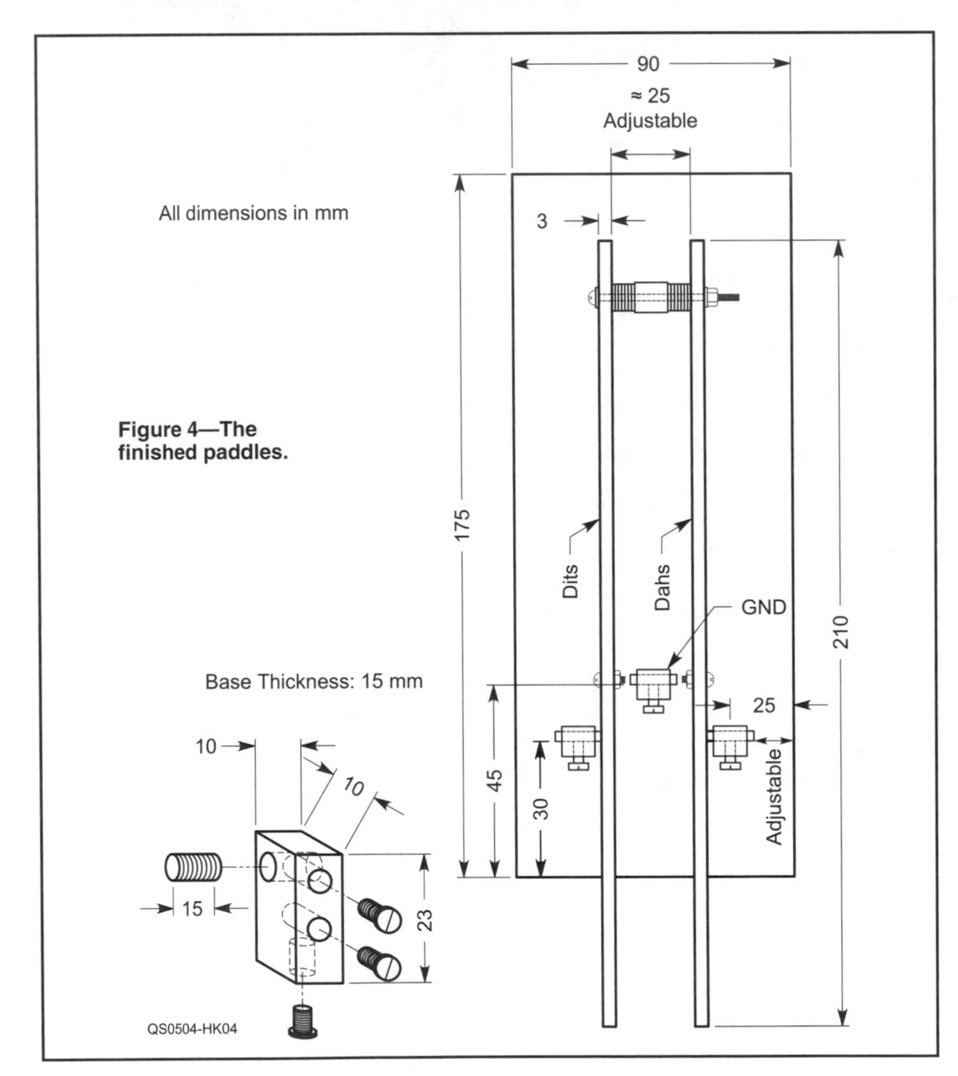


**Figure 4—The finished paddles.**

Although two screws are shown in the end posts, only one is actually required. I just left them there for the looks.

The post between the rulers serves as a contact support and the ground connection. The adjustable contact can be centered between the contacts on the rulers, providing smooth action. The movable contact is secured by tightening the top screw. The lower screw at the central post is used for the connection to the ground conductor of the cable.

The contacts on the rulers are made with two screws and nuts. The nuts are located at the "inner" side (between the rulers). The screws are filed flush to the nuts, so as to form a smooth surface to act as a contact. If you can locate small brass screws, so much the better; they won't rust and their golden color looks nice. In addition, they can be filed and smoothed much more easily than steel screws! The "dit" contact is usually on the left paddle. You may connect them any way you wish. Just wrap the wire once around the screw between the ruler and the nut and tighten the nut. For the connection to the radio, I use miniature shielded audio coax cable, connecting the braid to ground. In my design, I routed the coax through a hole in the back post and secured it with a tie-wrap at each side of the post (see Figure 2). The shielded cable provides EMI protection for some RF-sensitive keyers. Use the plug required by your setup.

After the building and wiring are complete, the paddles must be adjusted. I found it best if the contacts on the center (ground) post are first adjusted to protrude symmetrically from the post. Then, the contacts on the left and right posts are adjusted to push the rulers to the required distance for a smooth action. The distance between contacts is actually very small! It's roughly three tenths of a millimeter (0.012 inch) in my paddles. Take your time with the adjustments and you'll be rewarded with paddles that work wonderfully!

In case you're wondering, the two shiny screws at the back of the base, visible along the right paddle in Figure 1, serve only to cover two holes that existed there!

I am not a terribly proficient CW operator. For the last 20 years, I have used straight keys, exclusively, for my relatively few CW contacts. Now, however, my faithful old J-37 will be getting many more days off, because my paddles are soooo comfortable.

My fellow *QST* authors, please accept my sincere congratulations on the technical projects in *QST* —in this era of "off-the-shelf" equipment, such articles prove that creative radio amateurs can still build much of their equipment and have a ton of fun! Happy homebrewing!

I would be happy to answer any questions regarding this project! Just drop me a line.—*Tasos Thomaidis, SV8YM, 9-11, Therianou str, Zakynthos, GR 291 00, Greece;* **sv8ym@raag.org**

## DX CLUSTER MACROS

◊ I want to share this with *QST* readers. It is a kind of "how to" thing that was new to me. There are AR-Cluster commands for creating and managing macros that are pretty cool. They work great on my Mac and should work just as well on a PC. (I don't own a PC, so I haven't tried macros there.)

First, I use NI5V's *NetDXCluster* software on a Mac G4 to connect to the AB5K-2 AR-Cluster Telnet port in Bee Cave, Texas, via a DSL connection. I use some embedded macros over and over to check for RTTY and PSK spots. The macros are commanded with a mouse click. All I need do is pull down the

**Table 1**
**A List of AA5VU's *DXcluster* Macros**

```
AA5VU de AB5K-2 28-Jun 1354Z arc >_
Macro0 set to: sh/heading_
Macro1 set to: SH/ART
Macro2 set to: _
Macro3 set to: _
Macro4 set to: _
Macro5 set to: _
Macro6 set to: sh/dx/10 where comment like '%PSK%'_
Macro7 set to: sh/dx where comment like '%PSK%'_
Macro8 set to: sh/dx/10 where comment like '%RTTY%'_
Macro9 set to: sh/dx/30 where freq > 14060.0 and freq <14120.0_
```

Filter menu and select the command I want with a mouse click. (The current version of NI5V's *NetDXCluster* does not provide for user defined macros, so I use the Filter options as a work-around macro template.)

When I mentioned using embedded RTTY and PSK macros to Terry, AB5K, he suggested I try the new macros added to the A*R*-Cluster software. I was not aware that the AR-Cluster software had built-in macro support until Terry pointed me to the docs at: **www.ab5k.net/ArcDocs/UserManual/ArcPersonalization.htm#macros**.

Terry then provided the following explanation and an example using one of my cluster commands:

"————*AR-Cluster* has 10 programmable macros. Macros are user defined shortcuts. The 0-9 keys are defined as macro keys. If the first character of a user command is a 0 -9, the character is replaced with its macro definition.

In Dick's example, rather than type the command 'sh/dx/30 where freq > 14060.0 and freq < 14120.0' he can set up a macro by typing 'set/macro 9=sh/dx/30 where freq > 14060.0 and freq < 14120.0' and the next time simply type a 9 (followed by the Enter key)

————"

I tried Terry's tip and find it is much easier to type the number than to pull down a menu and click a macro to start it.

For your information, Table 1 shows the results of a "sh/macro" command to display my current macro settings.

If you are an AR Cluster user, you may want to experiment with macros. They are pretty cool and save typing and mousing effort. The macros I use most are 0, 6 and 8 (see Table 1).

For those unfamiliar with NI5V's *NetDXCluster*, it has a data entry (type in) box for user interaction with the cluster. One of the standard commands I use most is SH/HEADING, to find the name and location of a strange spot on the cluster display. With the cluster macros, all I need do is push the zero key, type in the prefix or call (for example, 7Q7PF) and push ENTER or RETURN. The cluster response is: `Country: 7Q = Malawi 81 deg (LP 261) 9279 mi (14,929 km) from AA5VU_`

The "ART" command (my macro 1 above) does a call sign lookup. If I push the "1" key, type in a call sign and push ENTER or RETURN, the cluster responds with my station data.

To find a AR-Cluster node near you, go to **www.ab5k.net/ArcNodeList.aspx**. Contact your local AR-Cluster sysop for help, if necessary.—*73, Dick Kriss, AA5VU, 904 Dartmoor Dr, Austin, TX 78746-5163;* **aa5vu@arrl.net**

◊ Here's a useful tip for Web surfing. When you paste or enter a complete (street, city and zip) address into Google's search box, Google will return a link to a map display of that address.—*AA5VU*

## A MEMORY FREQUENCY STORAGE SCHEME

◊ I have an Alinco DR 235. My only complaint was that there is no easy way to check someone on the reverse. Most radios have a button that you can press but not the Alinco. My solution was to program all the repeaters into even numbered memory locations. That left the odd numbers unused. The repeater in location number two can have the input frequency programmed into location number one. It would be a simple matter to rotate the knob one step to check the input. I put the odd numbers into skip mode so only the repeaters are in scan. —*Pete Ostapchuk, N9SFX, 59425 Apple Rd, Osceola, IN 46561-9393;* **n9sfx@aol.com**

## KEEP THAT LIST READABLE AND READY

◊ 3M Corporation business-card protectors are handy for keeping a list of repeaters and simplex frequencies programmed into your handheld transceiver, as reference or backup. Laminate the list (folded to hold more data), punch a hole, and keep it on your ID lanyard. For more information about the products see **www.3m.com/office**. — *John Griswold, KK1X, 34 Cambridge St, Ayer, MA 01432*

## LEFT-HANDED KEYER PADDLE

◊ Here may be an idea of interest to left-handed CW operators — a simple adapter I wired when teaching a left-handed person how to send CW with a keyer.

For right handers, the convention is dots on the thumb, apparently because of its agility. Some "southpaws" learn on right-handed paddles, but when teaching a local ham to use a keyer, I thought I would give him the advantage of the thumb on the dot paddle. Changing the wires at the paddle required disassembly. To get around this problem I wired an adapter with a three-wire shielded phone plug on one end and a similar jack on the other end. The ground is common, but the tip of the plug connects to the ring of the jack, and vice versa. By plugging the paddle into the adapter and plugging the adapter into the keyer, the dot and dash wires are interchanged. Now switching from right to left-handed keying is as simple as adding an adapter. — *Bill Conwell, K2PO, 6224 SW Tower Way, Portland, OR 97221;* **arrl@conwellpdx.com**

[I suggest that right-handed operators learn to send with their left hand and vice versa. That way you don't have to lay down your pen or pencil to reach for the keyer paddle. I had wired a DPDT switch into the line from my paddle to reverse the dot and dash leads, but this adapter is easier to wire. — *Ed.*]

## PADDLE HINT

◊ Following up on the editor's comment to K2PO in the July 2006 Hints & Kinks column about right-handed operators learning to send with their left hand to avoid dropping a pencil in the right hand — you already have the skill; you just don't know it.

Right-handed ops, take the paddle and turn it around 180° so the back of the paddle faces you. Place your left hand on top of the paddle, so your left thumb rests on the dit paddle and your left index finger on the dah paddle.

Start sending CW. Amazing, isn't it? You will be able do it with virtually no practice required. This is an ability now referred to as an obsolete contester skill. In the days before computer logging it would allow a CW contest operator who was paper logging to send with one hand and log with the other while minimizing hand movement. — *73, Scott Robbins, W4PA, 3225 Whittle Springs Rd, Knoxville, TN 37917;* **w4pa@yahoo.com**

## HEARING AIDS AS HEADPHONES

◊ Hearing loss is a fact of life for old hams like me. My spouse persuaded me to get hearing aids, but the hearing aids caused feedback when I used my regular headphones. How was I to operate the rig without annoying my spouse with sounds of CW emanating from speakers? It was too much trouble to remove my hearing aids every time I used headphones.

## T-Coil to the Rescue

My hearing aids came with T-Coils — magnetic pickups designed to enable telephone conversations. Would this work with my rig?

Large magnetic loops are part of many public address systems, so I decided to experiment with my own. My junk box contained a scrap of eight conductor rotator control cable about 20 feet long, which I configured as an 8 turn loop. By trial and error, I determined the best location for this audio amplifier fed loop — the ceiling of my shack. The input of the amplifier is plugged into my rig's phone jack.

My friends in the Tennessee Contest Group now call me "The Bionic Man" because I hear my rig without benefit of connecting wires. Also note that hearing aids with bluetooth technology are available and could be a better alternative.

## Special Considerations

1. T-coils in hearing aids are very directional so sometimes I must adjust the position of my head to avoid nulls.

2. Both ears respond to the same magnetic signal, so stereo (for two radios) is not possible with this configuration.

3. A CRT computer monitor also responds to a magnetic signal, so to avoid squiggles on my screen the output coil from the amplifier needs to be separated from the monitor and the power kept at a minimum.

4. Current-output type audio amplifiers work best for feeding wire loops — but these are expensive. I decided to use an amplifier I already had and not to worry about tuning and matching. — *73, Bill Hall, K3CQ, Apt 314, The Saint Paul, 5031 Hillsboro Rd, Nashville, TN 37215,* **k3cq@bellsouth.net**

## HANDY FREQUENCY LIST

◊I found that a simple frequency chart attached to the back of my microphone can help me remember where the various phone bands are while driving. It's very easy to make and very easy to use. Set up a table of the frequencies in a program like *Excel* and print it out. Cover it with clear packing tape, and then cut it out. You can then mount it on the back of the microphone with another piece of packing tape — simple as that. — *73, Andrew Thall, K2OO, 3709 Calle Chiapas, Laredo, TX 78046,* **thall@thall.net**

## QSL WIZARD AND GLOBAL QSL

◊My previous *QST* article discussed the free *QSL Maker* software to print excellent QSL cards at home.[1] This is still a great program but I have continued my search for the perfect way to QSL and I want to share two new methods.

There is an outstanding new shareware program to print QSL cards at home called *QSL Wizard* available from Alpine software. *QSL Wizard* lets you create fully customizable QSL cards with ease. This program utilizes point and click techniques and has a much different feel to the user than *QSL Maker*. Both programs import *ADIF* electronic logbook data to print your contact information directly on your cards. Anyone who wants to print their own QSL cards should evaluate both the *QSL Maker* and *QSL Wizard* software to see which program you prefer. Both will allow you to print cards at home that can rival a commercial QSL card. The *QSL Maker* program is available at the ARRL Web site **www.arrl.org/files/qst-binaries** and at **http://qslmaker.mi-nts.org**. *QSL Wizard* can be found at **www.alpinesoft.com/asqslwizard.html**.

Global QSL is a new way to send QSL cards via the DX bureau. Global QSL uses electronic submission of contact data such as eQSL and Logbook of The World (LoTW). Instead of an electronic QSL confirmation, Global QSL will print a beautiful custom double-sided full-color glossy QSL card with your specific contact information. This is done on heavy cardstock that is nicer than I can print at home. Global QSL not only prints your card but they also send it directly to the foreign DX bureau so you don't have to mail anything to your outgoing DX bureau.

To use Global QSL, download their free *QSL Graphic Editor* software to develop your custom QSL card. Their software includes a variety of images such as the ARRL logo and pictures of many radios that you can incorporate. You can also use your own images. You can update your card whenever there is a change in your station. There are online tutorials that teach you how to use the software. If you don't want to design your own QSL card then they will be happy to design the card for you. The Global QSL Web site is **www.globalqsl.com**.

Sending your cards from Global QSL could not be easier. You upload your *ADIF* logbook file (or you can manually add contact data like eQSL) to their Web site after logging in and they will print your cards. Your QSL card design is kept on their computer and your contact data is inserted when they print the cards. They print the cards when they get 500 cards for a specific DX bureau but they will not wait more than 2 months to print your cards. An active DX bureau may get cards printed and mailed every week. I think the cards go out as fast or faster than via the outgoing bureau. You can send the cards to specific people such as a QSL manager by using the "QSL via field" in the *ADIF* data.

Global QSL is used for DX cards since the cards are sent to a DX bureau. Global QSL should not be used by US hams to send cards to other US hams since the ARRL incoming bureau is not designed to handle QSL cards from US hams. Global QSL will not print a card from a ham in the continental USA to another ham in the continental USA unless a foreign call sign is involved such as a foreign QSL manager or a DXpedition. They will print cards that do not involve the continental 48 states, that is, Hawaii, Alaska, Puerto Rico, Guam, etc.

Global QSL costs $12.50 for 100 QSL cards, which includes printing and mailing. I can't buy the cardstock, print the cards using color ink and mail them for this price. A thousand cards cost $99 for printing and mailing. The 100 card offer is a great way to try the service.

I think you will find that the *QSL Wizard* software, the *QSL Maker* software and the Global QSL service all make it easy to send out beautiful QSL cards that are likely to cause the recipient to return the courtesy. These techniques can create a beautiful card that anyone would be proud to display. — *73, David Rabin, W9PH, 1330 Nyoda Pl, Highland Park, IL 60035,* **w9ph@arrl.net**

## HELP FOR THE LOW VISION HAM

◊I am writing this article to help and encourage my fellow hams with vision problems. I was first licensed in 1956 as KN8DIN in Gladwin, Michigan. I currently live in Minnesota with the call WØPSH.

I was diagnosed with macular degeneration in 1997. The doctor informed me that I was now legally blind and I started feeling a little down. As working DX was a big part

LARRY LENNON, WØPSH

**Figure 5— An eyeglass loupe made by Bausch & Lomb. It's called a jeweler's double eyeglass loupe p/n 81-41-78. There are many types of loupes available.**

LARRY LENNON, WØPSH

**Figure 6 — The telescope is an accessory affixed to the eyeglass lens by an optometrist who will be able to recommend the best one to meet your needs.**

[1]D. Rabin, W9PH, "QSL Maker," *QST*, Jul 2008, p 68.

of my ham life, (only three more countries to work). What will I do now?

First I tried hand magnifiers for reading but that took one hand to hold the magnifier and the other hand to hold what I wanted to read. I wanted a better way to read anywhere I happened to be.

I searched around and found a jeweler's loupe that you clip onto your glasses (see Figure 5). The loupe I selected has two lenses — one is 3× power and the other is 6× power. With the 3× lens I can see my computer screen (I'm using it to type this article) and I use it to take pictures with my digital camera. With both lenses I can read the newspaper and *QST*.

Now how do I see my transceiver? After doing some research, I contacted an optometrist and purchased a pair of glasses with a telescope mounted in the lens for my best eye (see Figure 6). Now I can see my radio to enjoy chasing DX and can fill out my log.

I still have some problems, as my ICOM IC-775DSP could use more contrast between the color of the case and the lettering for the buttons. Once in a while I forget to press the SPLIT button when chasing DX and I apologize for the problems this causes.

These are only a few of the things that visually impaired hams may do to enjoy life a little more. There are many places to find help for visual problems. The Internet is a good place to find low vision equipment. If you are a veteran, contact your local Department of Veterans Affairs and talk to the Visual Impairment Service Team (VIST) Coordinator. The VIST Program is designed to help veterans who are severely visually impaired and legally blind. Your vision problem does not have to be service connected. This is a great place to get help and talk to others with similar problems.

Don't let visual problems stop you from enjoying your life. I look forward to hearing you in the pileups. *73, Larry Lennon, WØPSH, 10725 Dahlia St NW, Coon Rapids, MN 55433*, **w0psh@arrl.net**

## USING A BROADCAST LOOP ON 160 METERS

◊The Terk Technologies AM Advantage tunable loop antenna does a great job as a receiving antenna for the 160 meter band by just setting the tuning dial as high as it will go. Along with an inexpensive Ten-Tec Model 1056 direct conversion receiver kit that I assembled for 160 meters, I was able to give our grandson a fully functional Amateur Radio receiving station, without an outdoor antenna or any of the "wires and things" that his parents might object to. The Saturday evening AM broadcast on 1860 kHz, nightly W1AW code practice and bulletins, and cross-country sideband roundtables, are all received very reliably and "Q5" with this desktop receiving system. Best reception is in the plane of the loop; the directivity helps eliminate noise. The antenna is available from Amateur Electronic Supply, C Crane and possibly others, for about $50. — *73, Dean Lewis, W9WGV, 1193 Azalea Ln, Palatine, IL 60074*, **w9wgv@arrl.net**

## KEYER PADDLE BASE FOR THE HEAVY-HANDED

◊I must confess that in the middle of a CW contest, when I have to use the keyer paddle, I am heavy-handed. So much so that I often slide the keyer paddle sideways, because the base is not heavy enough to resist the first hit.

I have an N2DAN Mercury paddle, which is a dual lever type for iambic keying. I learned electronic keying in the early '60s when there were no iambic keyer squeeze paddles, only single lever paddles that required you to push the lever back and forth to make dits and dahs. A pal who learned the same way told me he went back to an old Vibrokeyer single lever paddle (**www.vibroplex.com**) to take advantage of his hard-learned muscle memory, because he never quite got the hang of iambic keying. I felt the same way.

I tried an old Vibrokeyer single lever paddle, but it slid around the table, too. I bought a nice new Begali Simplex (single lever) paddle (**www.i2rtf.com**), which works great, but in the heat of battle I started sliding that one, too. I had to do something to stabilize the situation.

I had created a circular steel base for my portable low power rig's mini-paddle. I found an online supplier of die-cut steel disks about ¼ inch thick and 4 inches in diameter. The disk was a little rough, so I took it to a local powder coating place and had them sand blast and powder coat it red. I thought this base might be heavy enough (13 ounces) to stabilize the Begali key.

The little magnets on the bottom of the mini-paddle started me thinking: What if I got stronger magnets to mount the Begali atop the red steel disk? The Begali's base is ferrous metal (that is, magnetizable) so I looked online and found a supplier of neodymium magnets (K&J Magnetics, Inc; **www.kjmagnetics.com**) slightly smaller than quarter coins, which exert a force of 46 pounds each when placed between two steel plates.

I flipped the Begali paddle over, used quarter coins to see how many circular magnets would fit on the bottom, then ordered five neodymium circular magnets ⅞ inch diameter × 3⁄16 inch thick (the Begali could have accommodated six magnets). At 3⁄16 inch thick (your keyer paddle may be different), the paddle's height would not be raised too much by the magnets. The magnets' thickness was just enough to raise the Begali's four corner feet above the base, so the magnets would be making contact between the Begali and the circular base, exerting the magnets' full force between them.

I also wanted to put a thin layer of cork under the red circular steel base, so I ordered a small sheet of 3⁄64 inch thick cork from a music store (**www.MusicMedic.com**), traced it around the circular base, cut the cork circle out slightly smaller so it couldn't be seen at the edges of the red base and secured the cork with strips of double-sided plastic carpet tape (ACE Hardware #50106).

Ta-dah! It worked like a charm. Now I can slam the keyer paddle as hard as I want and it doesn't budge. If you are even heavier handed then I, you can look to **www.americanmorse.com**, which sells a rectangular, black steel base that weights 32 ounces. That, plus the original weight of the keyer paddle should satisfy the heaviest-handed of us.

Note, the neodymium magnets are *very* strong and come with all kinds of warnings about not getting your fingers caught between them and keeping them away from ferrous metal objects, magnetic computer media and all electronic appliances, TVs, radios, DVD players, VCRs, etc. The entire steel mass of the key paddle and the steel base become strongly magnetized, so keep them away from everything else in the shack. — *73, Larry Serra, N6NC, 750 B St, Ste 3300, San Diego, CA 92101-8188*, **n6nc@arrl.net**

## ACCURATE ZERO BEATING USING THE THREE-OSCILLATOR METHOD

◊When an oscillator is adjusted so it zero beats with WWV or other standard frequency transmissions, much comment has been made over the ability to approach true zero-beat. When the harmonic is directly zero-beat to the standard, the stated accuracy is generally in the 1-5 Hz range.

There is a technique that allows one to repeatedly zero-beat to a much higher accuracy. The method is called the "Three-oscillator Method" and dates back to the 1930s, or earlier. The earliest discussion I have found was on page 47 of Bulletin 10, *Frequency Measurements at Radio Frequencies*, published by the General Radio Company in February 1933. The bulletin states that the "method has been in use for a number of years…" The technique is also presented in the 1956 Technical Manual (TM11-2665) for the AN/URM-18 Frequency Calibrator Set. More recently, Alan Melia, G3NYK, reports an accuracy of 0.1 Hz using the same technique, **www.alan.melia.btinternet.co.uk/freqmeas.htm**.

The three oscillators are the standard, the unknown and either another less accurate oscillator or a receiver beat frequency oscillator (BFO). The AN/URM-18 and the General Radio 1100-A frequency standards utilize regenerative receivers. Using reception of WWV as an example, in normal practice the unknown signal is adjusted to zero-beat with WWV by injecting a sample of the unknown into the antenna of an AM receiver tuned to a WWV transmission. As the unknown is adjusted to match WWV, a beat frequency will be heard that approaches 0 Hz or zero-beat with the WWV transmission.

Unfortunately, the audio passband of the receiver and the observer's ear limit hearing a beat frequency much below 10 Hz. It is possible to reach closer (lower) beat frequencies by listening to the background noise wax and wane but the results are not readily repeatable.

Now, a third source is introduced when the receiver BFO is turned on or the regenerative receiver is adjusted to oscillate. With the unknown source temporarily disconnected, the receiver is tuned to give a nominal 1 kHz beat frequency while receiving the WWV transmission. When the unknown source is once again added, the 1 kHz beat will wax and wane at a rate equal to the beat between the unknown source and the WWV transmission. Changing the BFO or receiver tuning only changes the frequency of the tone that waxes and wanes. The waxing and waning rate is determined solely by the beat between the WWV transmission and the unknown source. It is now easy to reliably adjust the unknown, or its harmonic, to within a fraction of a hertz of the WWV transmission. — *73, John M. Franke, WA4WDL, 4500 Ibis Ct, Portsmouth, VA 23703*, **jmfranke@cox.net**

## DUAL-CHANNEL CW FILTER

◊In the old days of very wide receiver passbands we used "Q Multipliers" to peak a CW signal. This gave nowhere near the steep-skirted selectivity of modern DSP (digital signal processing) or SCAF (switched capacitor audio filter) filters, but would lift the signal out of the noise background enough to copy.

Today, many of us are using a very selective filter, which may cause us to miss a call 1 kHz or less away. If you have a good outboard audio filter of any type, it is easy to have the best of both worlds using the technique that follows:

First, obtain a pair of used computer speakers at the local thrift store. Then plug a "Y" splitter into your rig's audio output jack. Connect one lead to the unpowered speaker, the other to your external filter. Plug the powered speaker into your filter.

Select a CW signal to copy and place it in the center of your audio passband. Adjust the volume of the powered speaker until the sound is centered.

You now have a highly filtered, steep-skirted channel and a 2.1 kHz unfiltered channel mixing in the ultimate audio device — your brain. The desired signal will be head and shoulders above the noise but you will still be able to hear off-frequency calls. — *73, Norman Sullivan, NZ5L, PO Box 55, Beach Lake, PA 18405*, **nz5l@arrl.net**

# Chapter 10 Around the Shack

## FEED LINE ENTRANCE TIPS

◊ Those of us who operate from cellars, attics or garages need cable entrances that meet the electrical and mechanical needs of the cable(s), are weatherproof and insect proof. Let's look at some old and one new way to do this. The new way is stealthy and the interior portion is easy to cover when you sell the house.

The most direct route is to simply drill a hole in the wall (see Figure 1 and the sidebar "What's in the Wall"). If you're willing to cut the cable should it need service, you can simply size the hole to pass the cable and install connectors *after* the cable is through the wall. To keep out weather and insects, leave a drip loop and seal the cable jacket to the edges of the hole with caulk. Some department stores sell small-diameter "through wall" tubes meant for TV and telephone cable. Those that I've seen are large enough for RG-6, RG-58, RG-59 or "8X" *without connectors.*

It is much better to make your installation so that cables can be removed without removing connectors. For PL-259s, that means the smallest opening diameter must be at least ⅞ inch. Figure 2 shows a simple technique used by many hams. It is a "1½" PVC pipe part that is normally used to connect a household sink trap to the waste system. I used this method in my first station. It was in a ground-floor bedroom with a closet against an exterior wall. The interior end of the pipe entrance was concealed inside the closet.

Larger pipes can pass more cables. Prefabricated dryer vents are useful for this technique. They are large in diameter, and they do not look out of place on the exterior of a building.

If you need to hide cables running up or across walls, consider running them inside metal or plastic conduit, or gutter downspouts. You can find gutter parts made from aluminum or vinyl at home-improvement stores. All of these can be painted to match your home; Rustoleum now makes a primer for vinyl and other plastics.

One drawback to the various "pipe through the wall" methods is that they will require wall finish repairs when they are removed someday. This, and then need for indoor and outdoor appearances acceptable to my spouse, led me to the solution in Figures 3-5.

It's a project for advanced do-it-yourselfers; see the sidebar "What's in the Wall?" The idea is to mount a dual electrical outlet box to a stud with a UHF bulkhead connector that extends from the box interior through the back of the box, the exterior wall structure and a section of PC board secured outside the house siding. The PC board can be painted to match the siding and blends in very well. This technique provides neat, serviceable connections both inside and outside of the house (see Figures 4 and 5).

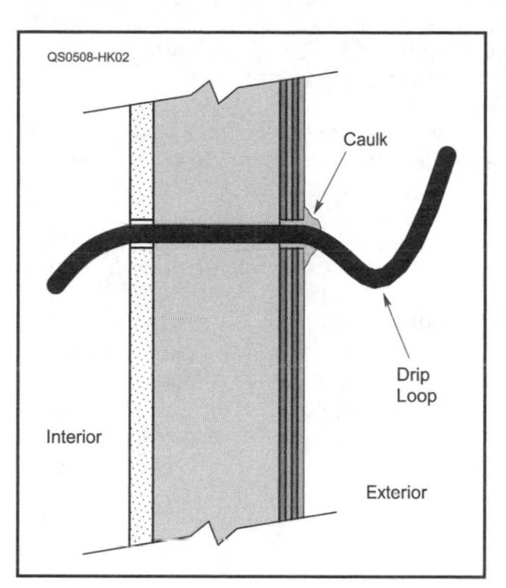


Figure 1

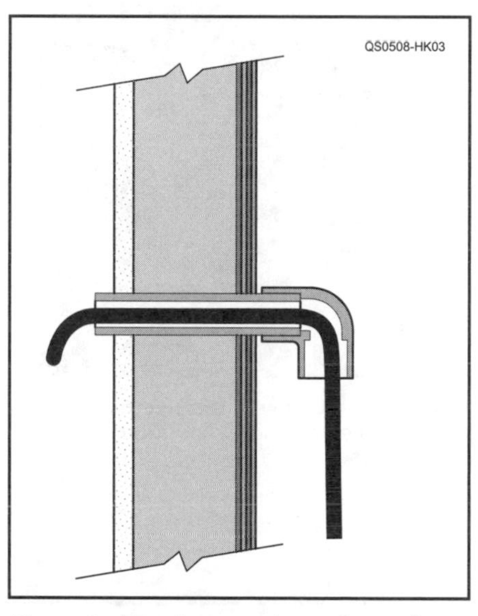


**Figure 2—The downward opening and drip loop prevent rain from entering. Pack the pipe with a few inches of steel wool to keep out drafts, insects and vermin.**

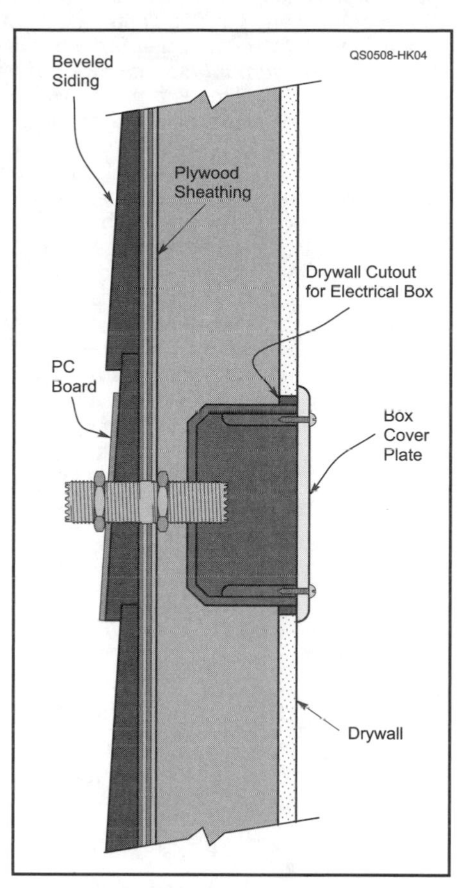


**Figure 3—An electrical junction box and a coaxial bulkhead connector provide a cable entrance that appears neat and professional from both inside (Figure 4) and outside (Figure 5) the house.**

Inside the house, a cable and PL-259 pass through a metal box cover and attach to the bulkhead connector inside the box.

If there is ever a need to remove the installation, the PC board can be replaced with a piece of aluminum flashing, and the box can be covered with a standard junction box trim plate.

1. Obtain a dual-gang electrical box. (Mine is plastic. It's cheaper and easier to machine.

Select an entrance location that is free of plumbing, and ac power, telephone and other wiring.

2. Using a hacksaw, remove the back wall from the box.

Locate the wall stud nearest your desired cable entrance. Place the side of the box on the line and mark the box outline on the wall. Cut out the drywall on the box outline.

Place box against stud face and secure the box to the stud with a couple of drywall screws.

3. Drill a hole large enough to pass a UHF bulkhead connector from the box interior through the siding and plywood. (It's a good idea to drill a small pilot hole first, so you can drill the finished hole from the exterior, which won't splinter the siding.)

## What's in the Wall?

Cutting or drilling into walls can be dangerous because they often contain plumbing, ac lines and telephone or other electrically "hot" lines.[A] How can we do it safely? Here are some ideas for modern stud-walls with drywall or thin paneling interior finishes and nonmetallic (Romex and similar) or wire in conduit construction. I wouldn't try this on plaster. Older homes may contain many sorts of other wiring construction techniques. Some of them are described in two do-it-yourself Electrical Code FAQs for the US and Canada at **ftp://rtfm.mit.edu/pub/usenet/news.answers/electrical-wiring/part1** and **ftp://rtfm.mit.edu/pub/usenet/news.answers/electrical-wiring/part2**.

Start with a good modern *electronic* stud locator. Some can sense metal and metal piping (rebar, plumbing, electrical conduit) or armored (bx) cable, electrical boxes, "hot" ac wiring and possibly non-metallic power cable within walls. For $30-40, they're inexpensive safety tools. Some also indicate *both* edges of each stud. I doubt that they detect telephone, intercom or other thin wiring.

Confirm the edge of the stud by driving a few small nails or screws through the drywall. It's easy to tell whether a screw is in air or wood once it's through the drywall, and screw holes are small and easy to repair.
(As the screw enters a soft wall material—don't try this with plaster—its threads displace bits of the wall material into a small mound around the screw. The hole can be repaired with "Spackle" or by squeezing a drop of white glue into it and carefully smoothing the mound from outside toward the center with your thumbnail.

Place the open face of the box on the wall with one edge on the edge of the stud. Mark the outline of the box on the wall.

With a keyhole saw, cut an observation hole slightly away from the stud edge and look to see if there is any wiring fastened to the stud.

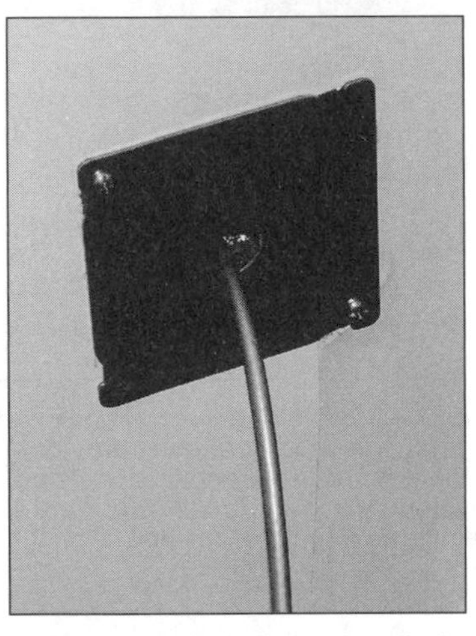

**Figure 4—A view of the electrical box cable entrance from inside the house. The apparently black rectangle is actually a silver box cover reflecting light away from the camera.**

**Figure 5—A view of the electrical box cable entrance from outside the house.**

4. Drill the PC board to pass the connector, secure the connector to the PC board and slip one edge of the PC board under the siding as shown. Secure the PC board to the siding with a couple of sheet metal screws and paint it to match the siding.

Drill an electrical box cover to pass a PL-259, and you can run the cable through the cover and attach it to the bulkhead connector inside the box in the wall.—*Bob Schetgen, KU7G, Hints and Kinks Editor*

### FINDING TRUE NORTH

◊ Numerous methods have been published for finding true north in *The ARRL Handbook* and elsewhere. While accurate, most require government maps, navigational textbooks and the knowledge to use them, or at least a good, dark sky and a star chart.

A rough sunny-day method involves sinking a straight stick vertically in level ground. (Such a stick is called a "gnomon.") Mark the tip of the stick's shadow on the ground with a rock or coin. Return in an hour or so and mark the shadow's new position. A line from the first mark to the second points roughly east. This is good enough for a general sense of direction when hiking, or for aiming antennas with very broad beamwidths, such as 90°.

We can learn more accurate methods from folks who install sundials. Get a piece of plywood and install a gnomon (nail, rod or dowel) at the center of one edge, perpendicular to the board (see Figure 6). Place the board in open sunlight with the gnomon to the north (in the Northern Hemisphere). Level the board and mark its edges on its support so that you can later return it to the same spot in the same orientation. Roughly an hour before local noon, mark the end of the gnomon's shadow as point B. Draw a circle around the gnomon base with its radius equal to the shadow's length (segment AB in Figure 6). Every 10 or 20 minutes, mark the tip of the gnomon's shadow on the board. The marks will define a curve that draws closer to the gnomon's base, then expands outward again to cross the circle. Once the curve is outside the circle again, no more measurements are necessary.

Sketch a fair curve through the marked points and label the point where it leaves the circle as point C. A line drawn through points B and C will be due East-West. A line from the midpoint of BC through the gnomon base (point A) is North-South. It should also cross the curve at the point where the curve comes closest to the gnomon base (local noon).—*William L. Fulton, PO Box 724, Sugar Land, TX 77487-0724*

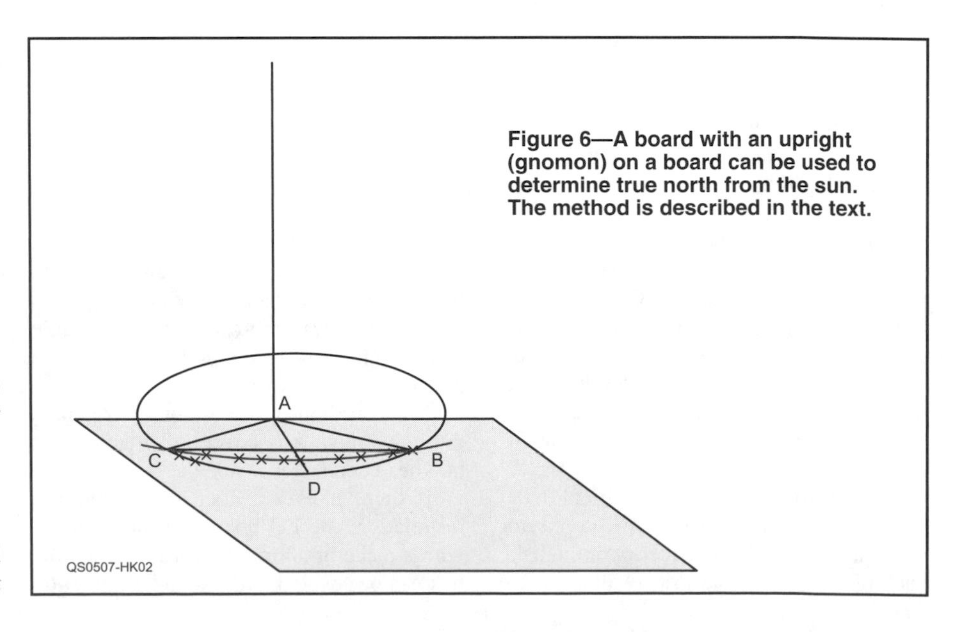


**Figure 6—A board with an upright (gnomon) on a board can be used to determine true north from the sun. The method is described in the text.**

## USE A TUNED PIPE TO "ZERO-BEAT" CW

◊ Recently, I became the proud builder/owner of a new Elecraft KX-1 QRP transceiver. I have fallen in love with this little rig, and the only fault I have found with it was the lack of a tone or other method of matching the frequency of a received CW signal. In other words, it lacks zero-beat tuning options. I know many hams tune to what sounds good to their ear, but I am obsessed with "zero beating" a station as close as I can.

There is an electronic signal indicator on the market that shows the received station is tuned for the strongest signal when an LED glows brightest.[3] I prefer matching a tone equal to the rig's CW offset to tune other stations. My K1 has a tone generator that I can switch on to match the tone of received signals, but the KX-1 lacks this feature.

I thought about it for a while and decided that I could use a pitch pipe like those used by groups singing a cappella. There are no pitch pipes tuned to 600 Hz, so I decided to make one. You have probably blown across the top of a soda or beer bottle and heard the low tone emitted. A vessel of smaller volume resonates at a higher tone.

Using this method, I first tried lengths of PVC pipe and found that a 5¼ inch piece worked okay. Then I thought to use one of my empty cigar tubes, and sure enough, it works even better. The thinner, lighter cigar tube seems to resonate better, and I've fine-tuned it against an electronic tone generator on my computer.

If you see me operating portable and blowing into an empty cigar tube (Figure 7), you'll know I am zero beating and not trying to play a tune.—*Dick Arnold, AF8X, 22901 E Schafer St, Clinton Township, MI 48035;* **af8x@arrl.net**

[3]When I searched the Web for such a product, I found several schematics and kits; just search for "zero beat."

**Figure 7—AF8X uses a tuned cigar tube as a pitch pipe to zero beat CW signals.**

## A PORTABLE SERVICE CART FOR EQUIPMENT

◊ My journey into Amateur Radio has been short by most standards, but the last four years have been full of hit-and-miss ideas putting together a usable ham shack.

I have been interested in radio since the late 1960s, listening to shortwave and AM band DX. I was always welcome in the living room or bedroom with a receiver. Then, in 2001, I took the Amateur exam and was able to make my own radio waves. As soon as the coax hit the living room windowsill and I sealed the gap with duct tape, I was sent to the garage like an old dog.

By that time, I had three HF rigs, an amplifier, 'scope and many other gadgets a ham needs to make a station work.

It was spring, so I set up on the workbench. That was fine—until the weather turned hot, and soon there was a new room in the corner of the garage.

When planning the size of the new radio room, I thought 6 by 8 feet would be fine, but when I moved in I found I could not move around easily. There was little room to service the rigs and wiring.

I made it one year in the tiny shack, with no good ideas to correct the problem of equipment access. I went to furniture and hardware stores but found nothing.

I work as an auto repair technician, and a mobile tool sales truck comes to our repair shop weekly to supply our tool needs.

One day I was on the tool truck, and there was the answer for my shack. Figure 8 shows a "Blue Point" portable service cart made by Snap On Tools (**www.snapon.com**). It has drawers, shelves and more importantly, wheels! These carts are very popular these days, for the computer scanners and laptops mechanics need.

I was on cloud 9 all day! This cart is the answer. As Figure 8 shows, everything for one complete station fits. The power supply is on the bottom shelf. The radio and amplifier are at eye level when the operator is seated.

DAVE MERRIAM, WØDKM

**Figure 8—A Blue Point equipment cart made by SnapOn Tools.**

DAVE MERRIAM, WØDKM

**Figure 9—WØDKM's roll-around station. The drawer has a thin plywood cover to serve as a desk. The cover lifts easily for access to tools and supplies inside.**

DAVE MERRIAM, WØDKM

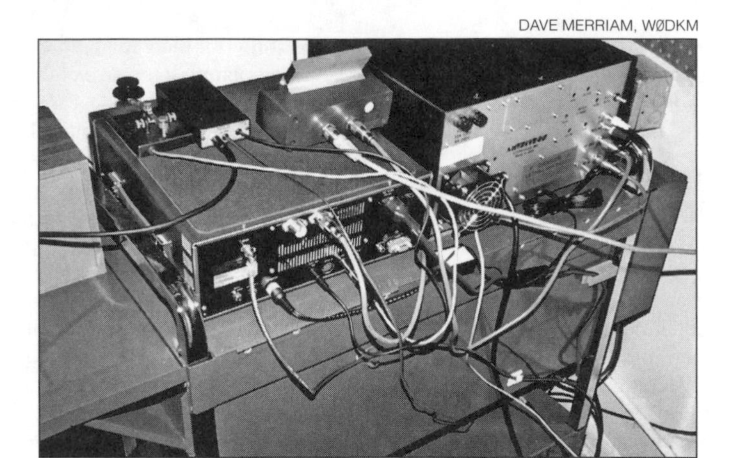

Figure 10—Oh happy day! Simply spin the cart around for easy access to rear-panel controls and connections.

STU COHEN, N1SC

Figure 11—Make any desk or table easily roll about with these tripod casters sold at many home-improvement stores.

STU COHEN, N1SC

Figure 12—Look where the legs meet the floor. All equipment is easily moveable at N1SC. Does Stu sign "mobile"?

The speaker is on a side shelf. The drawers hold hand tools, a soldering iron, hand keys and headphones.

In the top drawer, I added wood strips near the top to support a thin plywood insert used as a desktop for writing and logging contacts (Figure 9). This plywood can be lifted to access items in that drawer.

The best part of this cart is access to the rear of the equipment (Figure 10)—just unlock the wheels and spin it around. Oh happy day! *—Dave Merriam, WØDKM, 9308 Fort St, Omaha, NE 68134;* **merriam5@cox.net**

◊ Dave's wheeled cart is a great, professional looking way to achieve access to all sides of his radio gear. When I look at that sleek cart, however, I can feel my wallet cringe.

ARRL Technical Editor Stu Cohen, N1SC, has accomplished the same end with three-wheeled steel dollies (see Figure 11) sold at home improvement centers. Figure 12 shows Stu's shack with them in place. Any of the tables loaded with heavy equipment can be moved easily with the aid of these casters (which are permanently in place).

Other helpful devices are available. Look in your local mover's supply, hardware store or home improvement center for moving supplies. Although I cannot point to a supplier, I use some things called "furniture movers" (see Figure 13). They are essentially 6×10-inch sheets of heavy, flexible plastic sheet with ½ inch of high-density foam rubber fastened to one side. When a heavy object is placed on the foam side, the plastic slides easily over smooth or carpeted surfaces. If you find appropriate materials, you could easily make these yourself.*—Bob Schetgen, KU7G, Hints and Kinks Editor;* **rschetgen@arrl.org**

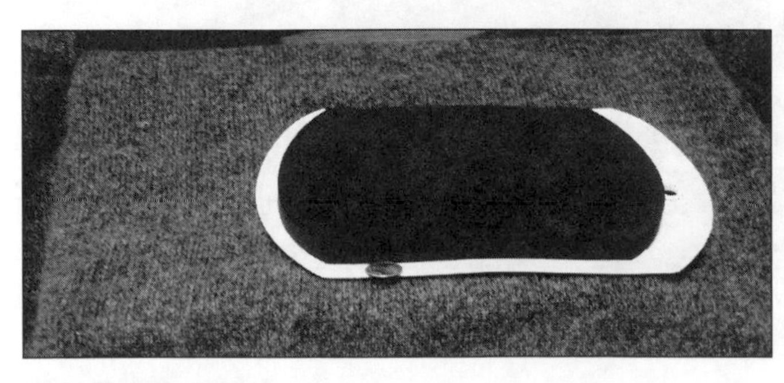

Figure 13—A "furniture mover" that helps heavy loads to slide easily.

## SOLVING "DROOPY BOOM" SYNDROME

◊ My wife and I each own a Heil Pro-Set Plus headset, and both have developed a similar problem I call "droopy boom syndrome." The Heil Pro-Set Plus is a popular communications headset with a boom microphone that contains two *switchable* mic elements. It also features a phase-reversal switch on the headphones.

The droop problem first developed after three or four contest weekends of use, and it gradually became worse and worse. As the headset warms up from being on your head for a while and the components inside expand ever so slightly, whatever holds the relatively heavy boom in place in front of your mouth begins to lose its grip.

At first, the boom droops a little when you move your head and you can push it back into place, where it would stay a while, but as a contest weekend goes on, the drooping becomes worse and worse. By the end of the 2004 IARU HF World Championship Contest, I was constantly holding the boom with one hand, lest it instantly drop down to its lower stop.

Both of the Pro-Set Pluses in our house-

WM5R

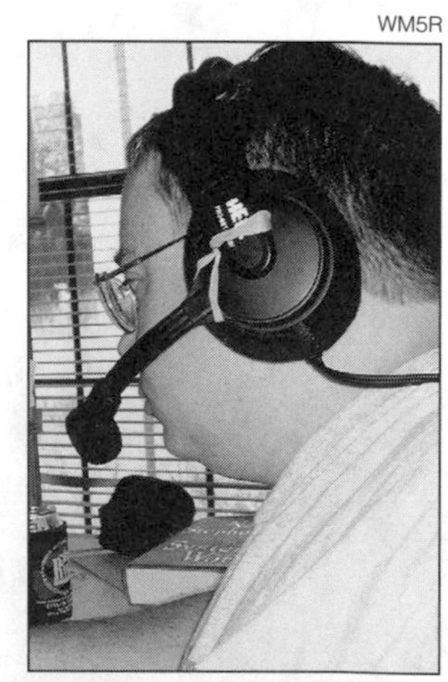

Figure 14—A rubber band helps prevent WM5R's microphone boom from drooping. Photo taken during the 2004 ARRL 10 Meter Contest at station K5TR.

hold now exhibit this problem, and I've heard similar reports from other local contesters. Yet there is a cheap solution.

While operating the ARRL November Sweepstakes (Phone) at W5KFT, I discovered that a rubber band wrapped around the headband post above the ear cup and then around the boom adds just enough friction to keep the boom from drooping, even when the headset warms up a little from normal use.—*Ken Harker, WM5R, 7009 Fireoak Dr, Austin, TX 78759;* **wm5r@arrl.net**

## HEIL PRO-SET PLUS — A BETTER SOLUTION THAN RUBBER BANDS!

◊ In Jan 2006 "Hints & Kinks" (p 63) we reported on WM5R's solution to insufficient tension of the mic boom of his Heil PRO-SET PLUS. We heard from Bob Heil, K9EID, the manufacturer, who let us know that the instructions that come with the PRO-SET PLUS describe the proper way to increase tension (see Figure 14):

*PRO-SET PLUS! Tension Adjustments*

*This first-ever competition headset has many adjustments that can be made by the user in the field. One of them is the tension adjustment for the microphone boom.*

- *Remove boom side cotton cover and ear pad*
- *Remove four Phillips head screws*
- *Carefully lift out the speaker plate*
- *This exposes the three small Phillips head adjustment screws 12, 4 and 8 o'clock position*
- *Tighten equally*
- *Re-assemble*

Bob goes on to note: "The first production run of 100 pieces were adjusted very lightly. After daily use here at K9EID, I determined the tension for production and was, of course, wrong. I apologize for that error but it is easily corrected, as are any of the adjustments. We, of course, corrected subsequent production runs and have not had one problem with this since. All of them were of that first 100 pieces. Please refer to the instruction sheet that was packaged with your Heil PRO-SET PLUS!

Please contact me, personally, if you have any problems. 73, Bob Heil, K9EID; **www.heilsound.com**."

Figure 14 — Close-up of the tension adjusting screws in the PRO-SET PLUS.

## TUNING FORK AS A CW TUNING AID

◊ There was a very good Hints & Kinks *QST* article recently that described an electronic musical instrument tuning device used to aid CW tuning.[1] A second Hints & Kinks item described blowing across the top of a hollow tube for the same purpose.[2] I fabricated a simple tool several years ago that I use. It is tuned to the default 600-Hz sidetone set on my radio.

I bought a tuning fork at the nearest music store for a few dollars. The fork for musical note D at 587 Hz is the one I purchased. I took it to work one evening and used our mighty HP spectrum analyzer with accelerometer input to "tune" the fork to 600 Hz. By grinding the tine tips equally, I lowered the mass, thereby increasing the frequency of the fork. See Figure 15. I ground the tines in small increments, with testing between each trip to the grinder, to get as close to 600 Hz as possible without overshooting. I overshot and quit at 601.2 Hz.

Use is quite simple: tune the radio to the received CW frequency you wish to contact (making sure your radio sidetone is adjusted to 600 Hz!), thump the fork on your hand, hold the base to the bone/skull behind your (good) ear and tune the radio to zero beat the fork with the received signal. (The fork frequency can be tuned to your favorite sidetone: just buy a fork lower in frequency than your desired sidetone and carefully grind to match that tone.) I have inscribed the frequency of the fork on the fork handle.

Most people probably won't have easy access to a spectrum analyzer and accelerometer. I did a bit of research and found a freeware program called *RightMark Audio Analyzer*, Version 5.5. This is a PC-based audio spectrum analyzer, which can be found at Web site **audio.rightmark.org/index_new.shtml**.

[1]Richard Lamb, KØKK, "CW Frequency Matching with Musical Instrument Tuners," Hints & Kinks, *QST*, May 2004, p 69.

[2]Richard Arnold, AF8X, "Use a Pipe to Zero Beat CW," Hints & Kinks, *QST*, Aug 2005, p 52.

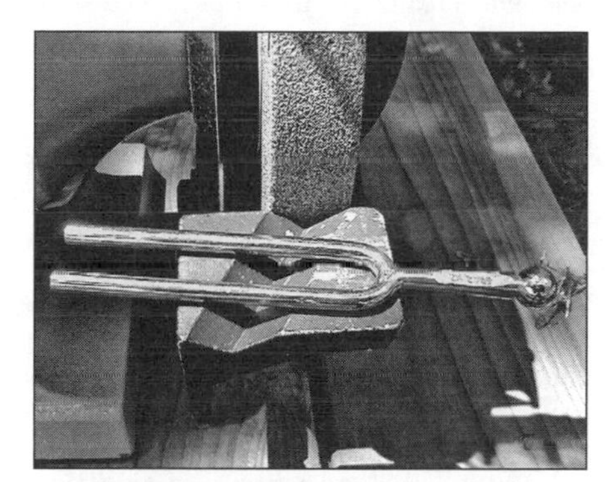

Figure 15 — Carefully grind both ends of the tuning fork equally, a bit at a time, to increase the tuning fork resonant frequency.

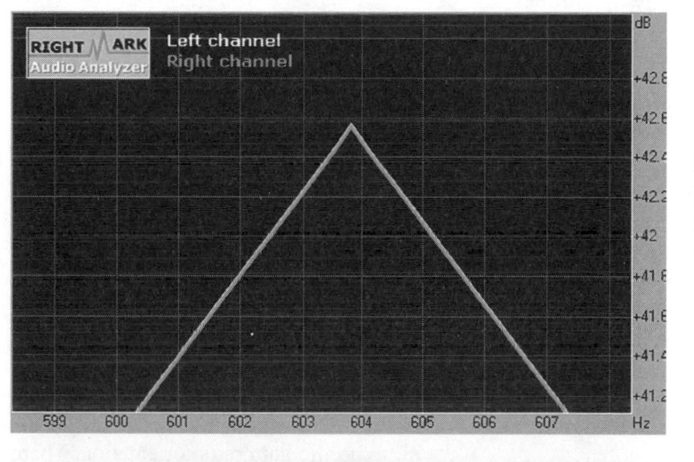


Figure 16 — Full bandwidth of the analyzer: note the peak at 600 Hz.

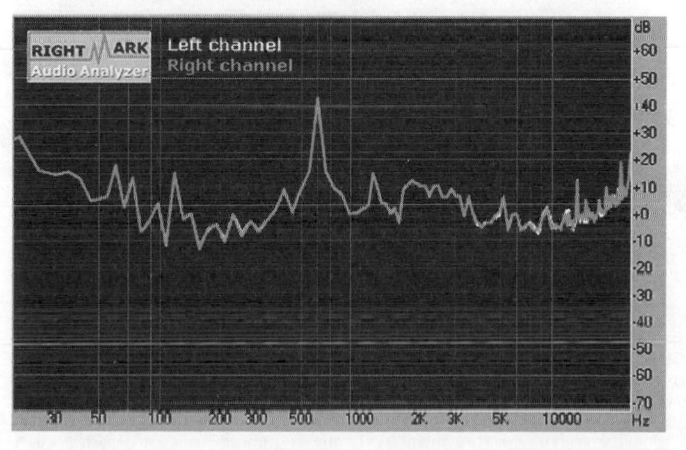


Figure 17 — Narrow bandwidth to resolve the fork frequency. I have estimated the peak at 603.8 Hz.

With a microphone as input to the PC sound card, you can determine the fork frequency and carefully grind metal off equally on each tine until you reach the desired frequency. Thump the fork and hold it in direct contact with the mike after you initiate the program. It will sample for a time then play back the results for analysis. The software is user friendly with several adjustments and analytical tools. Spectrum analyzer settings are critical so play with them to get the best presentation. See Figures 16 and 17 for software screens.

Using the newly installed program I determined my fork to be resonant at 603.8 Hz, pretty close to the 601.2 Hz determined by the $65,000 lab instrument with $600 accelerometer and charge follower as described above. (Normal wear may have increased the resonant point, which hints that the software and HP analyzer are in near exact agreement.) See Figures above for clarification. I wish I had the plot available I made on the HP instrument but I can't put my hands on it in time for this article. — *Ted Walker, KJ7V, 800 Palo Verde Pl, Sierra Vista, AZ 85635,* **kj7v@cox.net**

## TOILET PAPER TUBES IN THE HAM SHACK

◊Used toilet paper and paper towel tubes are handy helpers in the ham shack. They are great for storing test leads, scope probes, coax jumpers, and a myriad of other similar items that tend to clutter the workbench. Simply roll up the items and stuff them into a tube. Once placed in tubes, these items can be stacked in a box or a drawer. They are also useful in retaining power cords on test equipment that is stored on shelves, excess cord on "wall warts," and the power cords on soldering irons — heat guns — electric tools — and computer peripherals.

I use them behind equipment to keep the tangle of cords under control. Four or five tubes fitted into a coffee can make a handy holder for tuning tools, pencils or screwdrivers.

The cost is right, and the supply endless. — *Robert Tiffany, W1GWU, 102 Lockes Corner Rd, Alton, NH 03809,* **w1gwu@usadatanet.net**

## POWER STRIP LIBERATORS

◊I am writing in reference to the Hints & Kinks article "(Wall) Warts on Pigtails" in the April 2005 issue of *QST*. Zio Tek Power Strip Liberators are short extension cords, about 12 inches long. See Figure 18. They can be purchased at **www.cyberguys.com** for less than $10 plus shipping for a pack of five. I purchased two packs and they are very useful for using all of the outlets on a power strip. For that price it hardly seems worth the trouble of installing a new receptacle on the end of a printer power cord. — *Paul Bernhardt, WD4EBA, 480 Hartville Ave, Cocoa, FL 32926,* **wd4eba@arrl.net**

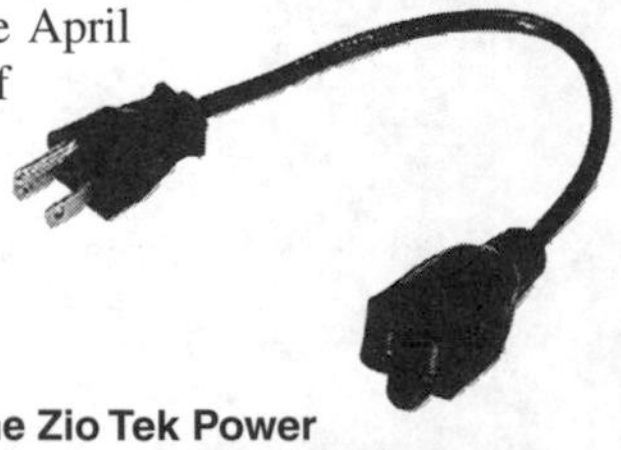

**Figure 18 — The Zio Tek Power Strip Liberator is a ready-made 1-foot extension cord that allows you to free up the extra outlets covered by wall warts on your power strip.**

## PROTECTING SILVER CONTACTS AND BUYING OLD ELECTRONICS PROJECTS BOOKS

◊I keep a lot of silver point contacts for testing and realized a trick I use could save a lot of trouble for others using boat anchors. I place the points and anything silver coated in Hagerty's tarnish intercept bags. You can find these bags at **www.hagerty-polish.com**. When the bags begin to turn black I trash the old one and transfer everything to a new bag. It works great.

I buy old electronics books for variations on projects. I found *99 Test Equipment Projects You Can Build* by *73 Amateur Radio* magazine, and on page 215 are the instructions for converting a Heathkit IG-102 Signal Generator to FET and battery operation. I've found these on eBay for $30 and the conversion parts cost $4. It makes a mighty fine signal generator for less than the cost of its open-air tuning capacitor. — *Clynton Yarter, ex WB6CSQ, 668 S Chipwood St, Orange, CA 92869*; **wb6csq@arrl.net**

## HOMEBREWING A TROPHY FOR YOUR HAM CLUB

◊One of the aspects of club membership that I enjoy most occurs usually once a year when the awards for the activities of the previous twelve months are doled out. Seeing the response from members who are rewarded for their efforts is very rewarding in itself.

Some clubs present certificates suitable for framing, others use generic trophies or wall plaques. In general these are marked for Amateur Radio only by the inscription on the award. It is very difficult to find trophies with a specific ham radio orientation. There are microphone-like tops for trophies, but all too often they have the letters "CB" on them. That, of course, may not sit well with many licensed amateurs.

My first club came up with a remarkable trophy for one of their top annual awards many years ago. It consisted of a miniature radio transmitter tower with a miniature "Yagi" or beam antenna on top. This award had been handcrafted by one of the members in the distant past, and was very sought-after by all the members. I do not remember what the award was for, exactly, but it was probably something like "Ham of the Year" or "Elmer of the Year." I do remember that everyone in the club would have given their favorite rig to be awarded that trophy. Each year another member's name and call would be added to the base and the winner would retain the trophy until next year.

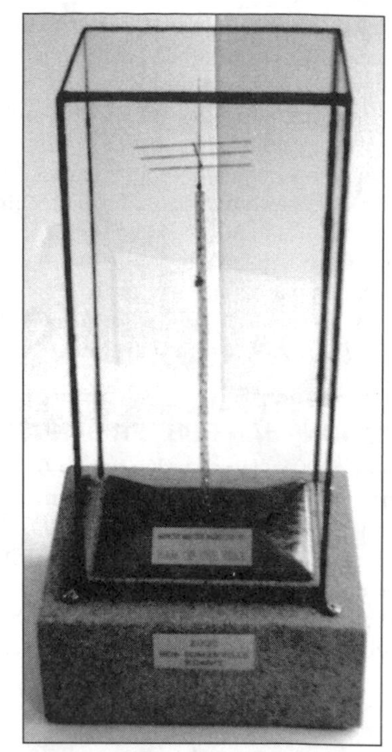

**Figure 19 — This photo shows the completed trophy that N1GY made for his Amateur Radio club.**

Recently, I became a member of the Board of Directors of another Amateur Radio club. Along with the position of Secretary came the responsibility of creating the certificates that our club awards at our annual Christmas party. With the publishing programs on my computer, that was no real problem, and all who received them seemed appreciative. I felt that the club ought to be able to do a little better than just a paper certificate for members who have really put their heart and soul into the activities that resulted in a top award. I remembered the tower trophy from years before and got to thinking.

Many years before I became a ham operator, I was heavily into model railroading, and I thought there might be a miniature tower available, so I went searching on the Internet. Sure enough, a company in California called Showcase Miniatures offers an N-scale radio tower for about $20 that looks very much like the typical Rohn tower many hams have in their backyards. The scale tower stands about 10 inches tall and even comes with a small Yagi antenna. The antenna is probably more like a two meter beam than an HF version, but that is easily made also. Any model railroad hobby store that can order or stocks the model tower can also provide brass wire and tubing to make a creditable HF beam antenna in very short order. The tower and

**Figure 20 — Here is a sample design for the brass tags that can be engraved and attached to the**

**Figure 21 — You are only limited by your imagination when it comes to making a suitable trophy for a radio club award. This drawing illustrates the idea of attaching a straight key to a wood base with a suitable brass tag attached.**

antenna can be painted with gloss gray paint or other color to suit.

There are several ways to make a base for the tower. I chose to build my own from a piece of pine lumber. After several passes through the table saw and sander, and then a couple of hours spread over several days applying a glossy finish with stain and polyurethane, I had a suitable base. If you want to avoid the work, most trophy shops can provide a blank that will be suitable. You will have to go to the trophy shop anyway to get the brass plate and have it engraved with the appropriate name and award. By the way, this is a very fragile model. It should be protected with a clear acrylic plastic or glass enclosure if possible. My wife, Audrey, happens to be a stained glass enthusiast, so creating a glass "box" to enclose the tower and its base was her contribution to the project. The manager of the trophy shop pointed out a very significant idea when building a trophy. He said that the recipient of a trophy often judges the importance of the award by the weight of the trophy. Since the trophy I designed was intended to be the Top Award for the club, I built a hefty base out of wood, which I then painted with a faux stone finish. You could also purchase a base from your local trophy shop.

When all is said and done, your club will have an outstanding ham radio trophy to award to the outstanding recipient. You could make a new one every year, or make the trophy a perpetual type with space on the base for multiple winners over many years. This is the way our club intends to go. The winner each year will get to keep the trophy for one year and will also get a framed certificate to keep forever. The choice is yours. Figure 19 is a photo of the trophy I made. Figure 20 is an example of the brass name tags that can be engraved and attached to the trophy base. The only limit is your imagination. You could use a desk microphone or Morse code key in place of the tower model. See Figure 21. Another possibility would be a large vacuum tube, modified to include an LED to run off a small battery in the base to light it up internally for display.

The club members will appreciate the trophy. Who knows, It may even increase the participation in activities as members see that there is recognition for hard work and enthusiasm throughout the year. — *Geoff Haines, N1GY, 708 52nd Ave LN W, Bradenton, FL 34207*; **n1gy@arrl.net**

## ARRL L/C/F CALCULATOR TIP

◊ Here is a tip for buyers of the ARRL L/C/F calculator: remove the slider and coat both sides of the top and bottom half-inch portions (mask off the printed areas) with a good quality silicone spray. Let it dry and then note the much smoother operation. One of my electronic calculators stores all the equations, but the "Type A" is faster. — *James Olsen Jr, W3KMN, 5905 Landon Ln, Bethesda, MD 20817*; **w3kmn@aol.com**

## MORE ON THE TUNING FORK AS A CW TUNING AID

◊ Ted Walker, KJ7V, has a good idea about using a tuning fork as a CW tuning aid. (See Hints & Kinks, Aug 2006 *QST*.) There are a number of audio tone generators and spectrum analyzers available on the Web, but the tuning fork is portable and doesn't require any external power.

Another option for tuning the tuning fork would be to cut the legs a bit short and drill and tap the ends for a 4-40 machine screw. Using screws with jam nuts would allow for tuning the fork down in frequency by lengthening the screws. — *73, Ed Bailen, N5KZW, 10405 Lime Creek Rd, Leander, TX 78641-6031*; **n5kzw@arrl.net**

## RECYCLED DISK CASES AS QSL CARD HOLDERS

I found a use for old 5 inch floppy disk cases. I found an old case in my attic and was about to put it into the trash when I thought of a better use for it. I now store my blank QSL cards and Mobile Response Cards (MRC) for the County Hunters' net in it. See Figure 22. The old dividers are also useful to separate different categories of cards. — *73, A. Robert Pantazes, W2ARP, 1 Madison Ave, Northfield, NJ 08225-1071*; **w2arp@arrl.net**

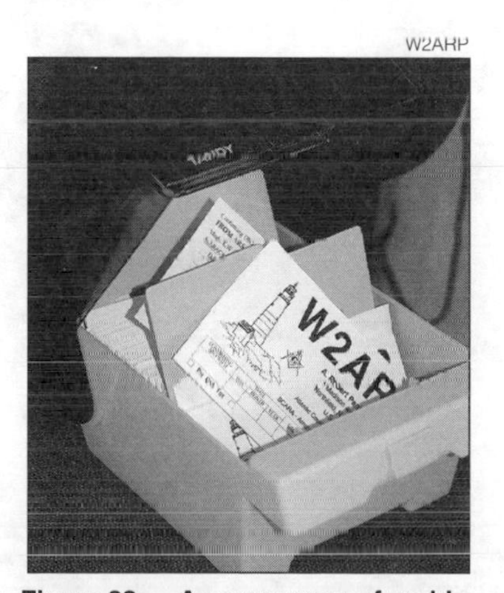


**Figure 22 — A new purpose for old 5½-inch floppy disk cases.**

## REMOTING A WIRELESS SPEAKER

◊ Perhaps you have a problem similar to mine. My shack is not in the house, but is attached to the garage. As both an active net operator and a good family man, I do not wish to spend 2 hours each evening closeted in an isolated and often cold shack, waiting to check into my 75 and 2 meter nets.

My first thought was to hard-wire a remote speaker in our kitchen to monitor the progress of the net while I was dining or working. The layout of the house and shack made this both impractical and aesthetically undesirable. My second thought was to install a UHF link from the garage to the house. The cost of this as well as the necessity of identifying with my call sign on the link ruled this out as well.

While rummaging through my box of extra speakers, I came across a pair of Emerson "wireless" audio speakers that I had once used in a non-ham application. These speakers were RF and not infrared activated. I connected the remote output of my receiver to the wireless speaker's transmitter. I then only had to find a good place to put the wireless speaker in the house. I used a small plug-in transformer to operate the wireless speaker, so I wouldn't have to rely on batteries.

Of course, I could not now hear the speaker to my rig when I was in my shack, since the Emerson transmitter was plugged into the rig's speaker jack. A simple ⅛ inch duplex plug in the wireless transmitter's jack allowed me to use an external speaker in the shack in parallel with the wireless transmitter. Now, I can listen to nets from anyplace in the house, and I can return to the shack to check in or to answer a call.

There may be other solutions to this problem. If you have a wireless intercom or a baby monitor, locating it near the speaker of your ham receiver may work. — *John Mollan, AE7P, 10901 Madrona Dr, Anderson Island, WA 98303*; **ae7p@arrl.net**

## USE YOUR MP3 PLAYER FOR CODE PRACTICE

◊ It's been a while since I operated CW, and I've wanted to brush up on my skills. I've tried listening to cassette tapes while walking or running but my movement always caused the tape to play erratically. This made for rough copy — there had to be a better way!

I decided to try using a digital audio player instead. I purchased a 1 GB flash memory MP3 player, sold online for about $50. See Figure 23. This is a big improvement over

KYLE YOKSH, KØKN

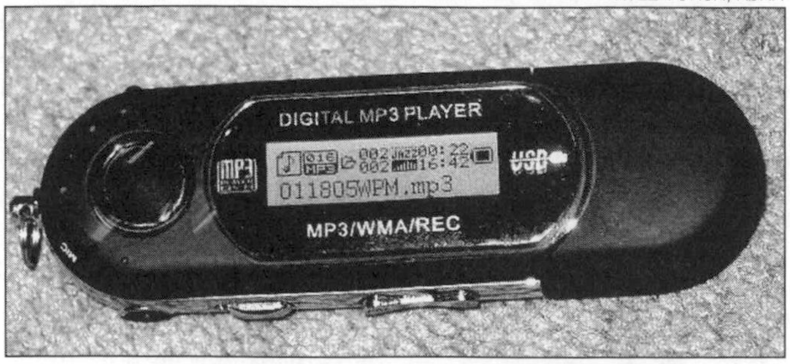

Figure 23 — Digital MP3 player loaded with W1AW code practice audio and text.

my old Walkman tape recorder, and is very inexpensive compared to some of the high-end, hard-drive players on the market today.

In addition to playing standard MP3, WAV and WMA files, this device features a voice recorder (similar to a micro cassette recorder but with no tapes and tens of hours of recording space) an FM radio, PC flash drive, telephone book and a text reader. I figured the voice recorder would come in handy for logging duties when operating my portable satellite station.

With the player ready to go, now I needed to load up some code to practice with. A Web search revealed that ARRL provides code practice transmissions in MP3 format at **www.arrl.org/w1aw/morse.html**. They even include a text file you can read later to check your copy. I routinely load new code practice files into my MP3 player, along with the associated text files. At any time I can choose a file, practice the code and immediately check my work, whether I'm in the lunchroom at work or on a park bench! — *73, Kyle Yoksh, KØKN, 125 N Chambery Dr, Olathe, KS 66061*; **k0kn@amsat.org**

JOHN J. ROESSLER, K6BX

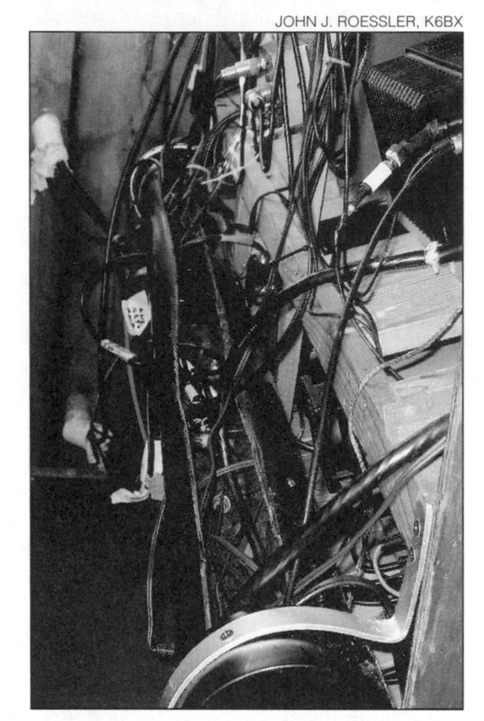

Figure 24 — Top view of 2 inch slit in the top of plastic sewer pipe used as a homemade wire tray.

## PLASTIC PIPE WIRE TRAY FOR BACK OF CONSOLE

◊Here's some info on the wire tray I made for the back of my equipment console. It works well and costs little. I had a 4 inch AWS Plastic Sewer Pipe that I had left over from installing a new sewer lateral in my house about 10 years ago. I cut it to be 5 feet long. I cut a 2 inch slit in the top for the wires to go into. See Figure 24. At the bottom of the pipe I cut five square holes to allow convenient access to the five ac power receptacles at the back of my console. I had some ¾ × 1⁄16 inch flat aluminum that I bent to make the end holders. See Figure 25. The only thing I had to buy was the two end caps for the pipe, which cost me $7.95 each. I think that PVC plastic pipe would work just as well. This installation has really neatened up the spider's web of wiring I had in the rear of my console! — *73, John J. Roessler, K6BX, 392 N Westwind Dr, El Cajon, CA 92020,* **r8382@sbcglobal.net**

## LABELING YOUR COAX CABLES

◊ I label my coax cables with a P-touch label marker (**www.brother-usa.com/ptouch/**). In Figure 26 you can see that I have placed my call sign and the length in feet on the coax cable (I also add the ¼ wave frequency, such as 146.250 MHz, on some coaxes). In addition, I place colored pieces of electrical tape on each end. The first color is for the near-end connector and the second color tells me what type of connector is on the far end. Blue = BNC, green = N, Red = PL259, white = other (such as SMC).

WB4PVT

Figure 26 — Labeling coax connector with call sign, length and color coding for connectors at either end.

I use this convention on all my coax cables, from test cables 2 feet long to spare 200 foot cables for Field Day. When I need a particular cable, say 150 feet long, I can just look in the junk box and from one end I can tell the length of the cable and what's on the other end. And when Field Day is over I know whether it's my cable or not. — *73, Charlie Stokes, WB4PVT, 494 Pamela Dr, Newport News, VA 23601*, **wb4pvt@arrl.net**

## PLANNING FOR SHACK DURING NEW CONSTRUCTION

◊ If you're having a new home built, a quick call to the builder can make your life as a ham radio operator much easier. When our home

JOHN J. ROESSLER, K6BX

Figure 25 — Bottom view of homemade wire tray

KA3MTT

Figure 27 — A PVC pipe run behind the wall from the attic down to the basement allows cables to be easily routed to the shack.

was being built, I had no idea where my shack was going to be, so I met with my builder and explained what I wanted accomplished. I wanted to be able to run feed lines to any floor of the house as easily as possible. I had the builder install a run of PVC tubing from the attic all the way into the basement in a straight run between the walls.

I did choose to have my shack in the basement, but my feed lines would be coming into the attic, especially since I live in a subdivision. Running them into the basement with a drop ceiling was easy, as you can see in Figure 27. — *73, Nathan Ciufo, KA3MTT, 6323 Cinnamon Ridge Dr, Burlington, KY 41005,* **ka3mtt@arrl.net**

## PAPER TOWEL TUBE CORD COVERS

◊ Another useful task for cardboard tubes used for 11 inch paper towels and 4½ inch tissue rolls is to secure many types of wires and cords. The rolls can easily be labeled to identify the wires. — *73, Robert Fairfield, K7RQN, 8325 W Daley Ln, Peoria, AZ 85388,* **jfairf1062@aol.com**

*And proving that great minds move in the same circles, with an additional twist on this topic, we have:*

## ANOTHER USE FOR PAPER TOWEL ROLLS

◊ Have you ever misplaced your DMM test leads on your workbench? For short leads, clip one end of your test lead to a toilet paper tube and simply drop the lead through the center of the tube. For longer leads, a used paper towel tube works best, with you wrapping the lead through the tube enough times to hold it neatly. Then clip the other end of the lead to the end of the tube.

Finally, you can fit five or six such tubes inside a large round Quaker Oats box, which has a convenient top cover to close up the whole box of leads. 73, *David Crown, K9SJN, 9386 Landings Sq, Des Plaines, IL 60016,* **k9sjn38@earthlink.net**

## REPAIRING 3-RING BINDER MANUAL PAGES

◊ Many service manuals are made up of loose pages in a three-ring binder. Over the years the pages may get torn from the binder, especially foldout schematics.

3M makes a tape known as Gentle Paper First Aid Tape ideal for repairing the pages. Simply put a length of this tough, transparent tape over both sides of the torn hole(s) and repunch the holes. The repair is almost invisible and the page is virtually impossible to tear again.

I found this tape at the local drug store in the first aid section. The cost was $1.99 for an 8 yard roll. Definitely better than those old donuts. I have used this tape to repair a very worn Collins KWM-380 manual using the above technique and it really works! — *73, Lee Craner, WB6SSW, PO Box 976, Agoura Hills, CA 91376,* **leeCraner@aol.com**

## TRANSCEIVER SUPPORT SHELF

◊ Recently I upgraded my main station antenna tuner to an autotuner, as this new autotuner is particularly well-suited for use with an amplifier. As a full legal limit autotuner, it is larger than my previous tuner. I resolved the space issue by building a transceiver support shelf that is high enough for the autotuner to fit under.

The main shelf consists of a leftover piece of 12 inch deep veneered particle board, two 4×1×12 inch pieces of oak and two 1×1×12 inch pieces of oak. I needed to stack the 4×1×12 inch and the 1×1×12 inch pieces of wood together since the autotuner is 4 inches high and the 4 inch high piece of oak is really 3.75 inches high. I held all the pieces together with countersunk #6 flat head wood screws.

Figure 28 shows the pieces prior to staining and assembly. Besides helping to keep my station compact, the new shelf also puts my transceiver closer to eye level, which makes operating even more of a pleasure! This is a good idea even if you don't have extra equipment to place under your transceiver, as you can also use this under-transceiver space for housing your logbook, key, microphone and anything else that normally clutters up your operating position. — *73, Phil Salas, AD5X, 1517 Creekside Dr, Richardson, TX 75081,* **ad5x@arrl.net**

PHIL SALAS, AD5X

Figure 28 — The parts used to make the shelf.

## NEAT HANDHELD TRANSCEIVER HOLDER

◊ Here is a neat little handheld transceiver holder that I created using a widely available computer mouse holder. You can use this to turn your handheld transceiver into a base station setup that is easily disassembled should you need to "grab it and go" or a holder for your radio in your car, kitchen, etc.

My handheld transceiver is an Alinco DJ-G5. I have a drop-in charger system that is not designed to trickle charge the battery. Therefore the radio cannot be left in it continuously. As a result, I was always charging the battery and hence connecting and disconnecting from my base antenna in the attic. I wanted a somewhat more fixed setup that I could break down quickly if I wanted to take my transceiver with me.

1. Purchase a mouse holder (I got mine from Ergo in Demand for about $4, **www.ergoindemand.com/EMH004.htm**). These are designed to hold a regular computer mouse and normally come with double stick tape to mount to the side of your PC monitor.

2. Remove the double stick tape. Odds are, like my handheld transceiver, double stick tape won't be enough to hold it up. I

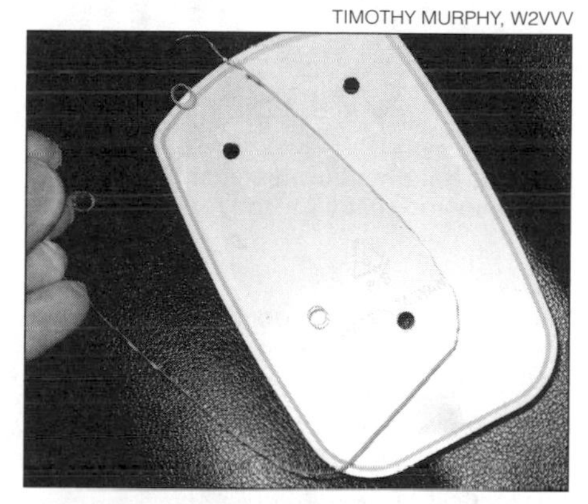
TIMOTHY MURPHY, W2VVV

Figure 29 — Cut the Plexiglas to fit and drill out it and the mouse holder to fit the mounting screws.

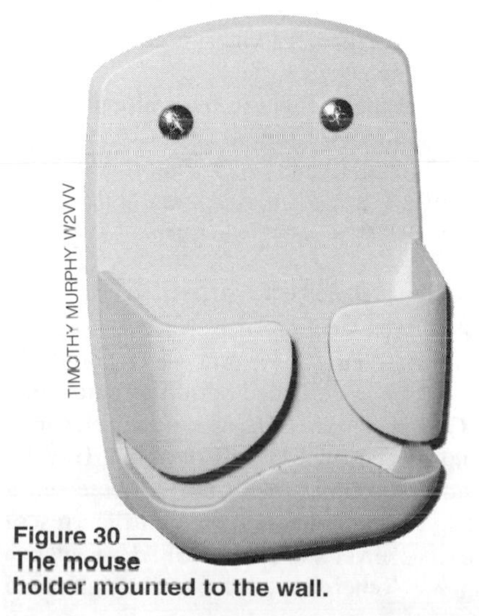
TIMOTHY MURPHY W2VVV

Figure 30 — The mouse holder mounted to the wall.

**Figure 31 — The final setup with the handheld transceiver wired up and ready for the repeaters.**

chose to use wall anchors.

3. Check for an extruded lip around the back of the holder that might cause it to crack if the screws are overtightened. Fill it in with a piece of ⅛ inch Plexiglas cut to fit in the recessed space (see Figure 29).

4. I then marked and drilled out three holes on the holder and Plexiglas to mount this to the wall (see Figure 30).

5. I placed the two items against the wall and leveled it prior to marking the spots for the anchors.

6. Put the anchors in the wall and screw in the holder (see Figure 31).

You're all done! — *73, Timothy Murphy, W2VVV, 25 Magnolia Ave, FL2, Dumont, NJ 07628-3019,* **w2vvv@arrl.org**

## MARK THOSE HEIL HEADSET ADAPTER CABLES!

◊ Before I leave my car at the airport parking lot for an extended period, I remove my ICOM IC-706MKIIG's detachable head from the dashboard and the Heil Traveler headset and leave them in the home shack along with its antenna and mount. The rest of the transceiver and the ICOM AH-4 antenna tuner are otherwise hidden out of sight and I leave them in the car.

When I returned home this last time, the 3B7C (St Brandon Island) DXpedition[2] was winding down and I wanted to get them in my log before they went off the air, so I pulled the transceiver from my car and set it up at my home station. I hooked the adapter cable to my headset, plugged it into the rig and proceeded to make a call. I was surprised when the headset's microphone did not work in either the push-to-talk (PTT) or voice operated transmitting (VOX) modes while the dual earphones worked just fine.

The headset had worked the last time I used it in the mobile. The frequency "SCAN UP" button on the switch pad worked fine, but the SCAN DOWN button caused my '706 to shut down with a loud "pop!" I thought that maybe, in handling the headset, I may have pulled on and damaged the cables connecting the PTT/SCAN switch pad. A check inside the switch pad showed everything to be okay. There were no broken or shorted wires. I tried its operation once again with the same result — a loud pop and immediate shut down of the '706. I then switched modes from phone to CW and put 3B7C in the log; that was the important thing to me at the moment!

Later, I decided to try the hand microphone that came with the '706. That worked perfectly. What could be wrong? Then it dawned on me; the headset adapter cable that I picked up off of my bench was the one for my Yaesu FT-817ND (Heil part number HS-TA-YM)! It looks identical to the HSTA-IM (ICOM) adapter cable! The adapter cable for the '706 was still in my car, under the seat! I attached the correct cable and the headset was back in business! The Heil Traveler requires different adapter cables for the various transceivers and they are easily removable. My two adapter cables are now marked with a band indicating transceiver usage. I feel lucky, because I could have damaged my '706 with this mistake! — *73, Karl T. Schwab, KO8S, 30752 Ridgefield Ave, Warren, MI 48088-3174,* **ko8s@arrl.net**

## QUICK, GOOD-LOOKING HOMEBREW LABELS

◊ I have found a simple and aesthetically pleasing method of putting labels on the panels of my projects. Dymo embossing tape is very nice, but my new method is better.

First, compose and print the labels with a *spreadsheet* program. Select fonts, sizes, colors and alignments suitable for the project. Surround each label with a solid line border to make it very easy to cut out the individual labels. Step two is to use a drop of rubber cement to hold the label in place on the panel.

[2]"The FSDXA 3B7C St Brandon DXpedition," *QST*, Jun 2008, pp 65-68.

The *piece de resistance* (things always sound better in French), is to place a sheet of self adhesive lamination over the panel and trim off the excess with a single edged razor blade. The finished panel has a neat appearance and is simple and inexpensive to prepare. The lamination protects the labels and gives the project a nice finished look. — *73, Tom Hart, AD1B, 54 Hermaine Ave, Dedham, MA 02026,* **tom.hart@verizon.net**

◊ Another method is to use *Microsoft Word* to generate the text (or to imbed graphics, if necessary) and then print the label on a transparency sheet (the kind used in overhead projectors). Then cut the labels out, glue them onto the panel using a dab of clear nail polish and then paint over them with a thin coat of nail polish to protect the labels. Both methods are simple and quick, even though they don't look quite as professional as the rub-on transfer decals. — *73, Dean Straw, N6BV, 5328 Fulton St, San Francisco, CA 94121,* **n6bv@arrl.net**

## CQ RINGTONE

◊ Being an avid CW operator, I thought it would be real cool if my cell phone would ring CQ CQ CQ in CW. My BlackBerry cell phone came with a program called *APPS* used to go online to reach *Access Shop* and purchase various cell phone software downloads. I went to the index and clicked on PERSONAL PRODUCTIVITY and downloaded a program called *VR+Voice Recorder* for $20 from By Share Service, GMBH. The program allows me to record voice messages with my cell phone, then e-mail them in *MP3* format to anyone. The received message can be played back on a PC or any mobile device.

I opened up the *VR+ Recorder* program and set my cell phone near my memory electronic keyer and recorded a series of CQs in the keyer. I flipped the keyer SEND switch while, at the same time, activating the cell phone recorder. After saving the file I played it back to check the sound. The next step was to change my cell phone ringtone to the new one. Cell phones do this differently so you will have to check your phone's instructions. For my BlackBerry I simply went to MEDIA, clicked on RING TONES, clicked on ALL RING TONES and selected the recorded file. I am happy to say that when I hear other cell phones ring I know right away that it isn't mine. When I am out in public and my cell phone rings CQ CQ CQ I get a comment from any ham within range. — *73, Greg Tyre, WXØE, 26673 Oldfield Ln, Lebanon, MO 65536,* **wx0e@embarqmail.com**

## GRIPPER PADS ARE REDEPLOYED

◊ I enjoy operating CW on the HF bands. I don't use my straight key too much except for QRP (less than 5 W) operation. Instead, I rely on my electronic keyer and paddle combination. On occasion, I break out the old standby "bug." When using either the bug or the paddles, I find myself repositioning these keys during a QSO due to side-to-side movement, causing them to "walk" across the operating desk. This is due in part to their rubber feet becoming hardened over time. The solution to this problem lay in the kitchen "gadget" drawer.

During a battle with the cap on a jar of olives, I realized I had the solution "in hand" — the rubber gripper pad sold as an aid to open jars. I reasoned that if it could improve my grip on a jar lid, it should be able to solve the dilemma of the "walking bug and paddles." Sure enough, by placing one of these pads under the paddles, the paddle stays put and no longer walks around the desk.

I have found additional applications: a lightweight tuner stays in place when its button is pushed, too-tight PL-259s are easily unscrewed and it's a handy workbench pad, which protects the equipment and keeps small parts and hardware from rolling away. Also, equipment without rubber feet won't scratch table tops when you put one of these underneath. You can cut the pads to size with ordinary scissors. They are sold in larger sizes for lining kitchen shelves or as placemats and are made in a variety of colors. Look for them in department stores and $1 discount stores. — *73, Steve VanSickle, WB2HPR, 3010 Tibbits Ave, Troy, NY 12180,* **wb2hpr@arrl.net**

## REUSE YOUR CRT STAND

◊ Hams are known for being resourceful and in the true spirit of the three Rs (reduce, reuse, recycle), I came up with the following useful project. My old CRT computer monitor was due to be replaced by a modern LCD screen. After I had removed it, I began thinking about the rotating stand that the screen sits on. It could be put to good use as a shelf for my HF rig.

After separating the screen and stand, I used a hacksaw to cut off the plastic guides that the screen fits into, thereby making a nice flat surface (see Figure 32). There was a small lip at the rear that I left to stop the rig from sliding backward. My Kenwood mobile rig sits nicely on the stand and can now be rotated left to right and up and down for easier viewing (see Figure 33). Many of these stands are being sent to the disposal centers as they are replaced with newer LCD screens. Consider saving yours and making this handy stand. *All photos by the author.* — *73, Jeff Richardson, VA3QSL, 36 Crawley Dr, Bramalea, ON L6T 2S1, Canada,* **va3qsl@arrl.net**

## RECEIVE INDICATOR

◊ Working with emergency communications as the Nantucket county RACES Radio Officer and ARES® emergency coordinator I have found it difficult to identity which one of my several active radios is receiving a transmission. This is especially important when working with Police, Fire, Coast Guard or any of the first responders.

To help correct the confusion, speed up communications and avoid mistakes I have installed external speakers for each radio. These are placed near its associated microphone with this circuit installed inside the speaker case. Each circuit has a red LED mounted to the speaker's front. When a signal is received the LED turns on for about 2 seconds, turns off for a fraction of a second to check the received signal and then repeats the process if the signal is still present.

This circuit needs 13.8 V dc at about 20 mA. Some radios have this voltage available at the mike connector. If using this source, check the specifications to be sure this doesn't overload the radio's circuit. This wire should also be fused for about ¼ A.

Since most radios use one speaker lead as ground this lead will also work as the circuit ground. Simply connect the circuit in parallel with the speaker. The volume control must be advanced slightly to fire the LED.

The circuit (see Figure 34) operates as a one-shot multivibrator. In the resting state Q1 is off and Q2 is on. When audio is detected, Q2 is turned off, Q1 and the LED turns on until the 100 μF capacitor discharges. At that time Q2 turns on again and looks to see if the signal is still present. If the signal is still present the process is repeated. If no signal is detected, Q1 and the LED stay off. Note that the LED will flash off as it checks. The on-time can be changed by increasing the value of the 100 μF capacitor. With values shown, LED

Figure 32 — The modified CRT stand ready for mounting a rig.

Figure 33— The completed stand with the rig in place.

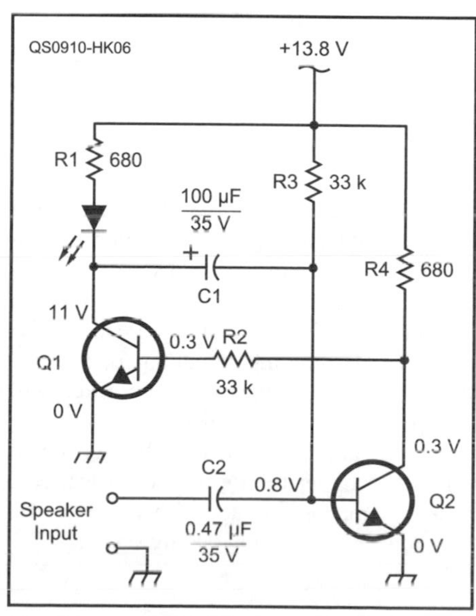


Figure 34 — The circuit for the receive indicator.

on-time is about 2 seconds, off is a fraction of a second.

### Parts List

| | |
|---|---|
| (2) | MPSA20 NPN transistors (ECG123AP) (NTE 123AP) |
| (2) | 680 Ω, ½ W resistors |
| (2) | 33 kΩ, ¼ W resistors |
| (1) | 0.47 µF 35 V capacitor |
| (1) | 100 µF 35 V electrolytic capacitor |
| (1) | LED (RadioShack 276-041) |
| (1) | circuit board |

*— 73, George Allen, N1NBQ, PO Box 727, Nantucket, MA 02554-0727,* **n1nbq@arrl.net**

## STATION GROUNDING

◊ After a few years' absence from ham radio, I have returned and am in the process of redesigning my test bench and operating desk space. I must admit that in the past I used the daisy chain method for interconnecting the various links in my transmitting and receiving equipment. That includes the usual collection of transceivers, power supplies, antenna tuners, amplifiers and so forth.

With the idea in mind to do away with the daisy chain mentality, I researched various sources and wound up in the "Safety" chapter of my *ARRL Handbook*. In that chapter appears a line drawing illustration showing a ½ inch copper pipe attached to the back of a typical test/operating bench.[1] All of the equipment is grounded through individual braided cabling to the common ground copper pipe.

I needed a method for attaching the pipe to my operating bench. It would seem a simple task to approach, but I was fresh out of practical solutions so I went looking for ideas. My patient spouse, upon hearing about my frantic search, simply pointed up on the living room wall over a front window and asked, "Why can't you use one of those curtain rod holders?"

She essentially "flung" her JC Penney housewares catalog at me and there it was. The catalog lists it as an "Ornamental Drapery Rod Extender Bracket" and carries a part number of R735-5287. [It's an adjustable metal bracket with a curved end (tray) meant to hold a tube. — *Ed.*] I would imagine that there might be a variety of such items available from other drapery stores.

The bracket will extend up to about 6 inches from the wall and has threaded locking screws for holding the curtain rod in place. The tray portion of the extender brackets will accept up to a ¾ inch diameter rod, so our ½ inch copper tubing fits nicely. The user should wrap the copper tubing with electrical tape where it makes contact with the extender brackets as well as the locking screws.

[1] *The ARRL Handbook for Radio Communications*, 86th Edition, 2009, Figure 3.8, p 3.8.

On my operating bench I mount my rig in a shelf array, which was originally intended as a tabletop entertainment center. The open-air arrangement of the shelving assures good cooling for the equipment without going to any great lengths of ducting and so forth. Also, stacking your radio equipment vertically versus "all over the tabletop" gives you a lot more room for your collection of *QST*s.

It consists of a shelf about 19 inches high, about 2 feet wide and about 16 inches deep with one movable middle shelf. This also needs a common grounding device, and a 2 foot length of ½ inch copper tubing was selected for that purpose. The tubing must be attached to the vertical sides of the shelf requiring some variation in the bracketing arrangement. For the shelf, I selected a pair of ½ inch suspension clamps, which come six pieces per pack and are described as a "pipe support system." They are used to hold ½ inch copper pipes onto wall studs or floor joists. The parts can be found in a hardware store. At the Orchard Supply Hardware (OSH) Web site (**www.osh.com**) they are called "Oatey #33914." The pieces are reasonably priced, made of durable plastic material and are predrilled for mounting nails or screws.

Attachment of the grounding braid to the copper tubing is accomplished with ⅝ inch adjustable stainless steel clamps (Orchard Supply Hardware 110000732799854389). I would suggest a minimum opening size of ⅝ inch; the ½ inch clamps don't leave much room for the grounding braid once the clamps are tightened.

### Grounding Wire

The *ARRL Handbook* suggests using AWG 6 solid copper wire for connecting the copper pipe to outdoor ground stakes. At the time of my project Hurricane Katrina reconstruction had raised the cost of solid copper wire; for this reason I opted for AWG 6 stranded THHN insulated copper wire. The use of tinned copper braid (in various widths) and/or flat copper strap material has advantages for connecting the copper tube to outdoor ground stakes or cold water pipes.

I prefer the convenience and availability of crimp-on electrical connectors, but I wasn't able to locate any that would accommodate AWG 6 wire. The reader should be aware that higher numbered wire gauges pertain to smaller wire diameters; thus AWG 6 is a much heavier wire than, say, AWG 12. The largest size crimp-on connectors my sources stock is for AWG 12 wire. Thus, some substitution may be in order to use AWG 6 wire in our grounding systems.

A workable substitute for interconnecting AWG 6 wire to another electrical conductor might be Gardner-Bender GSLU-35 Mechanical Lugs (OSH's #0000032076017187). These devices accept AWG 6 through 14 on one end and have a flat lug that will accept a #10 bolt on the other end.

For wire smaller than AWG 6 I suggest Calterm #65654. The package contains five pieces of gold-plated terminals; one end is a crimp-on which will accept an AWG 8 wire, while the other end accepts a ¼ inch bolt. *— 73, Arthur McAlister, KD6SF, 7570 Dartmouth Ave, Rancho Cucamonga, CA 91730-1534*

## DETERMINING BEARING FROM LONGITUDE AND LATITUDE

◊Strong winds this winter turned my beams and mast in the rotator. I discovered that my reference direction (a low Yagi pointing on a 135° bearing into the Caribbean) had also moved. I missed the spring equinox (where the sun rises at 90° and sets at 270°, or at least close enough for HF Yagi headings) and the North Star is too hard to see due to haze and city lights in my urban environment.

Fortunately, the Internet provided latitude and longitude data for my tower base and a nearby water tower; an *Excel* spreadsheet then computed the bearing.

To determine the bearing to a distant landmark: First, identify a landmark about 1-2 miles away that you can see from your tower. Searching Google for "lat-long finder" turned up **www.satsig.net/maps/satellite-photo-image-viewer.htm** and **www.sirzman.com/gc/llf.cfm**. There are others, but I used those two. It takes some playing, but you can find your neighborhood with the street maps, then switch to satellite imagery and put the cursor (or "pin") on the center of your tower base. That gives you the starting point latitude and longitude (LAT1/LONG1). Then find the water tower or other landmark by the same process. That is the end point latitude and longitude (LAT2/LONG2).

With a calculator or spreadsheet program like *Microsoft Excel*, compute intermediate values and the bearing to the end point using the following formulas. West longitude (what we have in North and South America) must be entered as *less* than zero. North latitude (USA, Europe) is greater than zero. Dallas TX is roughly 33° (north) latitude and –97° (west) longitude. Be aware that spreadsheets usually expect angles to be in radians, not degrees, so you have to multiply degrees by $\pi/180 = 0.017453293$ to get radians.

Enter the following into the spreadsheet:

LAT1/LONG1; LAT2/LONG2 [Enter the degree values first, then calculate the radian value; you need both.]

DLONG = LONG2 – LONG1 [Calculate in degrees. Convert to radians.]

COS(D) = [SIN(LAT1) × SIN(LAT2)] + [COS(LAT1) × COS(LAT2) × COS(DLONG)] [Calculate in radians.]

SIN(C1) = COS(LAT2) * SIN(DLONG) / SQRT(1-COS(D) * COS(D)) [Calculate in radians.]

**Table 1**

| | Degrees | Radians | | Degrees | Radians |
|---|---|---|---|---|---|
| LONG 1 | –96.64161 | –1.686714335 | LAT 1 | 33.05321 | 0.576887359 |
| LONG 2 | –96.620947 | –1.686353698 | LAT 2 | 33.05397 | 0.576900623 |
| **Intermediate Calculations** | | | | | |
| DLONG | 0.020663 | 0.000360637 | | | |
| COS(D) | | 0.999999954 | | | |
| SIN(C1) | | 0.999034228 | | | |
| COS(C1) | | 0.043938717 | | | |
| **Final Calculation** | | | | | |
| Bearing (radians) | | 1.526843459 | | | |
| Bearing (degrees) | 87.48168617 | | | | |

COS(C1) = (SIN(LAT2) – SIN(LAT1) × COS(D)) / (COS(LAT1) × SQRT(1-COS(D) × COS(D))) [Calculate in radians.]

$Bearing_R$ = ATAN(SIN(C1) / COS(C1)) [Calculate in radians.]

$Bearing_D = Bearing_R * 57.2957795$ [Convert radians to degrees.]

Example using the *Excel* spreadsheet calculator that can be found on the ARRL Web site is shown in Table 1.

If you are using a calculator, this may be in degrees. If using a spreadsheet, you will have to multiply by 180/PI = 57.2957795 to get back to degrees from radians. It may be necessary to add or subtract the result from 180 or 360°. Lay out the points on graph paper to get an idea of what the bearing should be. — *73, Timothy L. Bratton, K5RA, 4001 Trails End, Parker, TX 75002-6543,* **k5ra@verizon.net**

## INTERFACE CABLE MANAGEMENT

◊The interconnection of computers, GPSs, TNCs, interface modules and other equipment peripheral to our transceivers presents a problem with regard to management of the many specialized cables required. On the one hand, it is desirable to retain a certain lead length on the connecting cables so that various hookups can be accommodated in the shack without having to move the equipment very close together. On the other hand, cable lengths that are handy in the shack often lead to a "rats nest" setup in compact mobile or briefcase portable hookups. The active ham discovers that he has no sooner shortened and resoldered the interconnect cables for a compact setup when conditions necessitate lengthening the connecting cables to try out some alternate arrangement.

The widely available plastic food storage containers provide a handy solution to the problem. The interface cables can be coiled and layered in such containers with just the optimum length of connector ends passing out through a hole in the side or top of the container. Attaching the lid that comes with the container completes the assembly.

In my shack, one container captures the cables that interface a WiSys GPS/TNC unit to a Kenwood TM-241A for an under-the-seat mobile setup. The other container holds the interface cables that connect a computer to the Kenwood transceiver for a briefcase packet station. In both cases, a neat access hole was formed in the side of the plastic container using an old tube socket chassis hole cutter.

Fortunately, most of the signals that are transferred to and from transceivers to peripheral equipment are of fairly low frequency and the cables are almost always protected with ferrite cylinders to reduce RF pickup. We have found that if the setups did not suffer from RF pickup on the bench, they do not encounter similar problems when the connecting cables are coiled up in the plastic containers. Remember to operate the transceiver at the lowest reasonable RF power level. — *73, Ed Sack, W3NRG, 1780 Avenida Del Mundo, Apt 4, Coronado, CA 92118,* **esack@pacbell.net**

## ACCESSORY OCTOPUS

◊It used to be convenient to have a selection of jacks on the back of older transceivers that would provide easy access to a number of inputs and outputs. Those are now replaced by one or two DIN sockets to accomplish multiple functions. I find it inconvenient to hard-wire a DIN connector for a specific application, only to rewire it later when something new needs to be plugged into the radio.

CARL SOLOMON, W5SU

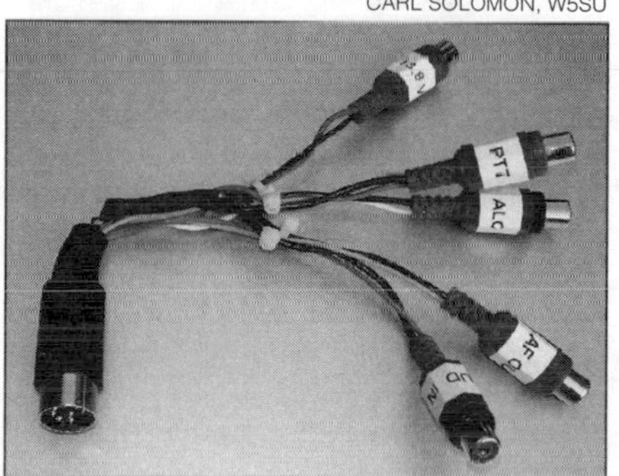


**Figure 35 — A simple homemade adapter for the accessory DIN socket found on most current transceivers.**

I came up with the following adapter that makes the process more versatile. Wire pairs from several female RCA connectors are soldered to the appropriate pins on a male DIN connector, with all of the black wires soldered together then to the ground pin. Each RCA connector is labeled, for example, ALC, PTT, 13.8V, audio in and audio out (see Figure 35). If only a couple of them are needed, the others dangle harmlessly but remain available for future use. I suppose some might argue that these be shielded, but I figured the short lengths wouldn't invite trouble and in my application there's been no problem. — *73, Carl Solomon, W5SU, 7110 Fernmeadow Cir, Dallas, TX 75248,* **w5su@arrl.net**

## HELP FOR THE MOBILITY IMPAIRED AMATEUR

◊I came across a simple and free keyboard alternative for amateurs who have limited dexterity, severely restricted hand motor skills or otherwise cannot use a keyboard.

My search for a modified keyboard arose from the needs of a dear friend, a long time CW man who can no longer manage a key or paddle and has great difficulty with a keyboard. Not only does this restrict his Amateur Radio activities but also any other computing that requires a keyboard. Simple things that we all take for granted like e-mail or surfing the net can be impossible for someone with limited motor abilities — particularly hand dexterity.

Fortunately, *Windows* has a built-in solution. It is called *On-Screen Keyboard* (*OSK*) (see Figure 36). *OSK* only requires the user to move a mouse or joystick a little. *OSK* will work with any application that requires input from a keyboard. It also works with all of the common e-mail clients, word processors and any Internet Web page.

*OSK* can be configured for "Hover to Select" mode. Rather than having to click on a key to produce a letter, this mode selects the letter simply by hovering the pointer over the key. It doesn't matter if you have a slight tremor, as long as the mouse pointer doesn't cross the boundaries of the key. You can set up the selection time delay to give yourself plenty of time to change your mind on the character or to move on to another character. *OSK* can also be set to "Click to Select" using the mouse or joystick button. Font color and size are also selectable.

So, how does this work for Amateur Radio? I have tested *OSK* with *DigiPan*, *MixW*, *CW Type* and *HRD DM780*. It works perfectly. In addition, macros can easily be used with *OSK* as all of these programs assign an F key to the macros. *OSK* can select up to 12 F keys.

To access *Windows OSK* click on START, then RUN and then enter osk into the OPEN box. Click on OK to open the *OSK* program.

JIM MATIS, K2TL

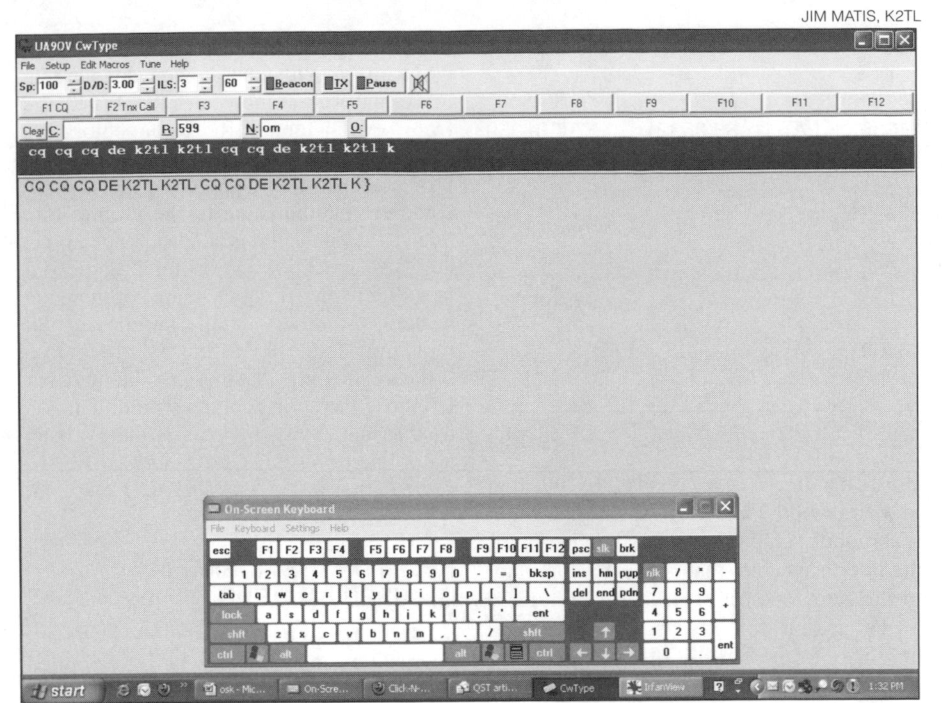


**Figure 36 — A screen capture showing *OSK* with *CW Type*, a free CW keyboard program from UA9OV (www.dxsoft.com).**

To create a shortcut on the desktop:

1. Right-click the desktop.
2. Point to NEW and a list will open. Click SHORTCUT.
3. Type osk then click NEXT.
4. Type a name for the shortcut and then click FINISH.

Some functions of any application require a mouse click and do not offer a keyboard alternative, so *Windows OSK* may not be a perfect solution for those who cannot maneuver a mouse and click. In this case, on-screen keyboard programs have a greater functionality. One such *free* program is *Click-N-Type* (**www.lakefolks.org/cnt**). This program makes it easier to get text into those uncooperative places, like browser address fields, e-mail address and subject fields, and other dialog box input fields. — *73, Jim Matis, K2TL, 11 Moss Haven Way, Howell, NJ 07731-2293*, **k2tl@arrl.net**

## STATIC AND LIGHTNING DANGERS

◊My spouse, Bonnie, KBØHTC, and I run a small Amateur Radio repair shop in a Denver suburb. I have had to change out some major components in a TS-480 and in two TS-2000s lately. Both had damage to the main circuit board, presumably from lightning or static surges. Both have multilayer boards; the missing signal or current paths just seemed to disappear into the board, not to reappear anywhere, making on-board repairs either too time consuming or not economical.

The repair bills for these rigs were extremely high due to the circuit board price and markup. Although the "commercial boys" would not even look twice at the bill, the average ham would have apoplexy. Even though you are getting a great many components replaced all at once, it's still hard to explain a $500 repair bill to an incredulous ham.

I would therefore offer this advice to any ham, especially someone running a newer (post 2000) radio, concerning the most destructive elements:

### *Static*

Always touch the side of the radio before plugging, unplugging or operating any switches or controls. Out west, we get static sparks on 90° days just from getting in and out of the car. It is not necessary to find a shiny spot on the case such as a screw head or something similar to touch; if your body charge is sufficient to cause damage, it will go through the paint on the covers.

### *Lightning*

The most commonly seen destructive path for lightning is through grounds and through data and audio lines. Antenna and dc lines are pretty heavy and can take a lot of surge current. The other signal lines in the new rigs are the width of a couple of human hairs and can only handle a few hundred milliamperes before they self-destruct. In addition, most of the traces are *in between* layers of the PC boards, not easily accessible; hence the need to change a whole board. So, we need an easy way to disconnect *all* data, audio and control lines from the back of the rig during electrical storms.

### *Power Supply Problems*

Get rid of that surplus 100 pound boat anchor 12 V, gazillion ampere power supply. Get a good, current limited supply or better yet, a switch mode supply. Switchers usually fail "dark," that is, their output just goes away. Most are quite inexpensive and easy to replace. Also install a good ac line filter. I recommend the Isobar Ultra series of line filters; they come with a $50,000 equipment guarantee.

If your rig has a negative lead fuse, be sure to check it from time to time, especially if you are having problems. The negative lead fuse in most installations can open but leave the rig operating, as it obtains its dc return to the power supply by other means, such as an antenna ground. Receive may be noisy and data or control errors may be noted. On transmit, a variety of problems will be apparent.

### *Interfacing*

A majority of problems in this area have to do with hams shorting the internal 8 V source during hook-up of audio or control accessories and cables to the rig. Extreme caution must be exercised if making your own cables. Your soldering skills must be above reproach; use pre-made or crimped connections wherever possible. Be very careful not to introduce "rogue" voltages or currents, even by way of ground loops to your rigs. Even very small fault currents can damage the radio or the accessory. Isolate wherever possible. When hooking gear together, always hook up the grounds first. — *73, Bill Leahy, KØZL (SK)*

## SPLIT BOLT

◊I'm embarrassed to say that it took me nearly 50 years to discover and fully appreciate the usefulness and versatility of the "split bolt" (see Figure 37) used so often by elec-

STEVE SANT ANDREA, AG1YK

**Figure 37 — Split bolts, commonly used in electrical work, can be used around the shack both inside and out.**

tricians. While they come in various sizes, I use primarily S8 and S6. Split bolts have many uses around the shack. They are useful for splicing additional wire onto the ends of a dipole to lower the frequency. Instead of soldering or tying knots, a split bolt will do a better job of attaching a wire through an insulator, it even allows for easier frequency adjustments since the wire can be quickly shortened and clamped again.

Instead of using standard ropes and lines to hold wire antennas up in trees, I now use 0.095 monofilament trimmer line since it doesn't fray in the branches when the wind blows. This heavier size is impossible to tie off but a split bolt holds it nicely. Of course, the best and most common use is to connect a pigtail from each piece of station equipment to the ground bus wire. Split bolts can be found in the electrical section of your favorite hardware store. — *73, Dick Hayman, WN3R, 15 Arlive Ct, Rockville, MD 20854,* **wn3r@arrl.net**

## LIGHTING FRONT PANELS

◊Although all rigs have dial backlighting, the front panel leaves a lot to be desired. Operate at night? Of course, we all do but the front panel of the rig is not lit. Visit your local hobby store and peruse the model train area. There will be a number of small lights available, from "grain of wheat" bulbs to LEDs. I found a package of six adhesive backed 12 V dc lights for just a couple of dollars. These can be easily mounted at the ham operating position to light up the front of the rig, an easy solution to a problem with most operating positions. — *73, Susan Meckley, W7KFI/MM, PO Box 1210, Pahrump, NV 89041,* **ussvdharma@yahoo.com**

## SILVER SHARPIES

◊Over the years, I've seen various methods of tagging wall wart type power supplies, those annoyingly prolific little black cubes that seem to fill up drawers and boxes as the devices they power become damaged and discarded or lost. Sometimes they get mixed together with others and you have to use a magnifying glass to read the voltage and amperage in an attempt to join the power cubes with their respective devices.

I discovered that a silver-colored Sharpie pen is ideal for marking directly on the exterior surface of these devices. You can write the name of the device the cube belongs to, the voltage and amperage rating, polarity information and whether it's an ac or dc device. Unlike attaching tags on the wire, the permanent Sharpie markings are difficult to remove once dry. You can also highlight the information molded into the device housing to make it easier to read by lightly rubbing the edge of the Sharpie point across the surface of embossed lettering.

Silver Sharpie markers are also ideal for restoring raised silver lettering and numbering on the faces of radios where the color has been rubbed off. I used one to color the MIC, PH and CW legends on the front of my Ten-Tec Orion II because the black background made the black embossed lettering difficult to read. Some radios have silver trim rings and the Sharpie is ideal for hiding small scratches and dings in these rings. Sharpie markers are also available in gold and other colors for custom applications. — *73, Webster Williams, WY3X, 4305 Fernwood Rd, Myrtle Beach, SC 29579,* **wy3x@arrl.net**

## GREEN ENERGY IN THE SHACK

◊We are all trying to move to a greener environment. Many hams have experimented with green energy solutions such as solar, wind and hydrogen fuel cell technologies. Most of these experiments have focused on Field Day and Emcomm applications. Up till now green energy devices for the shack have been expensive and complex.

I have happened upon a new green energy technology solution that is perfect for use right next to your rig. A company called Thermoelectric Ultra Seebeck Hybrid Equipment is marketing a thermoelectric charging unit (TCU) designed to be applied to the underside of the typical office chair.

As we all know, the human body has a temperature of 98.6°F (even higher during contests) and the average shack air temperature is about 68°F. The TCU utilizes the Seebeck effect [the direct conversion of a temperature difference into electricity. — *Ed.*] to generate about 4 V dc, a voltage sufficient to charge a lithium-ion (Li-ion) battery.

The TCU comes complete with the thermoelectric module, mounting components and a battery charging unit able to hold 4 AA Li-ion batteries (batteries not included). I found the installation to be simple and straightforward. Apply a coating of thermal grease to both the chair bottom and the TCU, mount the TCU to the chair using the adjustable brackets that attach to the chair edges. Plug in the battery charger and you are good to go. Once installed the TCU will silently charge your AA batteries, hour after hour, as you operate.

A couple of notes are in order. As explained in the instructions, charge time is gender dependent. Since heat transfer is related to surface area, and typically female hams have a smaller footprint when operating, a 37 percent reduction in power output can be expected for such operators. The TCU is designed to maintain a constant charging voltage for varying thermal input but charging current will drop and, therefore, charging time will increase.

Second, a problem I found in my early production model is that the battery charging unit was supplied without a polarized connector. This should be replaced with a polarized connector before use. While plugging in the battery pack with polarity reversed will not harm the batteries or the TCU, the nature of thermoelectric devices is such that, with polarity reversed, the TCU will draw power from the batteries and convert that power into heat, which will then be applied through the thermal grease to the bottom of the chair. This can become uncomfortable. — *73, Steve Sant Andrea, AG1YK,* **ag1yk@arrl.net**

## BALANCED LINES GROUNDING

◊I woke up at 6 AM in the middle of the December 27, 2010 Vermont blizzard and was in for a surprise. While in my shack I heard a zapping noise taking place 3-5 times per second. Outside the snow was flying sideways and the wind was howling from the north. A quick check revealed that the noise was coming from inside the MFJ-941D antenna tuner.

I have a number of coax antennas attached to this tuner as well as a 300 Ω balanced line twinlead feeding a 257 foot center-fed dipole for NVIS work on 160 and 80 meters.

Evidently, the WIRE/BALANCE position of the MFJ-941D tuner is not shorted to ground when another unused/unconnected position is chosen on the selector switch. The long dipole made a big capacitor with the earth and the snowflakes flying by charged it up. The noise was the dipole/capacitor discharging to ground inside the tuner.

The antenna twinlead is connected to the back of the tuner with a pair of banana plugs. Using insulated tools I disconnected the banana plugs and grounded them but not before discovering that the sparks could be as long as 2 inches and increased in frequency with wind speed.

If you have antennas connected to a tuner, you should check to see if the unused antennas are being grounded. If not, you need to create some kind of grounding system yourself since static can build to destructive levels very quickly. — *73, Jozef Hand-Boniakowski, WB2MIC, 465 Lamb Hill Rd, Wells, VT 05774-9707,* **wb2mic@arrl.net**

## VELCRO CABLE HOLDER

◊I made this storage device for adapters and cables because I could never find what I needed in a timely manner and everything was always getting tangled up. So, in the interest of speed and organization I came up with this idea (see Figure 38).

Take two pieces of sew-on Velcro and secure the loop strip to a piece of 1 × 12 × 24 inch pine with RTV. Then take the Velcro hook strips and cut them into 1½inch long pieces. With a toothpick, place a little bit of the Aleene's Super Fabric Textile Adhesive

JEANETTE DUNN, KI6AJJ

**Figure 38 — A simple solution for storing your cables, wall warts and sundry small accessories.**

JEANETTE DUNN, KI6AJJ

**Figure 39 — A completed set of flaps all ready to hold whatever needs holding. Note the short piece of string glued to the "open" end to make it easy to open the flap.**

STEVE LITTLE, AB9YN

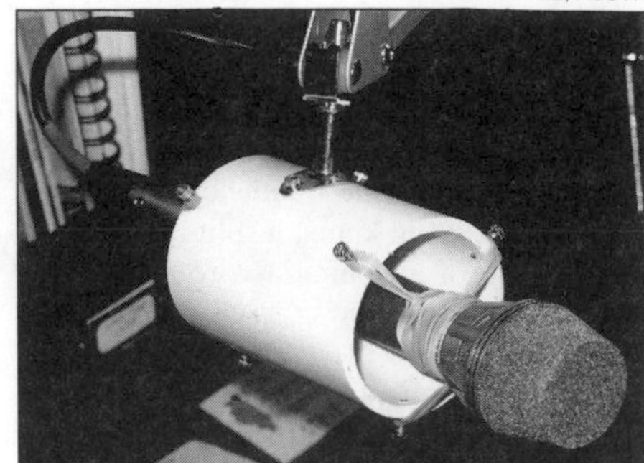

**Figure 41 — The finished microphone suspended in a makeshift holder on the end of the lamp "boom."**

JEANETTE DUNN, KI6AJJ

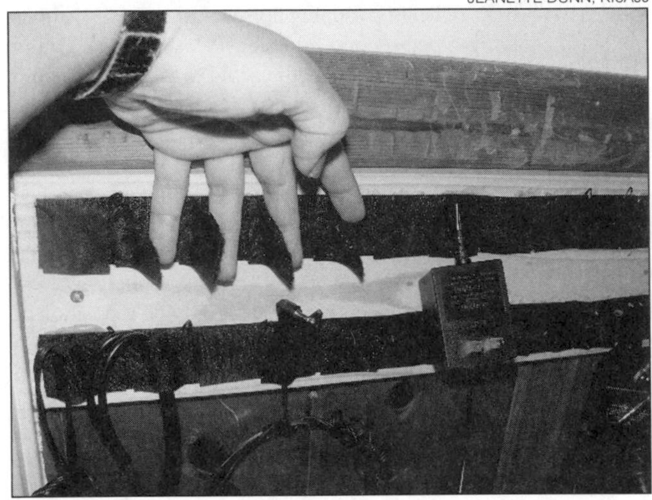

**Figure 40 — Each of the short Velcro hook strips is glued to the long loop strip at one end creating a flap that can be closed over whatever small accessory you want to keep available but out of the way.**

on the upper left corner and attach a small piece of string. Then put some glue on the upper right corner of each hook strip, place it on the loop strip and let it dry. This creates a flap (see Figures 39 and 40).

Now just place the cable inside the flap, fold it over and press. — *73, Jeanette Dunn, KI6AJJ, 3380 Malibu Dr, Santa Cruz, CA 95062*, **cell_safe@yahoo.com**

## NINE DOLLAR MICROPHONE BOOM

◊Having experience some years ago in professional broadcast stations, I got somewhat tired of having to hold the microphone in my shack when I transmitted. I knew what professional desk-style flexible microphone mounts looked like, but I also knew they were pretty expensive.

But things changed when my wife and I went to a local IKEA store. I was looking for a flexible desk lamp for my shack. That was a revelation because they offer a desk lamp for only $8.99. Now that's in my budget. So I went ahead and bought two of the lamps, one for the lamp and one for my microphone.

First, I removed the lamp cord that was threaded through the upper boom arm. Then I unscrewed the lamp shade and lamp socket to access the inside of the lamp assembly. Now, I could reach the two small screws that hold the lamp assembly to the lamp mounting plate. It turned out that the diameter of the lamp assembly was about the same as 2 inch PVC pipe. So, it's off to the hardware store for another very inexpensive part for the project. The pipe should be shorter than the microphone. This allows the rubber bands used to hold the microphone to extend outward and backward from the pipe and give better shock mounting. I particularly wanted a shock mount so moving the microphone wouldn't trigger the voice-operated transmit/receive switch.

I drilled and tapped 4-40 threaded holes for screws to mount the pipe to the curved lamp mounting plate. Then I drilled and tapped more screws to secure the rubber bands. Notice that I have two bands on the top side of the microphone (to hold its weight) and one on the bottom. The bands should not be pulled very tight. If they are a little soft, they absorb vibration better (see Figure 41).

I had a microphone on hand from my audio recording experience years ago. That was good news and bad. I didn't have to go out and buy a microphone, but this one happened to be a low-Z unit (AKG 190D) and my rig is designed for a high-Z microphone. So, check the impedance of the microphone input on your rig to see what you need. I could have just remounted the Kenwood MC-50 desk microphone that I have, but I wanted to keep that for portable operations.

I finished the project with a 2 × 4 inch plastic box from RadioShack. A metal box would provide better shielding against hum and buzz. I built in a red button as a PTT switch and the black toggle switch for long term transmit. I also added a 50 kΩ potentiometer for gain control mounted on the back side of the box. Once it is adjusted for proper sound level, it won't need to be adjusted again.

Using the MONITOR position on the rig you can hear the audio and adjust the gain, processor levels, etc. Check the manual for your rig for information on resistor values if the output of your microphone is too high. — *73, Steve Little, AB9YN, 749 Hunter Rd, Glenview, IL 60025-3402*, **ab9yn@arrl.net**

## EASY PTT

◊I needed a PTT switch on my desk that was easy to use with my boom microphone. I was in a Staples store to purchase some office supplies when I saw the perfect PTT switch: The Staples "That was Easy" talking button. I purchased an "Easy" button for $5 and returned home.

I placed the switch on my bench and opened it up. First, remove the battery cover and batteries. There are four small feet on the bottom; remove the feet to expose the screws. Before you open the case place a small mark with a pen to show yourself how to reassemble it. Remove the four screws and the case will come apart to expose the working switch and audio circuit. Next cut the circuit board as indicated (see Figure 42) with a small knife.

Solder a very small wire on the vacant solder pad near the right side of the switch. The other wire is soldered to a vacant solder pad near the white wire to the battery compartment. Run the wires out the battery

A ROBERT PANTAZES, W2ARP

Figure 42 — The inside of the "Easy" button. The point of the knife blade indicates where the foil must be cut.

A ROBERT PANTAZES, W2ARP

Figure 43 — The wires to your transceiver exit through the battery case.

compartment (see Figure 43) and test with an ohmmeter. The "Easy" red plastic button has to seat into two slots of the cover to work properly. Reassemble and retest making sure the switch moves freely. Wire to your rig's PTT and you have a desk PTT switch for $5. — *73, A Robert Pantazes, W2ARP, 1 Madison Ave, Northfield, NJ 08225-1071,* **w2arp@arrl.net**

## BALANCED LINES GROUNDING

◊I woke up at 6 AM in the middle of the December 27, 2010 Vermont blizzard and was in for a surprise. While in my shack I heard a zapping noise taking place 3-5 times per second. Outside the snow was flying sideways and the wind was howling from the north. A quick check revealed that the noise was coming from inside the MFJ-941D antenna tuner.

I have a number of coax antennas attached to this tuner as well as a 300 Ω balanced line twinlead feeding a 257 foot center-fed dipole for NVIS work on 160 and 80 meters.

Evidently, the WIRE/BALANCE position of the MFJ-941D tuner is not shorted to ground when another unused/unconnected position is chosen on the selector switch. The long dipole made a big capacitor with the earth and the snowflakes flying by charged it up. The noise was the dipole/capacitor discharging to ground inside the tuner.

The antenna twinlead is connected to the back of the tuner with a pair of banana plugs. Using insulated tools I disconnected the banana plugs and grounded them but not before discovering that the sparks could be as long as 2 inches and increased in frequency with wind speed.

If you have antennas connected to a tuner, you should check to see if the unused antennas are being grounded. If not, you need to create some kind of grounding system yourself since static can build to destructive levels very quickly. — *73, Jozef Hand-Boniakowski, WB2MIC, 465 Lamb Hill Rd, Wells, VT 05774-9707,* **wb2mic@arrl.net**

## ROUGHING UP MICROPHONE CONNECTORS

◊If you've ever had to install a new connector on a handheld microphone, you probably had a problem with obtaining a good grip on the outer jacket in the clamp of the connector. I had to replace the microphone on my 2 meter rig. After soldering the wires to the connector and tightening the clamp all the way so that the two pieces were fully merged, I figured I was done and could enjoy years of trouble-free service. Just a few contacts later I noticed that I was pulling the jacket out of the clamp and straining the conductors.

JIM KOCSIS, WA9PYH

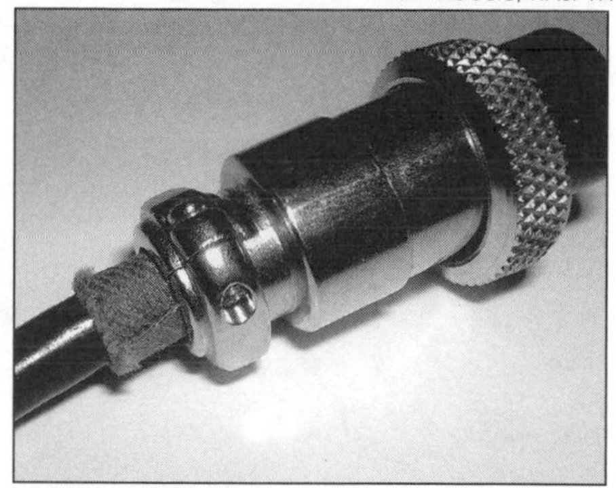

Figure 44 — A small piece of emery cloth helps to prevent the microphone cable from being pulled from the connector.

I tried adding a bit of outer jacketing to clamp the outer jacket more firmly. That didn't work. Clamping pressure wasn't the problem; the outer jacket was just too slippery. I tried some black electrical tape and that didn't work either — the outer jacket slipped right past the clamp and tape. I needed to add something that would grip the outer jacket very securely — something a bit abrasive.

After a brief scan of my workshop I saw the solution: emery cloth. (Note: This is not sandpaper. The material must be cloth backed abrasive. The paper backed abrasive will rip and disintegrate with time and pressure.) I cut a piece ½ × 1 inch and curled it to fit around the outer jacket where it went through the clamp. The long dimension is wrapped around the cable and the abrasive side must be on the inside so that it touches the outer jacket. I tightened the clamp so the two pieces of the connector were fully merged (see Figure 44).

Three years later the outer jacket is still where it was when I assembled it — it hasn't moved. A few times I've pulled so hard on the microphone cable that the rig began moving across the table — still the clamp holds the cable securely. Problem solved. — *73, Jim Kocsis, WA9PYH, 53180 Flicker Ln, South Bend, IN 46637,* **wa9pyh@arrl.net**

# Chapter 11 Interference (RFI/EMI)

## A HELP FOR COMPUTER MONITOR RFI

◊ Here's a useful tip that I didn't see on the ARRL computer RFI Web page (**www.arrl.org/tis/info/rficomp.html**). Sometimes it helps to change the refresh rate of the computer monitor. When I rebuilt the computer I use for the shack, I found I had a "new" source of noise in the HF transceiver. The process of elimination pointed an ugly finger at the monitor. Changing the monitor refresh fre-quency from 60 Hz to 70 Hz cured the problem.

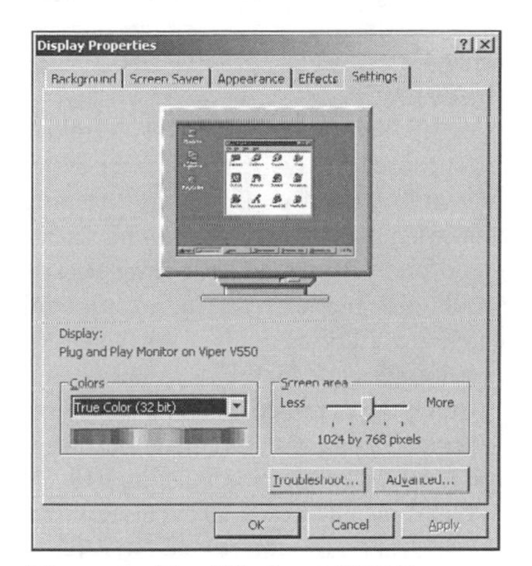


Figure 1—The *Windows 2000 Pro* screen properties dialog.

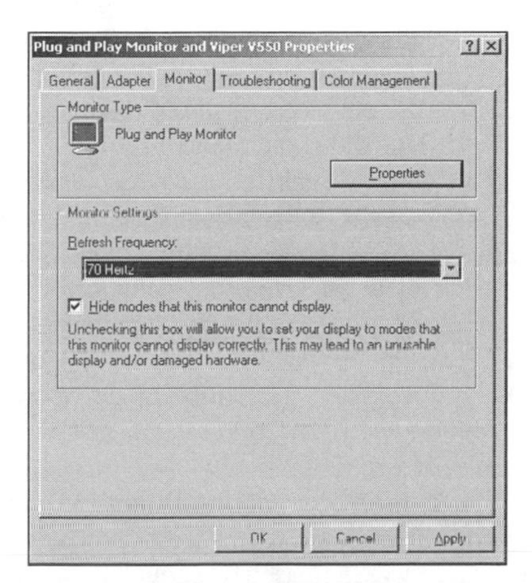


Figure 2—Make sure *Windows* won't show refresh rates your monitor can't handle.

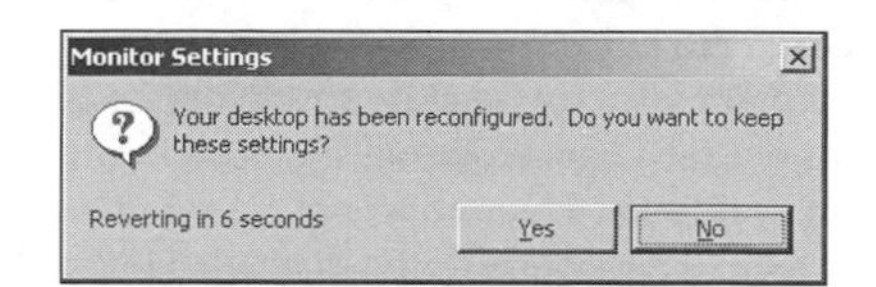


Figure 4—When the problem is cured,work your way out by accepting the changes.

In the *Windows 2000 Pro* operating system, right click on the desktop, choose PROPERTIES. (See Figure 1.) Click the ADVANCED button; then click the monitor tab and ensure that there is a check in the box for "Hide modes that this monitor cannot display." Then (Figure 2) select another REFRESH FREQUENCY value. Click the APPLY button. Make sure the display is acceptable (Figure 3), and check to see if the EMI is gone. Repeat this process as needed.

When the display quality is okay and the EMI is reduced to an acceptable level, click through OK buttons to the desktop (Figure 4). Other *Windows* operating systems use a similar process.—*73, Ed Bruette, N7NVP, Section Manager, WWA;* **www.arrl.org/sect/wwa**

## PUT A BRAKE ON HF RFI

◊ I had an RFI problem while operating on 20 meters in my 2002 Nissan Sentra. I was using a Lakeview TM-1 mount on the trunk and the radio worked fine with an SWR of less than 1.2:1. While the radio worked fine, the car didn't. Whenever I would transmit, the engine would buck and stall in time with my voice.

I first installed filters on the engine computer. That got rid of the stalling but not the bucking. A breakthrough came after I noticed that receiver ignition noise went away whenever the brake pedal was depressed. It appeared that there was RF coupling between the antenna and the brake light wiring. Could it be that it was causing the problem on transmit as well?

I installed snap-on ferrite chokes on the brake light wiring in the trunk as it disappeared into the fender well.

Now I have no further problems operating 20 meters from my car. If only the bands would come back! — *Jerry Turner, KØLSJ, 16298 Florida Way, Rosemount, MN, 55068;* **k0lsj@yahoo.com**

## KEEP THOSE DISHES CLEAN AND BRIGHT

◊ I just tamed my HF TVI with the help of *The ARRL RFI Book*.[1] The problem I had may be of interest to others using dish based TV systems.

My dish network satellite receiver feeds to my television through separate video and audio cables. Both are "shielded" cables connected with RCA plugs on each end. I had concluded, by systematically eliminating all other potential entry points, that my HF signals were getting in via these short cables.

I purchased some Amidon FT-240-77 ferrite toroid cores and passed about six turns of each cable through the toroids. When I checked out the results I found that the satellite video signals had a smeared appearance. I was surprised that a common mode choke on the outside of the shield could affect the signals inside a shielded cable.

I decided that it must be a problem with the cable shields and tried replacing the cables with some Monster Car Audio Interlink 101XNL cables (available from RadioShack). These are shielded with 100% Mylar foil shields, in addition to other features. I used the same six turns on each cable and both the TVI and smearing were gone. — *Joe Gore, WA4RTE, 203 Canterbury Dr, Wrens, GA 30833;* **joegore@firstate.net**

## PUT A BRAKE ON HF RFI

◊ This is in regard to my March 2006 Hints & Kinks item titled "Put a Brake on HF RFI." Several people have asked if the ignition noise also went away when I installed the snap-on ferrite chokes on the brake light wiring. Yes, it did, but the ferrite chokes may be overkill for that problem.

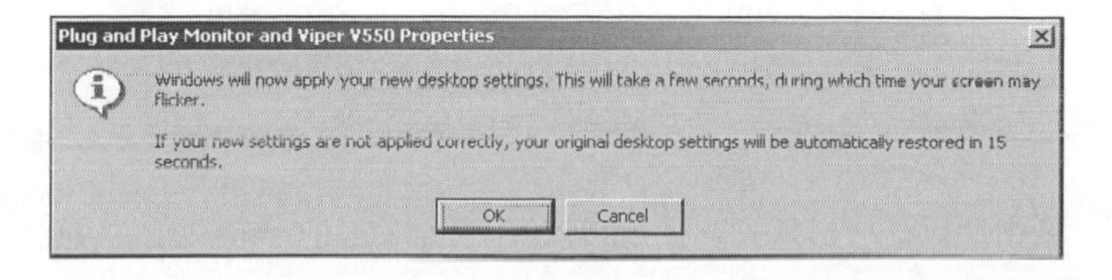


Figure 3—Try a rate and check both screen quality and EMI strength.

[1]Available from your ARRL dealer or the ARRL Bookstore, *The ARRL RFI Book*, ARRL order no. 6834. Telephone 860-594-0355, or toll-free in the US 888-277-5289; **www.arrl.org/shop/**; **pubsales@arrl.org**.

I would suggest putting capacitors across the brake light switch for EMI. The switch will appear to be shorted for RF, but still operate normally. Start with 0.1 µF capacitors and add more capacitance if necessary. You could try 100 pF, 1000 pF and a 1 µF tantalum electrolytic capacitor. The tantalum capacitor will conduct if you install it backwards, so be sure the positive lead connects to the positive voltage side of the switch. If you install it backwards, you may find that you have a dead battery on Monday morning. Reverse-connected tantalum capacitors can also explode and/or catch fire. Aluminum electrolytic capacitors do not work well at RF because they have a high "effective series resistance" or ESR. Tantalum capacitors have a much lower ESR, so they can be suitable for this application. *— Jerry Turner, KØLSJ, 16298 Florida Way, Rosemount, MN 55068*

## RF COMPUTER HASH SOLVED

◊ My problem started with the full-size tower PC in use at my station for many years. There was no detectable interference even though the radios, cabling and coax were close to one another. I was operating on 80, 40 and 20 meters and noticed the noise floor was much higher than normal, strong enough to wipe out weak signals. Also, there were birdies every 50 or 60 kHz up and down the band.

Using separate receivers and antennas, I concluded something in my station was causing the rise in the noise floor. First, I turned off every power-consuming device in the station, but this had no effect. Next, I shut down the computer, which eliminated the noise. Therefore, it was coming from the computer and I figured it was due to the proximity of the computer to the radio equipment and cabling.

Separating the cables and moving the antenna lead-in wire away from the equipment did not help. I then methodically unplugged one cable at a time from the back of the running computer until only the power cord was attached. No change! I then surmised the noise was coming down the power cord into the common power connection or radiating directly from the cable. I brought in an extension cord and plugged the computer into an outlet on a different circuit breaker. This reduced the hash somewhat but it was still there and very strong.

I then put ferrite slugs on the power line, but this had no effect either. At this point, I became somewhat resigned to the fact that I would have to put up with this situation by either shutting the computer off or locating it somewhere else. I also considered using a wireless hub, but I knew there had to be an easier way to solve this problem.

I sat back and thought, what has changed in the station in the past few months, when I did not have this problem? I remembered that the computer had failed and it needed a new power supply. Thinking the repair shop gave me a bad or dirty power supply, I decided to either return it to them or just change it on my own. Then during my attempts to move the computer, I accidentally bumped it hard and, lo and behold, the noise went away! Investigating further, I found that by tapping ever so slightly on the case of the computer the noise would disappear, but quickly come back. I said to myself: "Got cha!" I took the computer offline, opened it up and turned it on without the case. Noticing the noise or hash was still there, I started to poke around with an insulated tool and found that when I poked the power supply, the hash either resumed or disappeared. Unplugging the computer from the outlet, I pulled the power supply from the computer case, loosened four screws and opened it up.

What I found was interesting and satisfying at the same time. The ground lug from the power cord receptacle was very loose under one of the power supply circuit board screws. I also found the other three screws holding the circuit board to the case somewhat loose and tightened them as well. Replacing the supply back in the computer, I found the noise completely gone, no EMI at all, not an S meter flicker. While I still had the computer covers removed, I tightened every available grounding screw I could find including those on the computer mother board of which a few were also loose.

So if you are experiencing a sudden or unexplained raised noise floor, check out the switching power supply in your computer, it could be the culprit and it's an easy fix. Safety first though, unplug it. *— 73, Bill Gerhold, K2WH, 63 Goldfinch Ln, Hewitt, NJ 07421,* **k2wh@optonline.net**

## WIRELESS RFI DETECTION

◊ Most of us have gone through the hassle of trying to track down RFI sources around the house, including turning off the house circuit breakers and unplugging electronic devices one at a time to isolate the noise source. This usually involves tuning the rig to the offending noise and then going back and forth from the rig to remote parts of the house or yard de-energizing all possible noise sources one at a time. This back and forth process can be a hassle especially if you are dealing with an intermittent noise or one that is radiating or conducting from outside of your house.

This RFI noise detection process can be made a lot easier and more accurate by using an inexpensive pair of wireless headphones that plug into your rig's phone jack. Just tune in the offending RFI, put on the headphones and go anywhere within a 100 foot radius turning circuit breakers off and on, turning household electronics off and on, etc. You'll know immediately if you've isolated the culprit. You can also track noise sources with a sniffer such as a portable broadcast radio and simultaneously listen to see if the sniffer's noise is synchronized with what you are hearing remotely from your rig. If your neighbors are within 100 feet it can also help you track down RFI coming from their house.

I use a $20 wireless headphone set, model HO-900 made by Sentry Industries, Tarrytown, New York(**www.SentryIndustries.com**). I bought mine at a local department store. It comes complete, ready to plug into your transceiver (less batteries), has an approximate 100 foot range, operates on 49 MHz and has a signal/noise ratio of >50 dB. The fidelity is surprisingly good as is the construction quality. *— 73, Dick Goodwin, K4JJW, 2217 Caracara Dr, New Bern, NC 28560,* **rgoodwin41@embarqmail.com**

## ALARM SYSTEM RFI SOLVED

◊Most monitored systems have two keypads that are hard wired to the central alarm unit. The keypads are active, even when the system is in standby. I was experiencing 'beeping" from the keypads when transmitting on both the 160 and 80 meter bands. RFI was not a problem at higher frequencies, probably because the filtering built-in to the central alarm unit was adequate.

I tried several remedies including clamp-on toroids, bypass capacitors and single chokes, which failed to prevent the RFI. The RF chokes installed on the keypad terminals at the central alarm unit seemed to offer the most promise, but they only protected one band or the other. I did some impedance measurement on the chokes using the Autek RF-1 Analyst and found that using a large value choke in series with a smaller value choke exhibited a broadband high impedance range, sufficient to cover the frequencies of interest. This did solve the RFI problem.

The chokes I settled on were Fastron series 23 (**www.mouser.com/fastron**) with values of 680 µH in series with 47 µH. They are each about the size of a ½ W resistor.

A rule-of-thumb says an impedance of 1000 Ω or higher will usually provide good RFI suppression. The impedance of my chokes ranged from 2000 Ω at 1.5 MHz to 1250 Ω at 7.5 MHz.

Thus, high impedance is maintained over the frequencies of interest and overall resistance was increased by about 8 Ω, which did not affect the operation of the keypads. It is felt that combinations of chokes of different values in series can be an effective way of suppressing RFI in low current circuits where RFI is being experienced on several frequencies. *— 73, John Holliman, K5SEE, 11919 Bourgeois Forest Dr, Houston, TX 77066-3209,* **k5seeham@aol.com**

## FLOATING LEAD RFI

◊One of my neighbors had been experiencing some strange problems. His garage door opener ceased to operate after I transmitted with the beam pointing in his direction running 800 W on any band. The unit was on but would not respond to either the remote opener or the push button inside the garage. Only a power off/power on reset would restore operation. Ferrite chokes on the ac mains or the control line failed to make any difference.

I discovered that the cable from the opener to the push button control had three wires. Two were connected and one wire was floating. I thought that the floating wire might be acting as a decent antenna and might be inducing RF into the opener's main unit. I simply connected the third floating wire at both ends to one of the other wires and the problem was solved.

He also had a severe TVI issue on one of his two TV sets. Both sets are using the local cable TV service. Indications were that the TVI was unrelated to any particular channel and the symptom was fundamental overload. All the usual fixes were tried including an ac line RFI filter, ferrite chokes on the cable and ac mains, etc. I even tried a conventional high pass filter (which is not suited for cable TV as some of the channels are shifted down in frequency from their "normal" broadcast frequencies) but nothing worked.

I noticed that there were four cables going to the room with the TV being affected. One was carrying the TV signals and three were unconnected and just floating. After the garage door opener experience, I again thought that these floating cables might be acting like antennas. We removed the three floating cables and the TVI problem was solved.

In 48 years of ham radio operation, this is the first time I have encountered this and thought it would be good to share the experience. — *73, Al Koblinski, W7XA, 2733 S Davis, Mesa, AZ 85210*, **w7xa@qwest.net**

## DRILL CHARGER RFI

◊Last weekend while operating on HF I noticed some RFI that I had not previously heard. The interference signals sounded like noise, were about 20 kHz wide and spaced approximately 11 MHz apart from 9 MHz to 30 MHz. The noise occurred at exactly 1 second intervals with ½ second duration and registered about S9 on my transceiver.

I checked for the interference from my mobile and determined that it was only present near my home. I then took my portable HF receiver and walked around the house holding it near my clocks that monitored WWVB, my computers, Wi-Fi routers, TVs and anything else that I could find that was electronic. I noticed that the RFI was loudest in my shack, which also contains a small workbench. After further investigation, I noticed that my new Hitachi cordless drill battery charger has an LED that blinks at a 1 second rate when the battery is fully charged or not in the charger. I unplugged the charger and the RFI disappeared.

The drill is a Hitachi DS18DSAL and the charger is a UC 18YGSL. Now, I will only plug in the charger when charging a battery. — *73, Jere Sandidge, K4FUM, 1770 Oak Ridge Cir SW, Stone Mountain, GA 30087-3286*, **k4fum@arrl.net**

## SOLVING AN EMI PROBLEM

◊Shortly after having my new combo Internet/telephone/television system installed, I discovered RF from my transceiver was scrambling the residential gateway (modem). As little as 25 W was enough to cause havoc. The detective work began by determining how the RF was getting into the modem. It was either entering via the power source, the CAT 6 input cable, the RG-6 feeding the televisions, the USB cable to my computer or via direct radiation.

One at a time I disconnected the RG-6 cables, the USB cable and the CAT 6 cable with no success. Since the modem had a battery backup, I then disconnected the 120 V ac power supply to the modem and let it run on battery backup. Now the modem was naked with no input or output wiring and running on the battery; a quick transmission on 80 meters still scrambled the modem. I therefore verified the modem was sensitive to direct radiation.

Since the modem was housed in a plastic case and there was no direct attachment available for an earth ground, I decided to construct a Faraday shield to surround the modem. I remembered I had saved some scrap aluminum window screen so I made a pattern and cut and folded the screen to form a box for the modem.

I attached two AWG 12 flexible pigtails to the screen box and attached the other end of each pigtail to the cover screw of a nearby 240 V ac junction box providing power to my linear amplifier. The neutral for the 240 V ac cable coming from the main breaker box is somewhere around an AWG 6, only about 15 feet long and is tied to earth ground. [*Warning* — The neutral wire in your home's electrical system, while grounded by the power company, is *not an electrical safety ground* but an active part of the ac power circuit. A safety ground should be wired directly to the green safety ground wire in the outlet box or directly to the earth ground point with a separate piece of wire. If you are unsure about proper safety grounding, contact a licensed electrician. — *Ed.*]

I slid the modem into the screen box, reconnected all the leads and was ready for an RF test. I incremented the RF power from 25 W to the legal limit and from 160 meters to 6 meters and the modem never missed a lick — problem solved. I used an available convenient earth ground, but depending on your individual situation, experimentation is the key to finding a safe and effective ground. I later made a more permanent shield using some aluminum sheet metal I had lying around. This solution will not only make you happy but also a nearby neighbor whose TV or Internet suddenly drops out when you transmit and he doesn't know why. — *73, Joe Vlk, W8DCQ, 3967 Shoshone Ct, Oxford, MI 48370-2933*, **w8dcq@arrl.net**

# Abbreviations List

**A**

a—atto (prefix for $10^{-18}$)
A—ampere (unit of electrical current)
ac—alternating current
ACC—Affiliated Club Coordinator
ACSSB—amplitude-companded single sideband
A/D—analog-to-digital
ADC—analog-to-digital converter
AF—audio frequency
AFC—automatic frequency control
AFSK—audio frequency-shift keying
AGC—automatic gain control
Ah—ampere hour
ALC—automatic level control
AM—amplitude modulation
AMRAD—Amateur Radio Research and Development Corporation
AMSAT—Radio Amateur Satellite Corporation
AMTOR—Amateur Teleprinting Over Radio
ANT—antenna
ARA—Amateur Radio Association
ARC—Amateur Radio Club
ARES—Amateur Radio Emergency Service
ARQ—Automatic repeat request
ARRL—American Radio Relay League
ARS—Amateur Radio Society (station)
ASCII—American National Standard Code for Information Interchange
ATV—amateur television
AVC—automatic volume control
AWG—American wire gauge
az-el—azimuth-elevation

**B**

B—bel; blower; susceptance; flux density, (inductors)
balun—balanced to unbalanced (transformer)
BC—broadcast
BCD—binary coded decimal
BCI—broadcast interference
Bd—baud (bids in single-channel binary data transmission)
BER—bit error rate
BFO—beat-frequency oscillator
bit—binary digit
bit/s—bits per second
BM—Bulletin Manager
BPF—band-pass filter
BPL—Brass Pounders League
BPL—Broadband over Power Line
BT—battery
BW—bandwidth
Bytes—Bytes

**C**

c—centi (prefix for $10^{-2}$)
C—coulomb (quantity of electric charge); capacitor
CAC—Contest Advisory Committee
CATVI—cable television interference
CB—Citizens Band (radio)
CBBS—computer bulletin-board service
CBMS—computer-based message system
CCITT—International Telegraph and Telephone Consultative Committee
CCTV—closed-circuit television
CCW—coherent CW
ccw—counterclockwise
CD—civil defense
cm—centimeter
CMOS—complementary-symmetry metal-oxide semiconductor
coax—coaxial cable
COR—carrier-operated relay
CP—code proficiency (award)
CPU—central processing unit
CRT—cathode ray tube
CT—center tap
CTCSS—continuous tone-coded squelch system
cw—clockwise
CW—continuous wave

**D**

d—deci (prefix for $10^{-1}$)
D—diode
da—deca (prefix for 10)
D/A—digital-to-analog
DAC—digital-to-analog converter
dB—decibel (0.1 bel)
dBi—decibels above (or below) isotropic antenna
dBm—decibels above (or below) 1 milliwatt
DBM—double balanced mixer
dBV—decibels above/below 1 V (in video, relative to 1 V P-P)
dBW—decibels above/below 1 W
dc—direct current
D-C—direct conversion
DDS—direct digital synthesis
DEC—District Emergency Coordinator
deg—degree
DET—detector
DF—direction finding; direction finder
DIP—dual in-line package
DMM—digital multimeter
DPDT—double-pole double-throw (switch)
DPSK—differential phase-shift keying
DPST—double-pole single-throw (switch)
DS—direct sequence (spread spectrum); display
DSB—double sideband
DSP—digital signal processing
DTMF—dual-tone multifrequency
DVM—digital voltmeter
DX—long distance; duplex
DXAC—DX Advisory Committee
DXCC—DX Century Club

**E**

e—base of natural logarithms (2.71828)
E—voltage
EA—ARRL Educational Advisor
EC—Emergency Coordinator
ECL—emitter-coupled logic
EHF—extremely high frequency (30-300 GHz)
EIA—Electronic Industries Alliance
EIRP—effective isotropic radiated power
ELF—extremely low frequency
ELT—emergency locator transmitter
EMC—electromagnetic compatibility
EME—earth-moon-earth (moonbounce)
EMF—electromotive force
EMI—electromagnetic interference
EMP—electromagnetic pulse
EOC—emergency operations center
EPROM—erasable programmable read only memory

**F**

f—femto (prefix for $10^{-15}$); frequency
F—farad (capacitance unit); fuse
fax—facsimile
FCC—Federal Communications Commission
FD—Field Day
FEMA—Federal Emergency Management Agency
FET—field-effect transistor
FFT—fast Fourier transform
FL—filter
FM—frequency modulation
FMTV—frequency-modulated television
FSK—frequency-shift keying
FSTV—fast-scan (real-time) television
ft—foot (unit of length)

**G**

g—gram (unit of mass)
G—giga (prefix for $10^{9}$); conductance
GaAs—gallium arsenide
GB—gigabytes
GDO—grid- or gate-dip oscillator
GHz—gigahertz ($10^{9}$ Hz)
GND—ground

**H**

h—hecto (prefix for $10^{2}$)
H—henry (unit of inductance)
HF—high frequency (3-30 MHz)
HFO—high-frequency oscillator; heterodyne frequency oscillator
HPF—highest probable frequency; high-pass filter
Hz—hertz (unit of frequency, 1 cycle/s)

**I**

I—current, indicating lamp
IARU—International Amateur Radio Union
IC—integrated circuit
ID—identification; inside diameter
IEEE—Institute of Electrical and Electronics Engineers
IF—intermediate frequency
IMD—intermodulation distortion
in.—inch (unit of length)
in./s—inch per second (unit of velocity)
I/O—input/output
IRC—international reply coupon
ISB—independent sideband
ITF—Interference Task Force
ITU—International Telecommunication Union
ITU-T—ITU Telecommunication Standardization Bureau

**J-K**

*j*—operator for complex notation, as for reactive component of an impedance (+*j* inductive; –*j* capacitive)
J—joule (kg $m^2/s^2$) (energy or work unit); jack
JFET—junction field-effect transistor
k—kilo (prefix for $10^{3}$); Boltzmann's constant ($1.38 \times 10^{-23}$ J/K)
K—kelvin (used without degree symbol) absolute temperature scale; relay
kB—kilobytes
kBd—1000 bauds
kbit—1024 bits
kbit/s—1024 bits per second
kbyte—1024 bytes
kg—kilogram
kHz—kilohertz
km—kilometer
kV—kilovolt
kW—kilowatt
kΩ—kilohm

**L**

l—liter (liquid volume)
L—lambert; inductor
lb—pound (force unit)
LC—inductance-capacitance
LCD—liquid crystal display
LED—light-emitting diode
LF—low frequency (30-300 kHz)
LHC—left-hand circular (polarization)
LO—local oscillator; Leadership Official
LP—log periodic
LS—loudspeaker
lsb—least significant bit
LSB—lower sideband
LSI—large-scale integration
LUF—lowest usable frequency

**M**

m—meter (length); milli (prefix for $10^{-3}$)
M—mega (prefix for $10^{6}$); meter (instrument)
mA—milliampere
mAh—milliampere hour
MB—megabytes
MCP—multimode communications processor
MDS—Multipoint Distribution Service; minimum discernible (or detectable) signal
MF—medium frequency (300-3000 kHz)
mH—millihenry
MHz—megahertz
mi—mile, statute (unit of length)
mi/h (MPH)—mile per hour
mi/s—mile per second
mic—microphone
min—minute (time)
MIX—mixer
mm—millimeter
MOD—modulator
modem—modulator/demodulator
MOS—metal-oxide semiconductor
MOSFET—metal-oxide semiconductor field-effect transistor

MS—meteor scatter
ms—millisecond
m/s—meters per second
msb—most-significant bit
MSI—medium-scale integration
MSK—minimum-shift keying
MSO—message storage operation
MUF—maximum usable frequency
mV—millivolt
mW—milliwatt
MΩ—megohm

**N**

n—nano (prefix for $10^{-9}$); number of turns (inductors)
NBFM—narrow-band frequency modulation
NC—no connection; normally closed
NCS—net-control station; National Communications System
nF—nanofarad
NF—noise figure
nH—nanohenry
NiCd—nickel cadmium
NM—Net Manager
NMOS—N-channel metal-oxide silicon
NO—normally open
NPN—negative-positive-negative (transistor)
NPRM—Notice of Proposed Rule Making (FCC)
ns—nanosecond
NTIA—National Telecommunications and Information Administration
NTS—National Traffic System

**O**

OBS—Official Bulletin Station
OD—outside diameter
OES—Official Emergency Station
OO—Official Observer
op amp—operational amplifier
ORS—Official Relay Station
OSC—oscillator
OSCAR—Orbiting Satellite Carrying Amateur Radio
oz—ounce ($^1/_{16}$ pound)

**P**

p—pico (prefix for $10^{-12}$)
P—power; plug
PA—power amplifier
PACTOR—digital mode combining aspects of packet and AMTOR
PAM—pulse-amplitude modulation
PBS—packet bulletin-board system
PC—printed circuit
PD—power dissipation
PEP—peak envelope power
PEV—peak envelope voltage
pF—picofarad
pH—picohenry
PIC—Public Information Coordinator
PIN—positive-intrinsic-negative (semiconductor)
PIO—Public Information Officer
PIV—peak inverse voltage
PLC—Power Line Carrier
PLL—phase-locked loop
PM—phase modulation
PMOS—P-channel (metal-oxide semiconductor)
PNP—positive negative positive (transistor)
pot—potentiometer
P-P—peak to peak
ppd—postpaid
PROM—programmable read-only memory
PSAC—Public Service Advisory Committee
PSHR—Public Service Honor Roll
PTO—permeability-tuned oscillator
PTT—push to talk

**Q-R**

Q—figure of merit (tuned circuit); transistor
QRP—low power (less than 5-W output)
R—resistor
RACES—Radio Amateur Civil Emergency Service
RAM—random-access memory
RC—resistance-capacitance
R/C—radio control
RDF—radio direction finding
RF—radio frequency
RFC—radio-frequency choke
RFI—radio-frequency interference
RHC—right-hand circular (polarization)
RIT—receiver incremental tuning
RLC—resistance-inductance-capacitance
RM—rule making (number assigned to petition)
r/min (RPM)—revolutions per minute
rms—root mean square
ROM—read-only memory
r/s—revolutions per second
RS—Radio Sputnik (Russian ham satellite)
RST—readability-strength-tone (CW signal report)
RTTY—radioteletype
RX—receiver, receiving

**S**

s—second (time)
S—siemens (unit of conductance); switch
SASE—self-addressed stamped envelope
SCF—switched capacitor filter
SCR—silicon controlled rectifier
SEC—Section Emergency Coordinator
SET—Simulated Emergency Test
SGL—State Government Liaison
SHF—super-high frequency (3-30 GHz)
SM—Section Manager; silver mica (capacitor)
S/N—signal-to-noise ratio
SPDT—single-pole double-throw (switch)
SPST—single-pole single-throw (switch)
SS—ARRL Sweepstakes; spread spectrum
SSB—single sideband
SSC—Special Service Club
SSI—small-scale integration
SSTV—slow-scan television
STM—Section Traffic Manager
SX—simplex
sync—synchronous, synchronizing
SWL—shortwave listener
SWR—standing-wave ratio

**T**

T—tera (prefix for $10^{12}$); transformer
TA—ARRL Technical Advisor
TC—Technical Coordinator
TCC—Transcontinental Corps (NTS)
TCP/IP—Transmission Control Protocol/ Internet Protocol
tfc—traffic
TNC—terminal node controller (packet radio)
TR—transmit/receive
TS—Technical Specialist
TTL—transistor-transistor logic
TTY—teletypewriter
TU—terminal unit
TV—television
TVI—television interference
TX—transmitter, transmitting

**U**

U—integrated circuit
UHF—ultra-high frequency (300 MHz to 3 GHz)
USB—upper sideband
UTC—Coordinated Universal Time (also abbreviated Z)
UV—ultraviolet

**V**

V—volt; vacuum tube
VCO—voltage-controlled oscillator
VCR—video cassette recorder
VDT—video-display terminal
VE—Volunteer Examiner
VEC—Volunteer Examiner Coordinator
VFO—variable-frequency oscillator
VHF—very-high frequency (30-300 MHz)
VLF—very-low frequency (3-30 kHz)
VLSI—very-large-scale integration
VMOS—V-topology metal-oxide-semiconductor
VOM—volt-ohmmeter
VOX—voice-operated switch
VR—voltage regulator
VSWR—voltage standing-wave ratio
VTVM—vacuum-tube voltmeter
VUCC—VHF/UHF Century Club
VXO—variable-frequency crystal oscillator

**W**

W—watt (kg $m^2s^{-3}$), unit of power
WAC—Worked All Continents
WAS—Worked All States
WBFM—wide-band frequency modulation
WEFAX—weather facsimile
Wh—watthour
WPM—words per minute
WRC—World Radiocommunication Conference
WVDC—working voltage, direct current

**X**

X—reactance
XCVR—transceiver
XFMR—transformer
XIT—transmitter incremental tuning
XO—crystal oscillator
XTAL—crystal
XVTR—transverter

**Y-Z**

Y—crystal; admittance
YIG—yttrium iron garnet
Z—impedance; also see UTC

*Numbers/Symbols*

5BDXCC—Five-Band DXCC
5BWAC—Five-Band WAC
5BWAS—Five-Band WAS
6BWAC—Six-Band WAC
°—degree (plane angle)
°C—degree Celsius (temperature)
°F—degree Fahrenheit (temperature)
$\alpha$—(alpha) angles; coefficients, attenuation constant, absorption factor, area, common-base forward current-transfer ratio of a bipolar transistor
$\beta$—(beta) angles; coefficients, phase constant current gain of common-emitter transistor amplifiers
$\gamma$—(gamma) specific gravity, angles, electrical conductivity, propagation constant
$\Gamma$—(gamma) complex propagation constant
$\delta$—(delta) increment or decrement; density; angles
$\Delta$—(delta) increment or decrement determinant, permittivity
$\varepsilon$—(epsilon) dielectric constant; permittivity; electric intensity
$\zeta$—(zeta) coordinates; coefficients
$\eta$—(eta) intrinsic impedance; efficiency; surface charge density; hysteresis; coordinate
$\theta$—(theta) angular phase displacement; time constant; reluctance; angles
$\iota$—(iota) unit vector
K—(kappa) susceptibility; coupling coefficient
$\lambda$—(lambda) wavelength; attenuation constant
$\Lambda$—(lambda) permeance
$\mu$—(mu) permeability; amplification factor; micro (prefix for $10^{-6}$)
$\mu$F—microfarad
$\mu$H—microhenry
$\mu$P—microprocessor
$\xi$—(xi) coordinates
$\pi$—(pi) $\approx 3.14159$
$\rho$—(rho) resistivity; volume charge density; coordinates; reflection coefficient
$\sigma$—(sigma) surface charge density; complex propagation constant; electrical conductivity; leakage coefficient; deviation
$\Sigma$—(sigma) summation
$\tau$—(tau) time constant; volume resistivity; time-phase displacement; transmission factor; density
$\phi$—(phi) magnetic flux angles
$\Phi$—(phi) summation
$\chi$—(chi) electric susceptibility; angles
$\Psi$—(psi) dielectric flux; phase difference; coordinates; angles
$\omega$—(omega) angular velocity $2\pi F$
$\Omega$—(omega) resistance in ohms; solid angle